Studies in Systems, Decision and Control

Volume 379

Series Editor

The series "Studies in Systems, Decision and Control" (SSDC) covers both new developments and advances, as well as the state of the art, in the various areas of broadly perceived systems, decision making and control–quickly, up to date and with a high quality. The intent is to cover the theory, applications, and perspectives on the state of the art and future developments relevant to systems, decision making, control, complex processes and related areas, as embedded in the fields of engineering, computer science, physics, economics, social and life sciences, as well as the paradigms and methodologies behind them. The series contains monographs, textbooks, lecture notes and edited volumes in systems, decision making and control spanning the areas of Cyber-Physical Systems, Autonomous Systems, Sensor Networks, Control Systems, Energy Systems, Automotive Systems, Biological Systems, Vehicular Networking and Connected Vehicles, Aerospace Systems, Automation, Manufacturing, Smart Grids, Nonlinear Systems, Power Systems, Robotics, Social Systems, Economic Systems and other. Of particular value to both the contributors and the readership are the short publication timeframe and the world-wide distribution and exposure which enable both a wide and rapid dissemination of research output.

Indexed by SCOPUS, DBLP, WTI Frankfurt eG, zbMATH, SCImago.

All books published in the series are submitted for consideration in Web of Science.

More information about this series at http://www.springer.com/series/13304

Vassil Sgurev · Vladimir Jotsov · Janusz Kacprzyk
Editors

Advances in Intelligent Systems Research and Innovation

Springer

Editors
Vassil Sgurev
Institute of Information
and Communication Technologies
Bulgarian Academy of Sciences
Sofia, Bulgaria

Vladimir Jotsov
University of Library Studies
and Information Technologies
Sofia, Bulgaria

Janusz Kacprzyk
Systems Research Institute
Polish Academy of Sciences
Warsaw, Poland

ISSN 2198-4182 ISSN 2198-4190 (electronic)
Studies in Systems, Decision and Control
ISBN 978-3-030-78123-1 ISBN 978-3-030-78124-8 (eBook)
https://doi.org/10.1007/978-3-030-78124-8

This Springer imprint is published by the registered company Springer Nature Switzerland AG
The registered company address is: Gewerbestrasse 11, 6330 Cham, Switzerland

Contents

Balancing Exploration and Exploitation in Forward Model Learning

Alexander Dockhorn and Rudolf Kruse

Abstract Forward model learning algorithms enable the application of simulation-based search methods in environments for which the forward model is unknown. Multiple studies have shown great performance in game-related and motion control applications. In these, forward model learning agents often required less training time while achieving a similar performance than state-of-the-art reinforcement learning methods. However, several problems can emerge when replacing the environment's true model with a learned approximation. While the true forward model allows the accurate prediction of future time-steps, a learned forward model may always be inaccurate in its prediction. These inaccuracies become problematic when planning long action sequences since the confidence in predicted time-steps reduces with increasing depth of the simulation. In this work, we explore methods for balancing risk and reward in decision-making using inaccurate forward models. Therefore, we propose methods for measuring the variance of a forward model and the confidence in the predicted outcome of planned action sequences. Based on these metrics, we define methods for learning and using forward models under consideration of their current prediction accuracy. Proposed methods have been tested in various motion control tasks of the Open AI Gym framework. Results show that the information on the model's accuracy can be used to increase the efficiency of the agent's training and the agent's performance during evaluation.

A. Dockhorn (✉)
Queen Mary University, London, UK
e-mail: a.dockhorn@qmul.ac.uk

R. Kruse
Otto von Guericke University, Magdeburg, Germany
e-mail: rudolf.kruse@ovgu.de

V. Sgurev et al. (eds.), *Advances in Intelligent Systems Research and Innovation*,
Studies in Systems, Decision and Control 379,
https://doi.org/10.1007/978-3-030-78124-8_1

1 Introduction

Forward Model Learning describes the process of learning a model of a priori unknown environments. This enables an agent to anticipate the outcome of planned action sequences and use simulation-based search techniques to optimize its behavior.

This chapter builds upon our recent work on forward model learning in games [1] and directly extends our work on forward model learning for motion control tasks [12]. Previous studies have shown that reliable forward models can be learned by observation. Since no information on the environment is available, we have chosen to implement an uninformed training process in which the agent uses random actions to explore its environment. This results in many interactions being wasted due to (1) repeating a known interaction and (2) neither focusing on improving the model nor improving the agent's performance. As a result, we have observed that the model may be unreliable since large amounts of the state space remain unobserved. In return, the agent's performance during the evaluation will be limited by the prediction accuracy of the trained forward model.

Motivated by these shortcomings, we study methods for improving the efficiency of the agent's training process. Therefore, we formulate two learning goals (1) the exploration of the environment to train a reliable forward model, and (2) the exploitation of promising action sequences to become more proficient in the given task. The main contribution of this work can be summarized by:

- **Taxonomy of Learning Algorithms**: We present a unifying view on reinforcement learning, search, and forward model learning algorithms in the context of the agent-environment interface.
- **Decomposed Differential Forward Model**: We review the definitions of previously proposed model building heuristics and their generalization in form of the decomposed forward model. As a result, we propose the decomposed differential forward model which will be used throughout this study.
- **Risk Awareness**: We propose several methods for measuring the agent's confidence in the learned forward model's predictions. Furthermore, we propose several optimization goals to balance exploration and exploitation during the agent's training process.

The remainder of this chapter is structured as follows: In Sect. 2, we compare the concepts of reinforcement learning, search-based algorithms, and forward model learning for decision-making. The following sections (Sects. 3 and 4) focus on the introduction of forward model learning and types of forward model representations. Section 5 exemplarily shows the usage of Gaussian process regression and ensemble regression models such as a random forest regression. For both, we propose methods for incorporating the variance of made predictions in the agent's decision-making process (Sect. 5.2). We evaluate proposed methods based on their resulting performance in three simple motion control tasks in Sect. 6. For a more detailed analysis, we compare the impact of proposed measures on the agent's training process in Sect. 7. We conclude our results in Sect. 8 and discuss opportunities for further studies.

2 Taxonomy of Learning Algorithms

The agent-environment interface represents a general description of a learning scenario. Here, the agent is in continuous interaction with an environment. Each of the agent's executed actions has the potential to update the state of the environment. In return, the agent can be offered a numerical reward that is typically associated with the given task.

In reinforcement learning, the agent focuses on picking actions to maximize its expected reward over time. Given observations of all its previous interactions, the agent tries to estimate the expected value of state-action pairs. Algorithms such as Temporal Difference Learning (TDL) [32], Q-learning [35], and the Monte Carlo (MC) method [30], update the expected value after a reward signal has been observed or after the result of an episode has been observed respectively. These techniques are called model-free since they do not try to build a model of their environment.

In contrast to these simpler reinforcement learning algorithms, which are storing the value of each state-action pair, methods in deep reinforcement learning use a neural network to approximate the value based on the input, thus drastically decreasing the storage used for the model [37]. While deep reinforcement learning algorithms have resulted in an impressive performance in the context of many game-related benchmarks, the number of required training examples to fit all the network's parameters is often high and thus require much training time.

Simulation-based search algorithms, such as minimax [23] or Monte Carlo tree search (MCTS) [5], utilize the environment's forward model to simulate the outcome of planned action sequences. Each action sequence represents a candidate solution. Its simulation is called a rollout and the result of such a simulation can be used to estimate the value of actions in the simulated sequence. In contrast to reinforcement learning algorithms, simulation-based search algorithms estimate the value of an action at run-time. Therefore, these algorithms require knowledge of the current state and the game's model, and return the action with the highest expected value regarding the current state. In return, they can be applied without prior training. Due to their capability of being applied without training, search algorithms such as MCTS and Open Loop Search [26] have performed well in general game-playing tasks [27].

Dynamic programming algorithms [21] require knowledge of the environment's forward model to calculate the true value of any state-action pairs. This optimization process can yield a perfect policy for small state and action spaces. However, the high breadth or depth of the search tree often renders this method infeasible.

Next to traditional search schemes, the rolling horizon evolutionary algorithm (RHEA) [15, 16], uses mutation and crossover to optimize the agent's action sequence. A heuristic value of simulated action sequences can be used as a fitness measure to guide evolutionary optimization. The design of such a heuristic can drastically impact, the resulting behavior.

In the following, we compare the reviewed methods based on their knowledge of the environment. Given the agent's action and the current state, the agent will observe the next state and a reward. Both updates can be encoded in a separate model, namely the forward model producing the next state and the reward model which in turn provides the agent with a reward. However, since the accumulated reward (also known as return) is much more useful for the agent's action-selection process, the following comparison will use the agent's knowledge of the return to measure the value of a state-action pair. Therefore, we focus on the agent's knowledge of two of the environment's components, (1) the environment's forward model, and (2) the value of the environment's states and actions.

Considering the agent's knowledge of the return, reinforcement learning and simulation-based search methods represent two extremes. The eager learning process of reinforcement learning methods results in a return model. While TDL and the MC method do this without needing knowledge about the forward model, the computations done in dynamic programming need the forward model for its iterative update routine. Deep reinforcement learning replaces the need for storing the expected return for every state by learning a network that approximates the return function. The reduction in complexity can arguably be achieved through an understanding of the game's state-space.

In contrast, simulation-based search methods do not need to store the expected return of each state, since they approximate at run-time using the forward model. The rolling horizon evolutionary algorithm seems to be a slight exception to this since it is also making use of a heuristic function, which approximates the value of a state. This heuristic function is being used to rate a rollout's outcome and, in the best case, approximates the expected return.

The class of forward model learning algorithms represents a new approach that enables the application of search algorithms in case the environment's forward model is unknown to the agent. Therefore, the agent builds a model to predict upcoming states based on made observation during previous interactions. At the time, the agent can also learn to predict the reward of the environment or approximate the value of a state-action pair.

This process can be compared to recent experiments on world modeling [18] and imagination-based deep reinforcement learning [36]. In these, the agent uses a deep or recurrent neural network structure that handles the selection of the agent's actions according to predicted future states. Agents using these model-based deep reinforcement learning approaches have shown to be capable of playing games based on their visual state representation [20]. Similarly, they achieved improved performance in several visual control tasks in comparison to model-free reinforcement learning agents [19]. However, the sheer number of parameters to be tuned and the eager learning of the state's value function results in the agent requiring lots of training data, e.g. 10^8 training steps for learning to play the game Sokoban [36]. The amount of required iterations makes this process infeasible in case the model's training time is limited.

In contrast, a prediction-based search can be implemented which determines an action's value according to simulations of the trained forward model. Furthermore, reductions of the feasible model space could be achieved by assuming independencies among observed sensor values. Figure 1 presents a summary of the discussed

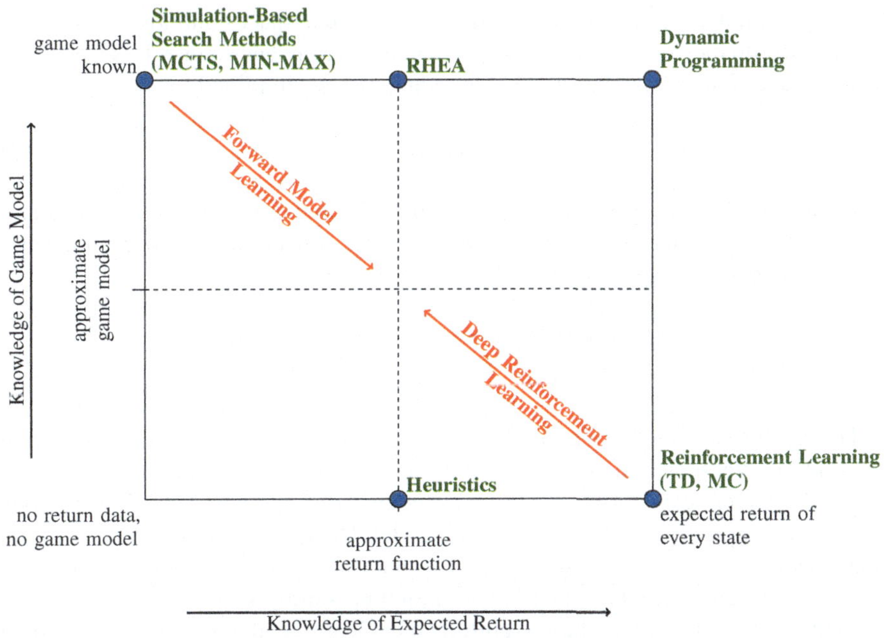

Fig. 1 Comparison of general game-playing and -learning techniques based on the agent's knowledge of the return function and the game's environment model [8]

methods, based on the two dimensions: knowledge of expected return and knowledge of the game model.

3 Forward Model Learning

In this section, we will take a closer look at the forward model learning process. First, we will define how the environment's model can be represented. The next section will cover the forward model's representation.

First of all, we assume that the environment model can be represented as a stochastic process that maps the sequence of previous actions $a \in A$ and states $s \in S$ to a probability of observing the next state and its associated reward $r \in \mathbb{R}$.

$$p(s_t, r_t \mid s_{t-1}, a_{t-1}, \ldots, s_1, a_1) \tag{1}$$

We consider each state $s_t \in S \subseteq \mathbb{R}^n$ to be a vector of real-valued observations $s_t = (s_t^1, s_i^2, \ldots, s_t^n)$, $s_t^i \in \mathbb{R}$ at time t. The same notation applies to the set of available actions and the action space: $a_t \in A \subseteq \mathbb{R}^n$.

For simplicity, we will assume single-value action spaces and an environment model that satisfies the Markov property. Therefore, the environment's response is independent of all but the latest interaction and can be simplified to:

$$p(s_t, r_t | s_{t-1}, a_{t-1}) \tag{2}$$

Up to now, the environment's response includes two components, the state and the reward. We further split the environment model into a state transition model and a reward model, both fulfilling the Markov property:

$$\begin{aligned} &\text{State Transition Model:} \quad p_{fm}(s_t \mid s_{t-1}, a_{t-1}) \\ &\text{Reward Model:} \quad p_r(r_t \mid s_{t-1}, a_{t-1}) \end{aligned}$$

During forward model learning, the agent will learn to approximate the state transition model based on observations of previous interactions. Given these observations, the agent can construct a training set and use supervised learning methods to create an approximation of the original model. Depending on the structure of the state-space a classification (discrete and nominal state-spaces) or a regression model (real-valued state spaces) can be used to predict upcoming states.

In recent works of model-based reinforcement learning, the state transition model has been represented by deep neural networks. The *World Models* agent learns a latent vector representation using LSTMs [18]. This allows the agent to keep track of previous events and further condense its action policy. Following up on this idea, the imagination-based *I2A* agent [28] adds a rollout phase for improving the agent's estimates of an action's expected return for long-time horizons. The *Dreamer* agent [19] has shown that latent state models can learn motion control behaviors of varying complexity based on high-dimensional sensory inputs.

In contrast to model-based reinforcement learning, forward model learning agents do not model the return of an action (accumulated reward) but estimate an action's value based on its simulated outcome. State-of-the-art planners such as Monte Carlo Tree Search [5] and Rolling Horizon Algorithms [15] have recently shown applications in game AI [15, 31] and motion control [6]. Since the accuracy of the planning process is dependent on the accuracy of the agent's prediction, the agent's goal is to improve the forward model's accuracy and not the agent's policy.

4 Forward Model Representation

Learning a model of the environment has been actively studied in model-based reinforcement learning. While many approaches rely on the generality of deep neural networks, they have not shown to be very sample efficient. Therefore, the model requires many iterations to be trained until a reliable prediction of the upcoming state can be made and the search-methods advantage to work without much training time is lost. Thus, finding a suitable state representation can drastically reduce the required training time as well as improve the model's final prediction accuracy.

Instead of predicting the environment's next state using a single model, the environment's response can often be split into multiple independent components. In return, we can decompose the forward model into independent sub-models. The reduced complexity of a sub-model's output space can result in a simpler learning task, hence, reducing the number of required model parameters and training examples.

4.1 Model Building Heuristics

Forward model decomposition has been shown to increase the sampling efficiency and resulting model performance in numerous applications. An example of a model decomposition heuristics is the local transition function that is used in local forward models [8]. Local forward models work similar to convolution neural networks, and predict the future state based on independent predictions of each cell, whereas the future state of each cell is predicted based on its current state and the state of its neighbors. Local forward models have been successfully applied to the game of life [22] and Sokoban [13] as well as several games of the GVGAI environment [1, 8].

In [14], the authors have shown that knowledge of the environment's state representation can be used to create an object-based decomposition. An extension of this work has been presented in [11], in which the authors proposed the use of stochastic independence tests to find a valid decomposition of the forward model without requiring background knowledge of the environment's state representation.

The local and object-based forward model assumes independency among the input and output variables of the forward model. Such dependencies among observable sensor variables may exist in motion-control tasks, but the underlying structure may be unknown. As a result, we have previously proposed the decomposed differential forward model [12]. For the decomposition, we assume that the next state's sensor values are dependent on the previous state (and possibly its predecessors), but independent of each other. This results in a decomposability of the state transition model into several sub-models, whereas each sub-model predicts the updated sensor value:

***i*-th Component Model:** $fm_i(s_{t-1}, a_{t-1}) \mapsto s_{i,\,t}$

The decomposed forward model can further be used to predict the future state by separately predicting the future state of each observable sensor-value (indicated by $\hat{s}_{i,\,t}$) and aggregating the result of each sub-model:

$$\begin{aligned} & fm(s_{t-1}, a_{t-1}) \\ =&(fm_1(s_{t-1}, a_{t-1}), \ \ldots, \ fm_n(s_{t-1}, a_{t-1})) \\ =&(\hat{s}_{1,t}, \ \ldots, \ \hat{s}_{n,t}) \end{aligned} \tag{3}$$

Alternatively, we can predict the changes of a sensor value:

***i*th Differentation Model**: $fm_{\Delta i}(s_{t-1}, a_{t-1}) \mapsto s_{i,t-1} - s_{i,t}$

Aggregating the predictions of all sensor values and the agent's reward, the agent can predict the outcome of an action. Similarly, a hierarchical structure can be built, in which the environment is first split into several units and the future state values of each unit are predicted independently [7, 9].

Since the result is another state observation, the agent can predict whole action sequences by repeatably applying the learned forward model. Using this, actions can be determined using forward planning methods or a policy can be learned using reinforcement learning on the simulated environment [25].

For motion control, the underlying modeling task is considered a supervised learning problem in which a regressor is trained to predict upcoming states. Several attributes of the model need to be considered upon model selection:

- **Model accuracy**: the trained model needs to be accurate for previous observations and future time steps. This is especially relevant for consecutive predictions since the error will propagate over multiple predictions.
- **Model speed**: the trained model needs to be applied very often during the search process. Studies on MCTS have shown that increasing the number or the quality of rollouts can improve the agent's performance [10].
- **Model size**: the model's size (in terms of parameter count) can impact the training time and the number of observations required for optimizing the model's parameters. While regression models such as linear regression are the most simple to train, their applicability is quite limited. In contrast, deep neural networks are flexible but can require large amounts of training data.
- **Model interpretability and reliability**: a characteristic that is often neglected in deep learning approaches is the model's interpretability. While deep reinforcement learning has shown great performance, the black-box nature of deep neural networks may not allow human interpretation of its results. This complicates to measure the reliability or risk of a trained model. In contrast, planning based approaches can transparently summarize the search path and its predicted states. This can be especially important in risk critical applications and allow the computation of confidence bounds.

5 Improving the Confidence of a Forward Model

The forward model allows us to anticipate the result of an agent's actions. However, the prediction may be inaccurate and therefore the predicted result of an action sequence may be desirable but unlikely. In the context of a simulation-based search agent which is using a learned forward model, we would like to incorporate the confidence in the model's prediction in the agent's decision-making process as well as improving it throughout the training process.

5.1 Measuring the Learned Model's Confidence

The following analysis is based on the Gaussian Process regression model [29]. In contrast to many other regression models, it does not return a single predicted value, but a full predictive distribution. Given this distribution, we can estimate the mean value and the distribution's variance to be used by the agent. Other regression models may not provide the same information but it may be inferred given their output.

In the case of ensemble regression models, we can measure the model's confidence by the agreement of all regression models included in the ensemble. Hence, the ensemble does not need to consist of probabilistic regression models but may be built using any type of regression model. A common representative is the random forest regression [3], which consists of multiple decision tree regression models each trained on a different slice of the training data set. The ensembles response will be the mean of predicted values. Similarly, we can measure the variance of made predictions to estimate the confidence of the ensemble model.

5.2 Learning Goals Based on the Model's Confidence

In previous studies, the training process was implemented as a random exploration [12, 13]. As a result, we have observed that the forward model's accuracy quickly degrades in situations that have not previously been observed. This is due to the missing diversity in the training set. Therefore, we want to actively approach states and actions that the forward model cannot accurately predict to improve its generality.

Given the task of predicting action sequences of length m, we will use a decomposed forward model to predict the upcoming states after each action of the action sequence. Starting with a state $s_t = (s_{1,t}, \dots, s_{n,t})$ we will predict the future states s_{t+1} to s_{t+m}. Similarly, we will learn to predict the rewards r_{t+1} to r_{t+n} the agent expects to receive, which can be modeled as an additional output of the environment's state observation. For simplicity, we will use s and r to denote the true state and reward, and $\hat{s}$ and $\hat{r}$ for the predicted state and reward respectively. The following heuristics will be used to guide the agent's action selection during training and evaluation.

First, we focus on improving the agent's performance by selecting actions that are predicted to be promising by the forward model. Naively, we can assume that we only need to take the sum of predicted rewards of an action sequence into account. Hence, the predicted discounted return Q of the agent's planned action sequence results in the weighted sum of rewards:

$$Q = \sum_{i=1}^{m} \gamma^{i} \hat{r}_{t+i} \tag{4}$$

for which $\gamma \in (0, 1]$ balances between favoring long-term and short-term rewards.

As we pointed out before, the regression model might be inaccurate in its prediction. Next, we want to incorporate information on the predicted reward distribution to make the agent aware of such inaccuracies. Hence, during training, we want to actively approach actions the current model cannot confidently predict. To motivate this behavior, we let the agent actively approach situations with a high variance. An optimistic return heuristic can be formulated as:

$$Q_{opt} = \sum_{i=1}^{m} \gamma^{i} \hat{r}_{t+i} + \sum_{i=1}^{m} \sigma^2_{\hat{r},t+i} = \sum_{i=1}^{m} \left[\gamma^{i} \hat{r}_{t+i} + \sigma^2_{\hat{r},t+i}\right] \tag{5}$$

During the evaluation, we want to avoid such situations to minimize the agent's risk. Taking a pessimistic (Q_{pess}) approach, we reduce the expected reward by the variance of its predictions.

$$Q_{pess} = \sum_{i=1}^{m} \gamma^{i} \hat{r}_{t+i} - \sum_{i=1}^{m} \sigma^2_{\hat{r},t+i} = \sum_{i=1}^{m} \left[\gamma^{i} \hat{r}_{t+i} - \sigma^2_{\hat{r},t+i}\right] \tag{6}$$

Thus we can assure that two action sequences with similar outcome but varying confidence in its predictions can be differentiated.

So far, both heuristics have focused on the predicted reward. To assure that the model will also become more confident in its prediction of future states, we introduce an exploration goal. For this purpose, we will use the information on the model's confidence in predicting upcoming state variables to guide the agent into states that cannot be accurately predicted yet. Once again, we measure the model's confidence by taking the variance of the prediction into account. Since the predicted states are dependent on each other, we would require additional simulations to identify the reward at each time-step of the predicted action sequence. This process would be infeasible for real-time applications because the number of forward model evaluations is limited. Instead, we will use the cumulative standard deviations to approximate the uncertainty of a whole action sequence:

$$\sum_{i=1}^{m}\sum_{j=1}^{i}\sigma^2_{\hat{s}_{t+j}}$$

Hereby, we punish action sequences in which the first actions cannot be accurately predicted more than action sequences in which the results of later actions are unsure. We further incorporate the defined optimization targets into a single heuristic for action-selection.

$$Q_{conf} = \sum_{i=1}^{m}\left[\gamma^{i}\hat{r}_{t+i} + \alpha\sigma^2_{\hat{r}} + \beta \underbrace{\sum_{j=1}^{i}\sigma^2_{\hat{s}_{t+j}}}_{\substack{\text{cumulative}\\ \text{state variance}}}\right] \tag{7}$$

for which α balances the agent's risk awareness considering the reward predictions and β balances the agent's drive to explore unknown states. This heuristic has been motivated by Bienaymé's identity but since the independency of variables cannot be assured here, we included α and β to configure the heuristic according to the user's requirements. Setting $\alpha > 0$ sets the agent to actively approach states in which the predicted reward is unsure, whereas values below 0 let the agent avoid those states. For $\beta > 0$ the agent is encouraged to explore action sequences the model cannot confidently predict and once again lets the avoid those states for values below 0. For training in a safe environment, we recommend setting both values to be positive, while during testing we recommend choosing negative values to let the agent explore its environment. Those values may be adapted throughout the training depending on the confidence of the model and the agent's success.

6 Evaluating the Agent's Performance

Our evaluation is split into two parts. First, we analyze the agent's ability to learn a forward model and using it for guiding its action-selection in three of the most common motion control tasks (Fig. 2). During this experiment, we will compare the agent with reinforcement learning-based approaches. To evaluate the proposed measures, we designed a second experiment series, to test their effects on the agent's training and its resulting performance in the aforementioned motion control tasks. The source-code to our experiments and our results are available for download.[1]

---

[1] https://github.com/ADockhorn/Balancing-Exploration-And-Exploitation-in-Forward-Model-Learning.

(a) Cart-Pole (b) Pendulum (c) Acrobot

Fig. 2 Three motion-control environments and their representation in the OpenAI Gym framework

6.1 Motion Control Environments

The *cart-pole* problem [2] requires the agent to balance a pole by moving a cart back and forth. The pole is attached to a cart by an un-actuated joint, which moves along a frictionless track. The agent can apply a discrete force of $+1$ or -1 to the cart. The pole is starting in an upright position and needs to be kept in the range of $[-15, +15]$ degree from vertical. If the pole is falling below that threshold or the cart moves more than 2.4 units from the center, the episode ends. A maximal reward is achieved after balancing the pole for a maximum of 500 time-steps.

The *pendulum problem* is a classic problem in the control literature. In this version of the problem, the pendulum starts in a random position, and the goal is to swing it up so it stays upright. The reward penalizes deviations from the equilibrium and the magnitude of the agent's applied actions.

The *acrobot problem* requires the agent to swing up a pendulum with two links. Similar to the pendulum problem, the agent can apply a discrete rotational force to the joint of the first link. Initially, both links hang downwards, and the goal is to swing the end of the lower link up to a given height indicated by the target line. The reward is the height of the tip of the pendulum. The implementation of these environments is provided by the Open AI Gym framework [4].

6.2 Experiment Setup—Agent Performance

The proposed decomposed differential forward model (DDFM) will be compared to several popular reinforcement learning algorithms. In case of a discrete action space (Cart-Pole and Acrobot) we chose to use Sarsa [33], DQN [24], and CEM [34]. Whereas for the Pendulum environment, which uses a continuous action space, we compared our approach with the NAF algorithm [17]. The hyper-parameters of each algorithm were tuned by a simple grid-search.

For each environment, we continuously train our agents for a total of 100,000 time-steps. During training, we record the agents' reward per episode. We repeat the training process 10 times to get more stable results. The final performance of each model is measured by the average return of the last 10 training episodes.

6.3 Results

Figure 3 shows the agents' performance in each of the three training environments. The results indicate that simple problems, such as the Cart Pole and the Pendulum environment, can efficiently be solved after just a few training steps. In these, the agent outperforms deep-reinforcement learning approaches in terms of sampling efficiency. Nevertheless, the proposed model performs worse in the Acrobot environment, since solving the task requires to plan longer action sequences. Due to the simplicity of all three environments, the learning curves are steep and converge quickly. The convergence level seems to be very much dependent on the agent's parameters, for which the planning horizon and the regression model have shown to have the most impact.

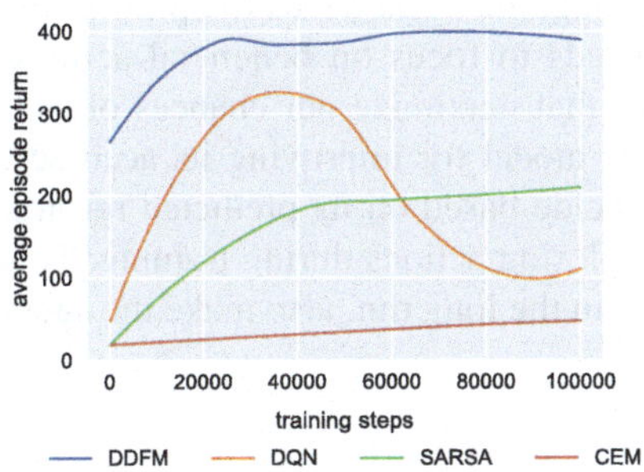

(a) Cart-Pole

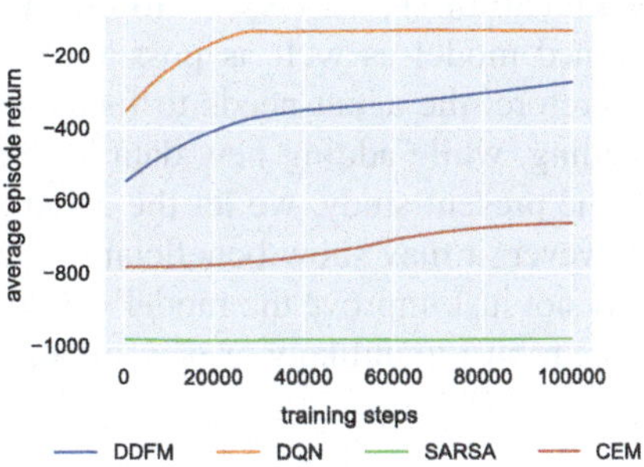

(b) Acrobot

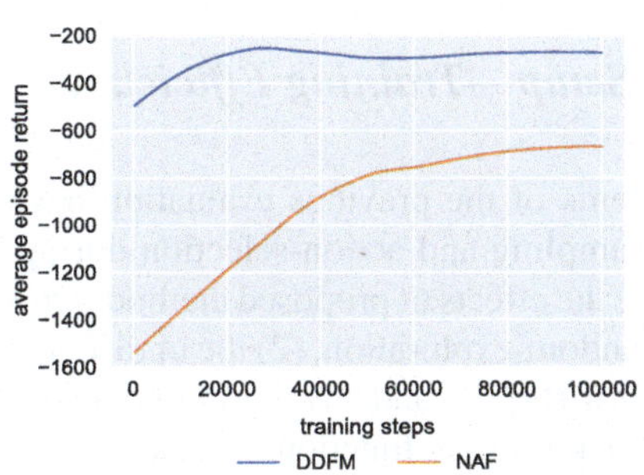

(c) Pendulum

Fig. 3 Training comparison of forward model learning and deep reinforcement learning agents. Graphs are showing the average episode return per training step smoothed using the local regression (loess). Each agent has been trained 10 times for 100,000 steps each

6.4 *Discussion*

Based on this evaluation and our preceding parameter optimization we have identified two problems of current forward model learning techniques.

Planning horizon dilemma: A common problem of planning algorithms is the choice of the planning horizon. While increasing the search can improve the accuracy of a planning agent's reward estimation, it also exponentially increases the number of possible states to be analyzed. In the case of forward model learning agents, the planning horizon is further restricted by the accuracy of the learned forward model. In contrast to using the true environment dynamics, predictions of a learned forward model become less accurate with increasing search depth. For choosing a suitable rollout length for a trained model we analyzed the model's prediction error over the length of multiple predicted action sequences. This can hint at a suitable parameter combination but has shown to be too costly to repeat throughout the model learning process. Dynamic measurements of the model's confidence in its prediction might help in improving the agent's performance. Similarly, algorithms like MCTS that do not use a fixed planning horizon may be beneficial.

Exploration versus Exploitation: During training, the agent needs to optimize the learned model as well as possible but also needs to focus on beneficial actions. Therefore, the agent needs to focus on actions that maximize our chances of succeeding, while adding new data to the forward model for improving its accuracy. In the present study, we let the agent solely decide based on its predicted reward. However, it may show beneficial to include exploring actions during training. This may not just improve the model's accuracy, but in the long run, also make the agent more robust to unlikely or new situations.

7 Evaluating the Effects of Confidence-Based Sampling

7.1 *Experiment Setup—Training Efficiency*

These observed problems of the previous evaluation have motivated the proposal of confidence-based sampling and action-selection during the agent's training and evaluation. To compare the effects of proposed methods we repeat the agent's training process using a (1) a random-exploration, (2) the undiscounted return (Eq. 4, $\gamma = 1$), (3) the optimistic return (Eq. 5), and (4) the confidence-based return (Eq. 7) with weights $\alpha = \beta = 1$. Each training function is used to train 10 models for 500 time-steps each. During training, we update the agent's forward model every 10 time-steps. The forward model will be represented as a differential decomposed forward model using Gaussian process regression with a radial basis function kernel.

After 500 time-steps of training, we store the agent's forward model and keep it constant for the remaining evaluation. Furthermore, we test the learned model's for

10 episodes using (1) a random agent, (2) the discounted return (Eq. 4, $\gamma = 1$), (3) the pessimistic return (Eq. 5), and (4) the confidence-based return (Eq. 7) with weights $\alpha = \beta = -1$. For a reliable action-selection, we simulate all action-sequences of length x and select the first action of the best-rated action sequence. In case multiple action sequences are predicted to yield the same value, we choose a random action sequence among the best action sequences.

7.2 Results

Figure 4 shows the results for each combination of train and test action-selection methods on the cart-pole environment. The box-plots show that the best performance is achieved by using the confidence-based return during the agent's training and evaluation. Even if the agent has been trained using random interactions, the agent can improve its performance using any of the proposed functions for action-selection.

The latter is especially interesting since it suggests that confidence-based sampling can benefit the action-selection of all existing applications of forward model learning without retraining the model. It allows the agent to easily balance between exploration and exploitation of the state-space by making it aware of action sequences that cannot be correctly predicted. This comes at the cost of estimating the prediction variance. Depending on the used regression method, this might result in just a small overhead, as we have demonstrated for ensemble regression models. Nevertheless, other regression models may require more work.

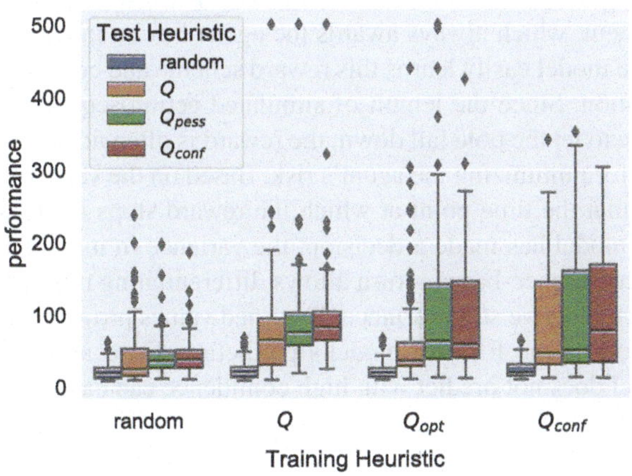

Fig. 4 Testing all combinations of train and test functions. Training has been done for 500 time-steps of the environment. For each trained model we performed 10 evaluations and measured the average length the agent kept the pole balanced. The experiment has been repeated 10 times to average the results of multiple training runs

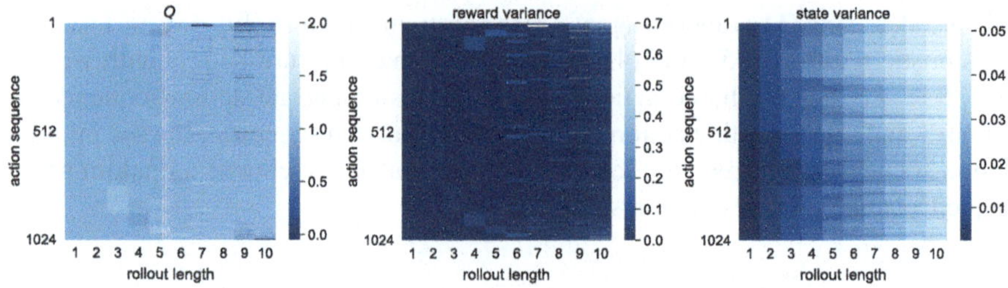

Fig. 5 Simulated action sequences an their predicted value given the proposed sampling methods

In our evaluation, we have only simulated action sequences of equal length. The application of confidence-based measures may allow using adaptive planning horizon by stopping the simulation as soon as the cumulative state confidence falls below a set threshold. This can be especially interesting for tree-search-based algorithms, such as MCTS, to cut off the current rollout in case the forward model cannot provide accurate predictions anymore.

To get a better picture of the effects of confidence-based action selection, we have devised a final test. In this experiment we have used a single state of the cart-pole environment, a trained model, and all simulated action sequences of length 5. We further used the (1) undiscounted return (Eq. 4, $\gamma = 1$), (2) the pessimistic return (Eq. 5), and (3) the confidence-based return (Eq. 7) with weights $\alpha = \beta = -1$ to evaluate all the action sequences.

Figure 5 show the respective values for all steps of all simulated action sequences. Looking at the values provided by the return shows that the agent can hardly differentiate action sequences based on their value. This is caused by the uniform reward of the environment, which always awards the agent 1 point in case the pole did not tip over yet. The model easily learns this reward scheme and correctly represents the reward distribution. Since the length of simulated action-sequences does in many cases not suffice to let the pole fall down, the reward is often not a sufficient signal to choose actions for minimizing the agent's risk. Based on the value of the pessimistic return, we see that the time point at which the reward stops is unsure, but as soon as the forward model has made a decision, the variance in its prediction returns to 0. Finally, the confidence-based return allows differentiating most action sequences based on their cumulative state variance. Returned values provide the agent with the most information on the forward model's prediction. Especially action sequences which the model does not predict with high confidence can easily be avoided.

8 Conclusion and Future Work

In this work, we have analyzed forward model learning algorithms and how they fit into the big picture of algorithms for computational intelligence in games and motion control. Based on our recent work on forward model learning, we have proposed the decomposed differential forward model. In its most general form, this model assumes that the environment's observable sensor values are independent of each other. This restriction can be loosened in case enough training data is available, such that the dependency structure can be inferred.

The model has further been tested in three classic motion control environments. Results have been compared to deep-reinforcement learning approaches. Our first experiments have shown that the performance is largely dependent on the accuracy of the learned forward model and its feasible prediction horizon. To further improve upon this, we have proposed several confidence-based sampling measures to make the agent aware of the accuracy of its predictions to judge the risk of an action sequence. A second experiment series has shown, that proposed sampling measures can improve the efficiency of the training process. Furthermore, they are capable of improving the agent's performance during evaluation even in case the model has been trained using random sampling.

Even if reinforcement-learning has shown to be able to solve many motion-control tasks, these methods still require large amounts of training data. Since the proposed planning based approaches have shown great learning performance during the first steps of acting in a new environment, the next step will be to combine these two algorithmic schemes in a single agent. Such an agent may dynamically choose to either trust the reinforcement learner's value model or to rely on a learned forward model for using a planning-based approach. Similarly, the planner could benefit from the reinforcement-learner's policy during the search phase. Since curiosity-driven learning and other intrinsic reward schemes have already shown promising results for improving reinforcement-learning agents, similar improvements could be achieved when being applied to forward model learning agents.

References

1. Apeldoorn, D., Dockhorn, A.: Exception-tolerant hierarchical knowledge bases for forward model learning. IEEE Trans. Games 1–14 (2020). https://doi.org/10.1109/TG.2020.3008002
2. Barto, A.G., Sutton, R.S., Anderson, C.W.: Neuronlike adaptive elements that can solve difficult learning control problems. IEEE Trans. Syst. Man Cybern. **SMC-13**(5), 834–846 (1983). https://doi.org/10.1109/TSMC.1983.6313077
3. Breiman, L.: Random forests. Mach. Learn. **45**(1), 5–32 (2001). https://doi.org/10.1023/a:1010933404324
4. Brockman, G., Cheung, V., Pettersson, L., Schneider, J., Schulman, J., Tang, J., Zaremba, W.: Openai gym (2016)
5. Browne, C.B., Powley, E., Whitehouse, D., Lucas, S.M., Cowling, P.I., Rohlfshagen, P., Tavener, S., Perez, D., Samothrakis, S., Colton, S.: A survey of Monte Carlo tree search methods. IEEE Trans. Comput. Intell. AI Games **4**(1), 1–43 (2012). https://doi.org/10.1109/TCIAIG.2012.2186810
6. Clary, P., Morais, P., Fern, A., Hurst, J.: Monte-Carlo planning for agile legged locomotion. In: Proceedings International Conference on Automated Planning and Scheduling, ICAPS 2018-June (Icaps), pp. 446–450 (2018)
7. Dearden, A., Demiris, Y.: Learning forward models for robots. In: IJCAI International Joint Conference on Artificial Intelligence, pp. 1440–1445 (2005)

8. Dockhorn, A.: Prediction-based search for autonomous game-playing. Ph.D. thesis, Otto-von-Guericke-Universität Magdeburg, Fakultät für Informatik (2020). https://doi.org/10.25673/34014
9. Dockhorn, A., Apeldoorn, D.: Forward model approximation for general video game learning. In: Proceedings of the 2018 IEEE Conference on Computational Intelligence and Games (CIG'18), pp. 425–432. IEEE (2018). https://doi.org/10.1109/CIG.2018.8490411
10. Dockhorn, A., Doell, C., Hewelt, M., Kruse, R.: A decision heuristic for Monte Carlo tree search Doppelkopf agents. In: 2017 IEEE Symposium Series on Computational Intelligence (SSCI), pp. 1–8. IEEE (2017). https://doi.org/10.1109/SSCI.2017.8285181
11. Dockhorn, A., Kruse, R.: Detecting sensor dependencies for building complementary model ensembles. In: Proceedings of the 28. Workshop Computational Intelligence, Dortmund, 29–30 Nov 2018, pp. 217–234 (2018)
12. Dockhorn, A., Kruse, R.: Forward model learning for motion control tasks. In: 2020 IEEE 10th International Conference on Intelligent Systems (IS), pp. 1–5. IEEE (2020). https://doi.org/10.1109/IS48319.2020.9199978
13. Dockhorn, A., Lucas, S.M., Volz, V., Bravi, I., Gaina, R.D., Perez-Liebana, D.: Learning local forward models on unforgiving games. In: 2019 IEEE Conference on Games (CoG), pp. 1–4. IEEE, London (2019). https://doi.org/10.1109/CIG.2019.8848044
14. Dockhorn, A., Tippelt, T., Kruse, R.: Model decomposition for forward model approximation. In: 2018 IEEE Symposium Series on Computational Intelligence (SSCI), pp. 1751–1757. IEEE (2018). https://doi.org/10.1109/SSCI.2018.8628624
15. Gaina, R.D., Liu, J., Lucas, S.M., Pérez-Liébana, D.: Analysis of vanilla rolling horizon evolution parameters in general video game playing. In: Lecture Notes in Computer Science (including subseries Lecture Notes in Artificial Intelligence and Lecture Notes in Bioinformatics), vol. 10199, pp. 418–434. Springer (2017). https://doi.org/10.1007/978-3-319-55849-3_28
16. Gaina, R.D., Lucas, S.M., Perez-Liebana, D.: Rolling horizon evolution enhancements in general video game playing. In: 2017 IEEE Conference on Computational Intelligence and Games (CIG), pp. 88–95. IEEE (2017). https://doi.org/10.1109/CIG.2017.8080420
17. Gu, S., Lillicrap, T., Sutskever, U., Levine, S.: Continuous deep q-learning with model-based acceleration. In: 33rd International Conference on Machine Learning, ICML 2016, vol. 6, pp. 4135–4148 (2016)
18. Ha, D., Schmidhuber, J.: Recurrent world models facilitate policy evolution. In: Bengio, S., Wallach, H., Larochelle, H., Grauman, K., Cesa-Bianchi, N., Garnett, R. (eds.) Advances in Neural Information Processing Systems, vol. 31, pp. 2450–2462. Curran Associates, Inc. (2018)
19. Hafner, D., Lillicrap, T., Ba, J., Norouzi, M.: Dream to Control: Learning Behaviors by Latent Imagination (2019). http://arxiv.org/abs/1912.01603
20. Henaff, M., Whitney, W.F., LeCun, Y.: Model-based planning with discrete and continuous actions (2017)
21. Howard, R.: Dynamic Programming and Markov Processes. Technology Press Research Monographs. Technology Press of Massachusetts Institute of Technology (1960)
22. Lucas, S.M., Dockhorn, A., Volz, V., Bamford, C., Gaina, R.D., Bravi, I., Perez-Liebana, D., Mostaghim, S., Kruse, R.: A local approach to forward model learning: results on the game of life game. In: 2019 IEEE Conference on Games (CoG), pp. 1–8. IEEE (2019). https://doi.org/10.1109/CIG.2019.8848002
23. Maschler, M., Solan, E., Zamir, S.: Game Theory. Cambridge University Press, Cambridge (2013). https://doi.org/10.1017/CBO9780511794216
24. Mnih, V., Kavukcuoglu, K., Silver, D., Rusu, A.a., Veness, J., Bellemare, M.G., Graves, A., Riedmiller, M., Fidjeland, A.K., Ostrovski, G., Petersen, S., Beattie, C., Sadik, A., Antonoglou, I., King, H., Kumaran, D., Wierstra, D., Legg, S., Hassabis, D.: Human-level control through deep reinforcement learning. Nature **518**(7540), 529–533 (2015). https://doi.org/10.1038/nature14236
25. Nguyen-Tuong, D., Peters, J.: Model learning for robot control: a survey. Cognit. Process. **12**(4), 319–340 (2011). https://doi.org/10.1007/s10339-011-0404-1

26. Perez Liebana, D., Dieskau, J., Hunermund, M., Mostaghim, S., Lucas, S.: Open loop search for general video game playing. In: Proceedings of the 2015 on Genetic and Evolutionary Computation Conference—GECCO '15, pp. 337–344. ACM Press, New York, NY, USA (2015). https://doi.org/10.1145/2739480.2754811
27. Perez-Liebana, D., Lucas, S.M., Gaina, R.D., Togelius, J., Khalifa, A., Liu, J.: General Video Game Artificial Intelligence, vol. 3. Morgan & Claypool Publishers (2019). https://gaigresearch.github.io/gvgaibook/
28. Racanière, S., Weber, T., Reichert, D.P., Buesing, L., Guez, A., Rezende, D., Badia, A.P., Vinyals, O., Heess, N., Li, Y., Pascanu, R., Battaglia, P., Hassabis, D., Silver, D., Wierstra, D.: Imagination-augmented agents for deep reinforcement learning. Advances in Neural Information Processing Systems 2017-Decem (Nips), pp. 5691–5702 (2017)
29. Rasmussen, C.E., Williams, C.K.I.: Gaussian Processes for Machine Learning. University Press Group Limited, New York (2006)
30. Rubinstein, R.Y., Kroese, D.P.: Simulation and the Monte Carlo Method. Wiley (2016). https://doi.org/10.1002/9781118631980
31. Silver, D., Huang, A., Maddison, C.J., Guez, A., Sifre, L., van den Driessche, G., Schrittwieser, J., Antonoglou, I., Panneershelvam, V., Lanctot, M., Dieleman, S., Grewe, D., Nham, J., Kalchbrenner, N., Sutskever, I., Lillicrap, T., Leach, M., Kavukcuoglu, K., Graepel, T., Hassabis, D.: Mastering the game of Go with deep neural networks and tree search. Nature **529**(7587), 484–489 (2016). https://doi.org/10.1038/nature16961
32. Sutton, R.S.: Learning to predict by the methods of temporal differences. Mach. Learn. **3**(1), 9–44 (1988). https://doi.org/10.1007/BF00115009
33. Sutton, R.S., Barto, A.G.: Reinforcement Learning, 2nd edn. The MIT Press, Cambridge (2018)
34. Szita, I., Lorincz, A.: Learning Tetris using the noisy cross-entropy method. Neural Comput. **18**(12), 2936–2941 (2006). https://doi.org/10.1162/neco.2006.18.12.2936
35. Watkins, C.J.C.H., Dayan, P.: Q-learning. Mach. Learn. **8**(3–4), 279–292 (1992). https://doi.org/10.1007/bf00992698
36. Weber, T., Racanière, S., Reichert, D.P., Buesing, L., Guez, A., Rezende, D.J., Badia, A.P., Vinyals, O., Heess, N., Li, Y., Pascanu, R., Battaglia, P., Silver, D., Wierstra, D.: Imagination-Augmented Agents for Deep Reinforcement Learning (2017). http://arxiv.org/abs/1707.06203
37. Yannakakis, G.N., Togelius, J.: Artificial Intelligence and Games. Springer, Cham (2018). https://doi.org/10.1007/978-3-319-63519-4

Universal Adversarial Perturbation Generated by Using Attention Information

Zifei Wang, Xiaolin Huang, Jie Yang, and Nikola Kasabov

Abstract Since the concept of adversarial attack was proposed, the vulnerability of Deep Neural Networks (DNNs) has attracted the interest of artificial intelligence researchers. Many state-of-the-art DNNs suffer the risk of adversarial perturbations, which can be divided into image-dependent and universal. In this paper, we review some traditional algorithms developed by others and introduce a novel method to create universal adversarial perturbations by attention. Our approach is the first to generate universal perturbations by attacking the attention heat maps with the interpretation method, Layer-wise Relevance Propagation. Our experiment results, when compared with traditional ones, showing an improvement both on fooling ratios and transferability with the ImageNet dataset. The goal of our research is to create less vulnerable DNNs to adversarial attacks.

1 Introduction

Deep neural networks (DNNs) have performed very well on many challenging tasks, such as image classification, semantic segmentation, and object detection. But DNNs are shown to be vulnerable under adversarial examples, which are artificial perturbed examples that are indistinguishable from benign examples but can lead DNNs to make

Z. Wang · X. Huang (✉) · J. Yang (✉)
Institute of Image Processing and Pattern Recognition,
Shanghai Jiao Tong University, Shanghai, China
e-mail: xiaolinhuang@sjtu.edu.cn

J. Yang
e-mail: jieyang@sjtu.edu.cn

Z. Wang
e-mail: wangzifei@sjtu.edu.cn

N. Kasabov
Auckland University of Technology, Auckland, New Zealand
e-mail: nk.kasabov@ulster.ac.uk; nkasabov@aut.ac.nz

George Moore Chair of Data Analytics, Ulster University, Coleraine, UK

V. Sgurev et al. (eds.), *Advances in Intelligent Systems Research and Innovation*,
Studies in Systems, Decision and Control 379,
https://doi.org/10.1007/978-3-030-78124-8_2

wrong decisions. Adversarial perturbation is first proposed by [28], and then various algorithms [4, 6, 14, 20, 23] are developed to confirm this phenomenon.

Adversarial perturbations can be divided into two categories: Image-dependent and Universal. Image-dependent perturbations depend on both the target networks and input images, which means that each image-dependent perturbation is crafted from one input benign image by attacking a specific network. What we traditionally call adversarial perturbation is just image-dependent perturbation. The perturbation is usually computed related to the gradients of the network output with the input image.

The attempt of universal perturbation is to make the adversarial attack get rid of the input images, which may reveal the essential weakness of the neural network. Moosavi-Dezfooli et al. [19] design a *universal* perturbation added on different benign images, which is thus called universal adversarial perturbation (UAP). Surprisingly, UAP could make most labels predicted wrongly by the attacked DNN. The goal of UAP is to develop universal perturbation v with small magnitude such that for most benign image x, $x + v$ can fool the per-trained network.

It is clear that adversarial attacks can reveal some weak aspects in deep learning. In spite of using adversarial attack to research the decision boundary of DNNs, there is an interpretation method which is another way to analyze the black-box behavior of DNN. Some novel neural network visualization methods can produce heat maps that manifest the attention of the network. Inspired by this, we develop our method for UAP from the aspect of attention.

Compared with image-dependent perturbations, the UAPs rely on image-independent weakness of DNNs. In [19], UAP is generated by attacking the classification output and in this paper. Different from [19], we propose a novel method to attack the attention heat maps. Heat maps have been widely used in interpretation and causal analysis for DNNs. Therefore, our UAP on heat maps is helpful to find interpretable image-independent weakness of DNNs. The proposed here UAP has higher attack success rates than that in [19], also showing that attack on attention heat maps is promising.

Specifically, we hypothesize that the heat maps produced by Layer-wise Relevance Propagation (LRP) [2] contain the correlations among different input images and are equally significant with the decision boundary. For example, the heat maps of "cat" images share some common statistics features, like magnitude, and semantic features, like eyes. Our algorithm seeks universal perturbation v with small magnitude such that for different benign images x, $x + v$ can change the amplitudes and structures of the heat maps. The UAPs generated by our method on ImageNet [25] can fool classification networks with high probability, more than 80%. Our main contributions can be summarized as follows:

1. We propose a novel architecture for creating non-targeted universal perturbations for classification networks, with L_∞ norm as the metric.
2. We achieve excellent performance in universal perturbations on challenging high-resolution dataset ImageNet [25] by attacking the heat maps produced by LRP iteratively.

3. We show that our universal perturbation has transferability across different networks, which is universal both for input images and target DNNs.
4. Our results demonstrate the correlation between the decision boundary and interpretation methods, encouraging a new view of adversarial attack.

The paper is organized as follows. In the next section, an introduction to the adversarial perturbation for deep neural networks is presented. In Sect. 3, the development of universal perturbation is discussed. In Sect. 4 interpretation methods for deep neural networks are divided into three categories. In Sect. 5, we introduce Layer-wise Relevance Propagation and our own method to generate universal adversarial perturbation based on LRP. In Sect. 6, the experiment results show our improvement on universal adversarial attack, both on the fooling ratio and transferability. The conclusions recapitulate our contributions and are displayed in the last section.

2 Adversarial Perturbation for Deep Neural Networks

Adversarial perturbation is the noise calculated artificially according to a certain attack algorithm for the specified neural network and data set. Different from random noises, the norm of the adversarial perturbation is extremely small. When added to the sample data, it will not interfere with human eye, but will turn benign samples into adversarial samples, making it impossible for a pre-trained neural network to perform correct classification.

Traditionally, the adversarial perturbation we usually say refers to the image-dependent perturbation. The image-dependent perturbation is a kind of noise obtained by artificial calculation, which works on a specific sample image. Adding the image-dependent perturbation to the benign image can get the adversarial example. The pre-trained neural network cannot correctly classify the adversarial example. But there is a limitation that the image-dependent perturbation only has an attack ability on specific sample image, and will not cause mistakes on other samples in the same data set. The definition of image-dependent perturbation is given below:

$$k(x_i + v_i) \neq k(x_i), \quad x_i \sim \mu, \tag{1}$$

while $k(x_i)$ donates the predicted category of data x_i in the distribution μ according to the pre-trained network, and v_i donates the perturbation related to x_i.

The main idea of generating adversarial examples depends on the gradient information of deep neural networks. After the benign image is feed into the deep neural network, the perturbation can be added to the image according to the direction of the gradients of the loss function with the input image. The perturbed image crosses the decision boundary, resulting in the misclassification of the network. Many methods for generating adversarial examples have been proposed, and there are also many classification standards. Generally speaking, the common classification standards are as follows:

1. White-box attack and black-box attack: If the parameters of the attacked model are known during the attack, then the attack is a white-box attack; if the parameters of the attacked model are unknown, then the attack is a black-box attack.
2. Target attack and untarget attack: If the adversarial sample is classified into the desired category, it is called a targeted attack; if the adversarial sample is only required to be misclassified, no specific misclassification is required, then it is called untarget attack.
3. Single-step attack and iteration attack: Single-step attack refers to the generation of adversarial perturbation in one iteration, while iteration attack means the adversarial perturbation is obtained after multiple iterations.

The most representative algorithm of adversarial attack is Fast Gradient Sign Method (FGSM) [6], The FGSM method fixes the model parameters, substitutes the sample x_i and the label y_i into the cost function $J(x_i, y_i)$, and then calculate the gradient of the sample x_i in the sign gradient direction $sign(\cdot)$ to generate the adversarial sample x_i^*. By adjusting the step η, the magnitude of the perturbation can be controlled and different attack success rates can be obtained.

$$x_i^* = x_i + v_i = x_i + \eta sign(\nabla_x J(x_i, y_i)). \tag{2}$$

The adversarial examples generated above are untarget adversarial examples. If the label y_i in the cost function is changed to y_i^*, and then subtract the calculated perturbation from the sample to generate a targeted adversarial sample classified as y_i^*:

$$x_i^* = x_i + v_i = x_i - \eta sign(\nabla_x J(x_i, y_i^*)). \tag{3}$$

FGSM is simple to calculate and can quickly generate a large number of adversarial samples. Researchers have proposed many other methods of generating adversarial samples based on FGSM, such as the multi-step iterative method i-FGSM [14] for generating adversarial examples:

$$\begin{aligned} &Initialization: \quad x_i^* = x_i, \\ &N\ times: \quad x_i^* = x_i^* + \frac{\eta}{N} sign(\nabla_x J(x_i^*, y_i)). \end{aligned} \tag{4}$$

It iterates from the original sample x_i and iterates N times to generate the final adversarial sample.

In the framework of adversarial attack, many other attack algorithms for adversarial sample generation are also proposed, such as Basic Iterative Method (BIM)/Projected Gradient Descent (PGD) [3], Carlini and Wagner (C&W) attack [4], Jacobian-based Saliency Map Attack (JSMA) [5] and DeepFool [6]. These attack algorithms are designed in the white box threat model. However, due to the transferability of adversarial examples between models, they are also effective in many gray-box and black-box settings.

3 Universal Adversarial Perturbations for Deep Neural Networks

Different from Image-dependent perturbations generated by methods [4, 6, 14, 20, 23], universal adversarial perturbations have the ability to generalize across different images. Universal adversarial perturbation is a kind of adversarial perturbation that can turn most of the samples in the data set into adversarial examples to attack the neural network. It has strong generalization ability and can cause serious attacks on the neural network, so generating universal perturbation is of great significance for detecting network robustness and improving network security. The definition of universal adversarial perturbation is as follows:

$$k(x_i + v) \neq k(x_i), \quad \text{for “most” } x_i \sim \mu, \tag{5}$$

while $k(x_i)$ donates the predicted category of data x_i in the distribution μ according to the pre-trained network, and v donates the universal adversarial perturbation not related to x_i.

Moosavi-Dezfooli et al. [19] first introduce the universal perturbation that can be added to any image to fool the DNNs with a high fooling rate. They create UAPs by adding and projecting small perturbations computed by DeepFool method [20] iteratively. For the benign image x_i sampled from the distribution μ, the perturbation v is calculated iteratively, so that $x_i + v$ tends to the decision boundary, and finally a perturbation with a universal attack effect is obtained.

Let us consider $X = x_1, \ldots, x_m$ as a data set sampled form the image distribution μ. We consider the image classification DNN as f, which is pre-trained on benign images from C different categories. For each input image x in the image distribution $\mu, f(x)$ produces a vector whose length is the number of image categories, and $f_c(x)$ is the cth element of $f(x)$. We report that f obtains a high accuracy on data set X. v donates the universal adversarial perturbation we wanted. Then the predicted category of data x_i mentioned above has such a relationship with DNN f:

$$k(x_i) = \arg\max_c f_c(x) \tag{6}$$

At the beginning of the iterations of the algorithm proposed by Moosavi-Dezfooli et al. [19], the universal perturbation vector v is zero-initialized. Then for each data-point $x_i \in X$, the algorithm seeks for the minimal perturbation Δv to make the current perturbed point, $x_i + v$, cross the decision boundary of the classifier. The new perturbation Δv is aggregated to the current instance of the universal perturbation. It's clear that the key point is an optimization problem to minimal the perturbation at the limitation of misclassification. The problem can be described as:

$$\Delta v_i \leftarrow \arg\min_r ||r||_2, \;\; s.t. \; k(x_i + v + r) \neq k(x_i). \tag{7}$$

Due to the high computational complexity of this optimization problem, Moosavi-Dezfooli et al. [19] choose to use Deepfool algorithm to provide an approximate solution. The idea of the Deepfool algorithm is to first find a number of vectors from the sample point to the decision boundary of the target CNN model f in the feature space near the sample point x. Then the closest hyperplane l of the boundary can be computed as follows,

$$l = \arg\min_{c \neq k(x)} \frac{|f_c(x) - f_{k(x)}(x)|}{\|\nabla f_c(x) - \nabla f_{k(x)}(x)\|_2^2} \tag{8}$$

The minimum perturbation r is the vector indexed by l,

$$r = \frac{|f_l(x) - f_{k(x)}(x)|}{\|\nabla f_l(x) - \nabla f_{k(x)}(x)\|_2^2}(\nabla f_l(x) - \nabla f_{k(x)}(x)) \tag{9}$$

Specific to the optimization problem mentioned above, we use $x_i + v$ to replace x in (8) and (9), and get the perturbation Δv_i we want.

Due to the requirement that the perturbation is imperceptible to humans, the magnitude of the final perturbation should be constrained. Usually, we limit $L_p (p \geq 1)$ norm of the perturbation to a constant L,

$$\|v\|_p \leq L. \tag{10}$$

In the algorithm, the updated universal perturbation is projected on the L_p ball of radius L and centered at 0. The projection operation defined as follows:

$$P_{p,L}(v) = \arg\min_{v'} \|v - v'\|_2, \ s.t. \|v'\|_p \leq L. \tag{11}$$

Then, we update the universal perturbation as

$$v \leftarrow P_{p,L}(v + \Delta v_i) \tag{12}$$

During the iteration, each image x_i in the data set X is feed into the algorithm above, and update the perturbation v. The iteration is terminated when the fooling ratio on the perturbed data set $X_v := \{x_1 + v, \ldots, x_m + v\}$ reaches our expectation.

There are also some other excellent algorithms about universal adversarial perturbations. [7, 24] train a generative neural network to generate universal adversarial perturbations by feeding random noises into the generative network. Once the training is finished, the generative neural network can transform random noises into universal perturbations. However, the real performance of the above methods relies heavily on the size of the training set and iteration numbers. To overcome this weakness, Khrulkov and Oseledets [12] proposes a new algorithm based on computing the so-called (p,q)-singular vectors of the Jacobian matrices of the DNNs' hidden layers.

Using only tens of images, Khrulkov and Oseledets [12] can construct universal perturbations with relatively high fooling rates.

To further get rid of the training set, [21, 22] present a training data independent approach to generate universal perturbations. Its target is to cause strong disturbances at the activation of the middle layers when the input is the UAP. [17] also proposes a Prior Driven Uncertainty Approximation (PD-UA) method to fully utilize the uncertainty of the DNNs, thus create UAPs without supervision.

To explore universal perturbations existed in other tasks, [3, 9, 15, 24] develop special algorithms in areas such as semantic segmentation, image retrieval, and natural language processing. These researches prove the widespread existence of universal perturbations in so many tasks, and call for researchers' attention to this phenomenon in both theory and practice.

4 Interpretation Methods for Deep Neural Networks

Traditional decision models such as linear models or decision trees are intuitive and easy to understand, but there are problems such as large deviations or high variance in performance. Linear models are prone to underfitting, and tree models are prone to overfitting. Deep learning has very high predict performance, but the end-to-end black-box model is difficult to interpret.

With the criticism of the black-box instinct of deep learning, various approaches to interpretation have been proposed to overcome this weakness. Neural network visualization methods can provide a heat map which donates the attention of the network, and is the main research area of interpretation. They can be separated into three categories: input modification methods, Class Activation Mapping (CAM) methods, and back propagation-based methods.

Among them, input modification methods analyze the changes of outputs according to the changes of inputs, such as noise [30], masks [29], or perturbed variants [32]. [29] puts masked patches at different places of the input image, which leads to the drop in the prediction probability. Then it creates heat maps based on this drop. However, these forward methods can be computationally costly due to the complex input modifications and bring up randomness.

Class Activation Mapping methods [16, 31] replace the fully-connected layers of DNNs with convolutional layers and global average pooling (GAP), thus achieving class-specific heat maps. The shortcomings of CAM are obvious. In order to obtain the weights in GAP, the model needs to be retrained after replacing the last classifier. And CAM is not applicable to all networks. Gradient-Weighted CAM (Grad-CAM) [26] combines the gradient information and the class activation maps of the DNNs to obtain heat maps. Grad-CAM can generalize on any targeted layer. Guided Grad-CAM [26] combine guided back propagation and CAM algorithm, and can produce pixel-wise heat maps.

Back propagation-based methods use the backpropagation mechanism of deep neural networks to propagate the importance information of model decision-making from the output layer neurons of the model to the input layer of the model to obtain the feature importance of the input samples. Gradient-based methods are usually able to generate pixel-wise correlation maps and obtain the visual interpretation of saliency maps.

In this paper, we concentrate on Layer-wise Relevance Propagation (LRP) [2], a back propagation-based method to "explain" a DNN with backpropagating relevance from the output to the input, which exploits the relationship between the input x and the output $f(x)$ with the DNN as f.

5 Universal Adversarial Perturbation Generated by Attention

5.1 Layer-Wise Relevance Propagation

Layer-wise Relevance Propagation is a Deep Taylor Decomposition (DTD) [18] based interpretation method. It redistributes the relevance of the output backwards through the pre-trained network using DTD and determines the contributions of the nodes to the classification. The relevance of each layer is obtained by propagating the relevance in the next layer based on the activation and network weights. The interpreter gives a pixel-wise heat map as the same dimensionality as the input image, thus visualizes the important regions of the input image which contribute to the selected category.

This method is derived from the law of conservation between layers, which means that the sum of the relevance of neurons in two adjacent layers remains the same. If $R_i^{(l)}$ represents the relevance between the ith neuron in the lth layer and the final output result, and $R_j^{(l+1)}$ donates the relevance between the jth neuron in the next layer and the final output result, then the conservation law requires:

$$\sum_i R_i = \ldots = \sum_i R_i^{(l)} = \sum_j R_j^{(l+1)} = \ldots = f(x). \tag{13}$$

If we use simple LRP rule to backpropagation the relevance:

$$R_i^{(l)} = \sum_j \frac{a_i^{(l)} w_{i,j}^{+(l)}}{\sum_i a_i^{(l)} w_{i,j}^{+(l)}} R_j^{(l+1)}, \tag{14}$$

$a_i^{(l)}$ is the post-activation output of node i in layer l. $w_{i,j}^{+(l)}$ and $w_{i,j}^{+(l)}$ donate:

$$w_{i,j}^{+(l)} = \max\left(w_{i,j}^{(l)}, 0\right),$$
$$w_{i,j}^{+(l)} = \max\left(w_{i,j}^{(l)}, 0\right). \quad (15)$$

According to these equations, we could produce a pixel-wise relevance map on a simple network. LRP can be further extended to more complex neural networks, and there are also some different propagation rules according to the layer type. For more details, refer to [2] to learn more (Figs. 1 and 2). The LRP interpreter g needs to be constructed on the target classification DNN f. It takes the image x and selected category label c as inputs, then generates the heat map $g(x, c)$ of the corresponding category. When universal perturbation v is added, the heat map of the perturbed image is defined as $g(x + v, c)$. It contains both positive and negative relevance values. In our work, we only use the positive values $\max(g(x + v, c), 0)$, donating how much a pixel contributes to predicting the category label c. For the sake of ease, we still use $g(x + v, c)$ to represent the positive heat map in this paper. In practice, it should be noted that the beginning of the back-propagation is the output of the last fully connected layer before softmax normalization layer, not the final probability vector. Figure 3 shows the visualization of LRP results. As we can see, LRP algorithm can provide the attention information of the neural network. We confirm that LRP method is able to isolate the relevance of the target class with high accuracy. There are less background noise and overlap. For instance, the monkey in the figure is highlighted in the LRP heat map, but the person beside is not highlighted.

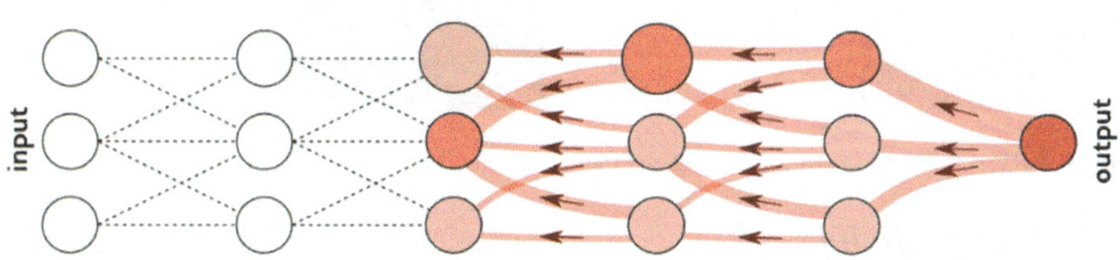

Fig. 1 LRP:Backpropagation the relevance

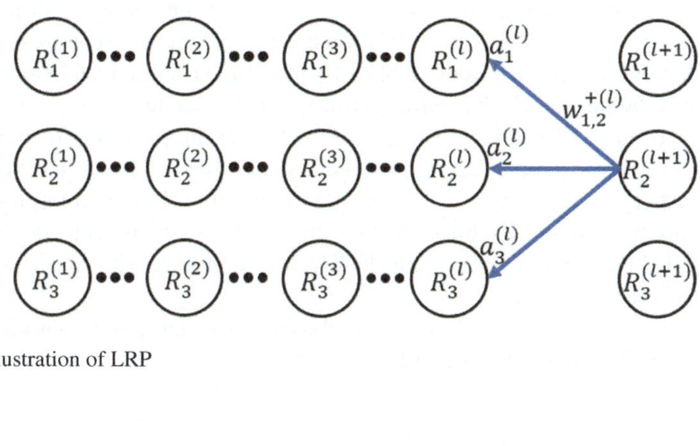

Fig. 2 Illustration of LRP

Fig. 3 LRP results visualization of original category

5.2 Attack Method

We consider the image classification DNN as a function f, which is pre-trained on benign images from C different categories. For each input image x in the image distribution $\mu, f(x)$ produces a vector of probabilities for image categories, and $f_c(x)$ is the cth element of $f(x)$. We report that f obtains a high accuracy on benign images and v donates the universal adversarial perturbation we wanted.

Different from [7, 19, 24] which immediately use the gradients of the DNN's outputs and inputs to create UAPs, our method chooses to attack the heat maps provided by LRP method instead. Due to the assumption that the interpretation of LRP contains the attention information of the DNN, we construct our algorithm to generate UAPs through destroying the DNN's attention. The overview of our architecture is shown in Fig. 4.

Suppose that we have a well-performed classification network f, our goal is that the UAP v can mislead the DNN f with small magnitude L, such that

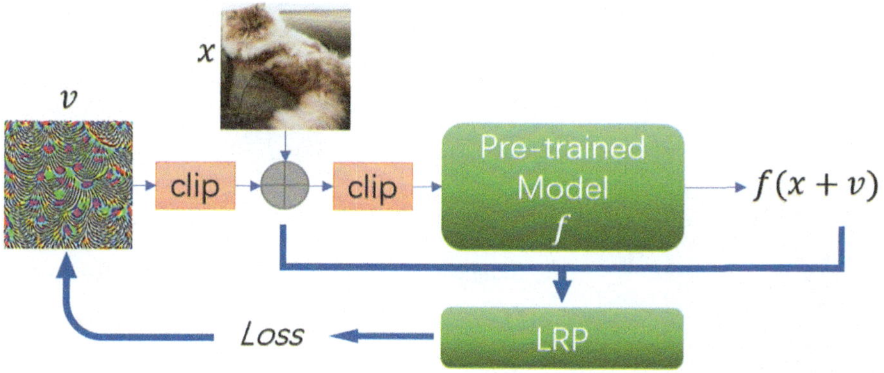

Fig. 4 Overview of the architecture for generating universal adversarial perturbations. A perturbation v is clipped and added to an image x, then the perturbed image is fed into a pre-trained model to get the output $f(x+v)$. LRP algorithm backpropagates the relevance of given labels to produces heat maps, which are for computing the loss function. Finally, the loss is used for backpropagation and optimization of the perturbation v

$$\begin{aligned} \arg\max_c f_c(x) &\neq \arg\max_c f_c(x+v), \\ |v\|_\infty = L, &\quad \text{for "most" } x \sim \mu. \end{aligned} \tag{16}$$

According to the objective of (16), we define two selected categories as c_{org} and c_{adv} to guide the attention of the DNN, such that

$$\begin{aligned} c_{org} &= \arg\max_c f_c(x), \\ c_{adv} &= \arg\max_{c \neq c_{org}} f_c(x+v). \end{aligned} \tag{17}$$

c_{org} donates the category with the highest prediction probability for $f(x)$, while c_{adv} donates the category with the highest probability except of c_{org} for $f(x+v)$. The basic idea of our attack is to transfer the DNN's attention from c_{org} to c_{adv}. To achieve such an objective, we choose to attack the L_p norm of the heat maps produced by LRP rather than the cross-entropy of the probability, such that

$$|g(x+v, c_{org})|_p \rightarrow \min, \quad |g(x+v, c_{adv})|_p \rightarrow \max. \tag{18}$$

We expect that the universal perturbation that changes the outputs of LRP can also change the predictions of the DNN. Our final loss function consists of the L_p norm of the two heat maps by LRP with respect to these labels, and is given by

$$Loss = |g(x+v, c_{org})|_p - |g(x+v, c_{adv})|_p, \tag{19}$$

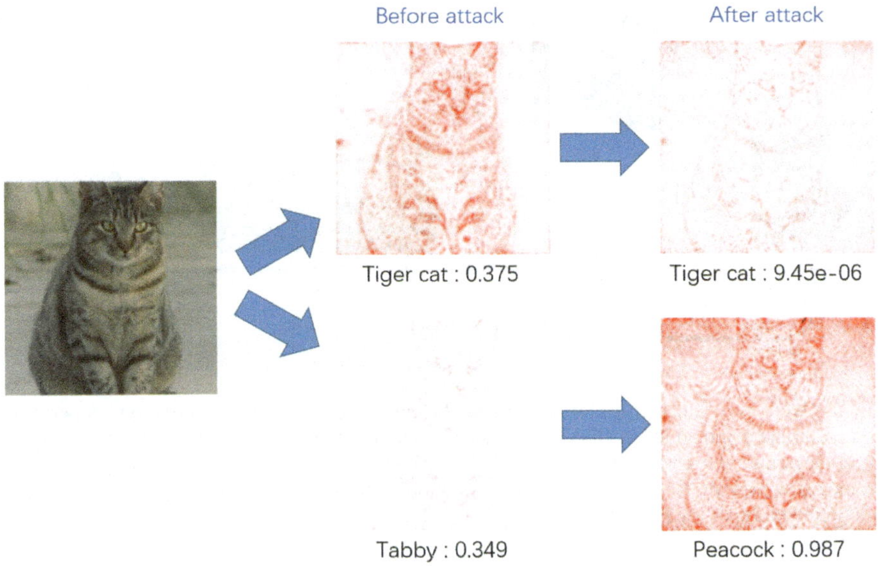

Fig. 5 LRP visualization of an example before and after attack. Left: benign image; Middle top: LRP of c_{org} before attack; Middle below: LRP of c_{adv} ($v = 0$) before attack; Right top: LRP of c_{org} after attack; Right below: LRP of c_{adv} ($v \neq 0$) after attack. The prediction probabilities are below them

where p is 1 or ∞. When p is 1, the loss function aims at decreasing all positive pixels of the heat map $g(x + v, c_{org})$, and increase all positive values of $g(x + v, c_{adv})$. When p is ∞, only the maximum values of heat maps is taken into consideration.

Figure 5 shows an example of LRP results before and after attack, which are scaled by diving the same factor for visualization. The heat map of the category "tiger cat" is decreased during attack, while the heat map of "peacock" is highlighted.

The summary of our algorithm is shown in Algorithm 1. The perturbation is zero-initialized at the beginning. We shuffle the training set X for each epoch. At the first of every iteration, we clip the perturbation v by $min(\varepsilon, v)$ in which ε is the maximum permissible L_∞ norm. Then the universal perturbation v is added to benign images to obtain the perturbed ones and clipped again to keep the pixel range [0, 255]. Before the LRP operation, the batch of benign images and perturbed images are fed to the pre-trained model f to acquire probability vectors, computing c_{org} and c_{adv} respectively according to (17).

After getting these dynamic labels, we re-feed the perturbed image $x + v$ to the model to get outputs $f(x + v)$ before the softmax layer, so that the LRP algorithm can produce heat maps with respect to c_{org} and c_{adv}. The loss function is designed as (19) to mislead the attention of the DNN. In all experiments, we use ADAM [13] optimizer to minimize the loss function and optimize the perturbation v by the gradient $\frac{\mathrm{d}Loss}{\mathrm{d}v}$. The final universal perturbation is created by iterating in batches of the training set. At the final of every epoch, the perturbation is scaled to $1/2$ for the optimization of the next epoch.

Algorithm 1 Generate UAPs by attacking LRP

Input: Training set X, classifier f, LRP g, max epoch numbers N, loss norm order p.
Output: universal perturbation v_{out}

Initialize $v \leftarrow 0$
while $k < N$ **do**
  Shuffle X
  for each data point $x \in X$ **do**
    if $f(x+v) = f(x)$ **then**
      Compute c_{org}, c_{adv} according to (17)
      Use LRP to get the loss function:
        $Loss = |g(x+v, c_{org})|_p - |g(x+v, c_{adv})|_p$
      Update v :
        $v \leftarrow v - \eta\Delta$ (ADAM approximate representation)
    end if
  end for
  $v_{out} \leftarrow v$
  $v = v/2$
end while
Return v_{out}

6 Experiment Results

In our experiments, we generate universal adversarial perturbations for the pre-trained classification networks on the ImageNet [25] dataset.

6.1 Dataset

In detail, all results are reported on the 50,000 RGB images of ImageNet [25] validation set consisting of 1000 categories. To attain sample balance and convincing results, we randomly select 40 images per class as our training set, and the other 10 images per class as our validation set. We first resize the shorter size of images to 256, then we crop 224×224 images in the center, such that the size of the input images is the same as the input size of the per-trained models.

6.2 Pre-trained Model

The pre-trained models we used are from Keras [5] applications module. We evaluate our attack on three widely-used DNNs: VGG-19 [27], ResNet-50 [8], DenseNet-121 [10], which have 71.55%, 74.77% and 74.37% top-1 accuracy respectively on our validation set. To match these models, we also utilize Keras [5] preprocessing functions provided by the application module [5].

6.3 Training Setup

For all three networks, we set the batch size to be 4, while the learning rate is varying. Similar to previous works, we set the perturbation's norm order $p = \infty$ in our experiments and limit the value ε of L_∞-norm as 10 when images are considered in [0, 255] range. The maximum number of epochs is 20. All the experiments are implemented in Tensorflow [1], Keras [5] with NVIDIA GeForce RTX 2080Ti GPUs.

6.4 Fooling Ratio

We first analyze the *fooling ratio* of universal adversarial perturbation based on state-of-the-art DNNs. *Fooling ratio* means the proportion of images that change the classification labels when added the perturbation v. Considering the validation set as D, we define the *fooling ratio* as

$$\frac{|\{x : \arg\max_i f_i(x) \neq \arg\max_i f_i(x+v), x \in D\}|}{|D|}, \tag{20}$$

which is like the evaluation setup in [19]. In our experiments, we assume that we have full access to the network to construct universal adversarial attack, which is non-target and white-box. Results are shown in Table 1.

As for all three networks, our method achieves quite high fooling ratios on the validation set. We discover that Vgg-19 is the most difficult to attack, and reports a fooling ratio of 82.42% when p in the loss function is 1. The robustness of ResNet-50 and DenseNet-121 is more fragile, and these two networks report higher fooling ratios, 91.15% and 91.50% respectively. The second line of the table introduces the fooling ratios when p is ∞, which are 81.30, 89.98 and 91.03%, a little lower than above.

It shows that our method can generalize onto various DNNs that LRP is applicable to, and produce satisfying fooling ratios. The magnitude and structure of heat maps based on relevance propagation can supply enough gradient information to help generate universal perturbations.

For a fair comparison, we retrain the algorithm from [19] on our pre-trained models using the same training set. The results are also listed in Table 1. Against this

Table 1 Fooling ratio comparison on the validation set

Method	Vgg-19 (%)	ResNet-50 (%)	DenseNet-121 (%)
Ours ($p = 1$)	**82.42**	**91.15**	**91.50**
Ours ($p = \infty$)	81.30	89.92	91.03
Moosavi-Dezfooli et al. [19]	77.80	84.52	84.68

algorithm, we outperform all networks. These results show that the LRP method can replace the classification output of the network, and provide better performance on generating universal perturbations.

6.5 *Visualization of Perturbations and Effect*

The perturbations generated by three networks are visualized in Fig. 6. For the UAP to be more striking, we scale perturbations from the range $[-10, 10]$ to range $[0, 255]$. From the horizontal comparison, we can notice that the perturbations generated by different models are different in texture and structure. From the vertical comparison, we can discover that UAPs generated by the loss functions with different norm order p are more similar, but still display different edges. In our experiments, different training initialization targeted on the same network will also lead to variable perturbations. It verifies that our method has the capacity to exploit adversarial perturbations in multiple directions.

We also claim that a single universal perturbation can destroy various LRP heat maps and cause many benign images to be misclassified with high probability. Figure 7 displays some perturbed images with those LRP results of the ground truth labels before and after attack. We can see that though the perturbations are imperceptible to humans, it makes the DenseNet-121 classifier to predict quite different labels. The LRP heat maps of the ground truth labels are all "cool down" comparing with benign ones. To be more interesting, the regions of interest also change from the object areas to corners or borders, showing the changes of the DNN's attention after our attack.

6.6 *Transferability*

Except for the attack ability on white-box models, we also concern about the *transferability* of the universal adversarial perturbations. *Transferability* is a property to evaluate the robustness of adversarial examples. An adversarial example generated based on its original network (e.g., Vgg-19) has *transferability* if it can also be misclassified by other networks (e.g., ResNet-50).

In our experiments, we generate perturbations based on Vgg-19, ResNet-50, DenseNet-121 respectively, and display their fooling ratios across different architectures. Table 2 summarizes the matrix which reports the fooling ratios when p in the loss function is 1. The rows indicate the networks for which the UAPs are computed, and the columns indicate the networks for which the fooling ratios are reported. We find that universal perturbations created by our method do generalize

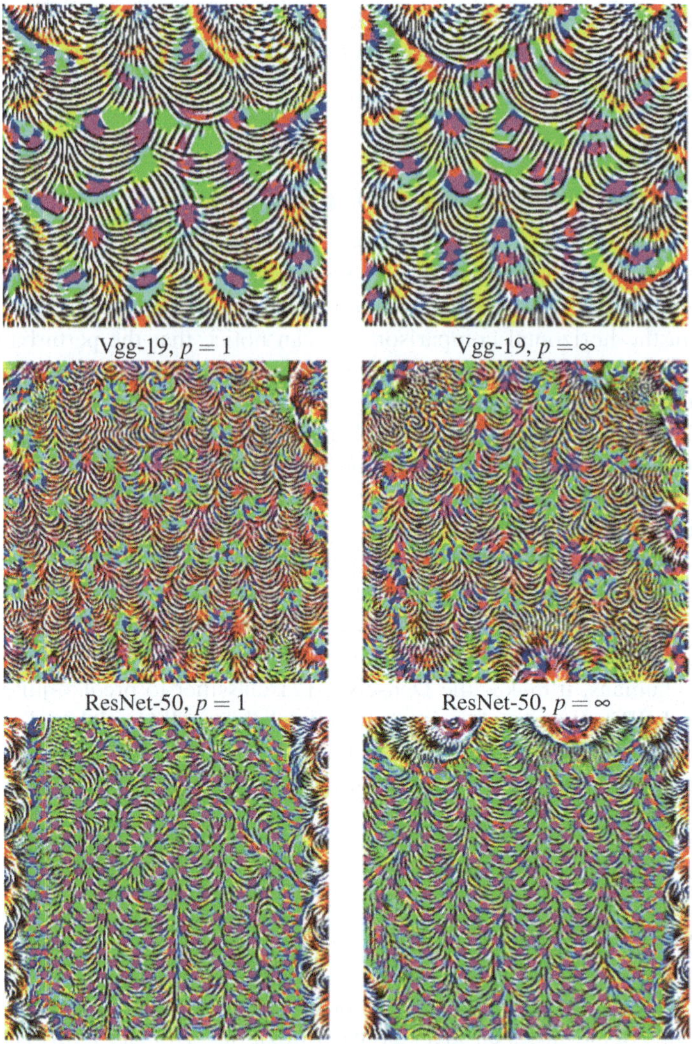

Fig. 6 Visualization results of universal adversarial perturbations generated by our method

well across different models, and show cross-model transferability. Among them, the universal perturbation generated by Vgg-19 reports better performance, and the fooling ratios are 62.17% and 62.18% respectively when the UAP is evaluated on ResNet-50 and DenseNet-121.

Fig. 7 Perturbed images ($p = 1$), corresponding labels and LRP results of the ground truth labels on DenseNet-121 (up: before attack; down: after attack)

Table 2 Transferability ($p = 1$)

	Vgg-19 (%)	ResNet-50 (%)	DenseNet-121 (%)
Vgg-19	**82.42**	62.17	62.18
ResNet-50	45.57	**91.15**	64.34
DenseNet-121	40.04	60.18	**91.50**

7 Conclusion

We propose a novel and well-performed universal adversarial attack that highlights the utilization of heat maps based on Layer-wise Relevance Propagation. In detail, we demonstrate that our optimization objective function based on the LRP method is effective to construct UAPs to fool the classification networks. Our algorithm achieves high fooling ratios on the ImageNet dataset with the small perturbation norm limitation. The perturbations can successfully transfer across different networks. In future work, we plan to explore the possibility of using other interpretation methods to craft adversarial attack, or to apply our method on other deep learning tasks. Future research is anticipated on UAP for spiking neural networks (SNN) that classify not only static images but dynamic spatio-temporal data using spike-time learning algorithms [11].

Acknowledgements This research is partly supported by NSFC, China (Nos.: 61876107, U1803-261, 61977046) and Committee of Science and Technology, Shanghai, China (No. 19510711200).

References

1. Abadi, M., Agarwal, A., Barham, P., Brevdo, E., Chen, Z., Citro, C., Corrado, G.S., Davis, A., Dean, J., Devin, M., Ghemawat, S., Goodfellow, I., Harp, A., Irving, G., Isard, M., Jia, Y., Jozefowicz, R., Kaiser, L., Kudlur, M., Levenberg, J., Mané, D., Monga, R., Moore, S., Murray, D., Olah, C., Schuster, M., Shlens, J., Steiner, B., Sutskever, I., Talwar, K., Tucker, P., Vanhoucke, V., Vasudevan, V., Viégas, F., Vinyals, O., Warden, P., Wattenberg, M., Wicke, M., Yu, Y., Zheng, X.: TensorFlow: large-scale machine learning on heterogeneous systems (2015). Software available from tensorflow.org
2. Bach, S., Binder, A., Montavon, G., Klauschen, F., Müller, K.-R., Samek, W.: On pixel-wise explanations for non-linear classifier decisions by layer-wise relevance propagation. PLoS ONE **10**(7) (2015)
3. Behjati, M., Moosavi-Dezfooli, S.-M., Baghshah, M.S., Frossard, P.: Universal adversarial attacks on text classifiers. In: ICASSP 2019—2019 IEEE International Conference on Acoustics, Speech and Signal Processing (ICASSP), pp. 7345–7349. IEEE (2019)
4. Carlini, N., Wagner, D.: Towards evaluating the robustness of neural networks. In: 2017 IEEE Symposium on Security and Privacy (SP), pp. 39–57. IEEE (2017)
5. Chollet, F., et al.: Keras. https://keras.io (2015)
6. Goodfellow, I.J., Shlens, J., Szegedy, C.: Explaining and harnessing adversarial examples. arXiv preprint arXiv:1412.6572 (2014)
7. Hayes, J., Danezis, G.: Learning universal adversarial perturbations with generative models. In: 2018 IEEE Security and Privacy Workshops (SPW), pp. 43–49. IEEE (2018)
8. He, K., Zhang, X., Ren, S., Sun, J.: Deep residual learning for image recognition. In: Proceedings of the IEEE Conference on Computer Vision and Pattern Recognition, pp. 770–778 (2016)
9. Hendrik Metzen, J., Chaithanya Kumar, M., Brox, T., Fischer, V.: Universal adversarial perturbations against semantic image segmentation. In: Proceedings of the IEEE International Conference on Computer Vision, pp. 2755–2764 (2017)
10. Huang, G., Liu, Z., Van Der Maaten, L., Weinberger, K.Q.: Densely connected convolutional networks. In: Proceedings of the IEEE Conference on Computer Vision and Pattern Recognition, pp. 4700–4708 (2017)
11. Kasabov, N.K.: Time-Space, Spiking Neural Networks and Brain-Inspired Artificial Intelligence. Springer, Berlin, Heidelberg (2019)
12. Khrulkov, V., Oseledets, I.: Art of singular vectors and universal adversarial perturbations. In: Proceedings of the IEEE Conference on Computer Vision and Pattern Recognition, pp. 8562–8570 (2018)
13. Kingma, D.P., Ba, J.: Adam: a method for stochastic optimization. arXiv preprint arXiv:1412.6980 (2014)
14. Kurakin, A., Goodfellow, I., Bengio, S.: Adversarial examples in the physical world. arXiv preprint arXiv:1607.02533 (2016)
15. Li, J., Ji, R., Liu, H., Hong, X., Gao, Y., Tian, Q.: Universal perturbation attack against image retrieval. In: Proceedings of the IEEE International Conference on Computer Vision, pp. 4899–4908 (2019)

16. Lin, M., Chen, Q., Yan, S.: Network in network. arXiv preprint arXiv:1312.4400 (2013)
17. Liu, H., Ji, R., Li, J., Zhang, B., Gao, Y., Wu, Y., Huang, F.: Universal adversarial perturbation via prior driven uncertainty approximation. In: Proceedings of the IEEE International Conference on Computer Vision, pp. 2941–2949 (2019)
18. Montavon, G., Lapuschkin, S., Binder, A., Samek, W., Müller, K.-R.: Explaining nonlinear classification decisions with deep Taylor decomposition. Pattern Recogn. **65**, 211–222 (2017)
19. Moosavi-Dezfooli, S.-M., Fawzi, A., Fawzi, O., Frossard, P.: Universal adversarial perturbations. In: Proceedings of the IEEE Conference on Computer Vision and Pattern Recognition, pp. 1765–1773 (2017)
20. Moosavi-Dezfooli, S.-M., Fawzi, A., Frossard, P.: Deepfool: a simple and accurate method to fool deep neural networks. In: Proceedings of the IEEE Conference on Computer Vision and Pattern Recognition, pp. 2574–2582 (2016)
21. Mopuri, K.R., Ganeshan, A., Venkatesh Babu, R.: Generalizable data-free objective for crafting universal adversarial perturbations. IEEE Trans. Pattern Anal. Mach. Intell. **41**(10), 2452–2465 (2018)
22. Mopuri, K.R., Garg, U., Venkatesh Babu, R.: Fast feature fool: a data independent approach to universal adversarial perturbations. arXiv preprint arXiv:1707.05572 (2017)
23. Papernot, N., McDaniel, P., Jha, S., Fredrikson, M., Berkay Celik, Z., Swami, A.: The limitations of deep learning in adversarial settings. In: 2016 IEEE European Symposium on Security and Privacy (EuroS&P), pp. 372–387. IEEE (2016)
24. Poursaeed, O., Katsman, I., Gao, B., Belongie, S.: Generative adversarial perturbations. In: Proceedings of the IEEE Conference on Computer Vision and Pattern Recognition, pp. 4422–4431 (2018)
25. Russakovsky, O., Deng, J., Hao, S., Krause, J., Satheesh, S., Ma, S., Huang, Z., Karpathy, A., Khosla, A., Bernstein, M., et al.: Imagenet large scale visual recognition challenge. Int. J. Comput. Vis. **115**(3), 211–252 (2015)
26. Selvaraju, R.R., Cogswell, M., Das, A., Vedantam, R., Parikh, D., Batra, D.: Grad-cam: visual explanations from deep networks via gradient-based localization. In: Proceedings of the IEEE International Conference on Computer Vision, pp. 618–626 (2017)
27. Simonyan, K., Zisserman, A.: Very deep convolutional networks for large-scale image recognition. arXiv preprint arXiv:1409.1556 (2014)
28. Szegedy, C., Zaremba, W., Sutskever, I., Bruna, J., Erhan, D., Goodfellow, I., Fergus, R.: Intriguing properties of neural networks. arXiv preprint arXiv:1312.6199 (2013)
29. Zeiler, M.D., Fergus, R.: Visualizing and understanding convolutional networks. In: European Conference on Computer Vision, pp. 818–833. Springer (2014)
30. Zhou, B., Khosla, A., Lapedriza, A., Oliva, A., Torralba, A.: Object detectors emerge in deep scene CNNs. arXiv preprint arXiv:1412.6856 (2014)
31. Zhou, B., Khosla, A., Lapedriza, A., Oliva, A., Torralba, A.: Learning deep features for discriminative localization. In: Proceedings of the IEEE Conference on Computer Vision and Pattern Recognition, pp. 2921–2929 (2016)
32. Zhou, J., Troyanskaya, O.G.: Predicting effects of noncoding variants with deep learning-based sequence model. Nat. Methods **12**(10), 931–934 (2015)

Framework and Development Process for IoT Data Gathering

Mika Saari, Petri Rantanen, Sami Hyrynsalmi, and David Hästbacka

Abstract The Internet of Things (IoT) is a growing area in everyday life. New applications under the umbrella term IoT are being developed continually. This development has raised the need for framework definitions for different purposes. This research introduces a special software/hardware framework for data gathering systems to be used in IoT related systems. The purpose of the research is to show the usability of a certain software/hardware combination in prototype development. The software/hardware framework has been developed during several research projects by following the same prototype development process. This is proposed as a descriptive model for the prototyping process. The main contribution of this research is the framework itself. The framework consists of a model of the system with selected components. The placement of the sensor network is also presented. The purpose of the framework is to guide and assist the construction of data gathering prototypes. Furthermore, the advantages of the framework are to support re-usability, portability, and interchangeability. This research introduces the framework, its main components, and their interconnections. In addition, the prototype development process used is presented.

M. Saari (✉) · P. Rantanen
Computing Sciences, Tampere University, Pori, Finland
e-mail: mika.saari@tuni.fi

P. Rantanen
e-mail: petri.rantanen@tuni.fi

S. Hyrynsalmi
Department of Software Engineering, LUT University, Lahti, Finland
e-mail: sami.hyrynsalmi@lut.fi

D. Hästbacka
Computing Sciences, Tampere University, Hervanta Campus, Pori, Finland
e-mail: david.hastbacka@tuni.fi

V. Sgurev et al. (eds.), *Advances in Intelligent Systems Research and Innovation*,
Studies in Systems, Decision and Control 379,
https://doi.org/10.1007/978-3-030-78124-8_3

1 Introduction

The Internet of Things (IoT) is a growing area in everyday life. New applications under the umbrella term IoT are being developed continually. The IoT paradigm is the integration of several technologies and communications solutions [1]. This development has raised the need for framework definitions for different purposes. For example, the draft of the IEEE standard [2] defines an architectural framework for the Internet of Things (IoT).

This article introduces a special software/hardware (SW/HW) framework for data gathering systems to be used in IoT related systems. The research question can be stated as follows: How to generalize the prototyping of IoT data gathering in a framework of required software and hardware components?

The research question was formulated during previous data gathering prototype system development projects. The main purpose of these prototypes is to gather data, for example environmental data such as temperature, humidity, or carbon dioxide levels (Fig. 1). The aim was to focus on the reproducibility of components within the development process. This study presents guidelines for selecting the required software and hardware components. The purpose of the SW/HW framework is to guide and assist when constructing data gathering prototypes. Furthermore, the advantages of the framework are that it supports re-usability, portability, and interchangeability.

This study is part of the research related to the Internet of Things (IoT) carried out by the Software Engineering and Intelligent Systems (SEIntS) group at Tampere University, Pori. The SW/HW framework has been developed during several research projects. These projects have contained multiple iteration rounds. Many of these rounds have produced a research article, whose main target was to describe the working prototype. The first prototype system was introduced by Saari et al. [3] in 2015. That research introduced the initial idea of a framework and a working

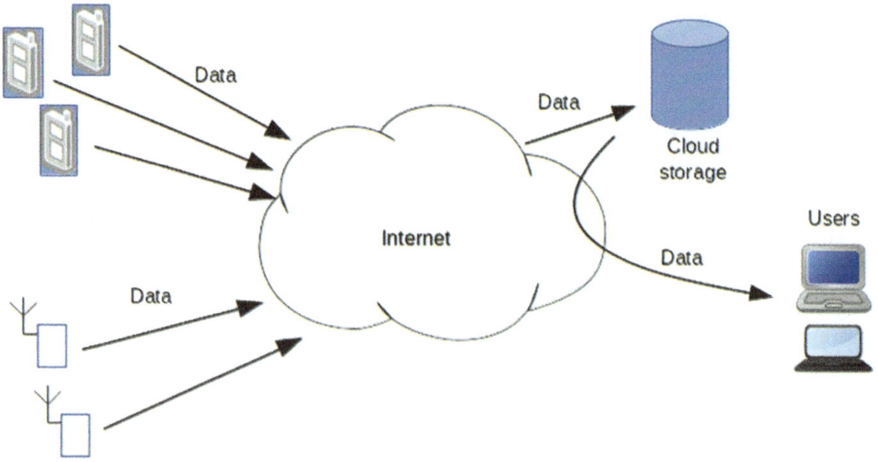

Fig. 1 The overall architecture of a data gathering IoT prototype system

implementation from it. Research on reducing energy consumption [4] presented the advantages of rapid prototyping with off-the-shelf devices and open source software.

The main idea of prototype development has been to start with off-the-shelf devices and open source software. These key software and hardware components are then modified in the desired direction and usually a working prototype system is produced.

The main result from our research is the framework itself. This has the ability to act as a guiding principle when developing new prototypes for gathering data. This framework also aims to represent the development of software and hardware usage in data gathering systems; in particular the evolution in the usage of both software and hardware is considered in different parts of the system. The framework could be used as a model when planning new data gathering prototypes for sensor networks.

The second finding made during the research is a descriptive model for the prototyping process (DMPP). This is a model of prototype development practices that have been applied in several research projects between university and enterprises [mostly small and medium-sized enterprises (SMEs)] in Finland. The purpose of the model is to introduce how academic research can conduct prototype development with regional enterprises.

The structure of this paper is as follows: In Sect. 2, we review the related research about prototyping, the development process, and related frameworks. In Sect. 3, we introduce an implementation of university-enterprise collaboration in prototype development described by means of process modeling notation. Section 4 introduces the SW/HW framework for IoT data gathering. Section 5 continues by describing the validation and testing of the framework. Section 6 includes a discussion and suggestions for future research on the topic and finally, Sect. 7 summarizes the study.

2 Related Research

The importance of prototyping embedded SW/HW systems was introduced by Buchenrieder [5]. The reason for this was because system differences had increased, and the product relied mainly on variations in software and system features. In addition, involving users in the specification process is important because more and more customers expect solutions and services tailored exactly to their particular needs. In a more recent study, Kruger et al. [6] developed a working prototype using off-the-shelf components. This also showed that a lengthy product development life cycle is not required when using a rapid prototyping process.

Rapid prototyping could be presented as a circle (Fig. 2). Rapid prototyping includes three stages: making a prototype, reviewing the result, and refining and iterating [7]. We used the idea of rapid prototyping in our projects. The working prototype solution in the context of IoT requirements is as follows: hardware to run the software; software to collect, save, and transmit the data; the right technologies for the use cases to make things easier for both developer and user [4].

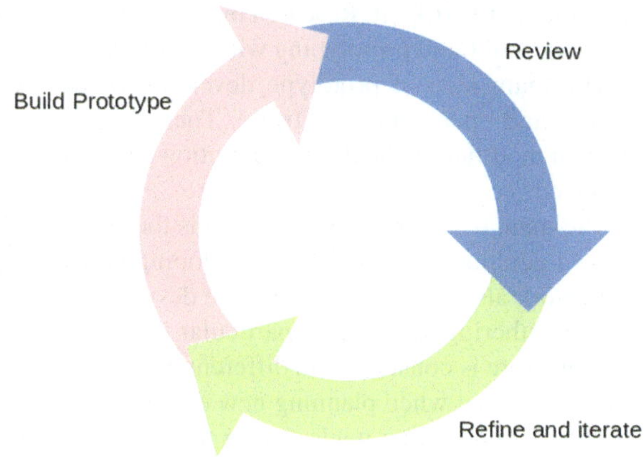

Fig. 2 The circle of the rapid prototyping process. Adapted from [7]

2.1 Development Process of the Prototypes

The IoT prototyping process can be viewed from two perspectives: a software development process and an embedded hardware development process. Furthermore, the authors have researched the IoT data gathering prototype development process by collecting data from several prototyping processes.

A development process from the area of software development is suitable for this study. The developed prototypes and the presented SW/HW framework include a lot of software development. The process model could be descriptive or prescriptive. A prescriptive model tells how the process should be performed, whereas a descriptive model tells how a process is performed in a particular environment. The third option, a proscriptive model, also describes the activities that could be done [8]. The descriptive process model (DPM) [9, 10] introduces an eight step approach for producing a process model. These steps are divided into two phases: the set-up phase and the execution phase. The eight step approach was used when the descriptive model for the prototyping process (DMPP) was developed. The DMPP is presented in Sect. 3.

The book "Introduction to Embedded Systems—A Cyber-Physical Systems Approach" by Lee and Sashia is based on the idea that designing and implementing an embedded system consists of three major parts of the process: modeling, design, and analysis [11]. The modeling phase specifies what a system does by defining the system model and the set of requirements. The artifacts, such as a combination of software and hardware components, are produced in the design phase. An artifact is a working system and it describes how the system works. In the analysis phase, information on the system is obtained to specify why the system works as it works [11].

2.2 *The Framework for Prototyping*

The SW/HW framework idea is not a new issue in the research field. For example, earlier studies [12–14] have addressed the framework subject from a real-time system perspective. In these studies, the design was at micro-controller level, whereas our prototypes use off-the shelf single-board micro-controllers. For example, Srivastava and Brodersen handled board level module generation, system software generation, and hardware-software integration in a unified framework [14]. Their study mentions a rapid prototyping method, but it was not explained further.

IoT architecture consists of several components, which can be divided into layers as follows: sensing layer, networking layer, service layer, and interface layer [15, 16]. The SW/HW framework is focused on the sensing and service layers. The networking layer exists but is not the focus of our research. The interface layer is described to the user in the SW/HW framework but is excluded from the study.

A wireless sensor network (WSN) can be used in various application areas. A WSN includes sensor nodes, which consist of sensing, data processing, and communicating components. A sensor network is composed of a large number of sensor nodes, which send data to the data storage. Since sensor nodes have data processing ability, the uploaded data can be either raw or pre-processed [17].

3 Descriptive Model for the Prototyping Process (DMPP)

We introduced our descriptive software process model for IoT prototyping in [18]. The purpose of this section is to present how the selected process model has supported the development of the framework.

The DMPP [18] could be extended to include hardware, because the model itself does not limit the type of artifact. Therefore, when the process is mentioned in this study, it generally means every kind of artifact, i.e., software or hardware, made in the prototype development process. The focus is on modeling the prototyping process in a research context, but its use in general is not restricted.

The DMPP was developed using the aforementioned descriptive process model (DPM) approach [10]. The basic concepts related to processes are role, activity, resource, and artifact (Fig. 3). The example is illustrated by the developer (role) involved in software developing (activity) using a programming tool (resource). The activity produces some software (artifact) used in the prototype system. The process data for the model is collected through interviews with the developers involved in the four different prototype development processes. These four prototype development projects and their outcomes are reported in several studies [19–22]. Common to all of the studies are that they present developed IoT prototype systems that gather data. Although the software and hardware components in the prototypes vary, overall they can be used to model the prototype development process.

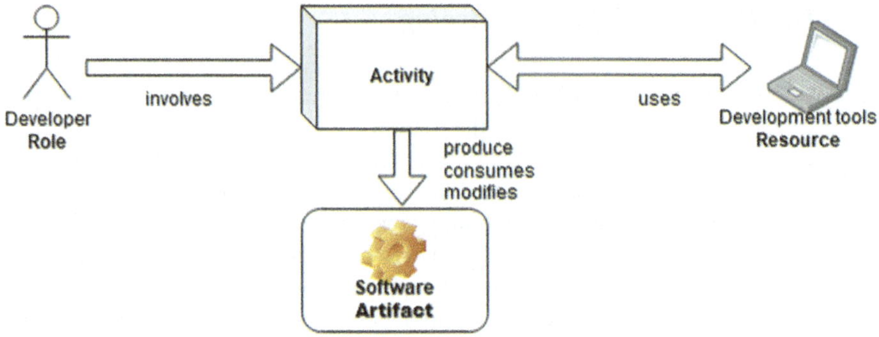

Fig. 3 Example of basic concepts related to the prototype development process. Adapted from [18]

Figure 4 presents the developed DMPP [18]. The model includes six steps. These steps support or use the SW/HW framework in the following ways:

1. The first step starts from the requirements definition, a collaborative discussion between the developers and the client. The client defines what kind of data would be useful. The developer group starts to define the hardware and overall architecture of system and how the data will be collected by the software. The selected hardware mostly determines the software environment and tools used.
2. The outcome of the discussion is the first artifacts: for example, the prototype system requirements within the discussion notes. The developer group constructs the first architecture model of the component interconnections.
3. The third step is the software/hardware prototype development made by the research group including the project manager and software/hardware developers. The clients' representatives are involved in the development process in the role of instructor. In this step the SW/HW framework is used as the guideline for selecting the components for the developed prototype.
4. The fourth step introduces the working prototype artifact. It contains the developed software and hardware components. Also, the interconnections of the components are tested. The testing process overall is usually only the functional testing of the

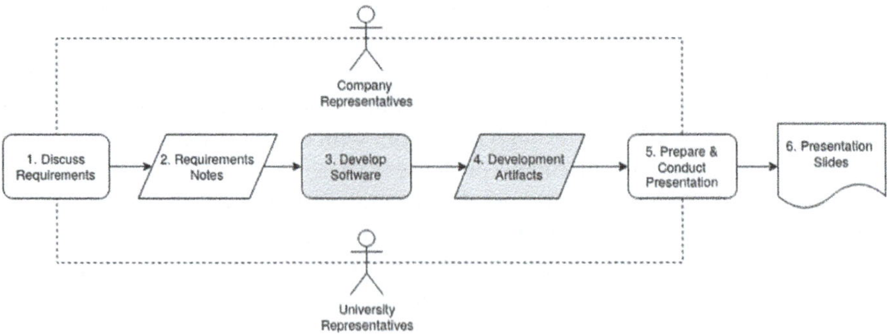

Fig. 4 Process model for prototype development. Adapted from [18]

prototype system. Additionally, the gathered data are inspected and if possible, compared to the expected results.
5. The fifth step includes preparing the outcome of the development process. The SW/HW framework can be complemented if necessary.
6. The sixth step is to publish the results, for example, the prototype system, the collected data, and the analysis of the project.

The process model in Fig. 4 is a simplified presentation of the prototype development process and therefore it does not mention common procedures such as iterations, testing, and customer testing.

Iterations are an efficient way to test and develop an idea. The first working prototype is made as simple as possible with basic components. For example, the hardware could be chosen at first only for testing the idea. If the idea works, the hardware is changed for more suitable hardware in the next iteration round. A good example of this is the application where we tested the use case [23]: Is it possible to take a photo in selected GPS coordinates automatically and send this photo to the cloud storage? This idea was tested with the Android application in a smartphone and the idea was found to be a workable solution. The smartphone had some limitations with the automation: The developed photographing application had to be started, it was not possible to be aware of the application crashing, and the possibility of remote control was not easily implemented. The second iteration round to solve the same use case was carried out with the following hardware: Raspberry Pi, camera, battery, GPS sensor, WiFi, and 4G modem. This time the Raspberry Pi OS made it possible to implement the automated operations and remote control with Linux tools.

4 SW/HW Framework for IoT Data Gathering

In this section we introduce the SW/HW framework for an IoT data gathering system. The framework consists of several hardware and software components. The purpose and advantage of the framework is to support re-usability, portability, and interchangeability. Another purpose of the framework is to guide and assist the construction of data gathering prototypes.

The definition of the SW/HW framework has been made in several academic research projects where the focus has been the collection of data using self-made prototype systems. Figure 5 illustrates the framework and its relationships. The previously presented model in Sect. 3 has connections to the framework and its development. DMPP step 1 gives guidelines for data handling. The client tells the developers what kind of data are useful. In step 3 the developers make the decisions on what kinds of software and hardware components are needed to fulfill the clients' data gathering expectations.

Furthermore, Fig. 5 clarifies the interconnections of the framework components. These interconnections guide the selection of software and hardware components when constructing a prototype. The framework relies on off-the-shelf devices,

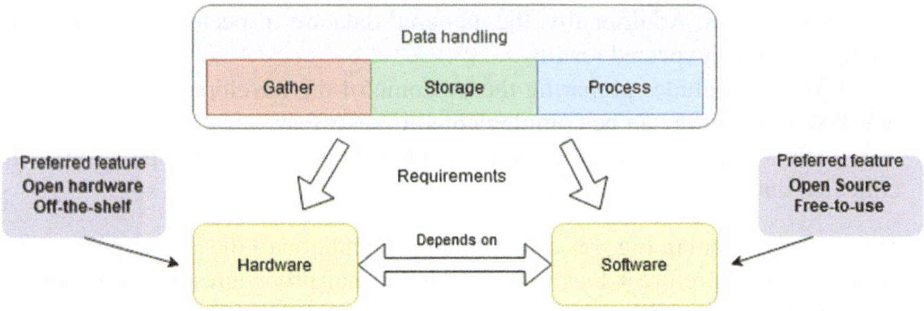

Fig. 5 Relationships of the SW/HW framework

because this speeds up development by minimizing hardware design and implementation [24]. The information that is to be collected determines the collection and structure used by the software and hardware components. Optional features are also used when selecting the components. In hardware, open hardware and off-the-shelf devices are preferable, because they are reasonably priced, quickly available, and have community support. The software components should have similar features: open source and community support. Please note that these criteria are not preferred when implementing the final application for production use.

The first question for the constructor of an IoT data gathering system when implementing the prototype is what information or data to collect. The answer to the question should be clear and it should also be the motivator of constructing the system. The next three questions are presented after the decision to collect data:

- How to gather data?
- How to store data?
- How to process data?

For an end user or client the availability of the data is important. Therefore the preferable place to store and process the data is in some cloud storage or server. Cloud storage could be some old Linux server or fully optimized commercial cloud computing service. However, before the data are in cloud storage, they have to be collected and stored temporarily in a sensor device or gateway device. Furthermore, the first data processing and storing should be managed by the sensor device or gateway device. For example, if the network has a communication problem, there is the possibility of losing data unless the data are forwarded rather than being stored temporarily. The SW/HW framework focuses on data storage and processing in the sensor or gateway layers.

Figure 6 shows the overall layer architecture of the IoT data gathering prototype system in a sensor network. The presented SW/HW framework is located in the lower part, containing the sensors, gateways, and preferable means of communication to operate with cloud storage. A few explanations of the figure are given below:

- The user and client layer utilizes the collected data.
- A processing layer is needed in most use cases. The data are processed in the way the clients require. The user can use the raw, unprocessed data.

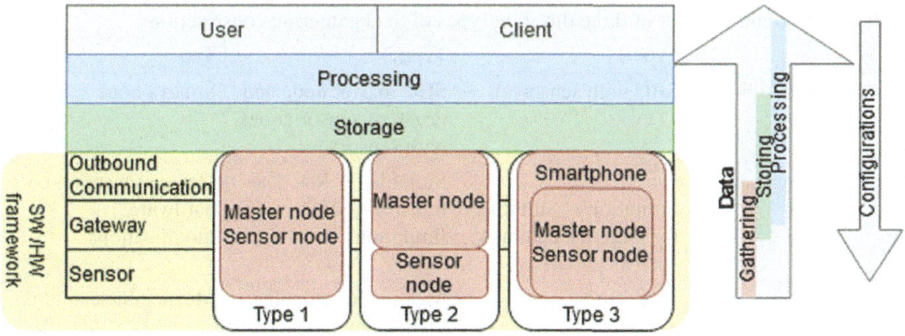

Fig. 6 Layer diagram of IoT data gathering and how the SW/HW framework is placed in it

- The storage layer collects and saves the data. The purpose of this level is to ensure data retention.
- The outbound communication layer belongs to the SW/HW framework. Its purpose is to offer a suitable data transfer method.
- Data gathering has three different hardware constructions: Type 1—master node-sensor node combination; Type 2—several sensor nodes collect the data in one master node; Type 3—Fully operational sensor device—a smartphone collects the data.
- The data flow from the sensors to the client or user. The data could be temporarily stored on the gateway level. The data processing could be also done on then gateway level if it is necessary and possible.
- Configurations and monitoring are enabled to ensure faultless operation and for testing purposes.

The division of the master node–sensor node into three different types of construction supports the versatility of the SW/HW framework. The main idea is that these three types make it possible to collect a wide range of data.

4.1 Hardware of the SW/HW Framework

The data gathering hardware can be divided on a higher level into three type of constructions, as can be seen in Fig. 6. The framework uses off-the-shelf hardware and devices. This limitation accelerates prototype development as at least partially tested devices can be used. The hardware can be categorized in two parts:

- Sensor node—Sensor hardware consisting of a combination of sensors and control device.
- Master node—data gathering and storing device that has the capability of collecting, storing, and processing data.

The division into two parts is enough for hardware when compared to the three types of data gathering devices in Table 1. On the hardware side, types 1 and 3 are

Table 1 The main features of three different types of data gathering constructions

	Type 1	Type 2	Type 3
Basic construction	SBC with sensor(s)	SBC master node and group of sensor nodes with sensors	Smart phone
Data gathering	No limitations—suitable for large data chunks such as photos	Suitable for low data transfer—SBC limitations	Device sensors—no hardware modifications
Data processing	SBC limitations	Mean value calculus, visualizing	Mean value calculus
Data storage	No limitations temporary storage	No limitations database storage	No limitations temporary storage

embedded together—the sensors and processing capabilities are in one device. In type 2 the master node can control several sensor nodes.

The **sensor node** collects the data with sensors. The data could be simple data, such as temperature and humidity. On the other hand, the data could be more complex, such as photos. The hardware is selected according to the data requirement. The separation of sensor node is made because of type 2 where the idea is to use several sensor nodes with one master node. In type 2 construction sensors are connected to a single-board micro-controller, such as Arduino or similar, which can handle a lot of simple data. Simple data are numerical values. Type 2 sensor nodes are simple, low-cost devices. These are connected to the master node and the amount of transferred data should be in bytes or kilobytes. The preferred communication methods include Bluetooth, ZigBee, and LoRa for short distance, low rates, and low power consumption [25, 26]. Types 1 and 3 are similar to each other; both are physically one entity: Type 1 consists of sensors and an SBC such as Raspberry Pi. The collected data could be complex and may need processing power, for example, photos. Type 3 is smartphone based solutions. A typical smartphone has several sensors, e.g., gyroscope, accelerometer, and ambient light sensor [27]. For example, Android phones could handle a lot of simple data from their own sensors, or complex data such as photos from the phone's own camera.

The **master node** collects data from the sensor nodes. The master node has the capability to pre-process or temporarily store data if needed. The master node has a communication channel for a larger data transfer rate and long distance, for example, 3G/4G/5G, or WiFi are suitable. Depending on the master node's communication channel, remote control and configuration are possible. For example, with the Raspberry Pi the remote configuration is easily constructed with a suitable communication channel and Linux OS tools. The idea of a master node in type 3 smartphone based solutions is implemented with a self-made application, which handles data collection, storage, and processing. The application limitations come from the phone's OS and the fact that hardware changes or modifications have not been made.

This SW/HW framework relies on communication to the public Internet. The collected data is transferred via the Internet to the data storage devices. These could be cloud servers with a database or dedicated open source Linux servers for saving data. There are several database models for storing sensor data and each of these has a special use case where they are best. The SW/HW framework can be applied to all of these techniques.

4.2 *Software of the SW/HW Framework*

The hardware of the SW/HW framework also requires software. The software used is mostly open source. In this way the selected software is community tested and the source code is freely available. Open source software is also free to use. Therefore, several software combinations can be used for testing purposes without extra costs.

The software components are divided into three parts:

- Sensor software—gets sensor data from sensors
- Data gathering and pre-processing software
- Data storage software

Regarding how these three parts work together, the diagram in Fig. 7 is a guideline for dividing software components between the master node and sensor node on the abstract level. This kind of approach supports the modular development and interchangeability of components. The diagram also illustrates the input from the sensors and the output to the cloud.

The list of software components starts from the sensor node with sensor software and ends with data storage software for the master node. The sensing and sensor software typically have low-level programming with C++, Python, or Perl scripts, which are dedicated to do a few tasks, for example reading sensor data, reviewing the data, and storing the data. Sensor devices, such as Arduino with sensors, are

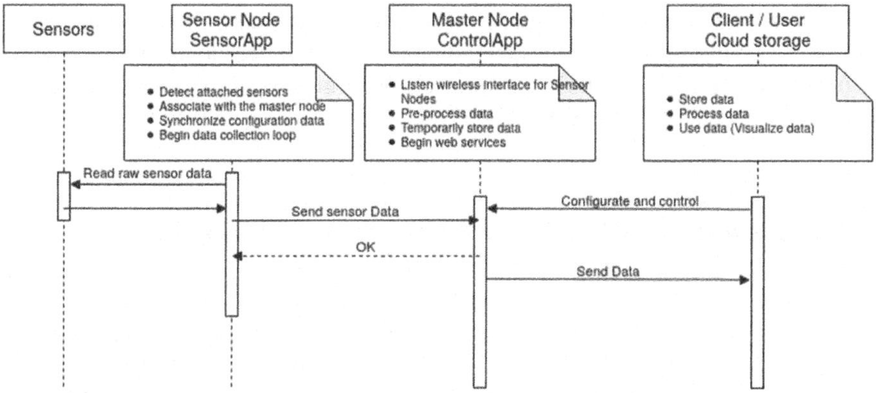

Fig. 7 Diagram of software component interconnections in the SW/HW framework

typically a micro-controller board that runs a dedicated program. Types 1 and 2 in Fig. 6 represent this kind of approach.

The Raspberry Pi based SBCs can perform both sensing and data gathering. In type 1 for example, the sensors could be connected to the Raspberry Pi and the sensor software reads the values from the sensors. If the construction is type 2, the gathering software handles the data collection from sensor nodes. The data gathering and preprocessing software are more complex and usually type 2 SBC devices are equipped with the full Linux operating system (OS). In the data storing and preprocessing phases, the SW/HW framework utilizes pre-made software and libraries. For example, Raspberry Pi could offer the gathered data to the Internet with a server application. Preprocessing in this scenario could be image recognition with image recognition software and library.

The gathered data are stored in the cloud server—this is the assumption of the SW/HW framework. Temporarily, the data could be saved to the master node using a suitable database. If the data meet the definition of a time series: "A sequence of numbers collected at regular intervals over a period of time" then a good choice is a time series database [28]. For example, the open-source time series database InfLuxDB is suitable for SBC hardware and is widely used in IoT solutions [29].

The other suitable database model for data storage is a relational model. The collected data could be stored locally in the sensor device, for example Raspberry Pi with Linux OS, MariaDB database, and a RESTful API combination. The RESTful API (Application Programming Interface) method allows remote control or management of a device over the network.

Smartphones equipped with the Android OS have been tested with this SW/HW framework. The Android OS has a software development kit (SDK), which enables the wide use of smartphone capabilities. For example, the SDK enables phone camera usage [23]. The SDK also enables usage of the smartphone's accelerometer sensor [30].

Data storage on the mobile phone is enabled by the OS. The SDK provides the capabilities to use files and databases for data storage. In terms of the SW/HW framework, the user should be able to use the data. The SDK also enables data transfer to cloud services.

Cloud storage for data is a better choice than local storage. Cloud storage could be, for example, a Linux server or maintained commercial cloud service such as Google Firebase. Both of these have more capabilities to store a larger amount of data than the local database in Raspberry Pi.

5 Validating the SW/HW Framework by Prototyping

The SW/HW framework has been developed and tested during several research projects. The majority of IoT data gathering prototype systems and their findings have been reported in different studies [3, 22, 23, 30–33]. The timeline of studies is presented in Fig. 8. The timeline also includes the first release of the SW/HW

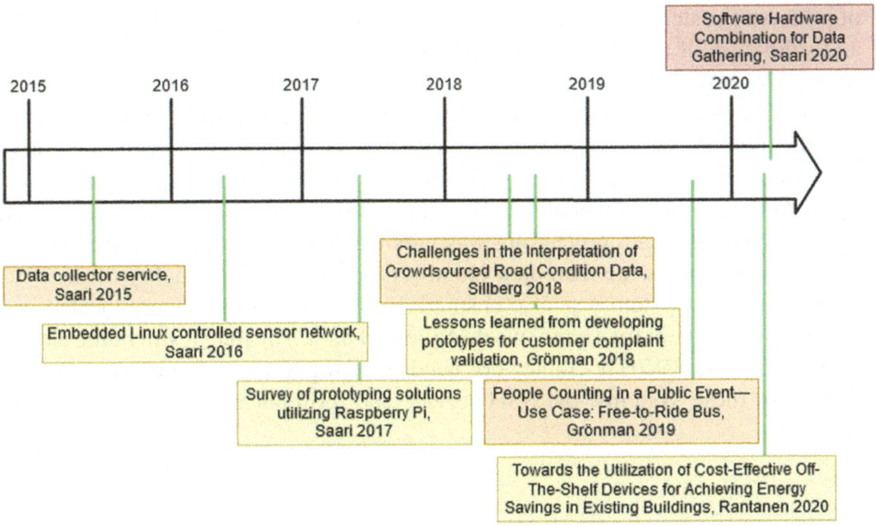

Fig. 8 Prototypes and validation in the SW/HW framework timeline

framework study [34]. These projects have contained multiple iteration rounds. Each iteration has produced a working prototype system. This section gives an overview of the systems and lists the main findings during the development of the system. The systems are divided into the previously presented types 1, 2, and 3.

5.1 Type 1 and Type 2 with SBC Related Prototype Systems

Types 1 and 2 are SBC and Linux OS based data gathering prototype systems. Type 1 usually contains one sensor connected directly to the SBC.

- A data collector service [3] was the first implementation. The SBC was a Beaglebone Black which gathered temperature, humidity, and brightness data. The data were stored in the SBC and there was a service to offer the data to users. The SBCs were connected with an Ethernet connection. The main goal of the study was to experiment how well a cost-efficient SBC could be used to gather sensory data, and how this data could be provided to the client over the public Internet. This goal was reached successfully, and the designed system was tested and found to work as planned [3]. The study proved that fully working data gathering prototypes can be developed with off-the-shelf devices and open source tools.
- The ability to use off-the-shelf devices prompted us to find out how others have used these devices in academic research. A survey of prototyping was made to find out the benefits and limitations of Raspberry Pi [32]. Also, we searched for the testing methods of these prototypes. The study showed that the Raspberry Pi is a widely used device in research implementations of different kinds. Some testing methods

of prototype systems were found: software testing, software performance testing, and validation of data tests [32]. This study clarified the operating environment for the SW/HW framework.
- The third [23] and fourth [22] studies were similar to each other. Both prototypes presented in the studies were based on a Raspberry Pi and camera combination in a vehicular environment. The power supply was a battery backup and both used a 3G/4G communication channel. The data, photos, were transferred to the cloud storage where the data processing was handled. The focus of the studies was to analyze and process the data on a cloud server but the functionality of the SBC based prototype system was also ensured. Both of the systems were located in a client's environment for several weeks to collect data. The long testing time clearly showed that the SW/HW framework needs a configuration channel from user to device. In these systems, the SSH service and terminal were used as a remote channel.

5.2 Type 2 with SBC Related Prototype Systems

The type 2 configuration has been tested in three real-world projects. The common construction of type 2 data gathering prototypes is that one master node SBC works with several sensor nodes. The communication is one-way from sensor nodes to the master nodes. The master node with a group of sensor nodes was tested in several research cases:

- The first was [31] where one SBC master node, an Intel Galileo Gen 2 development board, communicated using wireless XBee technology with several sensor nodes, on an Arduino development board. The collected data were environmental. The targets of the study were threefold: to test the model, to determine how well cost-efficient SBCs could be used to gather sensory data from several sensor nodes, and how to deliver this data to clients over the public Internet [31]. The study showed that SW/SW frameworks need a separate sensor node–master node architecture. This is useful when several sensors are required in a small area.
- The second use case [33] presented a wireless sensor system for monitoring indoor living or working conditions. The study expanded the range of sensors by using wireless LoRa technology in communication between the master node and sensor nodes. In the main construction, the master node Raspberry Pi received environmental data from several Sodaq sensor nodes. The data were stored in the master node for analysis and processing. The primary purpose of the tests was to validate the sensor system construction. Based on the experiments, we found that LoRa was a good choice for sensor applications within concrete buildings [33].
- The third use case used commercial, but open hardware and open source, RuuviTag sensor nodes [35]. These cost-efficient sensors collected temperature, humidity, pressure, and motion information. The data were sent using Bluetooth communication to the Raspberry Pi master node. The data were stored on an InfluxDB

database and visualized with Grafana visualizing software. One of the RuuviTag experiments is documented in a study [36]. This prototype system is quickly configurable, because RuuviTags do not need a configuration; only the Raspberry Pi needs a setup. These prototype systems showed that even though the gathered data were small in quantity, the Raspberry Pi limited the visualization. The second issue raised was a limitation in the amount of memory operations with a Raspberry Pi memory card. Therefore the prototype system experiment showed that data should be transferred and stored on a cloud server. From the perspective of the SW/HW framework, this prototype system showed the usability of modular development and the fact that the sensor nodes and master node are interchangeable.

5.3 Type 3: Smartphone Related Prototype Systems

The smartphone is an excellent WSN sensor node. It has a working hardware package: power source, communication skills, and sensor devices. It also has a suitable OS, which allows the wide usage of the hardware. Our first ideas of using the smartphone as a sensor were presented in [37]. The main question presented was "How to utilize mobile technology to supply disaster information to both mobile terminals and desktop computers?" [37].

In the recent type 3 prototype systems, we used Android smartphones. These two documented data gathering prototypes are next discussed from the perspective of the SW/HW framework.

- The first data gathering prototype implementation was presented in study [30], which utilized data collected by smartphone sensors, such as an accelerometer and GPS, to detect variations in road surface conditions. The use of a smartphone was preferred because the data were obtained from a group of users driving on roads in western Finland and most of these people owned a smartphone that was good enough for prototyping purposes. The data gathering device was therefore an off-the-shelf Android smartphone without any hardware modifications. The software was a combination of user interface application and background service. The results of the study were that the basic programming task of creating a simple application for tracking the user's location and gathering data from the basic sensors embedded in a mobile device is a straightforward process. Minor difficulties arose because of variations in the phones' basic software and hardware depending on the manufacturer. In relation to the SW/HW framework, the prototype showed the ability to use a smartphone as a WSN sensor node.
- The second data gathering prototype was a solution for idea testing. The question asked was: Is it possible to take a photo if the phone is in a selected GPS position? Study [23] presents the working prototype solution for this problem. For the SW/HW framework, this prototype again showed the ability to use a smartphone as a WSN sensor node. The program code utilized an implementation of the haversine formula, which determines the great circle distance between two

locations and is relatively simple to implement yet accurate enough for our use cases. The application can be installed on any reasonably new Android device and takes advantage of the built-in sensors and camera of the device [23].

6 Discussion

This study aims to resolve the research question: How to generalize the prototyping of IoT data gathering in a framework of required software and hardware components?

To answer the question, we made a reasonable number of data gathering prototypes and reported on them in academic research papers. Thus, our earlier research answered the question. The studies highlighted several aspects in constructing a SW/HW framework. The framework describes the main findings from developed prototypes. The framework brings out three different approaches for different use cases. The research papers presented how prototyping development can be made cost efficiently. This was enabled by using off-the-shelf embedded devices such as smartphones and Raspberry Pi SBCs. The devices have the type of OS that can be modified for sensor usage.

Furthermore, the prototype highlighted the knowledge we have about the process of prototyping. The model can be used as guidance when designing a new prototyping project together with a customer who wants to obtain information about some target environment.

The development of the SW/HW Framework raised several new ideas for research topics. These topics are briefly discussed here.

Each of the prototypes discussed has sensor software: software that reads the sensor, a temperature sensor for example, changes the value form bytes to an integer with a reasonable formula, and sends or stores the value somewhere. The study mentions the sensor software several times, but its construction has not been discussed in detail. This low-level program is coded using C/C++, Python, Java, or a similar programming language. It should be noted that at this level the initial data processing could be done, e.g., the mean value of accelerometer sensor values within one second. Is it possible to get improved performance without data loss? How should this low-level software be programmed? These would be suitable questions for future study.

The topics of user and user experience are beyond the scope of this study. Our prototypes were developed due to the needs of some project partner. The prototypes were tested with use case testing and once the customer had received a reasonable answer to a certain need, the development was stopped (except for one example [21] where there was a long piloting period in a real usage environment). The project outputs and prototypes are freely exploitable by the project partners.

The sensor prototypes produce a large amount of sensor data. Data processing and data mining are important issues, which this study leaves for future research, as it is such an extensive area. The issues of data visualization have been handled in some studies [30]. In addition, sensor data will become more usable if merged with other publicly available data. This kind of data could be weather data or map data [21, 38].

Performance problems have not been extensively addressed in this study, but when using the SW/HW framework they have to be taken into account. In data gathering construction type 2, the amount of sensor nodes is limited, but no exact limit can be set. The limit is changed by the range, communication channel, data to be transferred and so on. The data processing can also affect the performance problems. For example, in this study photos are often mentioned as difficult regarding performance, especially motion detection. For example, photos should be transferred to a cloud server for processing.

The SW/HW framework does not set the quality criteria for components, but how can the selection of high-quality components be ensured? "The hardware quality depends on the price" is one claim, which in most situations makes sense. The second level for selection is "good enough". The SW/HW framework does not set these kinds of selection criteria for software or hardware components and therefore these decisions are left to the framework user.

The vulnerability of data is worth considering. Is the data critical and what happens if we cannot obtain the data? Is it possible to manipulate the data and what would the consequences be in that case? For example, what happens if somebody changes the data. This SW/HW framework does not take a position on the matter, but these are significant issues. Furthermore, security issues are important for IoT devices. Security vulnerabilities and attacks on IoT systems have been covered extensively by [39]. The SW/HW framework does not pay attention to security except for the communication channel. This concern was raised in [31] and the proposed, more secure, communication technology LoRa has been discussed by Saari et al. [40].

7 Summary

This paper introduced the software/hardware framework and a descriptive model for the prototyping process. The framework was developed during several research projects by following the same prototype development process. The model presented the process for constructing and testing a data gathering prototype with six steps, starting from discussion of requirements and ending with the presentation of collected data. The main aspects of these steps were presented briefly.

A sensor network consists of several layers, from data gathering devices to data users. The framework is placed in the data gathering layer. The three types of data gathering constructions were presented by introducing the software components, the hardware components, and their interconnections.

Research findings: The model and the framework were validated by presenting several previous research projects and studies.

Acknowledgements This work is part of the KIEMI ("Vähemmällä Enemmän—Kohti Kiinteistöjen Energiaminimiä", or "Less is More: Towards Energy Minimum of Properties" in English) project and has been funded by the European Regional Development Fund and the Regional Council of Satakunta.

References

1. Atzori, L., Iera, A., Morabito, G.: The internet of things: a survey. Comput. Netw. 54(15), 2787–2805 (2010)
2. IEEE approved draft standard for an architectural framework for the internet of things (IoT) (2019)
3. Saari, M., Sillberg, P., Rantanen, P., Soini, J., Fukai, H.: Data collector service—practical approach with embedded Linux. In: 2015 38th International Convention on Information and Communication Technology, Electronics and Microelectronics (MIPRO), International Convention on Information and Communication Technology, Electronics and Microelectronics, pp. 1037–1041. IEEE (2015)
4. Saari, M., Sillberg, P., GrÃ¯Â¿Â½nman, J., Kuusisto, M., Rantanen, P., Jaakkola, H., Henno, J. Reducing energy consumption with IoT prototyping. Acta Polytech. Hung. **16**(9, SI), 73–91 (2019)
5. Buchenrieder, K.: Rapid prototyping of embedded hardware/software systems. In: Proceedings. Ninth International Workshop on Rapid System Prototyping (Cat. No. 98TB100237), vol. 5, pp. 2–3. IEEE Comput. Soc. (2000)
6. Kruger, C.P., Abu-Mahfouz, A.M., Hancke, G.P.: Rapid prototyping of a wireless sensor network gateway for the internet of things using off-the-shelf components. In: 2015 IEEE International Conference on Industrial Technology (ICIT), pp. 1926–1931 (2015)
7. Babich, N.: What is rapid prototyping? (2019). Accessed 16 Dec 2020
8. Jensen, C., Scacchi, W.: Discovering, modeling, and re-enacting open source software development processes: a case study, pp. 1–20 (2006)
9. Becker, U., Hamann, D., Verlage, M.: Descriptive Modeling of Software Processes. IESE-Report No. 047.97/E (1997)
10. Becker-Kornstaedt, U., Webby, R.: A comprehensive schema integrating software process modeling and software measurement. IESE-Report No. 047.99/E (1999)
11. Lee, E.A., Seshia, S.A.: Introduction to Embedded Systems. A Cyber-Physical Systems Approach. Second Edition, vol. 195 (2017)
12. Kreiner, C., Steger, C., Teiniker, E., Weiss, R.: A HW/SW codesign framework based on distributed DSP virtual machines. In: Proceedings—Euromicro Symposium on Digital Systems Design: Architectures, Methods and Tools, DSD 2001, pp. 212–219 (2001)
13. Saha, D., Mitra, R.S., Basu, A.: Hardware software partitioning using genetic algorithm. In: Proceedings Tenth International Conference on VLSI Design, pp. 155–160. IEEE Comput. Soc. Press (1997)
14. Srivastava, M.B., Brodersen, R.W.: Rapid-prototyping of hardware and software in a unified framework. In: 1991 IEEE International Conference on Computer-Aided Design Digest of Technical Papers, pp. 152–155. IEEE Comput. Soc. Press (1991)
15. Da Li, X., He, W., Li, S.: Internet of things in industries: a survey. IEEE Trans. Ind. Inform. 10(4), 2233–2243 (2014)
16. Vakaloudis, A., O'Leary, C.: A framework for rapid integration of IoT systems with industrial environments. In: 2019 IEEE 5th World Forum on Internet of Things (WF-IoT), pp. 601–605. IEEE (2019)
17. Akyildiz, I.F., Sankarasubramaniam, Y., Cayirci, E.: A survey on sensor networks. IEEE Commun. Mag. 40(8), 102–114 (2002)

18. Saari, M., Soini, J., Grönman, J., Rantanen, P., Mäkinen, T., Sillberg, P.: Modeling the software prototyping process in a research context. In: Tropmann-Frick, M., Thalheim, B., Jaakkola, H., Kiyoki, Y., Yoshida, N. (eds.) Information Modelling and Knowledge Bases XXXII, vol. 333, pp. 107–118. IOS Press (2021)
19. Grönman, J., Rantanen, P., Saari, M., Sillberg, P., Vihervaara, J.: Low-cost ultrasound measurement system for accurate detection of container utilization rate. In: 41st International Convention on Information and Communication Technology, 2018. Electronics and Microelectronics (MIPRO). IEEE (2018)
20. Soini, J., Sillberg, P., Rantanen, P.: Prototype system for improving manually collected data quality. In: Budimac, Z., Grbac, T.G. (eds.) Proceedings of the 3rd Workshop on Software Quality Analysis, Monitoring, Improvement, and Applications, SQAMIA 2014, 19–22 Sept 2014, Lovran, Croatia, CEUR Workshop Proceedings, pp. 99–106. M. Jeusfeld c/o Redaktion Sun SITE (2014)
21. Soini, J., Kuusisto, M., Rantanen, P., Saari, M., Sillberg, P.: A study on an evolution of a data collection system for knowledge representation. In: Dahanayake, A., Huiskonen, J., Kiyoki, Y. (eds.) Information Modelling and Knowledge Bases XXXI, vol. 321, pp. 161–174. IOS Press (2019)
22. Grönman, J., Sillberg, P., Rantanen, P., Saari, M.: People counting in a public event-use case: free-to-ride bus. In: 42nd International Convention on Information and Communication Technology, 2019. Electronics and Microelectronics (MIPRO). IEEE (2019)
23. Grönman, J., Rantanen, P., Saari, M., Sillberg, P., Jaakkola, H.: Lessons learned from developing prototypes for customer complaint validation. In: Proceedings of the SQAMIA 2018: 7th Workshop of Software Quality, Analysis, Monitoring, Improvement, and Applications, vol. 2217, pp. 27–30. CEUR Workshop Proceedings (2018)
24. Liou, F.F.: Rapid Prototyping and Engineering Applications, 2nd edn. Taylor & Francis, CRC Press, Boca Raton (2019)
25. Centenaro, M., Vangelista, L., Zanella, A., Zorzi, M.: Long-range communications in unlicensed bands: the rising stars in the IoT and smart city scenarios. IEEE Wirel. Commun. 23(5), 60–67 (2016)
26. Raza, U., Kulkarni, P., Sooriyabandara, M.: Low power wide area networks: an overview. IEEE Commun. Surv. Tutor. 19(2), 855–873 (2017)
27. Satoto, K.I., Widianto, E.D., Sumardi: Environmental health monitoring with smartphone application. In: 2018 5th International Conference on Information Technology, Computer, and Electrical Engineering (ICITACEE), pp. 281–286. IEEE (2018)
28. Namiot, D.: Time series databases. Data Analytics and Management in Data Intensive Domains (DAMDID/RCDL2015), vol. 1536, pp. 132–137 (2015)
29. Bader, A., Kopp, O., Falkenthal, M.: Survey and comparison of open source time series databases. Lecture Notes in Informatics (LNI), Proceedings—Series of the Gesellschaft fur Informatik (GI), vol. 266, pp. 249–268 (2017)
30. Sillberg, P., Gronman, J., Rantanen, P., Saari, M., Kuusisto, M.: Challenges in the interpretation of crowdsourced road condition data. In: 2018 International Conference on Intelligent Systems (IS), pp. 215–221. IEEE (2018)
31. Saari, M., Baharudin, A.M., Sillberg, P., Rantanen, P., Soini, J.: Embedded Linux controlled sensor network. In: 2016 39th International Convention on Information and Communication Technology, Electronics and Microelectronics (MIPRO), pp. 1185–1189. IEEE (2016)
32. Saari, M., bin Baharudin, A.M., Hyrynsalmi, S.: Survey of prototyping solutions utilizing Raspberry Pi. In: 2017 40th International Convention on Information and Communication Technology, Electronics and Microelectronics (MIPRO), pp. 991–994. IEEE (2017)
33. Towards the utilization of cost-effective off-the-shelf devices for achieving energy savings in existing buildings. IEEE (2020)
34. Saari, M., Rantanen, P., Hyrynsalmi, S.: Software hardware combination for data gathering. In: Proceedings of 2020 IEEE 10th International Conference on Intelligent Systems (IS2020) (2020)

35. Ruuvitag technical specifications. https://ruuvi.com/files/ruuvitag-tech-spec-2019-7.pdf (2019). Accessed 23 Nov 2020
36. Saari, M., Grönman, J., Soini, J., Rantanen, P., Mäkinen, T.: Experimenting with means to store and monitor IoT based measurement results for energy saving. In: 43rd International Convention on Information and Communication Technology, 2020. Electronics and Microelectronics (MIPRO). IEEE (2020)
37. Sillberg, P., Rantanen, P., Saari, M., Leppäniemi, J., Soini, J., Jaakkola, H.: Towards an IP-based alert message delivery system. In: Landgren, J., Jul, S. (eds.) ISCRAM 2009—6th International Conference on Information Systems for Crisis Response and Management: Boundary Spanning Initiatives and New Perspectives, June 2015, 8 p. Information Systems for Crisis Response and Management, ISCRAM (2009)
38. Sillberg, P., Saari, M., Grönman, J., Rantanen, P., Kuusisto, M. Interpretation, modeling, and visualization of crowdsourced road condition data. In: Goncalves, R., Sgurev, V., Jotsov, V., Kacpzyk, J. (eds.) Intelligent Systems: Theory, Research and Innovation in Applications, pp. 99–119. Springer, Cham (2020)
39. Babar, S., Stango, A., Prasad, N., Sen, J., Prasad, R.: Proposed embedded security framework for Internet of Things (IoT). In: 2011 2nd International Conference on Wireless Communication, Vehicular Technology, Information Theory and Aerospace & Electronic Systems Technology (Wireless VITAE), pp. 1–5. IEEE (2011)
40. Saari, M., Muzaffar bin Baharudin, A., Sillberg, P., Hyrynsalmi, S., Yan, W.: LoRa—a survey of recent research trends. In: 2018 41st International Convention on Information and Communication Technology, Electronics and Microelectronics (MIPRO), pp. 0872–0877. IEEE (2018)

Automated Environmental Mapping with Behavior Trees

Francisco Garcia Rosas, Frank Hoeller, and Frank E. Schneider

Abstract Automated exploration, environmental measuring and mapping using robotic systems is highly relevant in post-disaster scenarios. Challenging conditions and presence of unknown hazardous substances might disturb rescuers, affecting their performance or even endanger their lives. Under these circumstances, the deployment of robotic systems can reduce the risk and workload of rescuers and improve the situational awareness of emergency services. To accomplish this, multiple algorithms need to be synchronized on the robots. Behavior Trees offer a simple and modular structure for task execution, and provide a deep insight into complex task flows at first glance. The main contribution of this work is the development and integration of methods for environmental measuring and robot exploration using Behavior Trees. These methods include robot navigation, complete coverage planning, 2D/3D obstacle mapping and localization of hazard substances. Our solution is tailored for post-disaster scenarios and focuses on providing remote situational awareness for rescuers. Our system was successfully tested at the European Robotics Hackathon 2019, where a hazardous materials incident response operation was staged. With this implementation, we aim to simplify the system design and application of robotic systems used for autonomous operation.

Abbreviation

BT	Behavior Tree
CCPP	Complete Coverage Path Planning

F. G. Rosas (✉) · F. Hoeller · F. E. Schneider
Fraunhofer Institute for Communication, Information Processing and Ergonomics FKIE, Fraunhoferstrasse 20, 53343 Wachtberg, Germany
e-mail: francisco.garcia.rosas@fkie.fraunhofer.de

F. Hoeller
e-mail: frank.hoeller@fkie.fraunhofer.de

F. E. Schneider
e-mail: frank.schneider@fkie.fraunhofer.de

V. Sgurev et al. (eds.), *Advances in Intelligent Systems Research and Innovation*, Studies in Systems, Decision and Control 379,
https://doi.org/10.1007/978-3-030-78124-8_4

EnRicH	The European Robotics Hackathon
FSM	Finite State Machine
GNSS	Global Navigation Satellite System
LEO	Liquid Exploration Online
MBF	Move Base Flex
MS	Measurement Server
PN	Petri Nets
ROS	Robot Operating System
RTK	Real-Time Kinematics
SLAM	Simultaneous Localization and Mapping
UGV	Unmanned Ground Vehicle

1 Introduction

Natural or industrial post-disaster scenarios are typical applications for mobile robotic systems. Such situations always pose an imminent threat to emergency personnel operating in the affected area. Even simple teleoperated robots can significantly reduce the endangerment of humans. In addition, mobile robots can be equipped with sensors, improving the understanding of a disaster environment. Previous experiences demonstrated that robots with such capabilities can successfully be deployed [22]. Further common applications include search and rescue (SAR) [21] and building inspection.

In SAR missions, robot teleoperation is the dominant approach [32]. However, this is a cognitively exhausting task that usually requires a two-man team for controlling just one robot: a robot operator and a problem holder [20]. Current research in SAR focus on improving the human-robot interaction to allow one operator to control one or more robots simultaneously [8].

Additionally, a data connection link is required when teleoperating robotic systems, but depending on the scenario this can not always be provided. Then, it is necessary that the robot is able to accomplish tasks autonomously. This is rather complex, because automated exploration, map building and hazard data collection have to be carried out in an unknown and uncooperative environment.

The complexity of such combined tasks can be reduced with a suitable model of computation, for example by using traditional Finite State Machines (FSM), Petri Nets (PN) or Behavior Trees (BT). A problem with both FSM and PN is, as the amount of tasks increases, they quickly become illegible due to the *state explosion* [37]. However, the number of applications implementing BTs in robotics have recently increased, due to its modularity and reusability, that allows for a more complex and richer behavior modeling. Additionally, BTs have also been used for improving the robustness of mission-critical components in robotic systems [33].

Environmental measuring is a key component in SAR operations, and it is within the capabilities of modern robotic systems. This work presents a method for collecting, storing and presenting environmental measurements, aiming to deliver situational awareness to rescuers without exposing them to potential risks. To achieve this, a series of tasks have to be jointly solved. We develop individual solutions for addressing the problems of navigation, exploration and mapping in unknown environments and sensor measurement localization.

For exploration, we developed a covering path planner algorithm specially suited for robots with differential-drive. As an alternative, a wall-following path planner was also integrated. For mapping, a 2D, a 3D and an environmental mapper were used in combination to address robot localization and improve visual representation. To combine these tasks into a global solution, we develop a BT-based execution framework.

The European Robotics Hackathon (EnRicH) [29] 2019 offered several staged disaster scenarios. One of the goals was to map and localize hazardous substances in an unknown cluttered post-disaster environment. Our system successfully completed this mapping task autonomously, using the Unmanned Ground Vehicle (UGV) *Fenrir* presented on Fig. 1.

Fig. 1 *Fenrir* UGV equipped with a rotating Velodyne and spherical 360° cameras for 3D mapping at the EnRicH 2019

2 Related Work

Environmental measuring and monitoring using robotic systems have been extensively studied [9, 15, 16, 36]. However, most of these approaches do not consider hazardous or risky environments that might affect the robots or operators.

Similarly, robotic platforms have also been used in post-disaster scenarios for measuring and collecting data [27]. These efforts were mainly motivated by recent nuclear disasters in which high radiation poses a direct threat to human work forces.

Using teleoperation approaches, Nagatani et al. [23] deployed mobile rescue robots for carrying out surveillance missions after the nuclear accident at the Fukushima Daiichi Nuclear Power Plant. Likewise, Martin et al. [18] presented a 2.5D radiation mapping system using unmanned aerial vehicles, also targeting the Fukushima accident scenario. The problem with teleoperation approaches is that they involve a heavy cognitive load for operators of the system [20].

Robotic systems proved to be useful for environmental measuring, but enhancing their degree of autonomy is essential for reducing the load to operators.

Automated tasks for exploration and measuring have to be modeled and monitored using a tasks execution framework. The BT-based approach proposed by Colledanchise and Ögren [7] appears to be quite suitable. Sprague et al. [33] presented a practical example. They proposed a BT-based framework for controlling mission-critical systems, tailored to the needs of autonomous underwater vehicles. Their solution contains sub-trees for safety checks, system preparation, mission synchronization, execution and finalization.

BT-based approaches have also been integrated in robotic industrial applications. Rovida et al. [28] combined extended-BTs with task planning for executing complex robotic tasks. Segura-Muros et al. [31] used BTs for task execution of plans generated by hierarchical task planners. Although being effective in their scenarios, these approaches require to model the environment beforehand, which cannot be done in a disaster scenario.

For heterogeneous robotic systems, Colledanchise et al. [5] extended BTs to enable multi-robot configurations and improve fault tolerance by using built-in fallbacks. The authors provide a theoretical view on BTs and their application, but the proposed approach was not tested in simulation or real robots.

Finally, in the context of robot exploration, Bird et al. [1] proposed an autonomous robotic platform for conducting radiological monitoring in nuclear facilities, using a sequential execution of frontier exploration and mapping. Frontier-based explorations determine goal points based on a hidden evaluation formula, whose solutions may seem random and are difficult to convey to the operator. Additionally, no fallback strategies were proposed in case of navigation or frontier planner failure.

3 Environmental Measuring and Mapping

Our robotic system aims to support emergency personnel in post-disaster scenarios, where hazardous substances might be emitted or leaked. Our robot carries one or more suitable sensors into potentially dangerous areas and transmits the sensor measurements to the operator, so no human has to be endangered. Since the operation of a robotic system is time-consuming, repetitive and requires a lot of attention, our goal is to automate this process as much as possible. Furthermore, advanced robots can do more than just carrying measurement units. With the appropriate sensors, mostly based on LIDAR, robots can create 2D and 3D maps of the environment they traverse. These can even be colored using 360° cameras and augmented with the sensor measurements. Altogether, this can display the situation to emergency forces even before the first human has entered the area.

This section presents the three key components for processing the measurements collected by the robot, not limited to hazardous substance sensors. The first component is called Robot Localization and is a requirement for measurement mapping and automated exploration. Furthermore, a measurement server has been developed. Its main task is to store measurement values and reference them to past robot positions to simplify the post-processing and long-term storage. Finally, a 3D SLAM system is integrated, which is able to create fully colored 3D models of the environment. This produces a very realistic impression of the incident area.

3.1 Robot Localization

The final expected result of our robotic system is a measurement map. This implicitly demands for localization and a coordinate system, because the measurements from the sensor(s) mounted on the robot need to be placed on that map. Among the methods available, GNSS-based localization and Simultaneous Localization and Mapping (SLAM) are the most common. The robot itself provides an odometry-based coordinate system. Since it is error prone and tends to drift we refrain from using it.

The GNSS coordinate systems are in general very stable, especially when aided by local RTK ground stations. In an unknown environment, the only requirement for GNSS is an open sky. In comparison, SLAM does not have this limitation, but it requires structures to be observed. Large, flat, and empty areas can be problematic for it. To perform best in our scenario, we decided to use a laser-based SLAM. The main reason is that we cannot exclude indoor locations. Furthermore, we plan to use 3D-SLAM to generate a 3D model from the start.

This leads to a new issue: More advanced SLAM solutions change the coordinate system, if a new variation of the map is discovered by using global optimization. This happens regularly if an area is traversed twice. If repetitive measurements are recognized, the coordinate system is adjusted to reflect this event. This is called

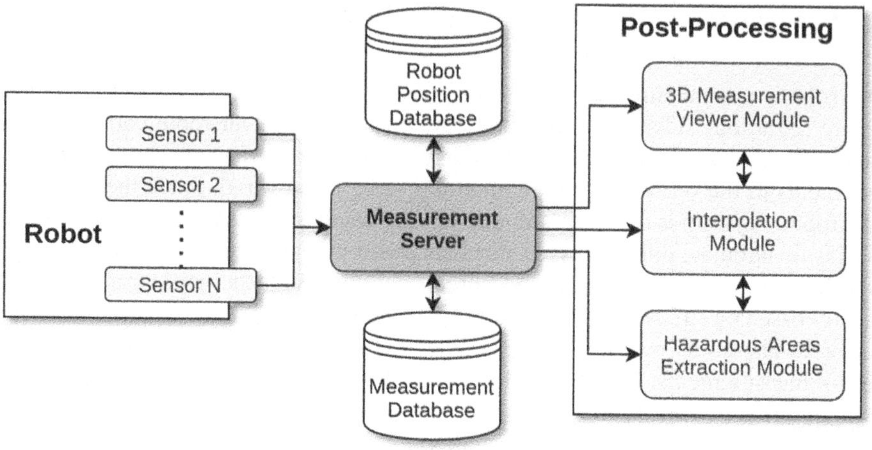

Fig. 2 The *Measurement Server* collects sensor measurements that are later transferred to post-processing applications or long-term storage

loop-closing and poses a new challenge to all connected software components. We designed some of our components to naturally handle this, e.g. the Measurement Server. Others, which do not explicitly handle coordinate system changes, are supported by additional helpers. The individual solutions to this challenge are addressed in the corresponding following sections.

For our robotic system, we employ the Google Cartographer [13] for 2D-SLAM because of its availability, capabilities and stability. The robot was equipped with an inertial measuring unit for improving the 2D mapping performance. Note, that we use a second SLAM method to generate the 3D map, but the 3D solution is not able to provide a real-time coordinate system. It is expected that future versions running on more powerful computers will overcome this limitation.

3.2 Measurement Server

As mentioned before, sensor measurements are collected and stored continuously using a *Measurement Server (MS)*. It supports different measurement types and multiple sensor devices, which is required when a heterogeneous set of sensors is used. This enables applications to map different sources simultaneously, e.g., temperature and humidity or 3G and 4G signal strength.

The MS provides an interface between different sensors and post-processing algorithms that require time-stamped measurement values. A connection overview is presented in Fig. 2.

Instead of referencing measurements with fixed poses they are stored with their measured time-stamp. For every examination of the measurements their locations are calculated online from the robot position history. Consequently, this enables us to handle coordinate system changes, especially coming from loop-closing events of the SLAM system. As result, this ensures that the local obstacle and measurement map are coherent at any time.

The MS can be deployed on the robot, so all measurements are stored locally. Alternatively, it is possible to deploy the MS as a central service, where many robots and operators can be connected to a global database.

The collected measurements can later be used to improve the situational awareness of operators. Post-processing steps such as measurement interpolation, extraction of hazardous areas or 3D visualization can help understanding the collected data. Section 3.4 presents some examples of such post-processing steps using collected measurements.

3.3 3D Mapping

Robots can also be equipped with LIDAR sensors and 360° cameras that delivers highly detailed distance data from the nearby physical structures. This data is not stored entirely, but instead processed and compressed into a single 3D map representation using SLAM algorithms. The map is presented to the operator in form of a colored point-cloud representation, which can be viewed interactively in 3D.

Considering the disaster scenario discussed in this work, it is a considerable improvement to be able to assess the current state of a building and its surroundings. Therefore, we have equipped our robot with appropriate sensors to perform this task.

Our approach gathers spatial data using the rotating platform presented on Fig. 1 while the robot explores the scenario. This platform includes a rotating *Velodyne VLP-16* 3D LIDAR and a spherical 360° *FLIR LadyBug5+* camera.

More specifically, 3D scans and updates on SLAM are triggered every time the robot reaches a goal point. Each scan contains the point-cloud from the LIDAR sensor, as well as RGB images from the 360° camera. A graph-based 3D SLAM [26] integrates them into a single graph representation of the environment. This graph is then exported as a colored 3D point-cloud map and saved locally. It can be transmitted to the control station, depending on the quality and bandwidth of the available communication network.

Our 3D mapping was successfully performed during the EnRicH 2019 competition. Excerpts from the generated 3D colored map are presented in Fig. 3. Additionally, a video of the full generated map is also available.[1]

[1] https://www.youtube.com/watch?v=rZFbHmJBtRY.

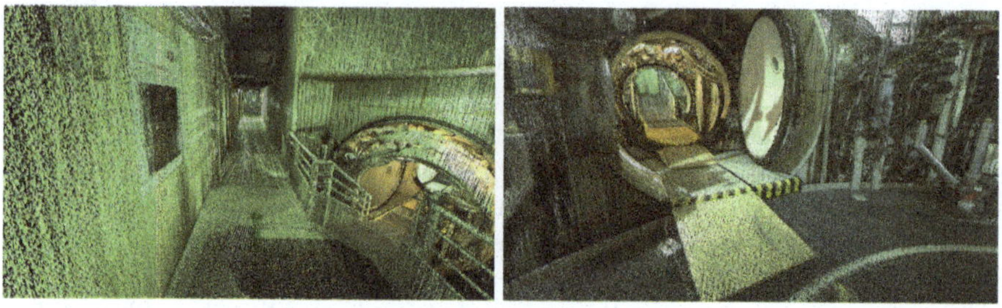

Fig. 3 Colored 3D map of the exploration scenario generated by the robot Fenrir at the EnRicH 2019. Entrance (left) to the reactor room (right)

3.4 Measurement Post-processing

As stated before, we want to improve situational awareness for operators and emergency personnel. It can be achieved by post-processing all the measurements that have been collected during the robot exploration. We want to improve the visual representation by translating the measurements into areas and visualize them as a measurement map to the operator. In addition, depending on the measurement type, there are certain value thresholds which indicate hazardous conditions for humans. We also highlight these risky areas on the map.

Depending on the scenario, the robot might not be able to visit the whole environment because of constraints such as time, battery capacity or obstacles. This creates holes of unmeasured space in the computed map.

Multiple approaches have been proposed [19, 24, 30, 35] for solving this problem, in which two of the most common solutions are the density estimation and the interpolation of measurements.

Density estimation aims to find the most likely localization of sources that influence the environmental measurements [30], such as hotspots or leaking pipes. It proves to be challenging, when both the number of sources and the environment are unknown. Some examples of single source [19] and multiple-source localization [24] of radioactive elements have been extensively discussed.

On the other hand, interpolation methods have been commonly used for extrapolating the collected measurements, and fill the spaces where no values are available [35]. Interpolation can only be used if the measurement values can be modeled as a continuous function. The general approach is to use the measurement covariance and its expected propagation pattern to predict the values for uncovered areas.

It is worth noting that both, density estimation and interpolation methods applied to environmental measuring, present some drawbacks and limitations that need to be considered:

- The method must be carefully tuned for the sensor and expected propagation pattern of the measurement type to be used. Some models might even require additional data, e.g., airflow information for the propagation of aerosols.

- External elements such as walls, pipes or obstacles influence the propagation and are not considered usually. Depending on the material they are composed of this might range from damping to reflection.
- Predicting environmental values always involves a high uncertainty. For example, a small but relatively strong source of energy can emit radiation which is virtually confined to a very small space. The risk of misinterpreting the estimation significantly grows with the size of the area to be predicted.

Some of these limitations can be addressed by properly modeling a dispersion model for every sensors and by including the 3D map into the estimation pipeline. Unfortunately, this exceeds the scope of this work, mainly because of the number of possible sensor models and the problem of on-the-fly material estimation. The operator has to be aware that propagation models must not be used for clearance measuring.

Nevertheless, using either interpolation or density estimation it is possible to extract areas where the expected measurements exceed a given threshold.

These areas can be extracted by searching and selecting local-maxima areas where the expected value is higher than a given safety threshold. As stated before, the approach does not considers additional physical structures such as walls, doors or obstacles, that might further reduce or even shield the expected measurement values.

Finally, after the areas have been extracted, a 3D visualization can be computed for operators to improve the understanding of hazardous areas. This method represents potentially dangerous areas as visual obstacles, creating a natural repellent, which can be perceived intuitively by human operators. An example of the method using two simulated sources is given in Fig. 4.

4 Exploration Path Planning

Although being monitored by an operator, the system is supposed to be automated without interactions as far as possible. While the BTs orchestrate the behavior of the robot, they do not fill the behaviors with actions. The calculation of specific target coordinates for the robot is done by path planner algorithms. They plan, based on the information collected so far, where to move the robot next. The following section will present two path planning approaches for exploration, which we used in combination.

The first approach solves the Complete Coverage Path Planning (CCPP) problem (Sect. 4.1). Its goal is to send the robot to every reachable cell of the scenario using a grid, to achieve data collection for the whole area. Considering the unknown scenario, the developed algorithm does not depend on any prior knowledge of the environment. Similar to other CCPP algorithms, our solution requires a stable coordinate system. As already stated, when using SLAM, loop-closing can occur, which might change the coordinate system.

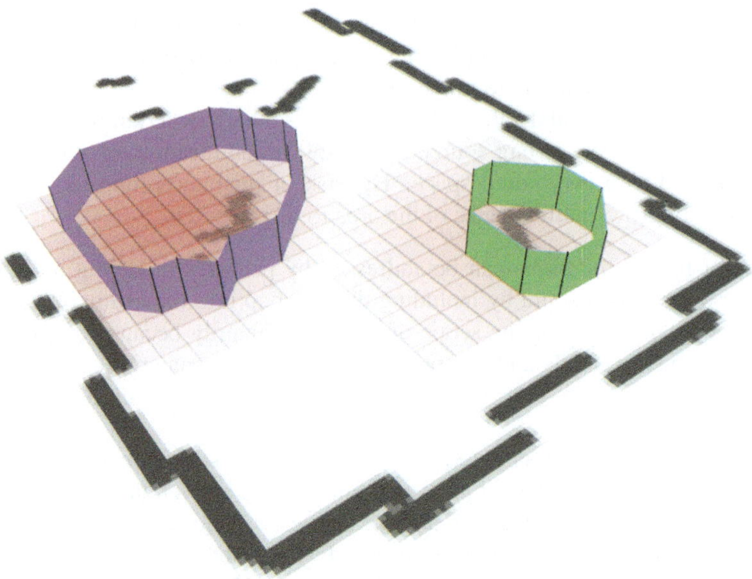

Fig. 4 Extraction and visualization of potentially hazardous areas using extrapolated measurements. It is used for improving situational awareness of operators and emergency personnel. Dangerous areas are represented as colored 3D polygons (green and purple). White-to-red grid shows the interpolated measurements and black cells represent obstacles

To stabilize the algorithm, we try to minimize such occurrences. For this, we use a contour-based wall-following mechanism (Sect. 4.2) before starting with CCPP. It makes the robot encircle the scenario, and initially feed the mapping components. The resulting, and probably still incomplete map, determines the scenario dimensions and should significantly reduce large-scaled coordinate system adjustments caused by loop-closing. Our CCPP algorithm can handle coordinate system changes because it is mostly a local-decision-based process. But these changes might create gaps or new unvisited cells in the grid, which would have to be covered by time-consuming backtracking. Having a stable coordinate system improves the overall performance and shows less unexpected behaviors.

Note that during the wall-following exploration the MS is already active. Therefore, the visited cells are not targeted by the CCPP algorithm again.

4.1 Complete Coverage Path Planning

As mentioned, the general problem to be solved is called Complete Coverage Path Planning (CCPP). It has been studied extensively in literature [11], and can roughly be divided into grid- and polygon-based approaches. Our scenario requires the robot to visit all cells in a grid to take measurements, thus we neglect polygon-based solutions. Unfortunately, most existing grid-based CCPP algorithms do not consider the robot rotation as a cost. This, in combination with the underlying rectangular grid, tends to result in excessive turning on-the-spot. This poses a problem to our Fenrir robot. It is propelled by a differential-drive system, meaning that on-spot rotations consume a high amount of power and put high stress on mechanical components. Considering the characteristics of our robot, conventional approaches are not suitable and therefore we developed our own algorithm. It is called *Liquid Exploration Online (LEO)* and reduces the amount of turning significantly. Nevertheless, it is still competitive regarding the total path length. The following section should give an inside to our algorithm. Please find a more detailed description in [14].

Algorithm Outline

The initial idea for the algorithm was taken from nature. When liquids are poured into receptacles, they fill up very efficiently the whole volume. We try to mimic this behavior without violating the completeness requirements of the CCPP. A big difference to water is that there is only one moving part, the robot. Especially the handling of dead ends is a lot more challenging for a robot than for a liquid.

LEO is a grid-based algorithm that fills an empty grid by visiting every cell. The core of the movement control is a local 3×3 matrix, which prioritizes the robot's neighboring cells. The matrix is not rotated with the robot, so its orientation is locked to the grid. In essence, the matrix generates local movement decisions, which result in boustrophedon motions [4] parallel to the grid, if not interfered by obstacles.

On a local scale, one of the currently neighboring cells has to be chosen as new target position. For this, we begin by marking and rejecting all neighboring cells already visited or occupied by obstacles. From the remaining cells, the one with the highest priority according to the prioritization matrix is chosen as new robot target. All other cells are marked as *to-be-visited* to enable backtracking and ensure completeness. This process is repeated, when the robot reaches the target cell until no more unvisited cells are known. If the robot reaches a dead end, and no neighboring cells are feasible, a backtracking mechanism is triggered as follows: From all cells marked *to-be-visited* the closest one is chosen. This is done by considering the data collected so far. Known obstacle cells are fed into a Dijkstra planner, which is then used to determine the closest cell to the robot position.

Similar to other CCPP solutions, *LEO* generates its own map of visited and *to-be-visited* cells. Note, that no information about the environment or the starting position is known beforehand.

Movement Prioritization Matrix

This section gives an insight into the prioritization matrix. Furthermore, some improvements to the naive matrix are shown, which drastically increase the performance of our approach.

For the sake of clarity, we consider only one preferred movement direction. To emphasize the comparison with a fluid, the south direction is used. In general and *w.l.o.g.*, the coordinate system for an unknown environment can be rotated as needed. In our measuring scenario, the preferred direction can be set accordingly to the cardinal direction of the MS.

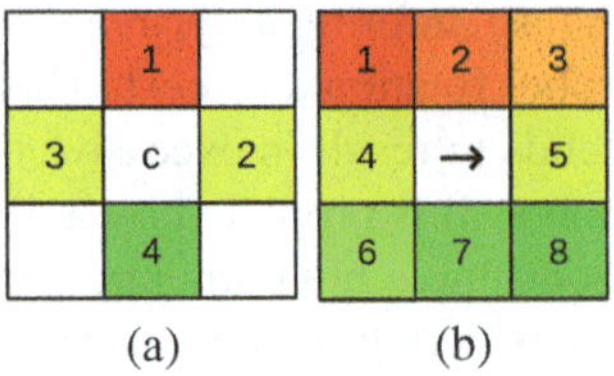

Fig. 5 **a** Naive prioritization matrix P, **b** improved prioritization matrix P_e

The naive prioritization matrix is shown in Fig. 5, and it already solves the CCPP problem. However, certain simulated trajectories show potential for improvement. Straight dividing movements into free space (Fig. 6a and lightly inclined walls (Fig. 6b) yield to unnecessary turnings.

Some of the latter is caused by dead-end cells created by gaps. These gaps were caused by the limitation to four movement directions. By moving from a 4-directional to an 8-directional prioritization matrix, we enable diagonal movements to counteract the occurrence of dead ends. This poses no problem to the robot's navigation system, which is not limited to cells.

Unfortunately, the detachment of robot movement direction and orientation of the matrix cannot be hold with 8 possible movement directions. The problem yields from the addition of two new upward directions (*up-west* and *up-east*). Selecting the best upward motion depends on the current direction of movement (west or east). Figure 5b shows the matrix P_e that is used when the robot is moving east/right. If the robot is moving west/left, the matrix is mirrored. The results from a straight forward implementation is shown in Fig. 6c. The resulting movement pattern generates considerably less dead-ends, less turning and an overall cleaner look.

The next problem to be addressed is the mentioned straight movements into free-space dividing the area to be covered. This causes unnecessary U-turns. Changing the starting behavior from a straight to a spiral movement is not an option, because it by definition would add more rotations and further would fracture the environment in a similar way. Because the division by a straight movement is a very common case, we propose a skipping mechanism for visited cells. This of course increased the number of revisited cells, but can potentially replace many 180° U-turns with

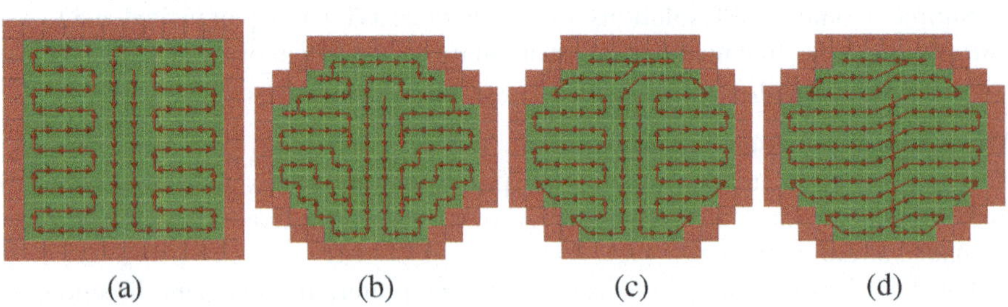

Fig. 6 Results of the evolution of the prioritization matrices. **a** Matrix P, **b** matrix P in non-trivial environment, **c** improvement with diagonal movements, **d** direction sensitive matrix P_e

small rotations of summed up to 55°. The cell skipping is implemented as follows: If the prioritization matrix chooses an already visited cell, the cell is not rejected. Instead the matrix is again applied at the visited cell, such that the cell's neighbors are now rated. The feasible cell with the highest priority, if any, is the replacement of the previously visited cell. Note that the skipping is not repeated if it fails, so that only one cell can be skipped. With this modification our algorithm shows a very clean, boustrophedon motion. It even recovers from disturbances caused by obstacles, as Fig. 6d shows.

LEO Performance Results

The *LEO* CCPP algorithm has been extensively tested and compared to other State-of-the-art approaches. Please find all of the results in [14]. An unknown simulated environment was configured for all algorithms considered. In accordance with the real-world exploration scenario of this work, the coordinate system during simulation was rotated between the experiment runs. This showed that many existing algorithms have problems with not-axis-aligned obstacles. Especially algorithms, which handle axis-aligned obstacles very well, struggle when looking at the amount of rotation. Since not-axis-aligned obstacles are to be expected in real-world application, it uncovers a serious weak point of these algorithms. Figure 7 shows this effect on four selected algorithms, including LEO.

Altogether, *LEO* handles not-axis-aligned obstacles very well. It fully fulfills its purpose of reducing the amount of turning, especially in-place rotations. It can even keep up with other algorithms when considering the total path length or revisiting of cells. Apparently, the trade-off made for cell skipping only has a minor effect.

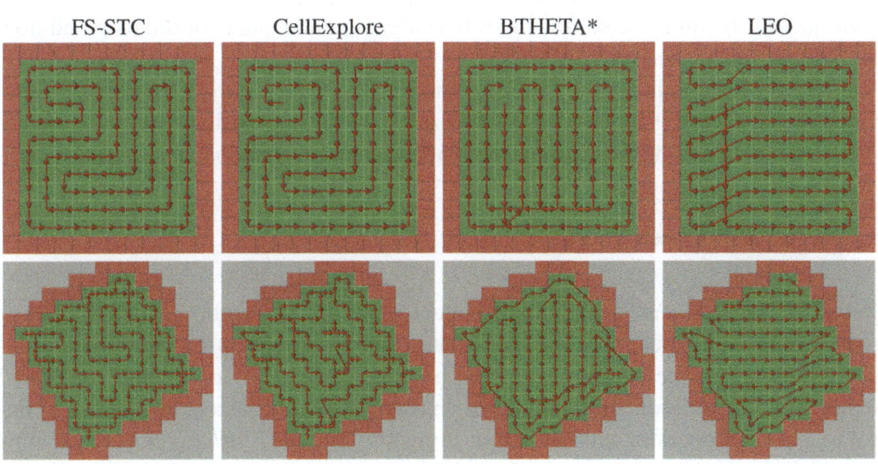

Fig. 7 Comparison of four CCPP algorithms running with axis-aligned and 30° rotated obstacles

4.2 Planning Using Contours

Considering an indoor scenario, it is beneficial to explore the outer-most boundary of a given environment. It stabilizes the coordinate system provided by SLAM in case loop-closing is supported. Furthermore, it also allows us to calculate an initial estimation of the exploration time by looking at the enclosed area.

To do this, we have developed a contour exploration mechanism that makes the robot circle the area using wall-following-like patterns. Our approach uses the map generated by SLAM and the current robot position as inputs. The general idea is to generate a polygon contour that guides the exploration, and then select the most appropriate goal based on this polygon.

Initially, the map is converted into an grayscale image by transforming the occupancy probabilities into pixel colors. The resulting gray-scale image is transformed again into a binary image using the Otsu Binarization [12]. Then, a topological structural analysis of the binary image using border following is performed as suggested by Suzuki [34]. This analysis returns a list of closed-contours modeled as polygons. They are sorted based on their location in the image, and contain closed-objects like obstacles, rooms or sections of unexplored areas. From the results, the biggest polygon containing the robot is chosen and its closest point to the robot is extracted. The point is used later as origin for searching a future goal located on the polygon, towards the unmapped space. While the robot is moving, SLAM is provided with new data. The map is updated and then used by the contour-follower planner in its next iteration. Once the space to be measured has been circled, the circumnavigation is completed and the exploration finishes.

As illustrative example, Fig. 8 shows snapshots of an exploration in a simulated environment. It can be seen how the robot discovers new areas of the map and how the polygons grow accordingly.

Finally, the contour-planner was also successfully tested during the EnRicH 2019 competition. Figure 9 shows a screenshot of the graphical feedback provided to the operator while running the exploration using the contour planner.

5 Behavior Trees (BTs)

So far, we have presented methods to collect, store and post-process measurements, as well as different techniques to automate the robot exploration using path planners. Considering the parallel execution of environmental mapping and automated exploration, the individual functionalities have to be synchronized, while giving control to a human operator. In scheduling-terms this is called *task execution* and aims to model the functionalities as tasks. Different approaches using State Machines,[2] PDDL Planning and Dispatching [2] or Behavior Trees[3, 4] have been presented.

[2] http://wiki.ros.org/smach.

[3] https://github.com/BehaviorTree/BehaviorTree.CPP.

[4] https://github.com/splintered-reality/py_trees_ros.

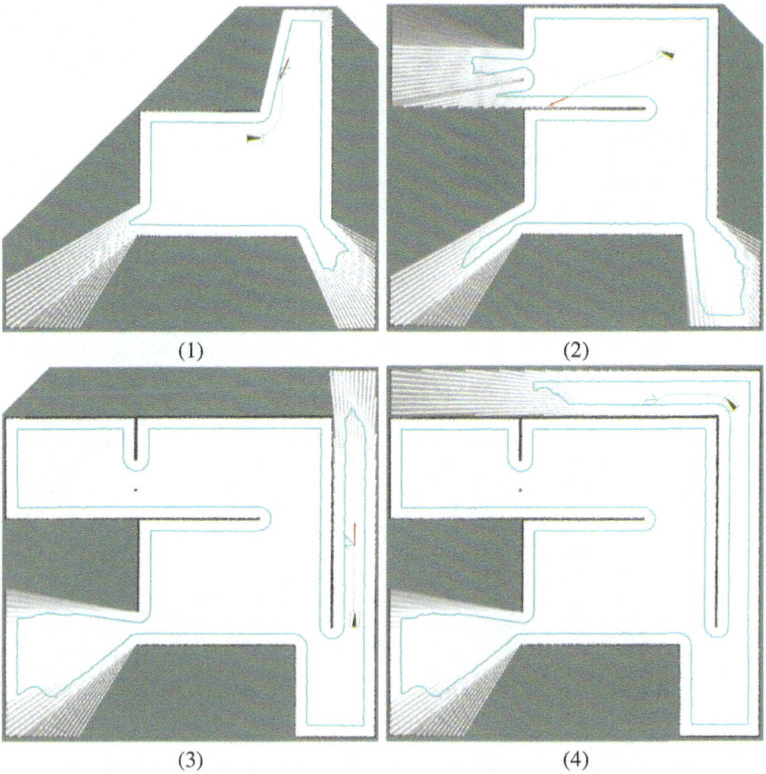

Fig. 8 Contour planner: four sequential snapshots of a simulated exploration. The dark green triangle represents the robots position, the blue line is the current contour and the red arrow marks the current goal pose

In general, tasks can be divided into *primitives* and *compounds*. *Primitives* are actions that can be executed directly, like accelerating or stopping motors. *Compounds* are a composition of partially ordered sets of further tasks, which can either be *primitives* or another *compound*. An example would be moving the robot to a new target location.

BTs offer modularity, flexibility and a simple syntax [33], which is why they are our choice for a task execution framework. The idea is to represent *compound* tasks as tree-branches and *primitives* tasks as the leaves of the BT.

This section presents a brief introduction to BTs together with an overview of a few implementation tools (Sect. 5.1). This is followed by a description of their integration on our robot (Sect. 5.2). Later, the two BTs used for managing automated exploration and mapping are described (Sect. 5.3).

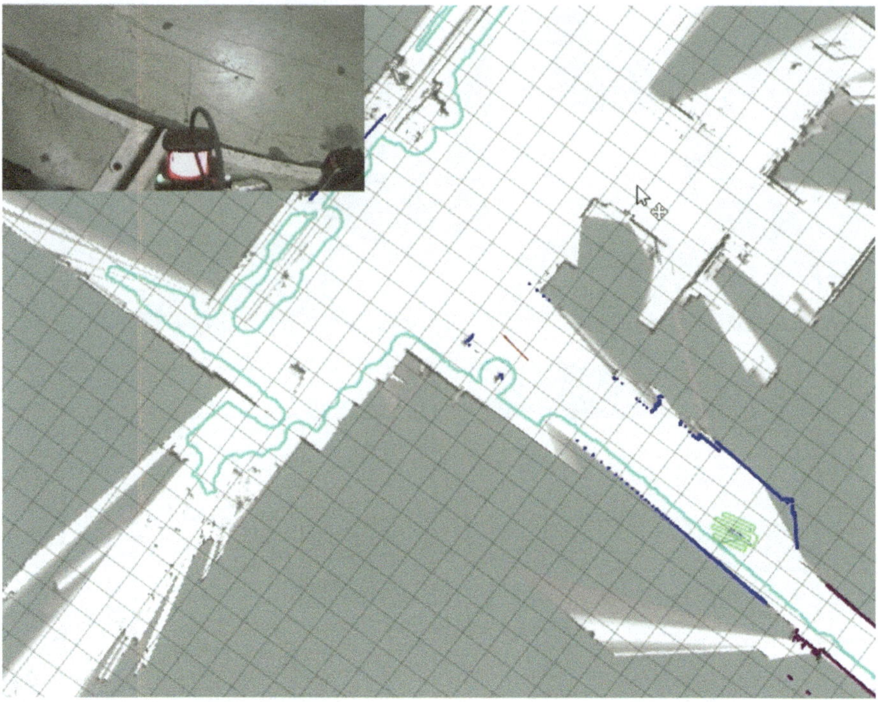

Fig. 9 Control station view of an active contour-following planner. The robot is in the lower right (green). The current contour is shown in cyan. The slim red arrow near the center is the current goal for the trajectory planner. On the top-left, an image captured by the frontal camera of the robot

5.1 Fundamentals and Tools

As the name implies BTs are modeled as trees. Internal nodes are called *control flow nodes* and leaf nodes (*primitives* and *compounds*) are denoted as *execution nodes* [7]. When a BT starts, the root node emits periodic signals, called *ticks*, which are sent to its currently active child. A node is executed if, and only if, it receives a tick signal. Once a node is executed, it might return *Success*, *Failure* or *Running*. In case a node is *Running*, it may propagate the tick to one of its children, based on its policy. The order of node switching inside the tree is from top to down and left to right.

Furthermore, control flow nodes determine if and in which order the child nodes are executed. In the classical formulation, a control node can represent one of the following four operators:

- *Sequence node*: Executes its children sequentially until one of them fails.
- *Fallback node*: Executes its children until one succeed.
- *Parallel node*: Runs simultaneously all its associated nodes.
- *Decorator node*: Typically used for looping or disabling sub-trees (a.k.a. conditionals).

Sequence, parallel and fallback nodes can have multiple children nodes, while decorators are limited to one.

Execution nodes represent the primitive tasks that the robot is able to execute. They are divided into *Actions*, that can change the environment, and *Conditions* that can not. An illustrative example is given in Fig. 11.

Different implementations of BTs are available since they have been extensively used in the computer game industry and robotics. Considering the features and requirements of the scenarios at hand, the open-source implementation *BehaviorTree.CPP* was selected. It offers a well-supported, community-driven BT execution framework, that integrates control flow nodes and some basic actions and conditions. For editing, debugging and monitoring BTs in real-time, we use the open-source graphical tool *Groot*.[5]

5.2 Integration in Robotic Systems

Our robot uses the Robotic Operating System (ROS) as middleware. ROS incorporates services[6] and action[7] interfaces for interacting with the robot's primitive tasks. We have developed the *BT Manager*, that connects the *BehaviorTree.CPP* library with the actions and services offered by ROS in the robot.

Primitive tasks in ROS are modeled as plugin nodes that can be loaded at any time or in any tree. They typically expose three components: a request or input data, feedback during operation, and a task result. The node execution is modeled using an XML tree, where only primitive tasks and control flow nodes are included.

The *BT Manager* ticks the current node and guarantees the continuous execution of the tree. It also provides monitoring and interaction with operators through a graphical interface, as well as control over the *compound* tasks (sub-trees or branches) on the tree.

Figure 10 presents an overview of how the different software components are interconnected. Two instances of the *BT Manager* control the mission execution (Sect. 5.3) and navigation (Sect. 5.4). It also shows the interaction between operator, control station, exploration planners, 3D mapper and robot.

5.3 Automated Exploration Using BTs

Following the findings from Colledanchise et al. [5], we use two different BTs running simultaneously: a *Mission Tree* and a *Navigation Tree*. The former holds the logical structure of the global task, while the latter handles all required tasks for moving the

---

[5] https://github.com/BehaviorTree/Groot.

[6] http://wiki.ros.org/Services.

[7] http://wiki.ros.org/actionlib.

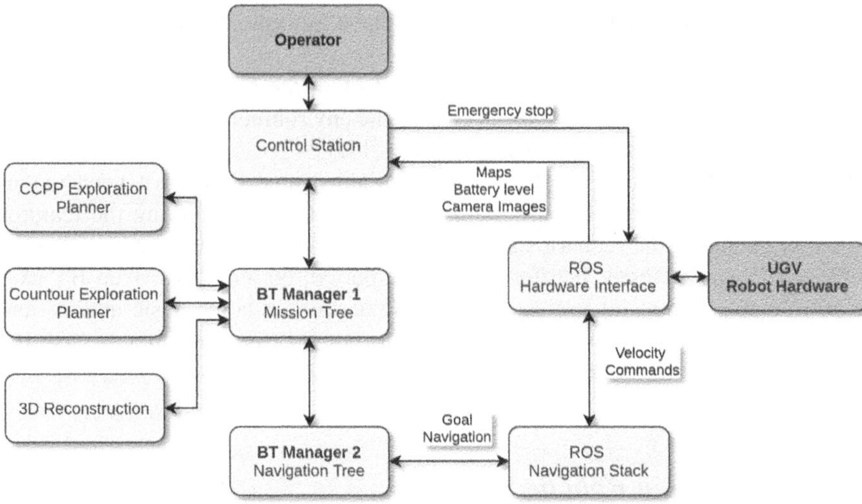

Fig. 10 Connection overview: the *BT Manager 1* synchronizes the tasks requested by the operator using the *Control Station*. Navigation commands are handled by the *BT Manager 2*, which uses the *Navigation Stack* to generate velocity commands

robot between two points reliably. In this way, it is possible to fully decouple the methods used for exploration from the methods used for robot navigation, simplifying the overall system design.

The *Mission Tree* is the main component of our framework and is presented in Fig. 11. It is composed of a sequence with three branches: *Goal definition*, *Navigate to Goal* and *3D Mapping*. First, the goal definition sub-tree ensures that a valid goal pose for the trajectory planning is received. This functionality is modeled as a fallback that prioritizes poses received from different sources. This includes manual pose commands given from the operator, goal poses generated by a contour-following planner (Sect. 4.2) or goals computed by the complete coverage planner (Sect. 4.1). Once a valid goal pose is received, the fallback node gives control to the Navigate to Goal node, which then interacts with the navigation tree. It transmits the goal pose and waits until it is reached by the robot. Finally, control is transferred to the 3D mapping sub-tree (Sect. 3.3), where a full sweep point-cloud scan is performed and combined with camera images. Once the process is finished, the tree returns to its initial state and waits for a new goal point.

It is worth noting that a measurement server (Sect. 3.2) is running in the background. It collects measurements continuously and its execution is not influenced by the current mission. Thus, it is not modeled in any BT. Similarly, the sensor data is continuously sent to the control station for visualization.

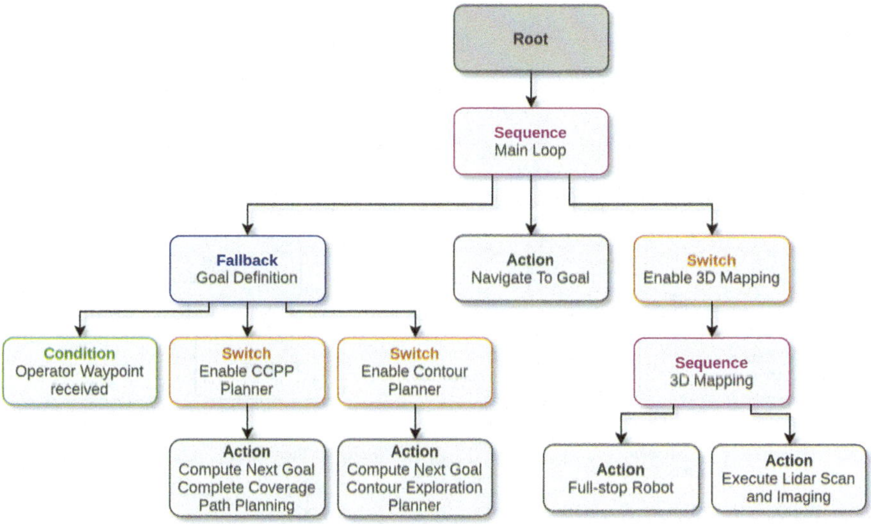

Fig. 11 The *Mission Tree* contains the logical structure of the global task. It consists of three main branches: goal definition, navigation and 3D mapping

5.4 Robot Navigation Using BTs

As described before, we use the secondary BT to model the navigation behavior presented in Fig. 12. Its purpose is to move the robot from its current position to a target goal reliably. For that, it starts with chaining the reception of goal poses with reception of the current SLAM map. The structure of the *Navigation Tree* follows a design approach similar to the one proposed by Pütz et al. [25] or Macenski et al. [17]. It integrates two main branches with altogether three blocks: global path planning, local navigation and recoveries. The path planning and execution is triggered via the Move Base Flex (MBF) [25] interface. At the end, two recovery behaviors are integrated as fallbacks in case the local navigation or collision avoidance fails.

Although BTs support parallel nodes, the classical formulation of BT requires tasks in a parallel operation to be independent of each other [6]. For our implementation, we decided to split off the navigation to an independent BT, such that both trees can run and trigger behaviors simultaneously. This is needed, if a smooth traverse of goal poses without stopping is desirable. Intermediate stopping is prevented by preparing and sending a new waypoint while the robot is still moving towards the current waypoint. MBF and its navigation plug-ins do not currently support intermediate waypoints, and therefore, always plan a deceleration while approaching a goal. This continuous navigation is used for the contour-follower and the CCPP planner. However, the current version of the 3D-mapping does require the robot to stop once a goal pose is reached, but the ongoing research efforts in the field of 3D-mapping is pushing towards continuous 3D-SLAM, and our BTs are prepared for this.

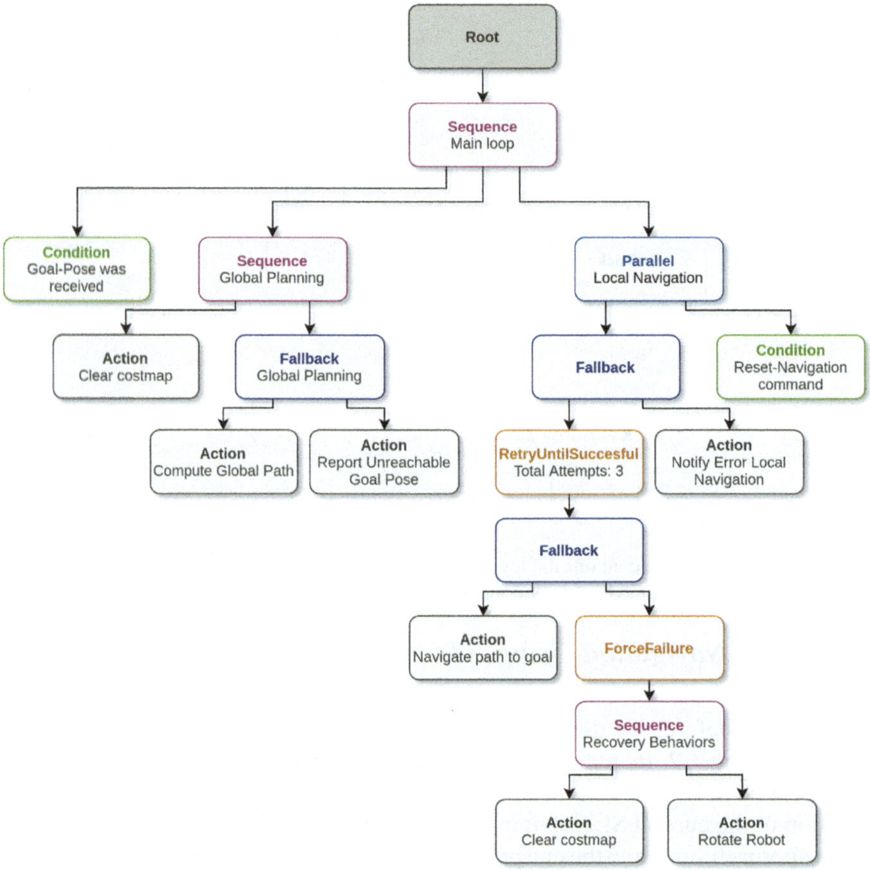

Fig. 12 *Navigation Tree* with two main branches: global path planning and local navigation with collision avoidance and recoveries, in case the navigation fails

The MBF was parameterized for our Fenrir UGV (see Fig. 1). The robot is equipped with four Pepperl+Fuchs R2000 2D LIDARs, two at the front and two at the back. One of each is inclined downwards to detect objects below the horizontal lasers. The output of the two horizontal LIDARs is directly fed into SLAM and MBF. Low and negative obstacles require a detection routine, because obstacle information cannot be directly derived from the raw distance data provided by the LIDARs. To classify each laser beam endpoint as drivable or obstacle, the measured position is compared with a reference scan. If the difference is significant, both the measured and the reference position are added to the set of obstacles. The reference point needs to be classified as inaccessible, because it references to the place where a negative obstacle is detected. The collected set of obstacles still cannot be sent to the MBF, because it (and also Google Cartographer) uses ray tracing to eliminate false positive obstacles. Since the data reception is not scheduled in ROS, it may happen that all low and negative obstacles are cleared by an incoming horizontal LIDAR scan. To prevent this from happening, all low obstacles are saved together with the time-stamp of their detection, similar to the hazard measurements. If a new SLAM map is generated, they are inserted into this map based on their time-stamps, before it is received by MBF.

The MBF was set to use a Dynamic Window Approach [10] local navigation plugin. The maximum velocity was set to 1.2 m/s, but because of the stop-and-go requirement of the 3D-mapping, it was rarely reached.

6 Performance at the EnRicH

The main scenario of the event was composed of three task categories: Exploration, manipulation and SAR. Eleven teams participated in this event and each team was allowed to take any combination of the available tasks. We were interested in building a 3D map of the area and detect and localize radiation sources with a 2D map. During the 30-minutes runs, nobody was allowed to accompany the robot, due to the active radiation sources present in the environment. This challenges the robustness and the autonomy of the participating systems, because the robot must not get stuck during the run. The organization team provided an area-wide WiFi network, so the operator could take direct control of the robot if required. Unfortunately, during handovers between the access points the connection might be interrupted during short periods of time. In such situations we decided not to stop the robot, because a WiFi handover is not distinguishable from a real, locally restricted, link outage. Thus allowing the robot to leave a potentially radio-uncovered area is a reasonable action to regain radio connectivity. Since all planning, navigation and mapping algorithms run locally on the robot, communication was only required for defining how the exploration had to be executed.

The operator can graphically choose between the contour-based or the complete coverage exploration, and add 3D scanning if wanted. As long as communication with the robot is available, the 2D measurement map, the 2D obstacle map and robot-specific data such as battery level and a video stream are transmitted to the control station for monitoring matters.

The navigation and collision avoidance system were set up to act safely, so the robot would not crash or get stuck while operating autonomously. Especially, the inclined LIDARs made sure the robot would not fall down stairs or into gaps.

The scenario included narrow corridors, negative obstacles, uneven surfaces and was partially cluttered with objects. Unfortunately, access to the circular reactor room was only possible via a ramp followed by a tubular corridor, as shown in Fig. 14. To access this area the robot was teleoperated. Yet, the remaining area was explored autonomously, without any manual intervention. The final 2D hazard map is presented in Fig. 13. No other team at the competition managed to autonomously generate a map with this coverage, detail or hazard detection accuracy. Overall, the robot was able to explore autonomously, localize radioactive sources and create the 2D measurement and 3D colored map of the given scenario.

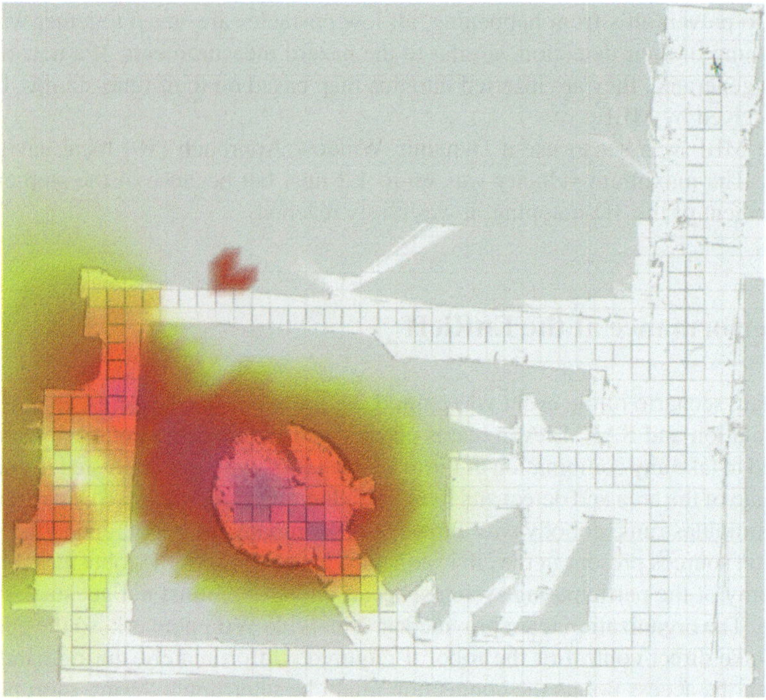

Fig. 13 2D measurement map of hazardous substances generated at EnRicH 2019. Cell size was 1.0 m. Results were padded using kriging [3]

Fig. 14 Entrance ramp and tubular corridor access to the reactor. These challenging conditions could only be overcome by teleoperation

7 Discussion

During the last years, robotic systems gained new capabilities and entered new domains of usage. In general, to be successful in a certain scenario, solutions for different problems are joined in single robotic systems. This introduces a need to describe and manage the global task. Depending on the application, this can be rather complex, because it may include different tasks, algorithms, choices and decisions. It has to be precisely defined when, and which subsystem performs a task. Wherever different solutions have to be chained, polled or switched to create complex behaviors, BTs are a well designed system for design and execution. For the application described here, automated exploration and environmental measuring and mapping of disaster scenarios, we adapted our existing subsystems to be controlled by BTs.

In this work, we present an UGV system to handle the given task. In detail, we describe the required subsystems and how we integrate BTs to manage their operation. The different subsystems are controlled by two BTs which run simultaneously: A navigation tree that executes the required algorithms for moving the robot between two points reliably, and an mission tree that defines how the robot explores and creates 2D/3D maps of the environment.

The exploration task is handled using two approaches: A contour-based planner for surrounding the scenario and a complete coverage planner for searching hazardous areas. Furthermore, three different methods of mapping were integrated. Two of them do SLAM, one in flat 2D, the other in 3D to generate a colored point-cloud representation of the environment. The third method augments the now available 2D map with a grid, which visualizes measurements of hazardous substances. The proposed system was successfully validated at the EnRicH 2019, where a hazardous materials incident response operation was staged.

At the beginning, our system requires an operator to define how the mission should be executed. But after that, it is capable of autonomous operation, reducing the operators cognitive workload noticeably. Of course the operator still can intervene and switch behaviors, like exploration or mapping, at any time. Additionally, it can greatly improve the situational awareness of operators by providing real-time detailed 2D/3D maps of the affected environment. Altogether, our system provides substantial remote situational awareness at a relatively low investment of manpower.

BTs proved to be very capable of capturing the complexity of real-world tasks, including the design, mission execution and monitoring. We designed our BTs in a modular fashion, so parts can easily be reused in other scenarios and applications.

Future work will focus on extending our approach to multi-robot systems, aiming to improve the exploration time and quality of measurement maps. Additionally, adding more post-processing methods such as object detection, classification or 3D reconstruction for hazardous sources might further improve the situational awareness for operators and rescuers.

References

1. Bird, B., Griffiths, A., Martin, H., Codres, E., Jones, J., Stancu, A., Lennox, B., Watson, S., Poteau, X.: A robot to monitor nuclear facilities: using autonomous radiation-monitoring assistance to reduce risk and cost. IEEE Robot. Autom. Mag. **26**(1), 35–43 (2019). https://doi.org/10.1109/MRA.2018.2879755
2. Cashmore, M., Fox, M., Long, D., Magazzeni, D., Ridder, B., Carrera Viñas, A., Palomeras Rovira, N., Hurtós Vilarnau, N., Carreras Pérez, M.: Rosplan: planning in the robot operating system. In: International Conference on Automated Planning and Scheduling (25th: 7–11 July 2015: Jerusalem, Israel). Proceedings of the Twenty-Fifth International Conference on Automated Planning and Scheduling, pp. 333–341. Association for the Advancement of Artificial Intelligence (2015)
3. Chiles, J.P., Delfiner, P.: Geostatistics: Modeling Spatial Uncertainty, vol. 497. Wiley, New York (2009)
4. Choset, H., Pignon, P.: Coverage path planning: the boustrophedon cellular decomposition. In: Field and Service Robotics, pp. 203–209. Springer, London (1998)
5. Colledanchise, M., Marzinotto, A., Dimarogonas, D.V., Ögren, P.: The advantages of using behavior trees in multi-robot systems. In: Proceedings of ISR 2016: 47th International Symposium on Robotics, pp. 1–8 (2016)
6. Colledanchise, M., Natale, L.: Improving the parallel execution of behavior trees. In: 2018 IEEE/RSJ International Conference on Intelligent Robots and Systems (IROS), pp. 7103–7110 (2018). https://doi.org/10.1109/IROS.2018.8593504
7. Colledanchise, M., Ögren, P.: Behavior Trees in Robotics and AI: An Introduction (2017)
8. Delmerico, J., Mintchev, S., Giusti, A., Gromov, B., Melo, K., Horvat, T., Cadena, C., Hutter, M., Ijspeert, A., Floreano, D., Gambardella, L.M., Siegwart, R., Scaramuzza, D.: The current state and future outlook of rescue robotics. J. Field Robot. **36**(7), 1171–1191 (2019). https://onlinelibrary.wiley.com/doi/abs/10.1002/rob.21887
9. Dunbabin, M., Marques, L.: Robots for environmental monitoring: significant advancements and applications. IEEE Robot. Autom. Mag. **19**(1), 24–39 (2012)
10. Fox, D., Burgard, W., Thrun, S.: The dynamic window approach to collision avoidance. IEEE Robot. Autom. Mag. **4**(1), 23–33 (1997)
11. Galceran, E., Carreras, M.: A survey on coverage path planning for robotics. Robot. Auton. Syst. **61**(12), 1258–1276 (2013). https://doi.org/10.1016/j.robot.2013.09.004
12. Garg, N., Garg, N.: Binarization techniques used for grey scale images. Int. J. Comput. Appl. **71**(1), 8–11 (2013)
13. Hess, W., Kohler, D., Rapp, H., Andor, D.: Real-time loop closure in 2D LIDAR slam. In: 2016 IEEE International Conference on Robotics and Automation (ICRA), pp. 1271–1278 (2016). https://doi.org/10.1109/ICRA.2016.7487258
14. Hoeller, F., Rosas, F.G., Schulz, D., Arya, D.: Leo: liquid exploration online. In: 2019 Third IEEE International Conference on Robotic Computing (IRC), pp. 17–24 (2019)
15. Jadaliha, M., Choi, J.: Environmental monitoring using autonomous aquatic robots: sampling algorithms and experiments. IEEE Trans. Control Syst. Technol. **21**(3), 899–905 (2012)
16. Lu, Q., Han, Q.L.: Mobile robot networks for environmental monitoring: a cooperative receding horizon temporal logic control approach. IEEE Trans. Cybern. **49**(2), 698–711 (2018)
17. Macenski, S., Martin, F., White, R., Clavero, J.G.: The Marathon 2: A Navigation System (2020)
18. Martin, P.G., Kwong, S., Smith, N., Yamashiki, Y., Payton, O.D., Russell-Pavier, F., Fardoulis, J.S., Richards, D., Scott, T.B.: 3D unmanned aerial vehicle radiation mapping for assessing contaminant distribution and mobility. Int. J. Appl. Earth Obs. Geoinf. **52**, 12–19 (2016)
19. Mascarich, F., Wilson, T., Papachristos, C., Alexis, K.: Radiation source localization in GPS-denied environments using aerial robots. In: 2018 IEEE International Conference on Robotics and Automation (ICRA), pp. 6537–6544. IEEE (2018)
20. Murphy, R.R.: Human-robot interaction in rescue robotics. IEEE Trans. Syst. Man Cybern. Part C (Appl. Rev.) **34**(2), 138–153 (2004). https://doi.org/10.1109/TSMCC.2004.826267

21. Murphy, R.R.: Trial by fire [rescue robots]. IEEE Robot. Autom. Mag. **11**(3), 50–61 (2004). https://doi.org/10.1109/MRA.2004.1337826
22. Murphy, R.R.: Disaster Robotics. MIT Press, Cambridge, MA (2014)
23. Nagatani, K., Kiribayashi, S., Okada, Y., Otake, K., Yoshida, K., Tadokoro, S., Nishimura, T., Yoshida, T., Koyanagi, E., Fukushima, M., Kawatsuma, S.: Emergency response to the nuclear accident at the Fukushima Daiichi nuclear power plants using mobile rescue robots. J. Field Robot. **30**(1), 44–63 (2013). https://onlinelibrary.wiley.com/doi/abs/10.1002/rob.21439
24. Newaz, A.A.R., Jeong, S., Lee, H., Ryu, H., Chong, N.Y., Mason, M.T.: Fast radiation mapping and multiple source localization using topographic contour map and incremental density estimation. In: 2016 IEEE International Conference on Robotics and Automation (ICRA), pp. 1515–1521. IEEE (2016)
25. Pütz, S., Santos Simn, J., Hertzberg, J.: Move base flex. In: 2018 IEEE/RSJ International Conference on Intelligent Robots and Systems (IROS), pp. 3416–3421 (2018). https://doi.org/10.1109/IROS.2018.8593829
26. Röhling, T., Mack, J., Schulz, D.: A fast histogram-based similarity measure for detecting loop closures in 3-D lidar data. In: 2015 IEEE/RSJ International Conference on Intelligent Robots and Systems (IROS), pp. 736–741 (2015)
27. Rosas, F.G., Hoeller, F., Schneider, F.E.: Autonomous robot exploration and measuring in disaster scenarios using behavior trees. In: 2020 IEEE 10th International Conference on Intelligent Systems (IS), pp. 469–474 (2020). https://doi.org/10.1109/IS48319.2020.9199973
28. Rovida, F., Grossmann, B., Krüger, V.: Extended behavior trees for quick definition of flexible robotic tasks. In: 2017 IEEE/RSJ International Conference on Intelligent Robots and Systems (IROS), pp. 6793–6800 (2017). https://doi.org/10.1109/IROS.2017.8206598
29. Schneider, F.E., Wildermuth, D.: Real-world robotic competitions for radiological and nuclear inspection tasks. In: 2019 20th International Carpathian Control Conference (ICCC), pp. 1–6 (2019). https://doi.org/10.1109/CarpathianCC.2019.8765680
30. Scott, D.W.: Multivariate Density Estimation: Theory, Practice, and Visualization. Wiley, New York (2015)
31. Segura-Muros, J., Fernández-Olivares, J.: Integration of an automated hierarchical task planner in ROS using behaviour trees. In: 2017 6th International Conference on Space Mission Challenges for Information Technology (SMC-IT), pp. 20–25 (2017). https://doi.org/10.1109/SMC-IT.2017.11
32. Sheridan, T.B.: Human robot interaction: status and challenges. Hum. Factors **58**(4), 525–532 (2016). https://doi.org/10.1177/0018720816644364. PMID: 27098262
33. Sprague, C.I., Özkahraman, Ö., Munafo, A., Marlow, R., Phillips, A., Ögren, P.: Improving the modularity of AUV control systems using behaviour trees. In: 2018 IEEE/OES Autonomous Underwater Vehicle Workshop (AUV), pp. 1–6 (2018). https://doi.org/10.1109/AUV.2018.8729810
34. Suzuki, S., et al.: Topological structural analysis of digitized binary images by border following. Comput. Vis. Graph. Image Process. **30**(1), 32–46 (1985)
35. Towler, J., Krawiec, B., Kochersberger, K.: Radiation mapping in post-disaster environments using an autonomous helicopter. Remote Sens. **4**(7), 1995–2015 (2012)
36. Trincavelli, M., Reggente, M., Coradeschi, S., Loutfi, A., Ishida, H., Lilienthal, A.J.: Towards environmental monitoring with mobile robots. In: 2008 IEEE/RSJ International Conference on Intelligent Robots and Systems, pp. 2210–2215. IEEE (2008)
37. Valmari, A.: The state explosion problem. In: Advanced Course on Petri Nets, pp. 429–528. Springer, Berlin (1996)

Multi-objective Automatic Clustering with Gene Rearrangement and Cluster Merging

Hongchun Qu, Li Yin, and Xiaoming Tang

Abstract As an unsupervised machine learning method, clustering is an important approach to understanding structural information in data. However, current adaptive clustering approach using multi-objective optimization framework have two apparent limitations. The first is that prior knowledge is needed to identify the correct cluster number. The second is difficulty in evaluating the best clustering solutions from the Pareto Optimal Front (POF) generated by a multi-objective optimization. These problems become severer in non-category datasets. Therefore, the primary goal of this research is to establish a genetic optimization based multi-objective clustering framework, in which multiple clustering validity indexes (CVIs) can be tested simultaneously to automatically obtain the optimal cluster number without knowing any sample label information in advance. In this effort, we will not only be able to consider clustering measurements such as cluster cohesion and separation, but also take other aspects, such as compactness, connectivity, variation among data elements, into consideration as well. Then, we aim to design a procedure to recommend three best solutions from the POF by using appropriate combination of CVIs without increasing computational cost. This procedure is expected to control the cluster number in a reasonable range and consequently decrease the difficulty in best solution recommendation. Finally, since we have the knowledge that using gene rearrangement in the genetic optimization does not affect partition, we take this advantage to merge clusters effectively and significantly speed the convergence of the algorithm. Our approach can outperform the state-of-the-art counterparts across diverse benchmark datasets in terms of partitioning accuracy and performance, as demonstrated in three experiments conducted on both artificial and typical real-world datasets.

H. Qu (✉) · X. Tang
College of Automation, Chongqing University of Posts and Telecommunications, Chongqing 400065, China

L. Yin
College of Computer Science, Chongqing University of Posts and Telecommunications, Chongqing 400065, China

V. Sgurev et al. (eds.), *Advances in Intelligent Systems Research and Innovation*, Studies in Systems, Decision and Control 379,
https://doi.org/10.1007/978-3-030-78124-8_5

1 Introduction

Clustering is currently one of the hottest topics in the field of machine learning because it is an important way to learn and understand structural information from a given dataset. The aim of cluster analysis is to find the potential distribution of data elements in high dimensional space, which is then used to classify the features presented in the data in an unsupervised way. As a popular data mining technique, it has been widely used in image segmentation [1], text analysis [2], gene analysis [3], social network [4] and intrusion detection [5], etc.

Technically, clustering algorithms attempt to divide data elements into several disjoint subsets, each of which is called a cluster. The guiding principle is that the elements classified into the same cluster should be as similar as possible and, the differences between the elements divided into different clusters need to be as great as possible. Classical clustering algorithms include prototype-based clustering algorithm, density-based clustering algorithm and hierarchical clustering algorithm. K-means [6] is a typical instance of prototype-based clustering. In K-means clustering algorithm, an initial center (representative) point of each cluster needs to be specified before the following procedures are applied. Firstly, it uses randomly specified representative points to divide datasets, and then re-determines the representative points according to the current partition results. K-means performs the above two steps until convergence is reached. Density-based algorithms can be interpreted as follows: the data space of clustering is a high-density connected region, which is separated from each other through low-density adjacent regions [7]. A typical density-based clustering algorithm can be referred to DBSCAN [8]. Hierarchical clustering algorithm generates nested cluster tree structure from bottom to top or vice versa, and partitions them at different levels, e.g., CHAMELEON [9] is a typical hierarchical clustering algorithm.

When the dimension of a dataset is high, the feature space is often very sparse, which makes it very difficult to distinguish high-density regions from the low-density ones by a density-based method [10]. Due to its high computational cost, hierarchical clustering algorithm is generally not suitable for processing dataset of huge size. On the contrary, prototype-based clustering algorithms, such as K-means, are simple and efficient, i.e., makes it one of the most commonly used clustering methods in real applications.

Most prototype-based clustering algorithms are based on measurements of distance among data elements. Given the number of clusters k, an initialization partition is first created. Then, they use an iterative repositioning technique to generate new partitions by moving objects from one group to another. For instance, many studies solve these problems by repeatedly running an algorithm with different fixed values k and different initialization [11]. However, in many cases, it is not realistic to have knowledge on the correct number of clusters. In order to achieve global optimum, prototype-based clustering may need to check all possible partitions, which leads to a considerable increasing in computation. In addition, when the number of clusters is larger, even for medium-size problem (i.e., the number of data elements is not too high), the procedure of clustering is not always computationally acceptable due to its NP-hard nature [12].

In recent years, multi-objective metaheuristic methods have gained increasing attention from the scientific community [13] not only because they are capable of automatically searching optimal clustering solutions from high dimensional space in an efficient manner, but also due to their ability to find multiple solutions simultaneously. These solutions usually form the optimal tradeoff between different conflicting objectives that evaluate the goodness of the clustering structure by a specific Cluster Validity Index (CVI) [14]. A CVI is often used as fitness function by metaheuristic algorithms to describe the relationship between cluster cohesion (i.e., the similarity among features of data elements within a cluster) and cluster separation (i.e., the dissimilarity among features of data elements between different clusters), i.e., as an estimation of quality of clustering solutions. Besides cluster cohesion and separation, other metrics such as compactness, connectivity, variation, etc., among data elements are also considered to be objectives optimized by metaheuristic clustering algorithms [15].

In comparison with single-objective clustering methods, multi-objective metaheuristic methods can optimize different CVIs simultaneously, which significantly promotes the clustering accuracy by evaluating solutions from various criteria at the same time. Nevertheless, one limitation is that the output of a multi-objective optimization method is a set of non-dominant solutions, which is usually denoted as Pareto optimal solutions or Pareto Optimal Front (POF). Mathematically, a POF is a scenario in which any given objective cannot be improved at the cost of deteriorating at least another one [16]. The number of non-dominant solutions in a POF is typically scenario dependent, i.e., the number of optimal solutions is usually indeterministic. For instance, the Multi-Objective Modified Differential Evolution based Fuzzy Clustering (MOMoDEFC) [17] takes Xie-Beni index and J_m (i.e., a CVI) as two objective functions and optimizes them simultaneously by using modified differential evolution based fuzzy clustering. However, its final output is a non-dominant front which is a set of pareto optimal solutions with uncertain size.

Another limitation is that the procedure of solution recommendation for a POF is generally computational cost because it is very difficult to tell the weight of each CVI in the holistic objective without necessary information. For example, the Multi-Objective Clustering with automatic K-determination (MOCK) [15] algorithm produces several solutions during the clustering phase but takes much more time to identify the best one. The final output of Fuzzy C-means-Non-dominated Sorting Genetic Algorithm-II (FCM-NSGA-II) algorithm is a single solution, but it needs to know 10% of the true sample labels when selecting the best solution, which makes it a semi-supervised learning method [18]. The Non-dominated Sorting Genetic Algorithm-Fuzzy Membership Chromosome (NSGA-FMC) algorithm can select the best solution without using sample information. It is, however, only effective for categorical datasets [19].

Fortunately, nature-inspired metaheuristics, such as genetic algorithms can be naturally combined with other algorithms to produce new hybrid framework that is useful to tackle these problems and has the ability of adaption to complex and changing environments [20]. For example, the Improved Fuzzy Relational Clustering-NSGA-II (IFRC-NSGA-II) [3] firstly uses NSGA-II to generate good clustering centers and partitions as an initialization. Then, the Fuzzy Relational Clustering (FRC) algorithm is performed on the initial partition obtained previously until convergence has been made. For both gene expression datasets and non-gene expression datasets, it achieves high performance in clustering, however the true sample label information still needs to be known. The Entropy C-Means-NSGA-II (ECM-NSGA-II) [21] algorithm only regards clustering as an optimization problem and does not construct a new hybrid algorithm by combining basic algorithms such as FCM and evolutionary algorithm. However, the algorithm still needs to preset the number of clusters.

To the best of our knowledge, there is currently no adaptive clustering approach based on multi-objective genetic algorithm which does not require any sample information and can also efficiently provide a fixed number of recommended solutions from non-category datasets. In this research, therefore, our primary goal was to establish a genetic optimization based multi-objective clustering framework, in which multiple CVIs can be tested simultaneously to automatically obtain the optimal cluster number without knowing any sample label information in advance. In this effort, we will not only be able to consider clustering measurements such as cluster cohesion and separation, but also take other aspects, such as compactness, connectivity, variation among data elements, into consideration as well. Then, we aimed to design a procedure to recommend three best solutions from the POF by using appropriate combination of CVIs without increasing computational cost. This procedure is expected to control the cluster number in a reasonable range and consequently decrease the difficulty in best solution recommendation. Finally, since we have the knowledge that using gene rearrangement in the genetic optimization does not affect partition, we hoped to take this advantage to merge clusters effectively and significantly speed the convergence of the algorithm, which can let it outperforms the state-of-the-art counterparts across diverse benchmark datasets.

2 Related Work

In this section, we briefly review the literatures that are relevant to our proposed approach in different aspects, such as multi-objective metaheuristics, genetic algorithms for automatic clustering as well as clustering validity indexes that are critical for optimization and evaluation of the clustering solutions.

2.1 Multi-objective Metaheuristics

For a given dataset, it is generally not realistic to define an unambiguous partitioning that can perfectly characterize the inherent distribution across the entire feature space, because the shape, size, connectivity, dimensionality and other characteristics of the dataset are not homogeneous to be identified by a single clustering criterion [22]. In contrast with single goodness measure of a partitioning, multi-objective metaheuristics therefore are more robust by simultaneously optimizing several cluster quality metrics and achieving optimal trade-off among multiple cluster validity indexes (CVI, see Sect. 2.3 for detail) under certain constraints [16].

In 2007, Handl and Knowles proposed a Multi-Objective Clustering to automatically determine the number of clusters (i.e., K), which is therefore also called MOCK [15]. The two-stage procedure employed in MOCK makes it to be able to find optimal clusters with good accuracy. It firstly uses a multi-objective evolutionary algorithm (MOEA) with graph-based representation to optimize two objectives that characterize compactness and connectedness of clusters. From the set of POF it produced, MOCK can then output a small and fixed number of clustering solutions by examining the tradeoffs between different validity indexes. The results showed that MOCK achieved much better performance than classical single-objective clustering techniques such as k-means on various datasets.

However, the study performed by Matake et al. [23] demonstrated that MOCK has apparent limitation to be applied in dataset with large size because it needs to iteratively compare different solutions until a promising solution can be output. They then proposed an updated clustering approach based on MOCK for web mining accordingly. The new algorithm can automatically find the optimal number of K in relatively smaller time scale. This advancement makes it to be suitable for large dataset in web information discovery.

Two years later, Suresh et al. [24] also developed a multi-objective optimization algorithm for clustering, in which two validity indexes, i.e., XB and FCM were employed as objective functions. The interesting feature of their work was that the differential evolution (DE) scheme was applied to simultaneously searching optimal solutions in a fuzzy manner. Their approach was named Multi-Objective Differential Evolution clustering (MODE). The algorithm was evaluated by two validity indexes, i.e., the adjusted Rand index and the SI index. The results showed that their algorithm had better clustering accuracy and performance than MOCK across six artificial and four real world datasets with different dimensions and structures.

After Suresh's work, Saha et al. [17] proposed an updated version based on MODE, but still used the fuzzy clustering method and the same clustering validity indexes, which is called Multi-Objective Modified Differential Evolution based Fuzzy Clustering (MOMoDEFC). Despite the fact that the output of MOMoDEFC is a non-dominant set of solutions with uncertain size, it outperformed a bunch of counterparts multi-objective DE-based fuzzy clustering, multi-objective GA-based fuzzy clustering, single-objective DE-based fuzzy clustering, and the fuzzy c-means algorithm in terms of clustering accuracy.

2.2 Genetic Algorithms for Automatic Clustering

Genetic algorithms (GA) have been widely used in multi-objective optimization framework for automatic clustering because they have demonstrated extraordinary searching ability for optimal solutions in complex feature space as well as excellent adaptiveness to various datasets with different cluster size, shape and dimensionality [25]. The most important GA-based approach was proposed by Deb et al. [26], which was named as the Non-dominated Sorting Genetic Algorithm II (NSGA-II). NSGA-II can overcome limitations of non-dominated sorting based multi-objective evolutionary algorithms (MOEAs) by improving the computational efficiency, elitism approach and avoiding specifying the sharing parameters. The performance of NSGA-II was demonstrated by simulation experiments conducted on a five-objective, seven-constraint nonlinear problem.

Liu et al. [27] studied to integrate multi-objective genetic algorithm and validity analysis for locating and ranking alternative clustering. In their research, they proposed the algorithm MOKGA based on NSGA-II to optimize two fitness functions that were respectively cluster number and partitioning error. They also studied a ranking strategy to shrink the population size during the genetic operation. The algorithm of MOKGA was tested on several open benchmark datasets and the accuracy and performance were validated.

Recent research has demonstrated that in contrast with typical gene crossover, gene rearrangement techniques can achieve better clustering results without increasing time complexity. For example, Liu et al. [28] proposed an automatic clustering approach based on genetic algorithm, which was named as AGCUK. In their approach, a noising selection and division-absorption mutation strategy were innovated to keep the balance between selection pressure and population diversity, and consequently obtained good performance. He and Tan [29] designed an automatic clustering approach which consists of a two-stage genetic algorithm (TGCA). The advantage of their approach is that the probability of genetic operation can be dynamically adapted to the number of clusters currently founded by the algorithm. Because of this technique, TGCA can effectively search for the best cluster number following with the adaption that allows it gradually moves to searching the globally optimal cluster centers. This method had not only higher accuracy, but also better performance, which were demonstrated in their experiments.

2.3 Clustering Validity Indexes

Clustering validity indexes (CVIs) are metrics for evaluating the quality of partitioning that best fits natural structure of a given dataset [16]. In general, CVIs can be classified into two distinct groups, which are external CVIs and internal CVIs respectively. External CVIs can be used to measure the goodness of a clustering procedure on a dataset if a priori information of the dataset is given, however, it is unlikely to obtain the priori information in advance for all situations [30]. This is why the external CVIs have very limited use in real problems, instead, they are often used in research purposes. On the contrary, internal CVIs do not need such priori information, but the intrinsic structure of the dataset to quantify the partitioning quality. Therefore, they are widely used to construct fitness or objective functions of clustering optimization algorithms. Typical internal and external CVIs used in this research are given below, other CVIs can be referred to literatures such as [31]. In this research, these internal CVIs were used to form fitness functions in the genetic algorithm. The external CVIs were then used as measurements for accuracy, performance comparisons among algorithms.

Internal clustering indexes

(1) Sum of square of errors

Suppose $X = \{X_i\}, i = 1 \ldots, n$ is a set of n D-dimensional vector sample points. x is divided into k clusters with $C = \{C_j\}, j = 1 \ldots, k$ and μ_i as the cluster center of cluster C_i. Each cluster has a covariance matrix, then there are K covariance matrices $S = \{S_j\}, j=1,\ldots,k$.

The square of errors of the sample points in C_j and μ_i is:

$$SE(C_j) = \sum_{x_i \in C_j} d(x_i, \mu_j)^2 \tag{1}$$

And sum of square of errors is:

$$SSE(C) = \sum_{j=1}^{k} \sum_{x_i \in C_j} d(x_i, \mu_j)^2 \tag{2}$$

Even if k = 2, minimizing this objective function is still a NP-hard problem [20]. SSE(C) is one of the most commonly used objective functions, but it has one drawback. That is, SSE always decreases with the increasing of cluster number k. This measure will make the selected k larger than the actual value. When SSE(C) is used to determine the adaptive clustering algorithm of k value by the algorithm, the k value will eventually equal to n, which is often inconsistent with the facts [21].

(2) Calinski-Harabasz index

Calinski-Harabasz index is based on intra-cluster compactness and inter-cluster dispersion. If all samples are considered as a cluster, its central point is:

$$\mu_c = \frac{1}{n} \sum_{i=1}^{n} x_i \tag{3}$$

BGSD (between-groups sum of square distances) is used to represent inter-cluster dispersion, which is defined as follows

$$BGSD = \sum_{j=1}^{k} n_j d(\mu_j, \mu_c)^2 \quad (4)$$

where n_j is the total number of samples in the j cluster. WGSD (within-group sum of square distances) is used to denote intra-cluster compactness, which is defined as:

$$WGSD = \sum_{j=1}^{k} \sum_{x_i \in C_j} d(x_i, \mu_j)^2 \quad (5)$$

Calinski-Harabasz index is given by the ratio between BGSD and WGSD:

$$Cal(C) = \frac{BGSD}{WGSD} \times \frac{n-k}{k-1} \quad (6)$$

For the same sample set, a higher Calinski-Harabasz index relates to a model with better defined clusters.

(3) Silhouette Coefficient

If one defines *a*(*i*) as the average distance from the sample point x_i to other sample points in the same cluster; and *b*(*i*) as the average distance from the sample point x_i to all the sample points in the nearest cluster.

Then the Silhouette Coefficient of single sample point x_i is

$$s(i) = \frac{b(i) - a(i)}{\max(a(i), b(i))} \quad (7)$$

The Silhouette Coefficient of cluster is the mean value of the Silhouette Coefficient of all samples:

$$Sil(C) = \sum_{i=1}^{n} s(i) \quad (8)$$

The value range of Silhouette Coefficient is − 1 ~ 1, the closer it is to 1, the better cluster result is.

(4) Davies-Bouldin index

The index can be expressed as the average similarity of each cluster C_j, $j = 1,\ldots,k$ and its most similar cluster $C_j, j \neq j'$. The average distance between all points in a cluster and the center of the cluster:

$$ds(C_j) = \frac{1}{n_j} \sum_{x_i \in C_j} d(x_i, \mu_j) \quad (9)$$

$d_{jj'}$ is used to express the distance between cluster centers μ_j and $\mu_{j'}$, then:

$$R_{jj'} = \frac{ds(C_j) + ds(C_{j'})}{d_{jj'}} \tag{10}$$

(11) is used to measure the similarity between cluster c_j and cluster $c_{jj'}$. Then Davies-Bouldin index is defined as:

$$Dbi(C) = \frac{1}{k}\sum_{j=1}^{k} \max_{j \neq 1} R_{jj'} \tag{11}$$

Its value range is a real number greater than 0. For different clustering results from the same dataset, the partition with the value closest to 0 is the best partition.

(5) Sum of generalized sample variance

Variance is a statistical measure of how much a set of observations differ from each other. In the case of a single variable, sample variance is usually used to describe the degree of variation of the observed value of the variable. When D variables are observed in each group, the variance is described by the sample variance–covariance matrix. Generalized sample variance = $|s|$, where s is the covariance matrix of the sample in the cluster, it is a generalization of the covariance matrix.

The generalized sample variance $GSV(C_j) = |S_j|$, S_j of samples in cluster S_j is the determinant of the sample covariance matrix S_j. When the number of samples in a cluster is less than 2, the covariance matrix does not exist and we set the generalized sample variance to 0. Geometrically, it is the volume of a superbox. For a fixed sample size, there are two factors affecting the variance of the generalized sample size. One is the variance within the sample features, and the other is the correlation between the sample features. sum of generalized sample variance:

$$SGSV(C) = \sum_{j=1}^{k} |S_j| \tag{12}$$

SGSV(C) is a general descriptor. When k clusters are obtained by the algorithm, the higher the density of clusters is, the greater the correlation between different features is, and the smaller the value of SGSV(C) is. It also has the defect that it always decreases with the increase of cluster number k. But compared with SSE(C), SGSV(C) takes into account the correlation between data.

External comparison indexes

(1) Adjusted rand index

Suppose we get the real label list T of a data set, and T is the label list generated by clustering results through a clustering algorithm. One of the following four relationships will occur between two samples x_1 and x_2:

R1: x_1, x_2 belong to the same cluster in T, and belong to the same cluster in P.
R2: x_1, x_2 belong to the same cluster in T, but not belong to the same cluster in P.
R3: x_1, x_2 do not belong to the same cluster in T, but belong to the same cluster in P.
R4: x_1, x_2 do not belong to the same cluster in T, and do not belong to the same cluster in P.

Let *a*, *b*, *c* and *d* represent the total number of R1, R2, R3 and R4 relationship pairs respectively. Where $a + b + c + d = c\binom{2}{n}$, then:

$$ARI(T, P) = \frac{2(ad - bc)}{(a + b)(b + d) + (a + c)(c + d)} \tag{13}$$

The value range of ARI(T,P) is 0 ~ 1, the closer the value is to 1, the better the cluster result is. If the label list P is randomly assigned, the value of ARI(T,P) is close to 0. If the predicted tag list P is equal to the real tag list T, the value of ARI(T,P) is 1.

(2) Mutual information index

Suppose *n* samples are divided into two distribution sets U and V. Their entropy is the uncertainty of a partition set, as follows:

$$H(U) = -\sum_{i=1}^{|U|} P(i)\log(P(i)) \tag{14}$$

$$H(V) = -\sum_{j=1}^{|V|} P'(i)\log\left(P'(i)\right) \tag{15}$$

where $P(i) = |U_i|/n$ is the probability of randomly selecting a sample from the sample set to be divided into the cluster U_i. Similarly, $P'(j) = \left|V_j\right|/n$ is the probability of randomly selecting a sample from the sample set to be divided into the cluster v_j. And

$$MI(U, V) = \sum_{i=1}^{|U|}\sum_{j=1}^{|V|} P(i, j)\log\left(\frac{P(i, j)}{P(i)P'(j)}\right) \tag{16}$$

where $P(i, j) = \left|U_i \cap V_j\right|/n$ is the probability that a sample is randomly selected from the sample set and divided into U_i and V_j at the same time. The normalized mutual information is:

$$NMI(U, V) = \frac{MI(U, V)}{mean(H(U), H(V))} \tag{17}$$

NMI(U,V) is usually used to detect the difference between the result of clustering and the real partition, and get the correct partition rate. NMI(U,V) is the similarity between the clustering results and the ground truth of the dataset. The value range is 0–1, the larger the value is, the greater the similarity between the clustering result and the ground truth of the data set is, so then the better the clustering result is.

3 Proposed Approach

In this section we propose a novel evolutionary clustering algorithm based on multi-objective genetic algorithm NSGA-II (Non-dominated Sorting Genetic Algorithm-II) [26]. The algorithm can obtain the pareto optimal solutions of partitions by iteratively evolve the population and finally three best solutions were selected as the recommended solutions. In this multi-objective genetic algorithm, two objective functions constructed by internal CVIs are optimized simultaneously (see Sect. 3.1: *Objective function*) as the fitness of each individual. For different purposes, the two objective functions can be replaced by different internal CVIs; An individual of the genetic population represents a partition, which is encoded by a chromosome (see Sect. 3.2: *Chromosome representation*); By calculating the Euclidean distance, all data points (samples, hereafter we use the two terms interchangeably) can be classified into different clusters that are recorded by chromosomes; When fitness of each individual in the population is calculated by the objective functions (see Sect. 3.3: *Individual fitness calculation*), they are sorted by NSGA-II, and then the next generation population is generated accordingly (see Sect. 3.4: *Non-dominated sorting and population selection*). In order to maintain the population size in a certain range and keep it diversified as well, new individuals are generated by the gene rearrangement technique incorporated into typical genetic operations such as crossover and mutation (See Sects. 3.5: *Gene crossover and gene rearrangement* and Sect. 3.6: *Gene mutation and chromosome variation*). Due to random effects, a few chromosomes generated by the above methods might be invalid because no data points can be classified into that cluster, the invalid chromosomes need to be removed before these clusters are merged, which is a new technique proposed by us to speed convergence of the algorithm (see Sect. 3.7: *Invalid chromosome check and cluster merge*). The genetic algorithm is terminated by outputting a pareto optimal solution set, then different internal CVIs are employed to select the best three solutions (see Sect. 3.8: *Best solutions recommendation*). Because the technique of Gene Rearrangement (GR) is used, our proposed algorithm is therefore name as: NSGAII-GR. Figure 1 gives flow chart of NSGAII-GR and the overall steps of NSGAII-GR are given in Algorithm 1.

Algorithm 1: NSGAII-GR

Input: dataset, population size, cycles, crossover probability, mutation probability

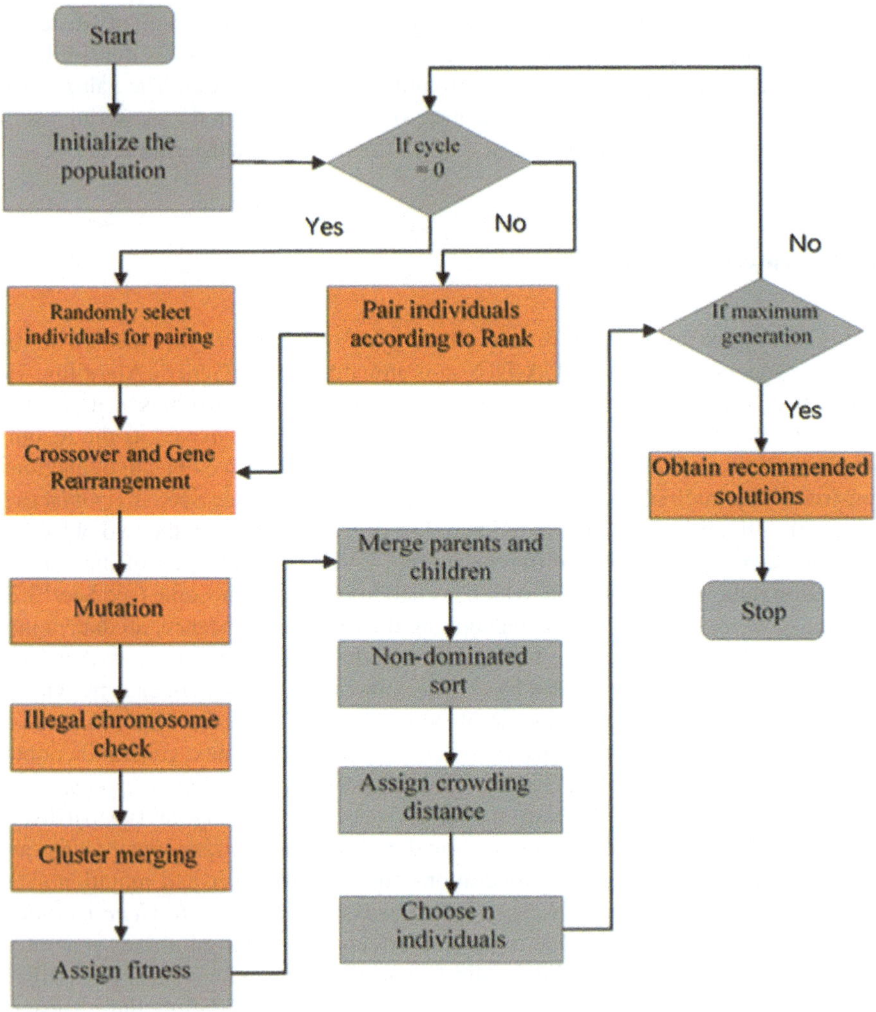

Fig. 1 Flow chart of NSGAII-GR. The novel operations introduced into this algorithm are colored in orange to differentiate ours from the traditional versions

Output: three recommended solutions

1. Initialize the population randomly.
2. Pare individuals in the population before chromosome crossover.
3. Generate new individuals by using gene rearrangement between matched individuals.
4. Mutate individuals.
5. Remove invalid individuals.

6. Merger clusters.
7. Calculate fitness for each individual.
8. Produce the new generation of population by screening the old individuals.
9. Repeat step 2–8 until the number of iterations is reached, otherwise, go to step 10.
10. Select and output the three best clustering solutions.

3.1 Objective Function

In NSGAII-GR, we use multi-objective function to optimize the accuracy of partitioning. In general, a multi-objective optimization problem can be desc-ribed as: a D-dimensional decision variable x and m objective functions $Obj_1(x)$, ..., $Obj_m(x)$. The objective functions are constructed by internal C-VIs, we will give details in the following sections. The goal of the multi-o-bjective problem is to maximize or minimize these objective functions sim-ultaneously. For the simplicity of description, we assume that the objective functions are minimized by searching the decision variable x in the solution space. If a decision variable cannot enhance any objective function without deteriorating at least another objective function, then this decision variable is regarded as a pareto optimal solution. This means that the results of part-itioning might not be non-exclusive in the context of multi-objective optimi-zation.

For example, in this research, we choose two objective functions min($Obj_1(x_1)$, $Obj_1(x_2)$) for the two decision variables x_1 and x_2 within the value range. If $Obj_1(x_1) < Obj_1(x_2)$ and $Obj_1(x_1) < Obj_1(x_2)$, we say that the decision variable x_1 dominates x_2, or the decision variable x_2 is dominated by x_1. When a variable x is not dominated by any other decision variables, then the variable x is a pareto optimal solution. In multi-objective problems, pareto optimal solution is often not a single value or point, but a set, which is called pareto optimal solution set or pareto optimal front, as shown in Fig. 2.

Specifically, the first objective function $Obj_1(C)$ used in NSGAII-GR is the sum of generalized sample variance (SGSV), which is able reflect the compactness within a cluster. However, considering its weakness that it tends to increase the chromosome length of individuals in the population during the evolution process. Therefore, we use the reciprocal of Calinski-Harabasz index as the second objective function $Obj_2(C)$ to control the length of the chromosome. Then the evolutionary objective of NSGAII-GR is to minimize the two functions.

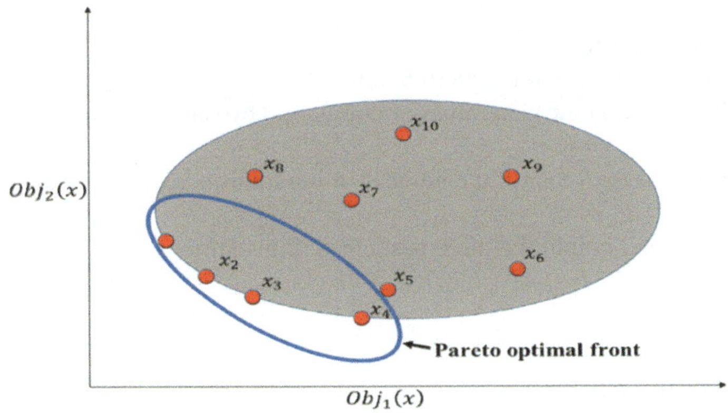

Fig. 2 Pareto optimal front. One should note that: x_1, x_2, x_2 and x_4 form a pareto optimal front because they are not dominated by any other solutions

3.2 Chromosome Representation

In a genetic algorithm, the coding of chromosome represents the way how problem is formulated and determines how the genetic operations are performed. In NSGAII-GR, alleles on the chromosomes are selected to represent clustering centers, and each gene is represented by a vector with dimension *D*. This coding method has been proofed by previous works that it can efficiently improve the computational performance of applied evolutionary clustering operators on the individuals [32]. The data points closest to a cluster center are classified into the cluster represented by the cluster center. Therefore, if the cluster center is determined, the cluster partition is obtained.

If a given dataset have *k* clusters, then a chromosome can have *k* alleles C_1, C_2, ..., C_k. The structure of its chromosome is shown in Fig. 3. An allele is expressed as a D-dimensional vector instead of a numerical value because an allele should be a meaningful expression (e.g., a gene controls a character). Because the number of clusters, i.e., *k*, is unknown in advance, the coding scheme are designed to be changeable, which means that the number of clusters is randomly generated within a predetermined length range $[K_{min}, K_{max}]$, where $K_{min} = 2$ and $K_{max} = round\ (\sqrt{n}\ )$ [33]. Another vector O is appended at the end of chromosome, but it is not involved in the genetic operations. O_m represents the fitness value of chromosome obtained

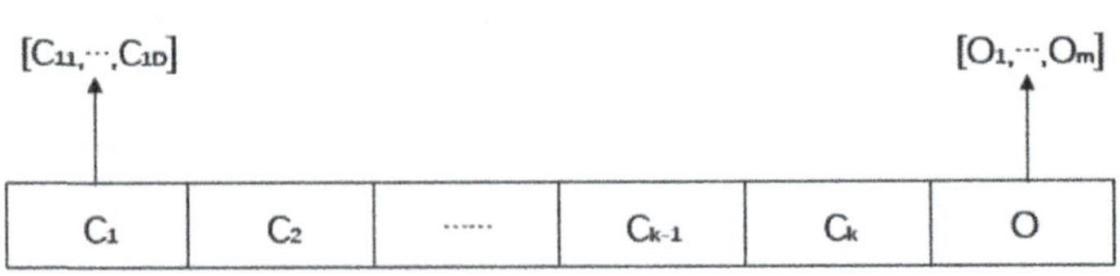

Fig. 3 Chromosome structure of individuals

in the m-th objective function as reference for genetic operations. Therefore, if each sample contains D features, and each feature is represented by a real number, the real coding length of the chromosome is $L = D * k + m$.

In the population, the chromosome length of the first generation of individuals is randomly selected from the length range. After the chromosome length is determined, a sample point is randomly selected from the dataset as the initial value of the gene.

3.3 Individual Fitness Calculation

Before calculating fitness value for individuals in the population, clusters are generated by allocating samples to their cluster centers represented by genes on the chromosomes. In this procedure, Euclidean distance is used to partition the samples into the nearest cluster.

The fitness value of an individual is calculated respectively by the two objective functions, i.e., $Obj_1(C)$ and $Obj_2(C)$, which are the sum of generalized sample variance and the reciprocal of Calinski-Harabasz index.

3.4 Non-dominated Sorting and Population Selection

After getting the fitness value of individuals in the population, the superiority among individuals can be distinguished. The original genetic algorithm, i.e., NSGA-II (Non-dominated Sorting Genetic Algorithm-II) divides individuals into different hierarchical subgroups through dominance relationship. The individuals in the higher-level subgroups therefore dominate the individuals in the lower-level subgroups.

Nevertheless, NSGA-II leaves a gap that the individuals in the same level of subgroups do not dominate each other. In our NSGAII-GR, we incorporate the method of Crowded-Comparison to sort individuals within the same subgroups. Specifically, if the fitness values calculated by $Obj_1(C)$ and $Obj_2(C)$ are used as horizontal and vertical coordinates, the corresponding individuals can be depicted in the two-dimensional coordinate system. The regions with higher point density are regarded as crowded area. Crowded-Comparison tends to ensure the diversity of individuals distribution in the population, which means that individuals in the more crowded areas are less likely to be selected. The application of non-dominated sorting (Fig. 4.) to chromosomes can generate ranks among them.

Finally, the first half (i.e., the 50% of the individuals with higher ranking) of the ranked chromosomes are reserved to be part of the next generation population, while the second half are discarded.

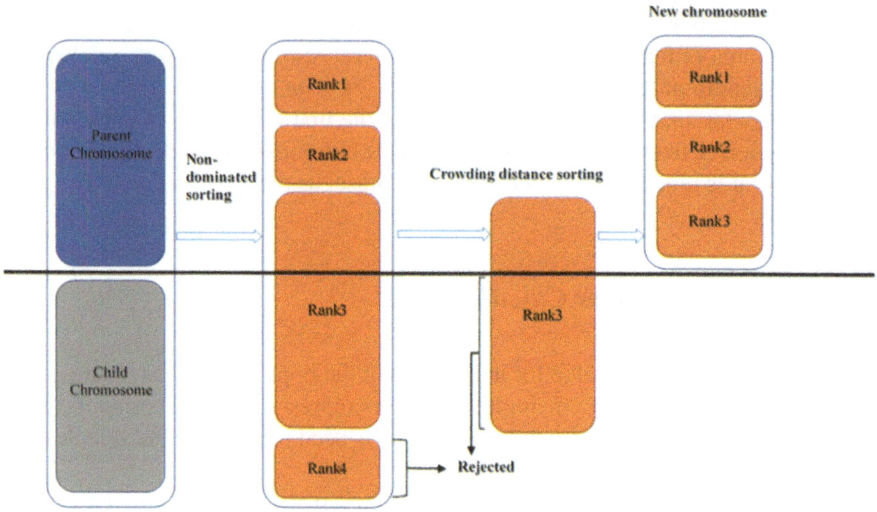

Fig. 4 NSGAII-GR working procedure for Non-dominated sorting

3.5 Gene Crossover and Gene Rearrangement

Crossover is a stochastic process (the probability of *probCrossover*) in which information is exchanged between a pair of parents to produce two offspring. In the original NSGA-II, individuals of the initial generation population are randomly matched. The individuals in subsequent generations are ordered by NSGA-II. However, in NSGAII-GR, we match all the neighboring individuals in order from the beginning. In this way, the individuals in the same level of subgroups are paired with each other, which makes the offspring of the higher-level individuals more likely to be higher-level ones. One exception of paring is that the two individuals located in tail and the real of a subgroup have the probability to be paired with other level of subgroups.

One problem is that the chromosome length of two matched individuals might be different. We use the following methods to direct exchange between two individuals A and B with different chromosome length.

Algorithm 2: Gene Crossover

Input: chromosome A, chromosome B

Output: chromosome A, chromosome B

1. Calculate the chromosome length(A) and length(B) of two individuals.
2. If length(A) > length(B):
3. Set indexArray = a list with size equals length(B).
4. Set choosedList = an empty list.
5. for $j = 1$ to length (A):

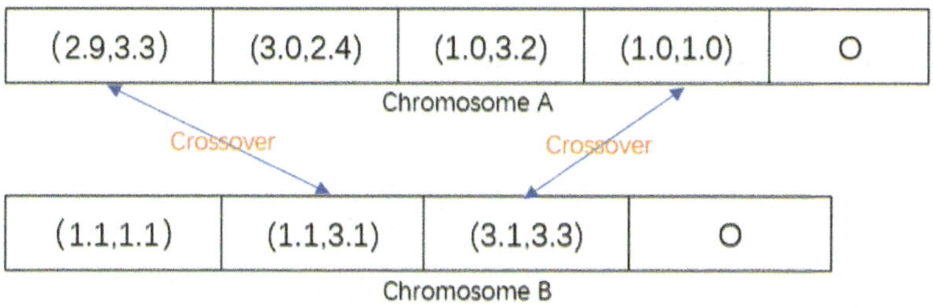

Fig. 5 Gene crossover operation in NSGAII-GR

6. Append element$j$ to the choosedList.
7. for$i$ = 1 to length(B):
8. indexArray[i] = a random element j in the choosedList.
9. remove element j from choosedList.
10. for i = 1 to length(B):
11. if a randomly generated number between 0 and 1 < the probability of Crossover:
12. exchange the i-th gene of chromosome B with the indexArray[i]-th gene of chromosome A.
13. 13.Do similar steps as 3 ~ 12 if length(A) < = length(B).

For example, if chromosome A = [(2.9,3.3), (3.0,2.4), (1.0,3.2), (1.0,1.0)] and chromosome B = [(1.1,1.1), (1.1,3.1), (3.1,3.3)]. If the *Indexarray* generated is [3, 1, 4], the NSGAII-GR gene crossover procedure between the two individuals is illustrated in Fig. 5. After gene exchanges, two offspring chromosomes, e.g., chromosome C = [(1.1,3.1), (3.0,2.4), (1.0,3.2), (3.1,3.3)] and chromosome D = [(1.1,1.1), (2.9,3.3), (1.0,1.0)] are generated.

Then, the two offspring chromosome C and chromosome D undergo gene rearrangement procedure, which is detailed in Algorithm 3. After gene rearrangement, chromosome C = [(1.1,3.1), (1.0,3.2), (3.1,3.3), (3.0,2.4)] and chromosome D = [(1.1,1.1), (1.0,1.0), (2.9,3.3)] are obtained. One should note that gene rearrangement proposed in NSGAII-GR does not affect cluster partitioning represented by individuals. Because the cluster centers obtained from chromosomes are independent to the order of genes, instead, the cluster partitioning is determined by distribution of samples to these centers. In addition, we use gene (i + 1) to present the most similar gene to gene(i) among all genes after gene rearrangement. The reason of doing so is justified in Sect. 3.6.

Algorithm 3: Gene Rearrangement.
Input: chromosome Ch

Output: chromosome Ch
1. Calculate length of the chromosome as length(Ch).
2. for $i = 1$ to length(Ch)-1:
3. In the remaining length(Ch)$-i$ genes, find the gene(j) closest to the gene(i).
4. change gene(j) with gene($i + 1$).

3.6 Gene Mutation and Chromosome Variation

The main purpose of mutation operation is to guarantee the global search ability of a genetic algorithm by introducing some additional variabilities to individuals. Gene mutation and chromosome variation are typical operations of genetic mutation. In NSGAII-GR, gene mutation is used to search for the optimal partition when the cluster number is determined, while chromosome variation is used to change the length of chromosome to determine the appropriate cluster number. For a chromosome containing k genes, each gene has the probabilities of *probMutation* to be mutated. The mutation procedure is detailed in Algorithm 4.

Algorithm 4: Gene Mutation
Input: chromosome Ch

Output: chromosome Ch
1. for $i = 1$ to k:
2. if gene mutations occur:
3. for $j = 1$ to D:
4. Set max(x_j) = the maximum value of the j-th feature in the sample
5. Set min(x_j) = the minimum value of the j-th feature in the sample
6. Set m_j = random number generated by uniform distribution on interval (-1,1)
7. if $m_j > 0$ then:
8. $x_{ij} = x_{ij} + m_j(\max(x_j) - x_{ij})$
9. else:
10. $x_{ij} = x_{ij} + m_j(x_{ij} - \min(x_j))$

In addition to gene mutation, chromosome variation might occur. The probability of changing the length of a chromosome is the same as *probMutation.* Chromosomes change only one gene at a time, that is, adding or removing one gene. When the chromosome length is equal to 2, the number of gene no longer decreases. When the chromosome length is equal to K_{max}, the number of gene no longer increases. The chromosome variation steps are listed in Algorithm 5:

Algorithm 5: Chromosome Variation

Input: chromosome Ch

Output: chromosome Ch

1. If chromosome variation occurs:
2. If length(Ch) < K_{max} and has 50% of chance do:
3. Randomly select a sampl x_i.
4. for $j = 1$ to D:
5. Set $\max(x_j) =$ the maximum value of the j-th feature in the sample.
6. Set $\min(x_j) =$ the minimum value of the j-th feature in the sample.
7. Set $m_j =$ random number generated by following a uniform distribution on interval (-1, 1).
8. if $m_j > 0$:
9. $x_{ij} = x_{ij} + m_j(\max(x_j) - x_{ij})$.
10. else:
11. $x_{ij} = x_{ij} + m_j(x_{ij} - \min(x_j))$.
12. Insert x_i into chromosome Ch.
13. else if length(Ch) > 2:
14. Randomly remove a gene from the chromosome Ch.

3.7 *Invalid Chromosome Check and Cluster Merge*

By using the Euclidean distance measurement, samples can be classified into the nearest centers that are recorded in one chromosome. Due to random effects, however, some clustering centers may not recruit any sample point, so such clusters are empty clusters which is regarded as invalid clusters. The chromosomes that contains more than one invalid cluster centers are denoted as invalid chromosomes. The invalid clusters should be removed from chromosomes to maintain their validity. In each iteration, NSGAII-GR needs to find out the invalid chromosomes, from which the genes representing invalid cluster centers are taken out to make the chromosome valid. The steps are defined in Algorithm 6.

Algorithm 6: Invalid Chromosome Check

Input: chromosome Ch

Output: chromosome Ch

1. Generate clusters $C = \{C_j\}, j = 1 \ldots, k$ based on center represented as gene(j).
2. Set $j = 1$.
3. While $j < = $ length(Ch) and length(Ch) > 2 do:
4. If C_j is an empty cluster:
5. Remove C_j from C
6. Remove gene((j) from the chromosome Ch.
7. $j = j - 1$.
8. length (Ch) = length (Ch) – 1.
9. $j = j + 1$

After invalid chromosomes are corrected, the clusters presented in these chromosomes are merged to speed the convergence of the algorithm. The steps of cluster merging are described in Algorithm 7. $GSV(C_x)$ shown in Algorithm 7 means the Generalized Sample Variance for cluster C_x.

Algorithm 7: Cluster Merge

Input: chromosome Ch

Output: chromosome Ch

1. if length(Ch) > 2 then:
2. Set j = 1.
3. While $j <$ length (Ch) and length(Ch) > 2 do:
4. $C_{merge} = C_j + C_{j+1}$
5. If GSV(C_j) + GSV(C_{j+1}) > GSV(C_{merge}) then:
6. Remove C_{j+1} from C.
7. C_j is replaced by C_{merge}.
8. gene(j) in chromosome is recalculated by C_j
9. length(Ch) = length(Ch) – 1.
10. Continue.
11. $j = j + 1$.

In this way, it is more likely to have individuals satisfied with the optimal objective function SGSV©, while Cal© (i.e., the Calinski-Harabasz index) is more likely to be a suboptimal objective. The principle is that as soon as the condition of SGSV© optimization is satisfied, the clusters will be merged without considering.

Cal(C). Therefore, one could observe that the value of Cal(C) might increase or decrease after cluster merging.

3.8 Best Solutions Recommendation

The NSGAII-GR finally outputs a set of solutions which are denoted as Pareto Optimal Front (POF). It is difficult to directly determinate the superiority of these solutions because the weights of Obj_1(C) and Obj_2(C) in solution evaluation are unknown. Our approach to three best solution recommendation is as follows. Firstly, the Davies-bouldin index (i.e., Dbi) is used to select the first recommended solution from the population. The length of the selected individual is regarded as the optimal cluster number K; Secondly, a subset of individuals that have the same chromosome length as the individual selected at the first step is selected; Thirdly, the individuals in the subset that have the highest fitness on Obj_1(C) and Obj_2(C) are respectively recommended as the second and third best solutions. In this way, we transform the set of non-dominating solutions with an uncertain number of individuals into three recommended solutions as output. The specific operations are defined in Algorithm 8.

Algorithm 8: Best solutions recommendation

Input: Pareto optimal front

Output: three recommended solutions

1. Set dbiList = an empty list.
2. Set length(front) = the size of Pareto Optimal Front (POF).
3. For i = 1 to length(front):
4. dbiScore = i-th individual's Dbi(C) score in the POF.
5. Append dbiScore to the end of dbiList.
6. Set dbiBestIndex = the index of min(dbiList) in the dbiList.
7. Set individualbestDbi as the dbiBestIndex-th individual in the pareto optimal front.
8. The best cluster number k_{best} is equal to the chromosome length (individualbestDbi).
9. Find out all individuals with chromosome length = k_{best} in the POF as a sub solution set.
10. Set length(sub) = the size of sub solution set.
11. Set calList = an empty list, sgsvList = an empty list.
12. For i = 1 to length(sub):
13. calScore = i-th individual's Cal(C)* score.
14. sgsvScore = i-th individual's SGSV(C)** score.
15. Append calScore to the end of calList.
16. Append sgsvScore to the end of sgsvList.
17. Set calBestIndex = the index of max(calList).
18. Set sgsvBestIndex = the index of min(sgsvList).
19. individualbestSGSV is selected as the sgsvBestIndex-th individual in the sub solution set.

20. individualbestCal is selected as the calBestIndex-th individual in the sub solution set.
17. Output the three recommended solutions (individualbestDbi, individualbestSGSV, individualbestCal).

*: Cal(C) denotes the Calinski-Harabasz index of cluster C.
**: SGSV(C) denotes the Sum of generalized sample variance index of cluster C.

4 Evaluation

4.1 Datasets

In order to verify and validate the proposed algorithm NSGAII-GR, we selected five two-dimensional artificial datasets, which are T9_1, Tb_1 from [34] and Square1, Square4, Long1 from MOCK; as well as five real-world datasets which are commonly used in clustering algorithm: Iris, Wine, NewThyroid, WBC and WDBC from UCI machine learning datasets. In addition, we selected ten datasets (10d-4c-No. (0–9) with features of 10 dimensions and 4 clusters, and ten datasets (10d-10c-No. (0–9) with features of 10 dimensions and 10 clusters from MOCK to feed the algorithm. Information of the selected datasets is described in Table 1

Table 1 Summary of datasets used for algorithm verification and validation, in which *D* represents dimension, *K* represents the number of clusters, *N* represents the sample size of the dataset, and C_i represents the sample size of the *i-th* cluster

Dataset	D	K	N	C_i
T9-1	2	9	450	9 × 50
Tb-1	2	11	550	11 × 50
Square1	2	4	1000	4 × 250
Square4	2	4	1000	4 × 250
Long1	2	2	1000	2 × 500
Iris	4	3	150	50, 50, 50
Wine	13	3	178	59, 71, 48
NewThyroid	5	3	215	150, 35, 30
WBC	9	2	683	444, 239
WDBC	30	2	569	357, 212
10d-4c-No. (0–9)	10	4	838–1482	–
10d-10c-No. (0–9)	10	10	2161–3789	–

4.2 Experiments

We conducted three experiments to verify and validate the proposed algorithm NSGAII-GR in terms of accuracy and performance with respect to the state-of-the-art counterparts as well as its original version. In the first experiment, we tested the ability of NSGAII-GR for automatically cluster number (i.e., *K*) finding without giving any prior information. The number of *K* was determined as the length of the chromosome that had achieved the highest fitness value given by the objective function on an internal CVI. To test the stability of the algorithm, we used two different internal CVIs to examine whether NSGAII-GR can obtain similar results. The two internal CVIs were the Davies-Bouldin index (Dbi) and the Silhouette Coefficient index (Sil), respectively. In order to minimize the random effect, furthermore, 20 replications were made for the experiment and the averaged *K* was used for comparison. The *K* number found by NSGAII-GR was compared with the actual *K* in the dataset. In addition, the range of *K* represented by the genetic population was also given as a reference.

In addition to *K* comparison, in the second experiment, we compared the partitioning accuracy of NSGAII-GR with two well-known multi-objective clustering algorithms, i.e., MOCK and FCM-NSGAII. Because the optimization objective functions of the candidate multi-objective evolutionary clustering algorithms are different, in order to be fair, we used two external clustering indexes, i.e., the adjusted range index (ARI(T,P)) and the mutual information index (NMI(U,V)), to measure the clustering results. In order to avoid the random effect, we averaged 20 runs of the three algorithms.

In the third experiment, the performance, i.e., partitioning accuracy and convergence speed of NSGAII-GR was tested with respect to its original version in which no gene arrangement and cluster merge were incorporated. In addition, the advantages of multi-objective optimization employed in NSGAII-GR were also demonstrated in contrast with a single-objective version that optimized one objective function at a time, i.e., the multi-objective version of NSGAII-GR, in which the Calinski-Harabasz index (Cal) and the sum of generalized sample variance index (SGSV) were simultaneously optimized, was compared with two single-objective version of NSGAII-GR, in which the Calinski-Harabasz index (Cal) or the sum of generalized sample variance index (SGSV) was separately optimized. Except the objective functions, other aspects and parameter settings of the two versions remained unchanged. The parameters of NSGAII-GR for all the three experiments were listed in Table 2.

Table 2 Parameter settings of NSGAII-GR for the three experiments. Parameters referred to previous work are annotated by references

Parameter	Setting
Population size	200
Cycles	100
Crossover probability (probCrossover)	0.8 [18]
Mutation probability (probMutation)	0.2 [18]
K_{min}	2 [33]
K_{max}	$\sqrt{n}$, n is the sample size [33]

5 Results

5.1 Automatic Clusters Number Finding

For all cases conducted on synthetic and real datasets, except for the one conducted on the NewThyroid dataset, e.g., 3.2 versus 3, the actual cluster number can all be correctly found by using the Davies-Bouldin index. Except for biases occurred in 10d-4c-no3 and 10d-4c-no5, the cluster number can be correctly selected in dataset 10d-4c and 10d-10c. Tables 3, 4 and 5 gave details of the comparison. Though NSGAII-GR cannot find the correct cluster number all the time in 10d-10c, its K value was very close to the actual K number. In this case, the accuracy of its partitioning was still acceptable (see Sect. 5.2).

It was shown in K-range in pop that the obtained range of cluster number k in the algorithm is less than $[K_{min}, K_{max}]$, but the correct cluster number we found was within in the range. In contrast with the method in which the number of clusters

Table 3 The value of K calculated by NSGAII-GR on synthetic and real datasets using Davies-Bouldin index (Dbi) and Silhouette Coefficient index (Sil) respectively. The range of K in the population is denoted by K-range in pop

Dataset	actual K	K(Dbi)	K(Sil)	K-range in pop
T9-1	9	9	9	9–20
Tb-1	11	11	11	11–21
Long1	2	2	2	2–29
Square1	4	4	4	4–31
Square4	4	4	4	4–31
Iris	3	3	3	3–12
Wine	3	3	3	2–12
NewThyroid	3	3.2	3	3–12
WBC	2	2	2	2–21
WDBC	2	2	2	2–20

Table 4 The value of *K* calculated by NSGAII-GR on 10d-4c datasets using Davies-Bouldin index (Dbi) and Silhouette Coefficient index (Sil) respectively. The range of *K* in the population is denoted by K-range in pop

Dataset	actual k	K(Dbi)	K(Sil)	K-range in pop
10d-4c-No. 0	4	4	4	3–34
10d-4c-No. 1	4	4	4	4–21
10d-4c-No. 2	4	4	3	3–26
10d-4c-No. 3	4	2	2	2–31
10d-4c-No. 4	4	4	4	2–29
10d-4c-No. 5	4	2	2	2–31
10d-4c-No. 6	4	4	4	2–30
10d-4c-No. 7	4	4	4	4–35
10d-4c-No. 8	4	4	4	4–26
10d-4c-No. 9	4	4	3	3–27

Table 5 The value of *K* calculated by NSGAII-GR on 10d-10c datasets using Davies-Bouldin index (Dbi) and Silhouette Coefficient index (Sil) respectively. The range of *K* in the population is denoted by K-range in pop

Dataset	actual K	K(Dbi)	K(Sil)	K-range in pop
10d-10c-No. 0	10	9.3	8.4	4–34
10d-10c-No. 1	10	9.1	9.2	2–37
10d-10c-No. 2	10	7	7	2–35
10d-10c-No. 3	10	7.7	7.4	4–25
10d-10c-No. 4	10	9.2	8.2	3–43
10d-10c-No. 5	10	9.5	9.5	2–60
10d-10c-No. 6	10	8.1	8	3–35
10d-10c-No. 7	10	7.3	7.9	2–46
10d-10c-No. 8	10	9.7	9.8	2–46
10d-10c-No. 9	10	9.5	9.5	2–44

is controlled by the objective function representing the separation between clusters, our proposed NSGAII-GR algorithm using Cal(C) as the objective function can significantly shrink the search space without affect the optimal solution. Cutting off the empty search space has been demonstrated to be very helpful for reducing the unnecessary computation and improving algorithm efficiency.

5.2 Partition Accuracy

For the automatic clustering algorithm, there might be such a scenario in which even the *K* value found by the algorithm equals to the real cluster number, but the partitioning accuracy is still very low. In such case, the right choice of *K* value might be just a coincidence. Because previous works found that in very few cases, the number of clusters do not equal to the correct number of clusters, but it is close to the correct value and the accuracy is still acceptable. This suggests that finding the number of clusters, i.e., *K*, is not enough to represent the goodness of partitioning alone. We still need incorporate accuracy to evaluate algorithms more robustly. In this section, two external clustering indexes, ARI(T,P) and NMI(U,V), were used to measure the accuracy of partitioning for the three algorithms.

From Table 6 one can observe that, for both T9-1 and Tb-1 dataset, NSGAII-GR and MOCK can adaptively get perfect correct partition. In Long1 dataset, NSGAII-GR performed better than the other two algorithms. In this case, it not only successfully determined the correct number of clusters, but also partitioned the samples correctly. All the three algorithms had relatively good performance in Square1, but the external clustering indexes in MOCK was slightly higher than others. In Square4 dataset, the overlapping degree between clusters was higher than the case in Square1, but NSGAII-GR had the best performance. Figure 6, 7, 8, 9 and 10 showed partitioning results generated by NSGAII-GR on the five 2D artificial datasets with normalized data distribution.

Except for the case in which FCM-NSGAII had a little higher score (less than 3%) on external clustering indexes than NSGAII-GR on WBC dataset, NSGAII-GR algorithm achieved significantly higher performance than the other two algorithms on the indexes of ARI(T,P) and NMI(U,V) in the remaining datasets. At the same time, it can be observed from Tables 6, 7 and 8 that NSGAII-GR can select good individuals no matter Dbi(C) or Sil(C) was used or not as the individual evaluation function. This indicated that the algorithm is able to find good partition structure in almost all datasets and is capable of distinguishing the good partitioning results from the solution set simply using the common internal clustering indexes.

The performance of all the three algorithms on the 10d-4c datasets were shown in Table 7. Except for the cases conducted on dataset 10d-4c-No. 3 and 10d-4c-No. 5 in which the number of clusters was not correctly found, NSGAII-GR algorithm can find adaptively better clustering partition in the remaining eight datasets, and their ARI and NMI were higher than that in FCM-NSGAII algorithm. On the 10d-4c-No. 0 and 10d-4c-No. 9 datasets, the values on the external indexes obtained by NSGAII-GR were slightly higher than that in MOCK algorithm, but slightly lower than MOCK algorithm on the other six datasets. Compared with FCM-NSGAII used in this experiment, individuals with the maximum ARI (T, P) value were selected from the whole population and the real label of the sample was needed to be known, however, NSGAII-GR did not need to know the sample information in advance. In the comparison with MOCK, we found that MOCK outputted more than three recommended solutions and its individual selection was more complicated than the

Table 6 Comparison of Adjusted rand index and Mutual information index over 2D-Synthetic and real datasets. Numbers in bold denoted the highest obtained on each dataset

Dataset	NSGAII-GR(Dbi)		NSGAII-GR(Sil)		FCM-NSGAII		MOCK	
	ARI	NMI	ARI	NMI	ARI	NMI	ARI	NMI
T9-1	1	1	1	1	0.9980	0.9982	1	1
Tb-1	1	1	1	1	0.9962	0.9971	1	1
Long1	1	1	1	1	0.3318	0.5705	0.8504	0.9098
Square1	0.9752	0.9589	0.9758	0.9597	0.9639	0.9447	0.9763	0.9610
Square4	0.8327	0.7953	0.8316	0.7935	0.8173	0.7824	0.7687	0.7448
Iris	0.8911	0.8749	0.8930	0.8778	0.7898	0.7861	0.5681	0.7611
Wine	0.8640	0.8495	0.8640	0.8495	0.4034	0.4381	0.3525	0.3921
Thyroid	0.6145	0.6017	0.6282	0.6026	0.5715	0.4544	0.2553	0.3305
WBC	0.8464	0.7478	0.8464	0.7478	0.8680	0.7758	0.7878	0.6774
WDBC	0.7325	0.6346	0.7434	0.6616	0.5838	0.4901	0.3815	0.4192

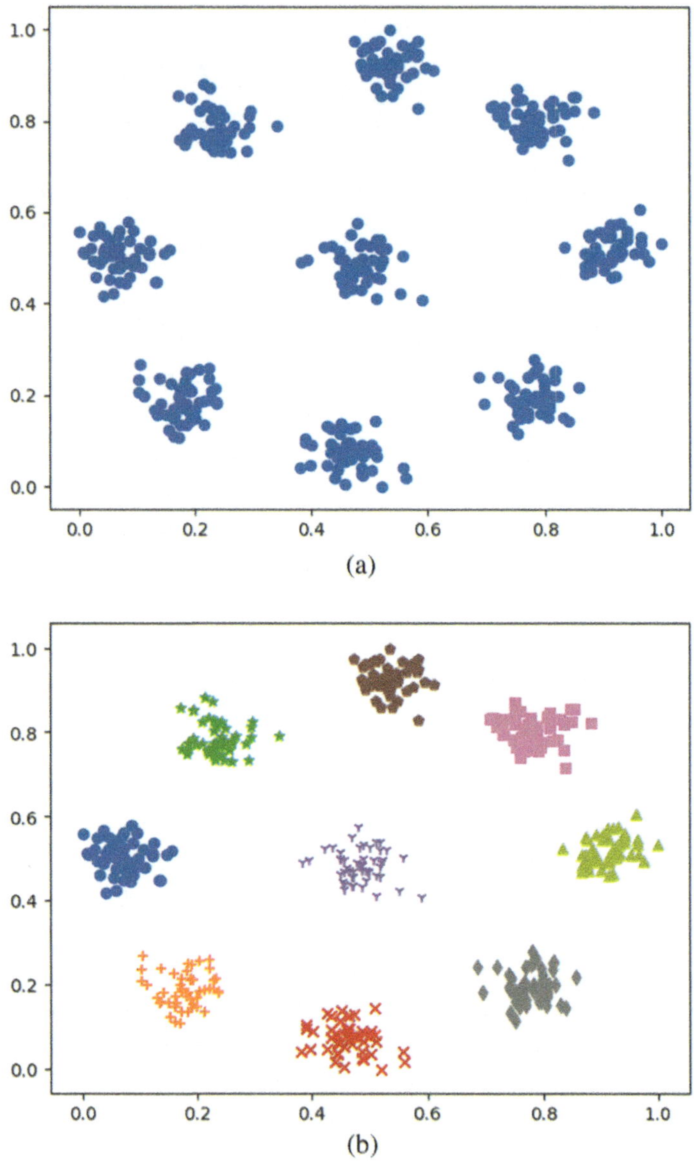

Fig.6 Comparison between the original distribution of T9-1 dataset (**a**) and the corresponding partitioning results (**b**) generated by NSGAII-GR with ARI(T, P) = 1.0

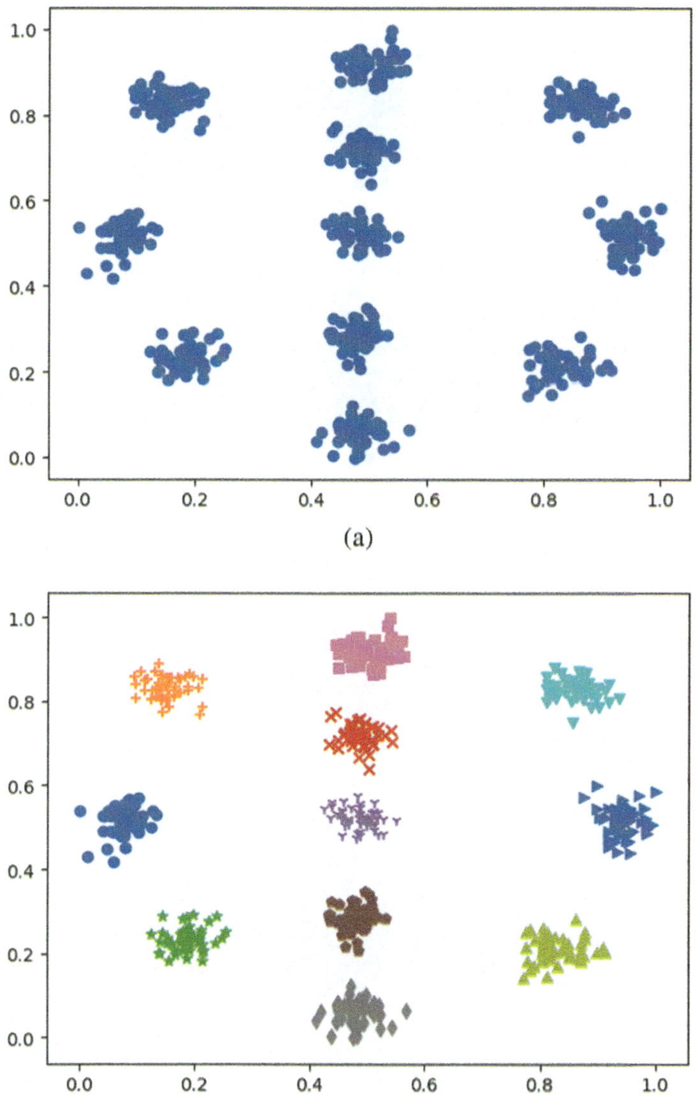

Fig. 7 Comparison between the original distribution of Tb-1 dataset (**a**) and the corresponding partitioning results (**b**) generated by NSGAII-GR with ARI(T, P) = 1.0

technique used in NSGAII-GR. According to the experimental results, NSGAII-GR was easier to operate, even though it did not perform well on the 10d-4c-No. 3 and 10d-4c-No. 5 datasets, overall, it gave promising results. In Table 8, the partitioning results of NSGAII-GR, FCM-NSGAII and MOCK algorithms on 10d-10c artificial datasets were listed.

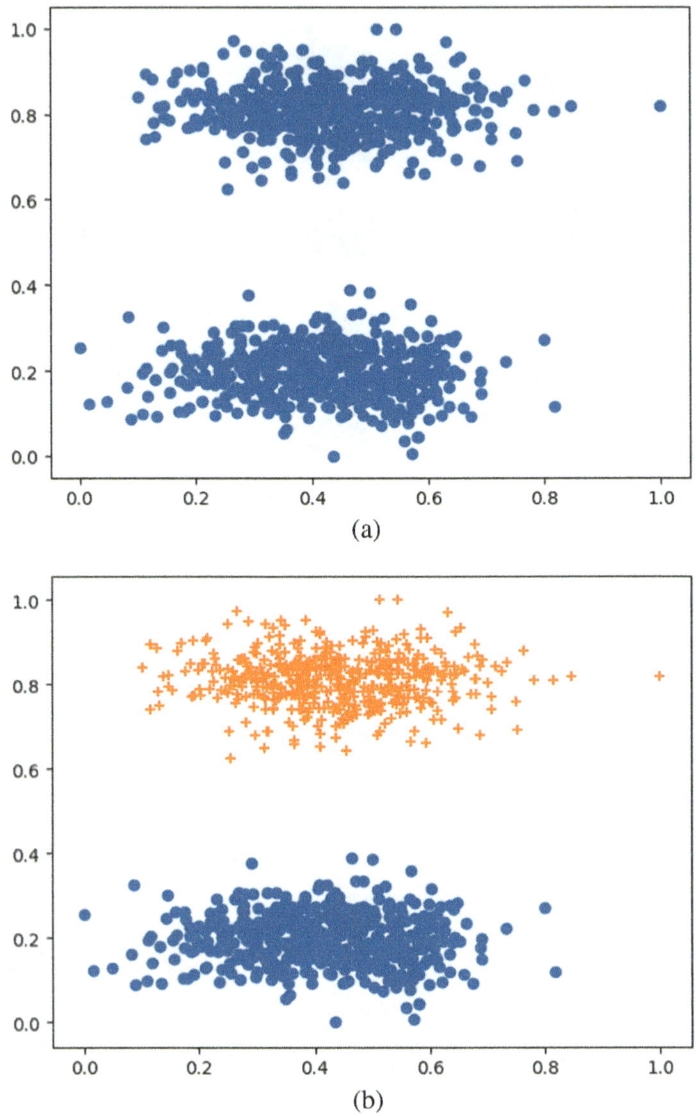

Fig. 8 Comparison between the original distribution of Long1 dataset (**a**) and the corresponding partitioning results (**b**) generated by NSGAII-GR with ARI(T, P) = 1.0

As shown in Sect. 5.1 that NSGAII-GR cannot always find the correct number of clusters on the datasets 10d-10c, but in general, the number of clusters it selected was very close to the actual number. Moreover, in the selected solution, we can observe that the accuracy was still good. FCM-NSGAII had the same algorithm time complexity as NSGAII-GR (see Sect. 5.3), but its accuracy on all 10d-10c datasets

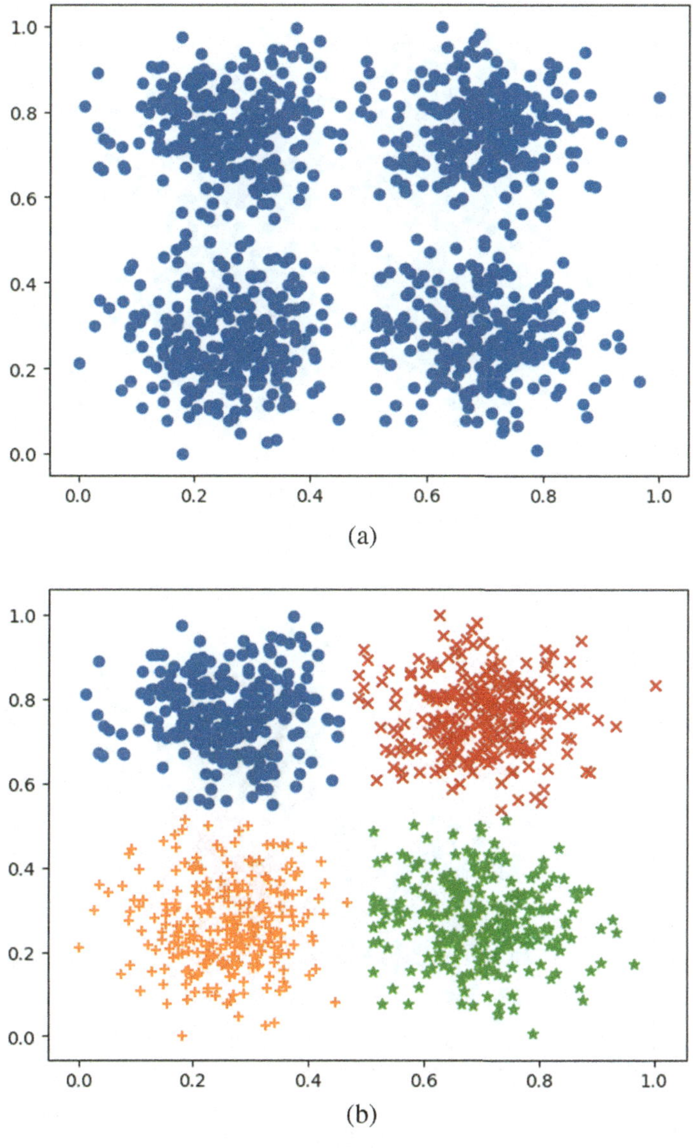

Fig. 9 Comparison between the original distribution of Square1 dataset (**a**) and the corresponding partitioning results (**b**) generated by NSGAII-GR with ARI(T, P) = 0.9763

were much lower than that in NSGAII-GR on both external clustering indexes ARI(T, P) and NMI(U,V). The external clustering indexes obtained by NSGAII-GR were slightly higher than that of MOCK on 10d-10c-No. 8 and 10d-10c-No. 9, but slightly lower than that of MOCK on other datasets. However, considering that the evolution and individual selection technique used in NSGAII-GR is overwhelmingly superior

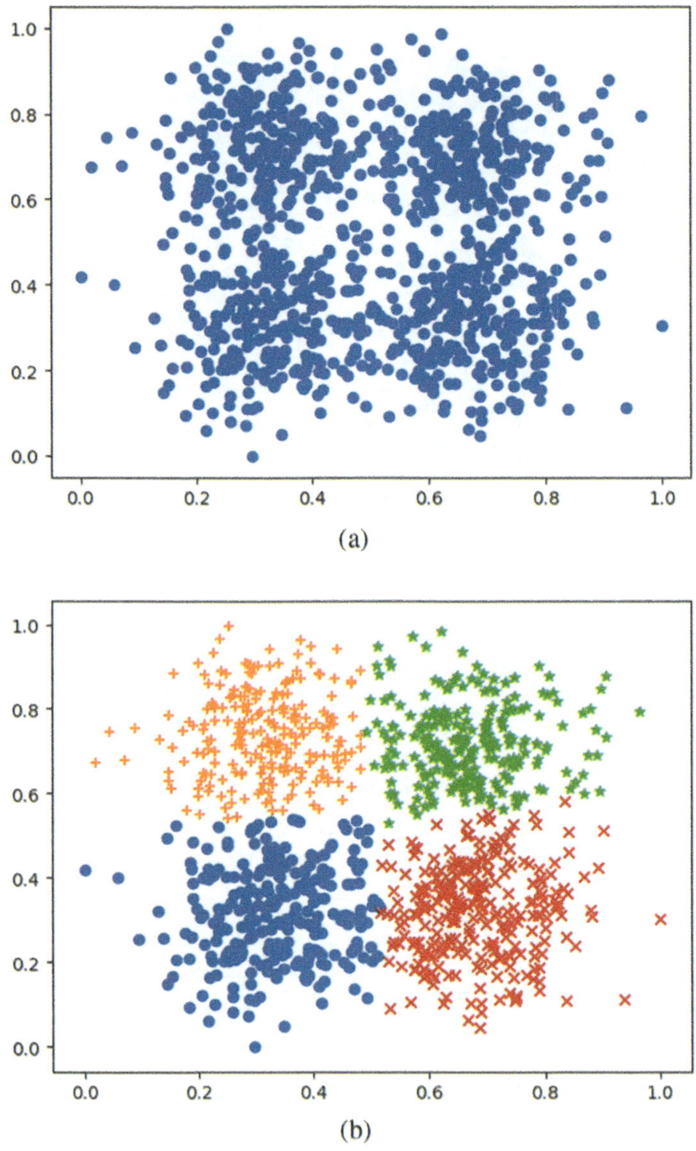

Fig. 10 Comparison between the original distribution of Square4 dataset (**a**) and the corresponding partitioning results (**b**) generated by NSGAII-GR with ARI(T, P) = 0.8373

Table 7 Comparison of Adjusted rand index (ARI) and Mutual information index (NMI) among NSGAII-GR, FCM-NSGAII and MOCK over 10d-4c datasets. ARI and NMI index were generated by two versions of NSGAII-GR, in which Davies-Bouldin index (Dbi) and Silhouette Coefficient (Sil) were respectively used as objective functions

Dataset	NSGAII-GR(Dbi)		NSGAII-GR(Sil)		FCM- NSGAII		MOCK	
	ARI	NMI	ARI	NMI	ARI	NMI	ARI	NMI
10d-4c-No. 0	1	1	1	1	0.9126	0.8655	0.9881	0.9785
10d-4c-No. 1	0.9917	0.9853	0.9928	0.9875	0.8996	0.8598	0.9972	0.9947
10d-4c-No. 2	0.9424	0.9158	0.8798	0.8932	0.8351	0.8046	0.9963	0.9929
10d-4c-No. 3	0.4957	0.6989	0.4961	0.7009	0.7346	0.7330	0.9894	0.9907
10d-4c-No. 4	0.9933	0.9872	0.9928	0.9868	0.8946	0.8455	0.9935	0.9899
10d-4c-No. 5	0.5594	0.6949	0.5594	0.6906	0.8806	0.8341	0.9474	0.9526
10d-4c-No. 6	0.9988	0.9974	0.9966	0.9929	0.8492	0.7969	1	1
10d-4c-No. 7	0.9940	0.9901	0.9940	0.9901	0.9536	0.9351	0.9971	0.9953
10d-4c-No. 8	0.9710	0.9630	0.9710	0.9631	0.8633	0.8467	0.9805	0.9769
10d-4c-No. 9	0.9937	0.9869	0.8565	0.8809	0.8191	0.8100	0.9594	0.9724

Table 8 Comparison of Adjusted rand index (ARI) and Mutual information index (NMI) among NSGAII-GR, FCM-NSGAII and MOCK over 10d-10c datasets. ARI and NMI index were generated by two versions of NSGAII-GR, in which Davies-Bouldin index (Dbi) and Silhouette Coefficient (Sil) were respectively used as objective functions

Dataset	NSGAII-GR(Dbi)		NSGAII-GR(Sil)		FCM- NSGAII		MOCK	
	ARI	NMI	ARI	NMI	ARI	NMI	ARI	NMI
10d-10c-No. 0	0.9377	0.9212	0.9179	0.9015	0.6291	0.6906	0.9611	0.9591
10d-10c-No. 1	0.8762	0.8917	0.8789	0.8915	0.6090	0.7286	0.9399	0.9556
10d-10c-No. 2	0.7628	0.8660	0.7644	0.8687	0.6300	0.7357	0.8703	0.9427
10d-10c-No. 3	0.8204	0.8378	0.8397	0.8537	0.6932	0.7548	0.9219	0.9492
10d-10c-No. 4	0.8433	0.8795	0.7868	0.8573	0.6027	0.7187	0.8518	0.9216
10d-10c-No. 5	0.8948	0.9044	0.8908	0.9029	0.6049	0.7095	0.9212	0.9574
10d-10c-No. 6	0.8832	0.8791	0.8838	0.8778	0.6799	0.7104	0.9153	0.9427
10d-10c-No. 7	0.7934	0.8628	0.8368	0.8863	0.6059	0.7155	0.8450	0.9319
10d-10c-No. 8	0.9262	0.9171	0.9082	0.9074	0.6487	0.7225	0.9230	0.9453
10d-10c-No. 9	0.9178	0.9104	0.9336	0.9196	0.6131	0.6924	0.9135	0.9423

to that of MOCK algorithm in terms of computational complexity, we thought that the above results are still acceptable.

5.3 *Performance*

We investigated the improvement in accuracy and performance of NSGAII-GR in contrast with the normal version without gene rearrangement and cluster merging on specific running cycles = {5,10,15,20,25,30,35,40,45,50}. The two versions were run for 20 times and the average values on external index ARI (T, P) were compared. Due to the limited space, only results on three data sets, i.e., 10d-4c-No. 0, 10d-4c-No. 1 and 10d-4c-No. 2 were selected for analysis and the results were shown in Fig. 11.

In Fig. 11a–c, it demonstrated that the version with gene rearrangement combined with cluster merging did accelerate the convergence and achieved better clustering

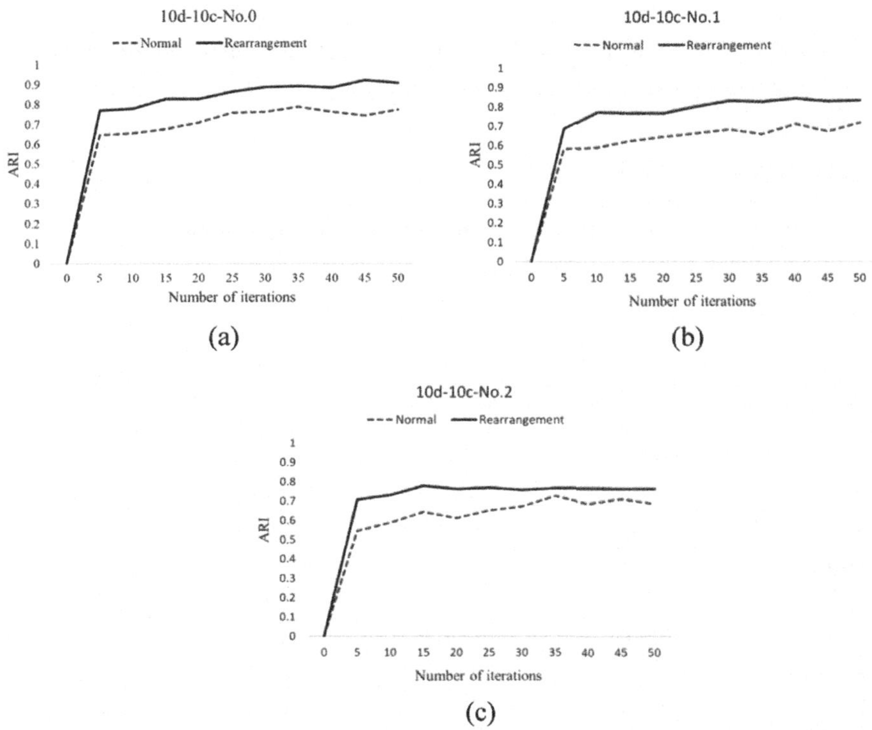

Fig. 11 Accuracy and performance comparison on the adjusted rand index (ARI) between the improved version (i.e., with the gene Rearrangement and cluster merging) and the normal version (i.e., without gene rearrangement and cluster merging) on dataset 10d-10c-No. 0 (**a**), 10d-10c-No. 1 (**b**) and 10d-10c-No. 2 (**c**)

results. For the selected datasets, i.e., 10d-10c-No. 0, 10d-10c-No. 1 and 10d-10c-No. 2, the adjusted range index (ARI) of the improved version was significantly higher than that in the normal version at the same iteration (cycle). One impressive advantage of the algorithm is that the operations of gene rearrangement and inter cluster merging do not increase the time complexity of the algorithm, see Sect. 5.4 for detail.

Here we gave the experimental results of adjusted rand index based on the comparison between the multi-objective algorithm NSGAII-GR and its single-objective versions in which the internal CVI of the sum of generalized sample variance (SGSV), Calinski-Harabasz index (Cal) and Davies-Bouldin index (Dbi) were used as fitness function on 10d-10c datasets. One should note that the single objective version did not employ the technique of gene rearrangement and cluster merging, and the individual selection method was also different from the multi-objective optimization method as well. Other than that, the steps in the single objective method were basically the same as its multi-objective counterpart. Algorithm 9 illustrated the detailed single-objective version. Table 9 gave the accuracy compassion between these versions on 10d-10c datasets.

Algorithm 9: Clustering based on Single-Objective Genetic Algorithm

Input: dataset, population size, cycles, crossover probability, mutation probability

Output: recommended solution

1. Initialize population.
2. Pair and crossover individuals to generate offspring without gene rearrangement.

Table 9 Comparison of Adjusted rand index between multi-objective algorithm NSGAII-GR and its single-objective versions in which the internal CVI of the sum of generalized sample variance (SGSV), Calinski-Harabasz index (Cal) and Davies-Bouldin index (Dbi) were used as fitness function on 10d-10c datasets

Dataset	NSGAII-GR	GA-SGSV	GA-Cal	GA-Dbi
10d-10c-No. 0	0.9377	0.4163	0.5725	0.0001
10d-10c-No. 1	0.8762	0.4313	0.1620	0.0001
10d-10c-No. 2	0.7628	0.5813	0.1949	0.0007
10d-10c-No. 3	0.8204	0.6556	0.4700	0.0003
10d-10c-No. 4	0.8433	0.4453	0.3493	0.0004
10d-10c-No. 5	0.8948	0.4557	0.1749	0.0001
10d-10c-No. 6	0.8832	0.4785	0.3796	0.0007
10d-10c-No. 7	0.7934	0.4719	0.1812	0.0010
10d-10c-No. 8	0.9262	0.4606	0.2057	0.0001
10d-10c-No. 9	0.9178	0.4813	0.2406	0.0003

3. Perform mutation operations on the newly generated individuals.
4. Correct invalid individuals.
5. Calculate fitness for individuals using a certain single objective function.
6. Rank fitness values for the individuals.
7. Select the first N individuals with the highest fitness into the next generation.
8. Repeat step 2–7 until the number of iterations is not reached; otherwise go to step 9:
9. Output the individual with the highest fitness as the recommended solution.

We observed that the multi-objective optimization method was overwhelmingly superior to any types of single objective optimization method using the same objective function, as shown in Table 9. Although the two properties, i.e., compactness and separation were considered in Calinski-Harabasz index (Cal) and Davies-Bould in index (Dbi) at the same time, they inherently regarded that the shape of clusters is spherical, (i.e., the correlation coefficient between different features is 0). The sum of generalized sample variance (SGSV) took correlations between samples into account. It regarded that samples in different clusters came from different distributions of which the covariance matrix was described by SGSV(C). By incorporating SGSV(C) and Cal(C) into NSGAII-GR, the structure of the solution can reach a balance between the two. The partitioning not only took into account the homogeneity within the cluster, but also the heterogeneity between the clusters, as well as the correlation between characteristics within the cluster. Moreover, we also found in our experiments that if SGSV(C) and Cal(C) were not used to select the partitioning structure, using Dbi(C) alone was unlikely to find good clustering results from the population.

5.4 Computational Time Complexity

Let G represents the number of evolutionary iterations, P represents the population size, N represents the sample size, K_{max} is the maximum cluster number, and D represents the dimension of the data, we analyze the time complexity of the algorithm as follows:

(1) The maximum number of genes in a chromosome is K_{max}. If there are D variables in each gene, it needs $O(K_{max}D)$ to initialize one gene. P chromosomes need to be initialized, so the initialization of the whole population needs $O(PK_{max}D)$.
(2) In one iteration, $O(PK_{max})$ is required for chromosome crossing. In one chromosome, gene rearrangement requires $O({K_{max}}^2D)$ times. Because $K_{max} \leq 2\,N$,

the chromosome crossover and gene rearrangement of the whole population are $O(PND)$.

(3) Each iteration of mutation step is performed $O(PK_{max}D)$ times.

(4) $O(NK_{max}D)$ is required for all samples to be assigned to a cluster.

(5) Two objective functions are used in this research. The first one is SGSV(C). If a cluster contains N_i samples, the covariance matrix of the cluster needs $O(N_i D^2)$, and the generalized sample variance needs $O(D^3)$. Because $\sum_{i=1}^{k} N_i = N$ and $N > D$, $O(ND^2)$ is required to calculate SGSV(C) once. The second is Cal(C) and requires $O(ND)$. Each iteration of the algorithm requires $O(PND^2)$ to calculate individual fitness.

(6) It takes $O(PK_{max})$ to test whether an individual is invalid or not and takes $O(PND^2)$ to test whether clusters need to be merged.

(7) The time complexity of NSGA-II algorithm is $O(MP^2)$, where M is the number of optimization objectives with $M = 2$. $O(P_2)$ is required to perform one non-dominated sorting.

(8) The time complexity of Dbi(C) is $O(ND + K_{max}{}^2 D)$, so the individual selection step needs $O(PND)$ times.

Overall, $ND^2 > P$, then NSGAII-GR has the worst-case time complexity of $max\{O(GPND^2), O(GPNK_{max}D)\}$.

6 Conclusion

In this research, an automatic clustering method based on multi-objective genetic algorithm, NSGAII-GR, is proposed. In multi-objective optimization, two internal CVIs, i.e., the sum of generalized sample variance and Calinski-Harabasz index are used to simultaneously optimize the partitioning structure. Then another internal clustering index, such as Davies-Bouldin index, is used to determine the correct number of clusters and select the first recommended solution. Because the structure of the solutions found by our algorithm is good enough, in another set of tests, we found that NSGAII-GR can also determine the number of clusters in the solution set and output recommended solutions by using the Silhouette Coefficient index.

For an automatic clustering algorithm, two aspects, i.e., proper search space and convergence, are critical to find a good solution. In the proposed algorithm, the partition procedure is limited to a convex region with the cluster center as the "inner point", in which all the sample points are regarded as a cluster. In this way, the search space is shrunk and the requirement that the cluster samples should be as similar as possible is satisfied. Then the cluster center is rearranged by genetic operations to achieve global optima. From here, the center of clusters could be any value within the sample value range. The second is the proper way of convergence. NSGAII-GR takes Cal(C) and SGSV(C) as objective functions, then NSGA-II is employed to select individual and the technique of gene rearrangement as well as cluster merging are applied to speed convergence.

Three experiments conducted on five 2D artificial datasets and five real world datasets demonstrated that NSGAII-GR achieved overall the best accuracy and performance of automatic clustering over the three state-of-the-art counterparts. The most attractive aspects of NSGAII-GR are that firstly, cluster coherent, separation and correlation are simultaneously considered to optimize the partitioning structure; and secondly, Davies-bouldin index is used to find the best solution from Pareto Optimal From without any prior information. These advantages lead NSGAII-GR to be one of the promising approaches in automatic clustering metaheuristics.

Future research efforts should be invested to address two possible challenges being observed during our investigations, which are: (1) occasionally, the individuals with the correct number of clusters cannot be correctly selected even they are found in the pareto optimal front; (2) individuals with the highest ARI score are not always selected as recommended solutions. Even though the two cases are far less likely to be observed in common datasets, we still think that the reasons behind deserve to be thoroughly investigated. Despite the random effects, these problems are probably sourced either from the traits of given dataset or from the design of internal CVIs. In some datasets, on one hand, the distribution of cluster size is extremely imbalanced and the uneven overlaps between them could let the algorithm unexpectedly merge two clusters into a single one, which might cause incorrect cluster number. On the other hand, not all internal CVIs are able to perfectly reflect the sample distribution patterns in a dataset with complex clustering structure. Except classical CVIs, novel measurements such as connectivity, overlap, entropy inference, etc., are emerging recently. We think that the refinement of CVI combinations in a multi-objective optimization framework could be a potential approaching the solution.

Acknowledgements This research was funded by the National Natural Science Foundation of China (61871061), which is gratefully acknowledged.

References

1. Ren, Y., Liu, X., Liu, W.: DBCAMM: A novel density based clustering algorithm via us-ing the Mahalanobis metric. Appl. Soft Comput. **12**, 1542–1554 (2012)
2. He, P., Zhu, J., He, S., et al.: Towards Automated Log Parsing for Large-Scale Log Data Analysis. IEEE Trans. Dependable Secure Comput. **15**, 931–944 (2018)
3. Paul, A.K., Shill, P.C.: New automatic fuzzy relational clustering algorithms using multi-objective NSGA-II. Inf. Sci. **448**, 112–133 (2018)
4. Said, A., Abbasi, R.A., Maqbool, O., et al.: CC-GA: a clustering coefficient based genetic algorithm for detecting communities in social networks. Appl. Soft Comput. **63**, 59–70 (2018)
5. Qu, H.C., Qiu, Z.L., Tang, X.M., Xiang, M., Wang, P.: Incorporating un-supervised learn-ing into intrusion detection for wireless sensor networks with structural co-evolvability. Appl. Soft Comput. **71**, 939–951 (2018)
6. MacQueen, J.B.: Some methods for the classification and analysis of multivariate observations. In: Proceedings of the Fifth Berkeley Symposium on Mathematical Statistics and Probability, pp. 281–297. University of California Press (1967)
7. Rodriguez, A., Laio, A.: Clustering by fast search and find of density peaks. Science **344**, 1492–1496 (2014)

8. Ester, M., Kriege, H.P., Sander, J., et al.: A density-based algorithm for discovering clusters in large spatial databases with noise. In: International Conference on Knowledge Discovery & Data Mining, Chiang Mai (1996)
9. Karypis, G., Han, E.H., Kumar, V.: Chameleon: hierarchical clustering using dynamic modeling. Computer **32**, 68–75 (2002)
10. Jain, A.K.: Data clustering: 50 Years beyond K-means. Pattern Recogn. Lett. **31**, 651–666 (2010)
11. Hruschka, E.R., Campello, R.J.G.B., Freitas, A.A., Leon, P.: A survey of evolutionary algorithms for clustering. IEEE Trans. Syst. Man Cybern. Syst. Part C (Appl. Rev.) **39**, 133–155 (2009)
12. Cowgill, M.C., Harvey, R.J., Watson, L.T.: A genetic algorithm approach to cluster an-alysis. Comput. Math. Appl **37**(7), 99–108 (1999)
13. Coello Coello, C.A.: List of References on Evolutionary Multi-objective Optimization (2010)
14. Theodoridis, S., Koutroumbas, K.: Pattern Recognition, 4th edn. Elsevier, Amsterdam (2009)
15. Handl, J., Knowles, J.: An evolutionary approach to multi-objective clustering. IEEE Trans. Evol. Comput. **11**(1), 56–76 (2007)
16. Kim, M., Ramakrishna, R.S.: New indices for cluster validity assessment. Pattern Recogn. Lett. **26**(15), 2353–2363 (2005)
17. Saha, I., Maulik, U., Plewczynski, D.: A new multi-objective technique for differential fuzzy clustering. Appl. Soft Comput. **11**(2), 2765–2776 (2011)
18. Wikaisuksakul, S.: A multi-objective genetic algorithm with fuzzy c-means for auto-matic data clustering. Appl. Soft Comput. **24**, 679–691 (2014)
19. Yang, C.L., Kuo, R.J., Chien, C.H., et al.: Non-dominated sorting genetic algorithm using fuzzy membership chromosome for categorical data clustering. Appl. Soft Comput. **30**, 113–122 (2015)
20. Engelbrecht, A.P.: Computational Intelligence: an Introduction, 2nd edn. Wiley, Hoboken (2007)
21. Gupta, A., Datta, S., Das, S.: Fuzzy clustering to identify clusters at different levels of fuzziness: an evolutionary multi-objective optimization approach. IEEE Trans. Cybern. 1–11 (2019)
22. Nanda, S. J, Panda, G.: Automatic clustering using MOCLONAL for classifying actions of 3D human models. In: Symposium on Humanities. Science and Engineering Research, Kuala Lumpur, pp. 945–950. IEEE (2012)
23. Matake, N., Hiroyasu, T., Miki, M., Senda, T.: Multiobjective clustering with automatic k-deter-mination for large-scale data. In: Proceedings of the 9th Annual Conference on Genetic and Evolutionary Computation (GECCO'07). ACM Press, New York, pp. 861–868 (2007)
24. Suresh, K., Kundu, D., Ghosh, S., Das, S., Abraham, A., Han, S.Y.: Multi-objective diff-erential evolution for automatic clustering with application to microarray data analysis. Sensors **9**(5), 3981–4004 (2009)
25. Xia, H., Zhuang, J., Yu, D.: Novel soft subspace clustering with multi-objective evolutionary approach for high-dimensional data. Pattern Recogn. **46**(9), 2562–2575 (2013)
26. Deb, K., Pratap, A., Agarwal, S., Meyarivan, T.: A fast and elitist multi-objective genetic algorithm: NSGA-II. IEEE Trans. Evol. Comput. **6**(2), 182–197 (2002)
27. Liu, Y., Ozyer, T., Alhajj, R., Barker, K.: Integrating multi-objective genetic algorithm an-d valid-ity analysis for locating and ranking alternative clustering. Informatica **29**(1), 33–40 (2005)
28. Liu, Y., Wu, X., Shen, Y.: Automatic clustering using genetic algorithms. Appl. Soft Comput. **218**(4), 1267–1279 (2011)
29. He, H., Tan, Y.: A two-stage genetic algorithm for automatic clustering. Neurocomputing **81**, 49–59 (2012)
30. Ding, Y., Fu, X.: Kernel-based fuzzy C-means clustering algorithm based on genetic algorithm. Neurocomputing **188**, 233–238 (2015)
31. Rendón, E., Abundez, I., Arizmendi, A., Quiroz, E.M.: Internal versus external cluster validation indexes. Int. J. Comput. Commun. **1**, 27–34 (2011)

32. Hancer, E., Karaboga, D.: A comprehensive survey of traditional, merge-split and Evolutionary approaches proposed for determination of cluster number. Swarm Evolution. Comput. **32**, 49–67 (2017)
33. Pal, N.R., Bezdek, J.C.: On cluster validity for the fuzzy c-means model. IEEE Trans. Fuzzy Syst. **3**(3), 370–379 (1995)
34. Hwei-Jen, L., et al.: An efficient GA-based clustering technique. J. Tamkang J. Sci. Eng. **8**(2), 113–122 (2005)
35. Qu, H.C., Liu, G.: Threshold optimized strategy based on improved flower pollination algorithm for unbalanced data. In: IEEE 10th International Conference on Intelligent System-s (IS), pp. 551–556 (2020)

Assessment of Deep Learning Models for Human Activity Recognition on Multi-variate Time Series Data and Non-targeted Adversarial Attack

Mahbuba Tasmin, Sharif Uddin Ruman, Taoseef Ishtiak, Arif-ur-Rahman Chowdhury Suhan, Redwan Hasif, Shahnawaz Zulminan, and Rashedur M. Rahman

Abstract Human activity recognition (HAR) is one of the leading research fields in ubiquitous computing working to integrate seamless technologies in our daily lives. The researches in this field focus on the technological advancement of fine activity recognition through using minimal technological deployments along with consideration of human factors. The domain specific knowledge of human activity recognition has shed significant importance on the activity data processing, due to the innate random nature of the activity data. The necessity for improved prediction accuracy of activities within limited infrastructural support and cost is necessary for the research works to be conducted at large. The data set utilized in this study is modified through tree-based feature-engineering method, which has significantly impacted the intrinsic time-series pattern associated with the original data set acquired from UCI machine learning repository. In this paper, the primary focus lies on improving classification accuracy of the human activities included in the data set through utilizing deep learning classifier models. The particular modified pattern of the data devoid of time-series pattern to major extent has resulted in lower classification accuracy by benchmark time-series classification models namely Keras-LSTM and RNN-LSTM. CNN based networks have demonstrated consisted improved training performance on the data set. The state-of-the-art classifier ResNet introduced by Microsoft Research in 2015 has achieved 99.9% classification accuracy on the data set of this study. Another important aim of this study is to infer the risk attached with adversarial attack on machine learning models. The high vulnerability of machine learning models in face of adversarial attacks has emerged as security concern of the deployment pipeline of machine learning models. To the author's best knowledge, there has been no prior study to find out the high insecurity associated with the human activity recognition data set and concerned benchmark models. In this study, two prevalent adversarial approaches namely Fast Gradient Sign Method (FGSM) and Basic Iterative Method

M. Tasmin · S. U. Ruman · T. Ishtiak · A.-R. C. Suhan · R. Hasif · S. Zulminan · R. M. Rahman (✉)
North South University, Dhaka, Bangladesh
e-mail: rashedur.rahman@northsouth.edu

M. Tasmin
e-mail: mahbuba.tasmin@northsouth.edu

V. Sgurev et al. (eds.), *Advances in Intelligent Systems Research and Innovation*, Studies in Systems, Decision and Control 379,
https://doi.org/10.1007/978-3-030-78124-8_6

(BIM) have been employed to carry out non-targeted attack on the pre-trained ResNet model. The accuracy loss by the adversarial attack over a small range of added perturbation is tremendous, nearly 96% accuracy is diminished in the process. Finally, we have demonstrated the comparative performance by FGSM and BIM on reducing the classification accuracy of pre-trained model. The findings shed light on the nature of the data influencing the classification accuracy and the resiliency of state-of-the-art classifier model on facing adversarial attacks.

1 Introduction

Human Activity Recognition (HAR) is an active area of research that has gained its pace to improve human interaction with technological presence at complex indoor environments. The Internet of Things (IoT) has radically changed the concept of using technology in indoor space, where almost all the appliances are now remotely controllable through the tap of smart phones by the resident users. The concept of Ambient Assisted Living (AAL) has flourished on the basis of providing sophisticated indoor services, specially to the elderly population, by integrating technology at usual daily activities.

There are benchmark deep learning models for human activity recognition with higher accuracy. Activity recognition requires collection of substantial data-set for training. The discrete nature of human activities makes the data set full of noisy and anomalous instances. Hence, training on the raw data with little to no feature engineering puts overhead on the model training accuracy and time. This leads to dependency on heavy deep learning models, which require expensive configuration setup to continue the training work. However, recent advances in adversarial attacks have broken the myth of deep learning models reign on accuracy just by the measure of training performance or accuracy score. As the nature of HAR data set is inherently sensitive and easily perturbable, the adversarial attack on HAR techniques can lead to analyze the vulnerability of the classification models and can bring insights in building robust models viable of handling adversarial threats.

Adversarial attack changes the original input data through adding slight perturbations to mislead pre-trained models on classification [44]. The added noise is unrecognizable by human observation and subconsciously still being processed by humans due to cognitive bias, leaving the noise undetected in the process. The completely different mode of analysis between human and machine creates the difficulty of detecting the potential security concern in machine learning models if not treated from the aspect of the machine. It leaves the machine learning model vulnerable to such tiny changes in data. For example, perturbations contained in 32-bit inputs of machine language but ignored in human readable 8-bit representation of the input value, can affect the human maintainers/users at the user-end of the model services [5, 16, 38, 44]. Adversarial attack is unique in the application to machine learning models without any prior access to the trained model and the data underneath. To add more to the questioned vulnerability of machine learning models, the perturbation are so subtle in amount and impeccably integrated with input data that the observer-end can quite possibly ignore the perturbation. Adversarial training is the tool to check the vulnerability of the neural networks and robustness of the models created [16, 44]. Researches on adversarial attacks have been initiated in the image processing field. Some of the advanced adversarial training on ImageNet is ensemble adversarial training [48]. The Basic Iterative Method implemented at randomly chosen starting point achieved best result on MNIST [31]. Existing research works on adversarial attack consider predominantly three sub fields: Non-targeted Adversarial Attack, Targeted Adversarial Attack and Defense against Adversarial Attack. Non-targeted adversarial attack has the objective of producing wrong classification by machine learning models through subtle change in the source data.

In this paper, the research work has been carried out on the benchmark data set collected by a distinct research group from Washington State University and currently available in UCI machine learning repository, namely "Human Activity Recognition from Continuous Ambient Sensor Data Data Set" [14]. The original data set has been collected on an ideal indoor setup, where the detecting sensors have been placed at locations specific to human activities indoors in the resident houses. In the paper [46], the preceding findings from our work showed the necessity of drastic feature engineering, while using baseline classifier models to classify selected five activities—"Cooking", "Eating", "Talking Over Phone", "Reading", "Watching TV".

The research findings highlight the necessity of data pre-processing and significant feature selection for achieving better classification accuracy score on noisy data of HAR instances. The classification is run on the data set for extrapolating the significance of exhaustive data pre-processing on the original source data from Washington State University and feature engineered data set from previous paper. In this paper, the classification accuracy of data set has been achieved through training of several deep learning models. The highest accuracy score of activity recognition on the data set is achieved by ResNet (approximately 99.9%). The adversarial attack is run on the data set by injecting noise through FGSM and BIM. Consequently, the pre-trained model is tested on the perturbed data set to record the misclassification score. To ensure the standardized benchmark implementation of the adversarial training, the adversarial attacks are implemented through using Cleverhans library [38]. The Cleverhans library is an open source contributing software licensed by MIT.

The working scheme of this paper is as follows:

1. Normalization and data preparation of instances for five selected activities obtained from original source and tree-based feature engineering data set
2. Accuracy improvement of the activities in the data set through application of several deep-learning models including ResNet.
3. Non-targeted Adversarial Attack (FGSM and BIM) to test the robustness of the model.

   The paper is structured as the following: the literature review on Human Activity Recognition, Feature Engineering, Deep Learning Models and Adversarial Attacks are included in Sect. 2. The original data source, feature engineered data set and the test data set are described in Sect. 3. Section 4 covers work flow of this paper, data preprocessing, ResNet model architecture and principal concepts of applied adversarial attack methods: FGSM and BIM. Following Sect. 5 includes performance evaluation of classification accuracy of ResNet, prediction score per activity and resiliency of the pre-trained ResNet model on facing adversarial attack. Section 6 comprises of discussion on our research insights, findings and related benchmark studies on feature engineering, activity classification through several deep learning models and adversarial attack on time-series data. Section 7 includes a proposed extension of the current work on a custom data collection setup for human activity recognition techniques. Section 8 concludes the present study with brief reflection on the findings and significance of the novel approach of this paper.

2 Related Work

2.1 Feature Engineering in the Field of Human Activity Recognition

Human activity recognition based tasks require data from embedded sensors. For this reason it is pretty obvious that the raw data will have lots of noise, null values and other environment related effects. The raw data needs to be cleaned and instances of each activity should be brought under a certain range for proper labeling and training. The data of this study was obtained from a well known data source [14], yet we had to address such issues in the data. Consequently, we chose to build a data-engineering pipeline constituting of feature engineering and normalization. Our previous work on this data set focused on the importance of feature engineering through significant feature selection from the acquired data set, where we had shown that the tree-based feature engineering techniques significantly improve models' training performance [46]. But this approach of course is not very new; lots of work in the domain of human activity recognition used feature engineering as a substantial step. Chung et al. presented such a work where sensors' data were collected from wearable devices, smart phones and indoor infrastructures of wireless sensors [12]. Gyroscopes and accelerometers are now being widely used to perceive gesture and motion pattern-based activity recognition. In all such work important feature selection and data dimensionality reduction have been given paramount importance. Karegowda et al. have used gain ratio and correlation as two important factors to reduce the dimensionality [21]. In their work they have determined the correlation with C4.5 tree, gain ratio and genetic algorithms. They chose filter method instead of wrapper method for gaining computational support. Tama et al. used tree-based classifiers and PSO-based feature selection in their intrusion detection system [45]. In a notable paper of time-series classification [25], the authors have demonstrated exemplary prediction score through employing FCN and ResNet based on exhaustive statistical feature extraction by sliding window principle.

2.2 *Deep Neural Network for Time-Series Data*

In paper [9] Carlini et al. have proved that defensive distillation cannot make a neural network robust against adversarial attacks. The writers have made three new adversarial examples which readily fail the accuracy of neural networks. Even if the neural network can fail the adversarial attack, transferability characteristics of adversarial attacks can let the neural network down. The tree adversarial example can inject noise on an easy-to-attack model; eventually transferring those examples to the hard-to-attack model. CNN was used in [53] with a view to automating feature learning from raw inputs of multichannel time series signals obtained from wearable inertial sensors. Automatic discriminative feature extraction was also performed in [55] through CNN that can capture local dependency. Partial weight sharing technique has been proposed there to accelerate signals. Walking, standing, sitting, running, falling, and climbing stairs were detected and classified in [39] with accelerometer data placed on the body. "Data-Driven" and "Knowledge-Driven" are the two fundamental techniques to classify HAR data set. But in [36] "Data-Driven" practice was applied with five classical supervised machine learning classifiers.

2.3 *Adversarial Attack*

One of the pioneering comprehensive studies of adversarial attack on time series classification conducted by H. Ismail Fawaz et al. demonstrated non-targeted adversarial attack on 85 benchmark data set published by UCR archive [19]. The paper presented a comparative view between several data set performances to two species of adversarial attack. The computer-vision and time-series based data set have showed significant differences in handling the perturbation added by adversarial attack. The results are standardized with Cleverhans API and it evidently shows the vulnerability of current state-of-the-art classification models to adversarial attacks in several human-sensitive application fields. Non-targeted adversarial examples generally transfer. But in [27] the authors have observed some targeted adversarial instances can also transfer with target labels which is rare. They trained their model on ImageNet and applied targeted adversarial examples on them.

A learning algorithm named Adversarial Classifier Reverse Engineering (ACRE) had been introduced in the paper [28]. The objective was to build an adversarial model capable of recognizing high quality instances rather than a generalized classifier. The learning efficiency of the applied adversary was tested by ACRE algorithm in linear reverse engineering classifiers. It measures the vulnerabilities of different classifiers with varied adversary applied on continuous and Boolean feature data set.

To make a classifier more robust learning, an adversarial evasion model consisting of feature cross-substitution attack is introduced in the paper [26]. Besides the adversarial model, they also applied effective and efficient feature usage during adversarial classification. Two more solutions were proposed by them, first one is Equivalence-Based Classification (EBC) which is a highly heuristic solution of using equivalence classes rather than primary features and second one is Stackelberg Game Multi-Adversary Model (SMA), a general approach of using mixed-integer linear programming. Both solve the problem that arises during feature reduction in adversarial classification. For stronger adversaries with sufficient feature, SMA brings a great balance between dimensionality reduction and evasion attacks.

3 Data Source

The original data set was collected from the UCI Machine Learning Repository [14] and in our previous work [46] we had demonstrated feature selection with tree-based feature engineering techniques on the obtained noisy data set. The tree-based feature engineering techniques helped to increase the accuracy scores of the applied baseline classifiers. Table 1 presents all the 36 features of the original data set collected from UCI machine learning repository. It was first published by the authors in 2012 [13]. Total number of activities covered by this data set are 12 (11 specific activities and another anonymous activity). The data set is collected from volunteer resident houses, where the location and motion detector sensors were placed at different parts of the house. Activities were collected and labeled based on the domain-specific knowledge of human activities and general locations in concern with the activities.

The features are divided into categories according to similar working principle for easy understanding. The "Time-related" features in Table 1 are the latest time stamps recorded on per activity instances. "Time Since Last Sensor Event" presents the gap time between occurrence of consecutive events. The "Sliding Window" categorized features include windowDuration (time duration of specific number of activities recorded in sliding window), two previous dominant sensors (prevDominantSensor1, prevDominantSensor2), local information about the last recorded sensor in the window (ID, Sensor Location, Motion Location), change of major sensor locations (areaTransitions) and distinct number of sensors in the window. The "Weight-count" category enlist features with specific sensors assigned to count the weighted average number of activities recorded by the sensor in concern, placed at different location of the house and named accordingly. The "Activity Location" category encompasses number of features collecting the last event recording time by the sensors in 24 h time. The activity feature is annotated manually and complexity contains the estimation of entropy in sensor counts.

For the purpose of this study, we have selected five specific activities from the original data set: Cooking, Eating, Talking over Phone, Reading and Watching TV. The activity instances are run through different feature-engineering approaches for selecting top features with most significant scores [46]. The feature engineered data

Table 1 Original features of the data set

Time-related	Sliding window	Weight-count	Activity location	Computed columns
^{L}Hours	lastMotionLocation	$^S C$-LivingRoom	$^S E$-OutsideDoor	Activity (label)
^{L}Seconds	lastSensorLocation	$^S C$-Kitchen	$^S E$-Chair	Complexity
^{L}DayOfWeek	lastSensorID	$^S C$-Bedroom	$^S E$-WorkArea	
timeSince^L	windowDuration	$^S C$-Bathroom	$^S E$-DiningRoom	
	prevDominantSensor1	$^S C$-Chair	$^S E$-Bathroom	
	prevDominantSensor2	$^S C$-DiningRoom	$^S E$-Bedroom	
	activityChange	$^S C$-Hall		
	areaTransitions	$^S C$-Office		
	numDistinctSensors	$^S C$-OutsideDoor		
		$^S C$-WorkArea		
		s$^S C$-Ignore		

^{L}lastSensorEvent; $^S C$sensorCount; $^S E$sensorElTime

set obtained from Extra-Tree classifier achieved consistent highest accuracy against all five baseline classifiers tested in previous work. In this paper, we have simultaneously run classification and adversary testing on this dataset.

Extra-tree classifier sorts all the columns according to their significance score and after the top 21 features, the scores were nearly zero. Hence, we have selected the top 21 features in the engineered data set. In Fig. 1, feature importance score is

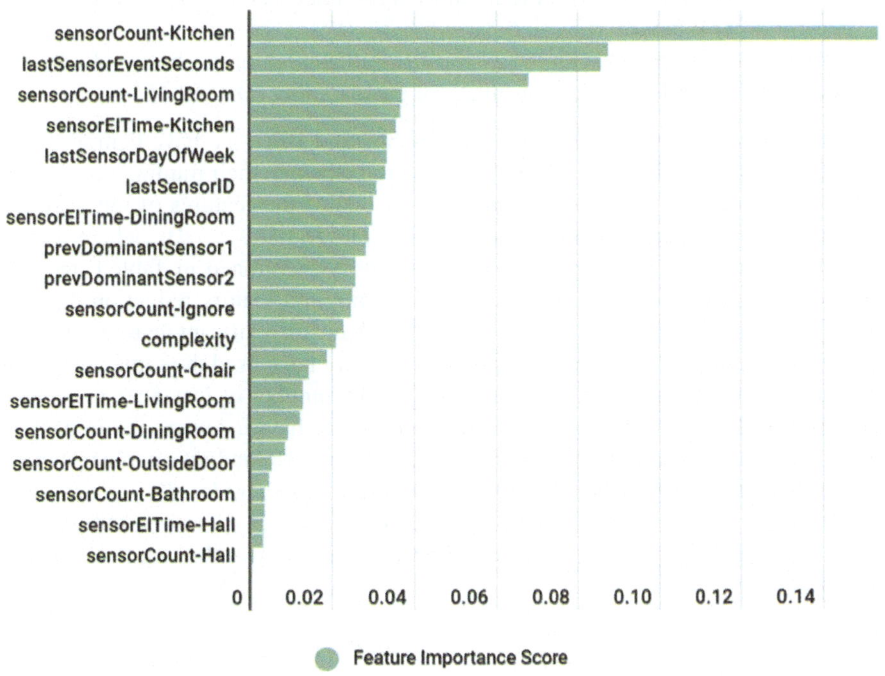

Fig. 1 Extra-tree classifier obtained features

Table 2 Data sets of the study

Source data set	Original no. of instances	Equal no. of instances
Original	Dataset 1	Dataset 2
Extra-tree classifier	Dataset 3	Dataset 4

presented along x-axis and name of the features are enlisted in y-axis. The feature "sensorCount-Kitchen" tops the list with significance score of 0.153. The scraped data set for five activities were run through feature engineering without trimming the number of instances per activity. The activity "Cooking" had nearly 300,000 instances recorded, for which the associated location sensor had topped the list.

In this research, we have modified the original data set into four versions, as listed in Table 2. Since the number of instances per activities have drastic differences in count and hence may affect the prediction accuracy per activity during classification, we have decided to extrapolate the prediction difference between the original number of instances and equalized number of instances. From both of the original data set and feature-engineered generated data set of five activities, two versions are rendered based on the number of instances. The different versions can be described as:

1. Normalized data points with the original 36 features (Dataset 1).
2. Normalized data points with the original 36 features, and with 9218 instances of each activity (Dataset 2).
3. Normalized data points with the 21 most significant features obtained by Extra-tree classifier (Dataset 3).
4. Normalized data points with the 21 most significant features, obtained by Extra-tree classifier, and with 9218 instances of each activity (Dataset 4).

After injecting adversarial attack with FGSM and BIM, we have demonstrated the effects of noise injection in such data set with respect to their training accuracies. Let, D = $x_1, x_2, x_3, \ldots, x_n$ be the instances of one of the features of the data set. We have such instances for other features from $D_1, D_2, D_3, \ldots, D_n$, where n = 36 in the actual data set obtained from UCI repository mentioned in Table 1 and n = 21 on the data set modified by feature engineering and feature reduction through Extra-tree classifier. The feature reduction through most significant feature selection was a part of our previous work [46]. In this work we have used the same version of the data set to make the comparative analysis. The number of instances in each of the feature columns is not even, which we found out to have a detrimental impact on prediction score per activity. We have decided to take 9218 random sample instances from each activity class. Consequently, each of the vectors $D_1, D_2, D_3, \ldots, D_{36}$ and $D_1, D_2, D_3, \ldots, D_{21}$ in our training data set had $x_1, x_2, x_3, \ldots, x_{9218}$ sample data points. The data points from each instance was normalized following the equation presented in Eq. 1.

$$x_i^j norm = \frac{x_i^j - \min(x_i)}{\max(x_i) - \min(x_i)}, \tag{1}$$

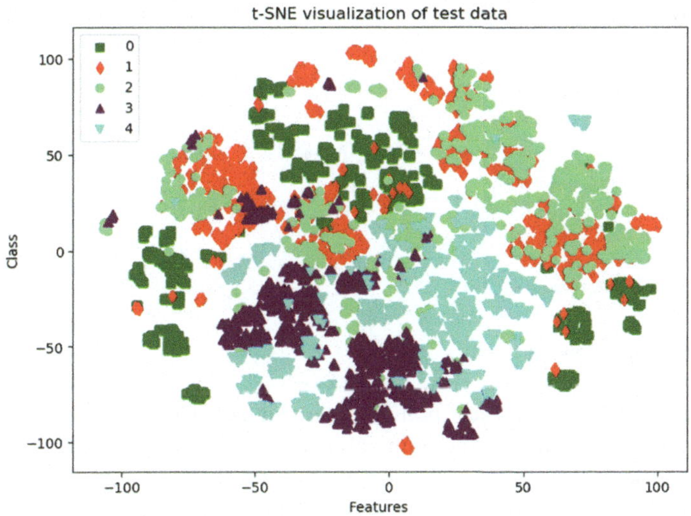

Fig. 2 Representation of the data set obtained by extra-tree classifier

where x_i^j is the ith feature of the data set and jth instance of the feature vector.

Figure 2 represents a visualization of the normalized version of feature-engineered data set, through t-Distributed Stochastic Neighbor Embedding (t-SNE) dimensionality reduction method [30]. The activity instances are dominantly overlapping after performing feature engineering and data preprocessing steps.

4 Methodology

The work discussed in this paper can be divided into two steps: firstly, training the data sets with different deep learning classifiers to achieve significant prediction accuracy and secondly, non-targeted adversarial attack. In this paper, we have demonstrated classification performance of the several deep learning models on both feature engineered and originally obtained data sets. The subsequent resiliency testing with adversarial attack by FGSM and BIM [19] has been implemented on the pre-trained model. The data sets are normalized before feeding into classifier models according to Eq. 1.

Among four different deep learning based approaches (Keras-LSTM, CNN, CNN-LSTM, RNN-LSTM, ResNet), the research team has come up with the conclusion that ResNet achieves the best result as presented in Table 3. As suggested in [19] time series data have not been largely explored to test the robustness of the classifier models concerning the risk of adversarial attacks. The highest classification accuracy on the data set of this study has been achieved by ResNet (99.9%). Few other relevant research works on benchmark data set by ResNet has also found to achieve excellent classification accuracy scores. This fact also motivated us to ana-

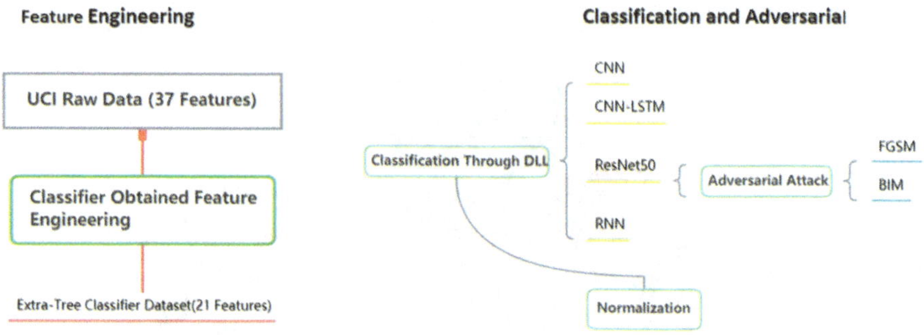

Fig. 3 Methodology flowchart

lyze the vulnerability of ResNet upon adversarial attack. Consequently, non-targeted adversarial attack has proven the vulnerability of such excellent performing model and reduced the accuracy score drastically. Several implementation of adversarial attack through comprehensive change in magnitude of perturbation have produced number of cases to analyze the robustness of the pretrained model. Figure 3 presents the Methodology workflow of this paper.

4.1 Data Set Preprocessing (Normalization)

The four versions of data set in this study consisted of time-series data collected from multiple sensors. In addition, there are columns derived from weighted counts of dependent variables as discussed in Sect. 3. Selection of five specific activities for this study from a total of 12 activities and subsequent feature engineering steps have significantly changed the original sequential data pattern. The data sets at this point consisted of input values with non-identical scales. Consequently, the data sets have been normalized feature-wise to prevent overfitting. Feature scaling of min-max normalization has been adopted to rescale the data sets [47]. The data sets are split into train, validation and test set according to the following ratio: 70, 15 and 15%.

4.2 Accuracy Improvement by Deep Neural Network Based Classifier

Residual Network (ResNet) introduced in 2015 [17] is suited to enable hundreds of convolutional layers, and is able to overcome the "vanishing gradient" problem of shallower neural networks through "identity shortcut connections". Compared to other Convolutional Neural Networks (CNN), Multi-channel CNN (MC-CNN) and Inception networks, ResNet has been proven to obtain better performance on time-series data [20]. ResNet for time-series data has been validated on 44 data sets from

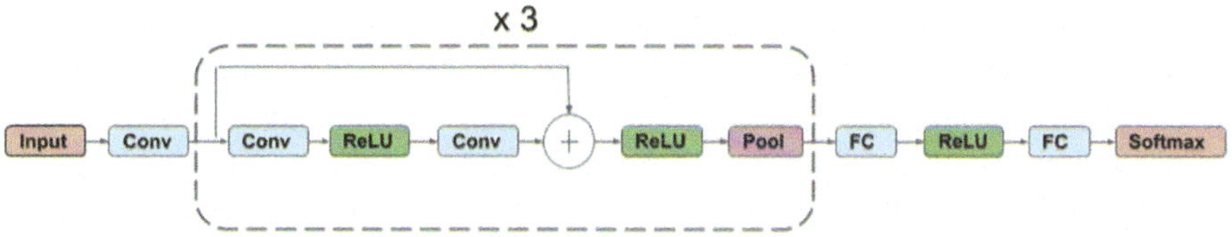

Fig. 4 ResNet architecture adopted in this study

UCR archive in [51] and it was the first recorded ResNet performance on time-series classification. In a recent empirical study on the UCR archive data set [20], ResNet achieved state-of-the-art performance for the time-series classification. The UCR archive data set is idealistic and considerably smaller in size, whereas the data set followed in this research is completely collected from real-life setup and required heavy denoising through feature engineering techniques for training the model.

The ResNet architecture adopted in this study is presented in Fig. 4 and is following the architecture introduced in the study by Wang et al. [51]. The network accepts the multivariate time-series data set and outputs a probability distribution over the predefined number of classes in the data set.

The first nine layers of the used ResNet architecture are 1D convolutional layers, each layer is followed by a batch normalization operation [18]. The first, second and third convolutions of each convolutional block consist of filters of length 8, 5 and 3 respectively. Number of filters in first, second and third convolutional blocks are 64, 128 and 128 respectively. The stride is set to 1 for all convolutional layers. The activation function employed in the first nine convolutional layers is Rectified Linear Unit (ReLU) [23].

Each convolutional layer derives the dimensions of the input multivariate time series from the feature maps employed at each layer. The large number of parameters in the deep network structure is substantially minimized through averaging the input data over the time dimension at the Global Average Pooling (GAP) operation at the tenth layer of the network. The learned features are evaluated by class activation map enabled at this layer. The output from GAP layer is fed to the softmax classification layer. The number of neurons at softmax layer is equal to the predefined number of classes in the data set. The network is trained for 100 epochs using Adam optimizer, and the cross-categorical entropy loss function is minimized.

4.3 Adversarial Attack on Pretrained Model

The open-source Cleverhans software library offers benchmark standardized implementations of adversarial examples and training construction techniques. To ensure a comparable result of adversarial training on the data set adopted in this study, Cleverhans API has been utilized for the execution of two adversarial training constructions: Fast Gradient Sign Method (FGSM) and Basic Iterative Method (BIM). Cleverhans is based on Tensorflow [1], and the interface allows it to accept pretrained models of any framework or models trained without specific abstraction. The core functionality of Cleverhans API is divided between two modules, namely attacks and model. The attacks module contains implementation of several adversarial attacks. The module defines a simple interface to execute adversarial attack on the pretrained models.

From the experimentations run by this study in order to obtain adversarial attack results on pre-trained models, the standard interface maintenance by the classification models became a crucial parameter for successful adversarial training. The attack module in Cleverhans accepts the input model and returns corresponding perturbed adversarial training output. The attack method follows two steps: FGSM and BIM are initialized with a wrapper version of the pre-trained model. Then the two methods independently produce the perturbed version of the input data set and the prediction of pretrained model on this adversary data set is recorded to evaluate adversarial accuracy. FGSM produces the perturbed version of the input data set through using ε (hyperparameter of perturbation), mentioned in Eq. 2. In case of BIM, it accepts ε, α (parameter of noise addition in small amount per iteration), number of iterations as its hyperparameters and consequently produces BIM version of adversary data set. In the following section, the two adopted adversarial algorithms (FGSM and BIM) are discussed in terms of mathematical interpretation.

4.3.1 Fast Gradient Sign Method (FGSM)

The Fast Gradient Sign Method (FGSM) was first introduced in [16] by Goodfellow et al. The algorithm succeeded to produce adversarial images which fooled GoogleNet model [2]. The scheme follows the target of training a model f around the locality of the target training point x, where the adversary is exerted on. The cost function is linearized in the process. The resulting adversarial output $\overrightarrow{x}^*$, originated from the training point x is presented in Eq. 2 as follows:

$$\overrightarrow{x}^* \leftarrow x + \varepsilon \cdot \nabla_{\overrightarrow{x}} J(f, \theta, \overrightarrow{x}) \tag{2}$$

where ε is the hyperparameter of magnitude of the perturbation enforced in adversarial attack. The ε value is carefully curated to adjust with any pretrained model interface in an open interval of values. The optimum ε value is significant for an expected execution of adversarial training. Depending on the nature of the data set and pretrained model, epsilon value significantly differs to produce optimum perturbation in the classification. Epsilon value adopted in this study ranges from 0.01 to as large as 10. The attack follows on a one-step gradient update at each time step along the updated direction of the gradient's sign. The perturbation process is computed through Eq. 3:

$$\eta = \varepsilon \cdot sign(\nabla_x J(X, \hat{Y})) \tag{3}$$

4.3.2 Basic Iterative Method

The Basic iterative method (BIM) extends the FGSM by repeated application with smaller step size. BIM was first proposed by Kurakini et al. [24]. The algorithm follows the intuitive clipping and restoring approach of Residual Network. It clips the value of intermediate results after completion of each step and ensures the adversary is in the ε-neighborhood of the target training point X. The repeated and significantly smaller perturbation in an iterative manner showcases intricate misclassification rate. The hyperparameters of BIM are ε (maximum perturbation), α (the per step small perturbation and number of iterations I).

5 Performance Evaluation

The working pipeline of the present study aims to implement a non-targeted adversarial attack on a pre-trained deep learning model with significantly better performance score of classification. The deep learning neural networks adopted for classification purpose demonstrated varied performance depending on the behavior of the input data set and different hyper-parameters of the networks. Depending on the accuracy score and time required to train on the data set, ResNet is chosen as the best performing model. The next step of the study is forwarded based on the pretrained model by ResNet. Non-targeted adversarial attack is later enacted on two chosen schemes of adversarial attacks: FGSM and BIM on the previously pre-trained models. The performance of the adversarial approaches and the resiliency of the state-of-the-art classification model revealed significant point of concerns in terms of the magnitude of perturbation added in the adversarial attack and performance of the model upon adversarial attack situation. This section will discuss the above mentioned performances in details.

5.1 Accuracy

Time-series forecasting with deep learning network has drawn attention of researchers for quite a-few years due to its applicability in various real-world aspects (financial technology, weather forecasting, agricultural analysis and many more). The concern with time-series data set remains rooted in the fine application of data-mining before running a classification model to achieve desired level of accuracy. Time-series data set are noisy in nature, with lots of missing values and seemingly unexpected pattern of data. Such randomness is occurred due to the mode of data-collection scheme. The time series data is mainly collected from real-life configuration setups and the users of such setup are always in their natural state. This close-to-life pattern of data-collection makes the data-mining step crucial to produce a data set on which machine-predictability can operate to reach expected output. Feature-engineering also ensures lesser requirement of training time by the model to train with less configuration dependency of the machine.

Through feature-engineering of the collected data set, four versions of compact data set are produced. While this step achieved the targets of considerable accuracy score by light-weight baseline classifiers, lesser time and configuration dependency, the intricate feature engineering also incurred loss to the original time-series pattern of the data set. In the previous work executed by our research team, the Multi-Layer Perceptron model achieved the highest classification accuracy. It was trained on the feature set obtained by Extra-tree classifier-based feature engineering and significant feature selection technique [46]. In the current work, the data points are normalized to ensure smooth acceptance by the models [3, 43]. We have run exploratory analysis on five different benchmark deep learning models, to comprehend the limitations of the data set on achieving significant classification score. The models are: Keras-LSTM, CNN, CNN-LSTM, RNN-LSTM and ResNet. ResNet has achieved the highest accuracy score compared to the other models. The architecture of ResNet is also not heavily dependent upon any specific data pattern. Hence, ResNet has been chosen as the selected model to test the resiliency against adversarial attack. Detailed discussion on the exploratory analysis of four other classification models are documented in Sect. 6.2.

The original setup of the data set collection from real-life resident users influenced in the number of instances for each activities in the data set. Some activities were lacking of comparable number of instances compared to some other activities, namely "Talking over Phone", "Watching TV" in comparison to "Eating" and "Cooking". The reason is the use of motion and IR sensors in the data-collection setup. The unevenly distributed data instances on the type of activities presents the domain-specific knowledge of activity pattern and how the related data can be better retrieved keeping the limitations in mind. Human activities are diverse and personalized in nature. Some of the activities like walking, running are better observed by motion sensors, similarly standing and sitting can be easily retrieved by IR sensors. While activities like talking over phone or watching TVs can be tricky to observe by static placement of sensors. In order to extrapolate the difference in the final classification score with the skewed nature of the data in consideration, the research team decided to run classification on four versions of the data set mentioned in Sect. 3. Table 3 encompasses the training performance by ResNet on the four data sets of this study.

In Fig. 5, the performance of ResNet on Dataset 3 and Dataset 4 are demonstrated. The trend of loss graph shows higher accuracy score for Dataset 3 than Dataset 4 by 0.1%. After normalizing the data points, higher accuracy scores are achieved in our current findings compared to the accuracies obtained from our previous work with the non-normalized and original data set. Loss of accuracy score for equally-distributed data set is coupled with the loss of time-series pattern fact. When the number of instances are arbitrarily fixed at a equal range, the original pattern of the data set is even more distorted. Hence, the skewed data set is dominating over the equally-distributed version. Feature engineering has increased the accuracy score for both skewed and equally-distributed data sets. On the other hand, equal data-distribution results in a faster execution of training over the data sets (by 22–41%).

Faster execution of neural network training is another significant output of careful feature engineering in terms of an optimum choice of machine learning model to maintain a significant accuracy score.

The skewed data instances on the type of activities presents the domain-specific knowledge of activity pattern and how the related data can be better retrieved keeping the limitations in mind. Human activities are diverse and personalized in nature. Keeping the skewed pattern of the data set in mind, the research team decided to run classification on four versions of the data sets mentioned in Sect. 3, in order to extrapolate the difference in the final classification score. In Fig. 5, the performance of ResNet on Dataset 3 and Dataset 4 are demonstrated.

The trend of loss performance shows higher accuracy score for Dataset 3 than Dataset 4 by 0.1%. The feature-engineered data sets achieved higher accuracy than original source data sets (Dataset 1 and Dataset 2). Loss of accuracy score for equally-distributed data sets is coupled with the loss of time-series pattern fact. When the number of instances are arbitrarily fixed at a equal range, the original pattern of the data set is even more distorted. Hence, the skewed data set is dominating over the equally-distributed version. Feature engineering has increased the accuracy score for both skewed and equally-distributed data sets. On the other hand, equal data-distribution results in a faster execution of training over the data sets (by 22–41%). Faster execution of neural network training is another significant output of careful feature engineering in terms of an optimum choice of machine learning model to maintain a significant accuracy score.

Figure 6 illustrates the predictive performance of the model trained by ResNet on Dataset 4. The total number of instances per activities is 9218 in the Dataset 4 before splitting. All the five selected activities are accurately predicted with significant accuracy. The classification performance of ResNet on the four test data sets shows an overfitting pattern. ResNet was built in order to classify image data sets with larger number of features [51]. Comparing to the benchmark image data sets where ResNet

Table 3 Performance evaluation of ResNet on the study data sets

Parameters	Dataset 1	Dataset 2	Dataset 3	Dataset 4
Accuracy (%)	99.889	99.783	99.913	99.900
Loss (%)	0.003716	0.007881	0.002843	0.004032
Test duration (min)	7.924612	6.15523	8.564008	5.02549

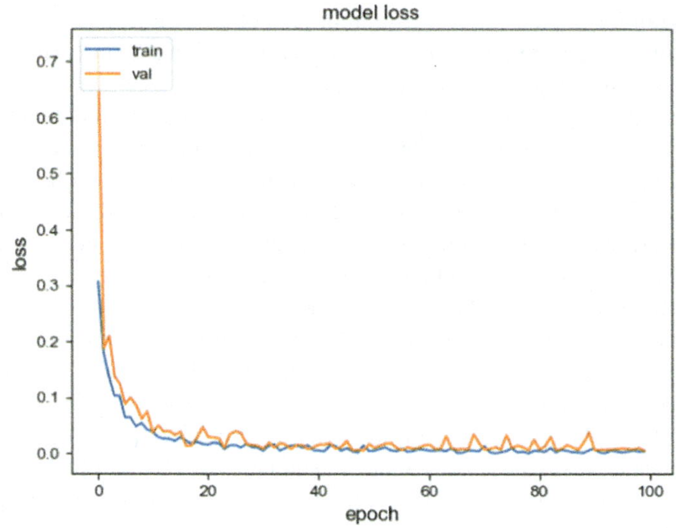

(a) Dataset 4

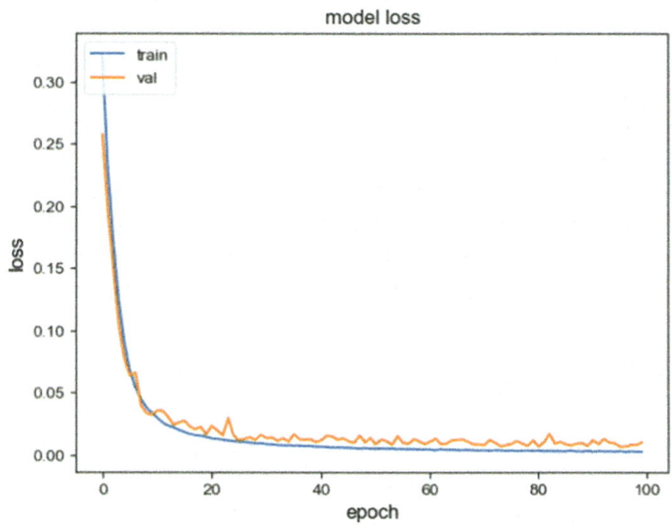

(b) Dataset 3

Fig. 5 Training loss over 100 epochs for ResNet trained on the data set obtained through feature engineering by extra-tree classifier

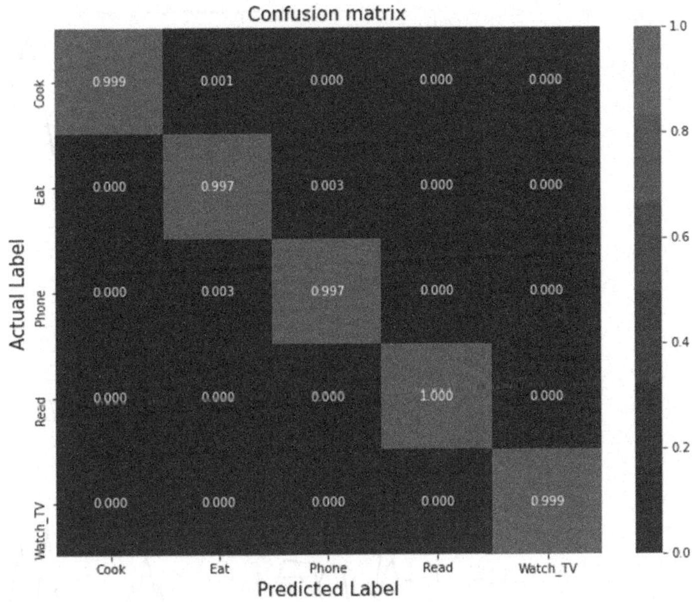

Fig. 6 Classification accuracy of ResNet

has achieved state-of-the-art performance, the smaller data set with less variance in this study has resulted in an overfitted trained model. Studies of similarly feature engineered time-series data classification using alike structure of ResNet validates the achievement of higher accuracy [29, 49].

5.2 *Adversarial Attack*

In this study, we have opted to implement non-targeted adversarial attack on pre-trained ResNet model. The benchmark adversarial classifier by Cleverhans API [37] has been utilized as the code base. Two well known adversarial schemes, namely Fast Gradient Sign Method (FGSM) and Basic Iterative Method (BIM) have been chosen to attack on Dataset 2 and Dataset 4. The perturbation is added to the input instances and the corresponding labels in test data set is removed beforehand, similar to the studies in [19, 54].

Applying the FGSM and BIM implementation on the data set obtained in this research work, the accuracies of the model ResNet decreased to different values depending on the value of the hyperparameter ε presented in Eqs. 2 and 3. The difference between original instances and perturbed instances is captured as the noise injected in adversarial attack, as depicted in Fig. 7. Augmented value of ε leads to distinct differences between original and perturbed instances. The x-axis in the

Fig. 7 Visualization of added perturbation at $\varepsilon = 0.075$ by FGSM

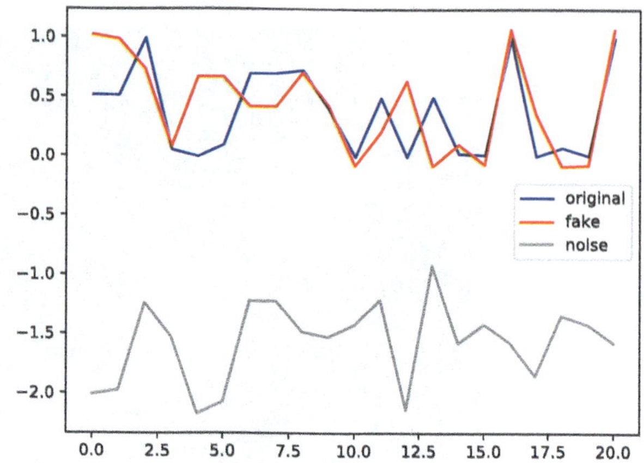

(a) FGSM On Dataset 4

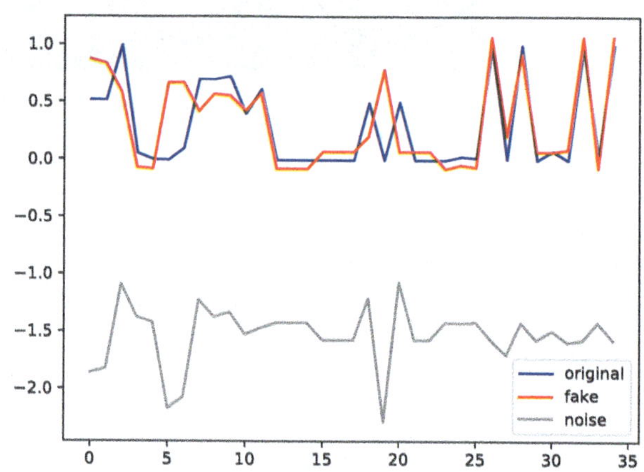

(b) FGSM On Dataset 2

figure represents the addition of perturbation on the number of feature columns in the concerned data set, while y-axis holds the perturbation amount.

5.2.1 Analysis of Adversarial Attack by FGSM and BIM

Figure 8 demonstrates performance of FGSM and BIM through the change of epsilon magnitude over two equally distributed data sets: Dataset 2 and Dataset 4. The wrong classification rates produced by BIM than FGSM at similar ε value are more acute and steep in nature. The injection of perturbation is random in nature. The algorithmic difference of the two methods creates dissimilarity in the perturbed classification.

It is noticeable in Fig. 8 that within a short range of $\varepsilon = [0.01, 0.025, 0.05, 0.075]$, newly generated accuracy scores by both attack schemes on both data sets have

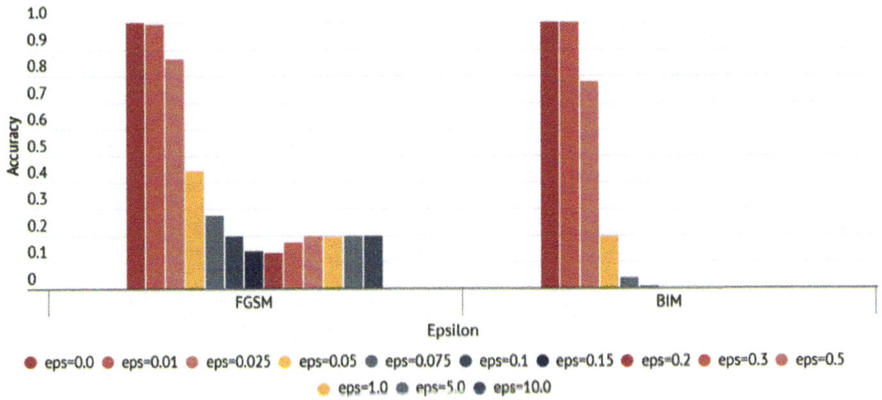

(a) Dataset 4

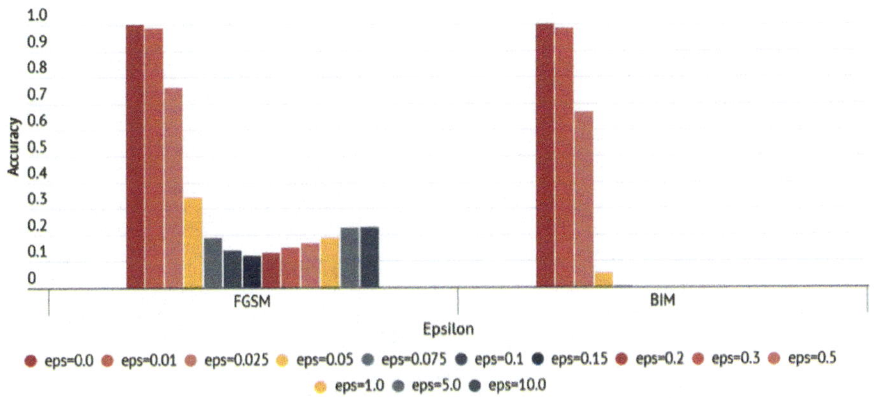

(b) Dataset 2

Fig. 8 Comparative performance evaluation FGSM and BIM adversarial attack on pretrained ResNet model trained on equally distributed normalized data sets

reduced significantly. At Fig. 8a and b, the pattern of accuracy loss is similar for FGSM and BIM. The rate of misclassification on Dataset 2 is steeper than Dataset 4. At $\varepsilon = 0.05$, FGSM reduces the accuracy score to 0.44 and 0.34 on Dataset 4 and Dataset 2 respectively as shown in Fig. 8. In Fig. 8b BIM drags down the accuracy score to 0.19 for the model trained on Dataset 4 and to 0.0026 for the model trained on Dataset 2 for the value of $\varepsilon = 0.075$.

In our current work, the non-targeted adversarial attack on the state-of-the-art classifier ResNet has reduced its score from 99.9% obtained accuracy to 19% as an adversarial outcome. This clearly states the vulnerability of ResNet classification performance over adversarial attack, which can turn an excellent model accuracy to a substantially lower value.

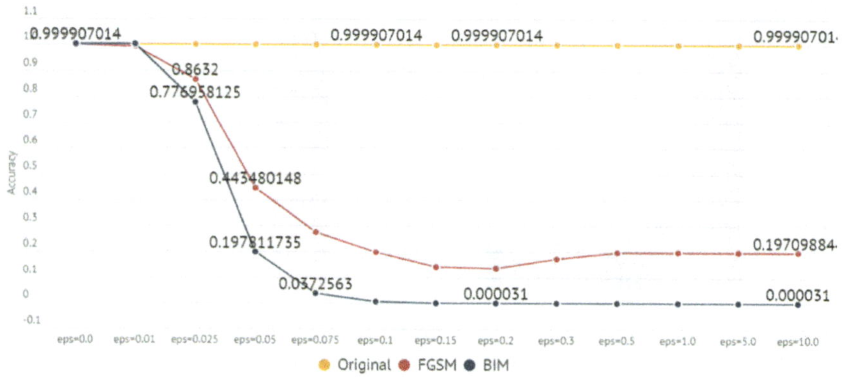

(a) Dataset 4

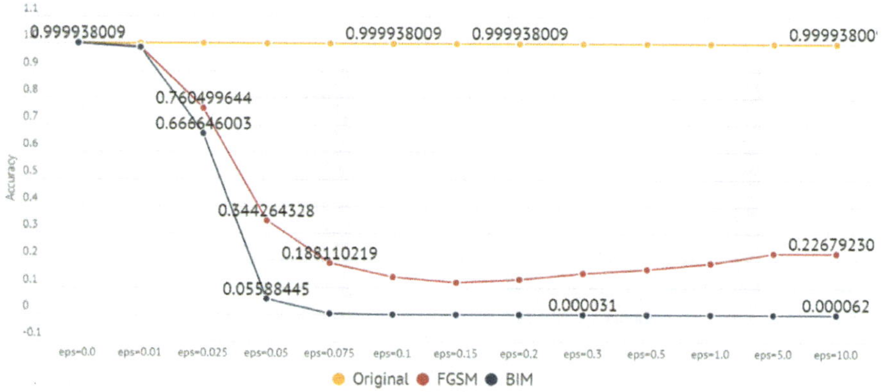

(b) Dataset 2.

Fig. 9 Comparative performance evaluation of adversarial attack with FGSM and BIM on ResNet

5.2.2 Resiliency of ResNet on Adversarial Attack

Both FGSM and BIM have significantly reduced the original accuracy of pretrained ResNet. Figure 9 shows the accuracy obtained from ResNet and accuracies obtained after FGSM and BIM attack. BIM has managed to fool ResNet network by multiple factors than FGSM. Increasing magnitude of perturbation augmented the strength of adversarial attack further and accuracy score also reduced consequently. At $\varepsilon = 0.075$, FGSM reduced the original accuracy by 72%, while BIM reduces the accuracy by 96%.

BIM has almost nullified the classification accuracy of ResNet at this value, depicting the subtle yet effective perturbation to input data instances. At $\varepsilon = 10$, FGSM accuracy reduces by 80.28%, a slow degradation from previous FGSM score at $\varepsilon = 0.75$. BIM accuracy score at $\varepsilon = 10$ is nearly vanishing, which demonstrates that the accuracy of ResNet is degraded to the situation of no training at all. The ResNet

Table 4 Wilcoxon signed-rank test between accuracy obtained by FGSM and BIM attacks

Data set	z	p-value	Two-tail	Hypothesis
Dataset 4 (%)	1.725822	0.04219	0.084379	No difference
Dataset 2 (%)	3.059412	0.001109	0.002218	BIM is better

network is trained on a large data set of approximately 4 millions of instances, and hence has easily become perceptible to the precise perturbations applied at large number of data points. Thereby, it reduces the accuracy of the network rapidly. BIM injects perturbation through multiple iterations, while FGSM operates on real-time generation of adversarial perturbation. The repeated addition of perturbation of small amount α by BIM at each iteration leverages the noise addition, as mentioned in Sect. 4.3.2. FGSM adds a constant ε magnitude of perturbation to the input data at run time. BIM outperformed FGSM in carrying out adversarial attack on the performance of the ResNet network.

Statistical significance testing on performance of FGSM and BIM on Dataset 2 and Dataset 4 revealed the better performing attack scheme through Wilcoxon signed-rank test [42, 52], as enlisted in Table 4. The considered value of alpha is $\alpha = 0.05$. There is no statistical difference recorded when the same attack (FGSM or BIM) method is applied on either Dataset 2 or Dataset 4. Interestingly, when the performance of FGSM and BIM on individual data set are tested, BIM performs significantly better on Dataset 2 than FGSM, at the same time on Dataset 4, there is no performance difference between the adversarial attacks.

6 Discussion

6.1 Importance of Feature Engineering

Feature engineering is a widely accepted technique for improving performance of machine learning models. Especially in the area of human activity recognition, the data are mostly noisy and environmental effects often occupy data extraction procedures. As a result, effective model training on raw/unprocessed data set cannot be trusted to achieve considerable classification. Feature engineering in such domain work requires data cleaning, important feature selection/redundant feature exclusion, uniform data pipelining through statistical analysis and so on. Among the abundance of available features, data dimensionality and other factors; feature engineering is a very effective way that can actually help more to understand the nature of the data for any machine learning project. Meaningful insights based on satisfactory statistical analysis can be helpful to choose the primary models to work on and also for hyperparameters selection. The original data set acquired for our work was no exception. Models training performance upon the baseline classifiers with the original unprocessed data set was very poor and in our previous work we had presented the usage of tree-based feature engineering techniques: random forest classifier [6], extra-tree classifier [15] to improve classification performance. We have demonstrated improved model performance after using the feature engineering techniques and also showed the significant set of features in the data set. Such feature engineering techniques also help avoiding unnecessary computational complexity and in the domain of human activity recognition from IoT/sensor-based data, these techniques should be useful to focus on training with the values obtained from important sensors and features selection for efficient classification.

6.2 *Time-Series Data Classification*

Time series-based data requires seasonality. In many real-life scenario based research; like stock market prediction, anomaly detection, medical studies, weather forecasting-based researches machine learning models are continuously being developed for fine-tuned classification or forecasting. The principal goal of time-series based data classification is to find the seasonality, behavior or order of the data through extensive statistical analysis. One of the most pivotal goals of our work was to find the presence of seasonality in the data. The loss of time-series pattern caused in lower classification accuracy obtained by classic Keras-LSTM [11] when trained on Dataset 1 and Dataset 3. On the other hand, Keras-LSTM obtained considerable accuracy on Dataset 1, as shown in Fig. 10a. The LSTM structure adopted following hyper parameters: number of neurons = 100 and dropout rate = 0.2.

While independent implementation of LSTM failed to achieve desired accuracy level, CNN demonstrated better performance on the Dataset 3 (85.563%). The CNN architecture is constituted of two convolution layers followed by one dense layer, illustrated in Fig. 11. The learning of smoothing parameters through convolutional and pooling layers in the first two layers of CNN leverages classification accuracy. The time-dependency of LSTM is limited in CNN learning architecture which provides better classification performance on the data set.

CNN-inspired LSTM also achieved a better accuracy score 95.248%, showing the clear significance of optimum hyperparameters for reaching accuracy target. We have also implemented classification through RNN-LSTM model [41]. RNN-LSTM is inspired by LSTM and it's architecture has two LSTM layers of 256 nodes and 128 nodes respectively, followed by an output dense layer. While accuracy score of RNN-LSTM is quite significant, the architecture of the model searches for time-series data comprehensively in a non-linear dynamical system, which makes the network dependent on temporal relationship between data instances through-out the complete time paradigm of the data set. This dependency of each instance on the former instance based on time stamp is a overhead given that CNN and CNN based deep learning networks can achieve similar accuracy without time dependency.

Instead of depending on consecutive pair of instances, CNN assumes time-dependency on previous N instances. Here, N is the hyperparameter used to determine the spatial extent of connection between each neuron to a specific local region of the

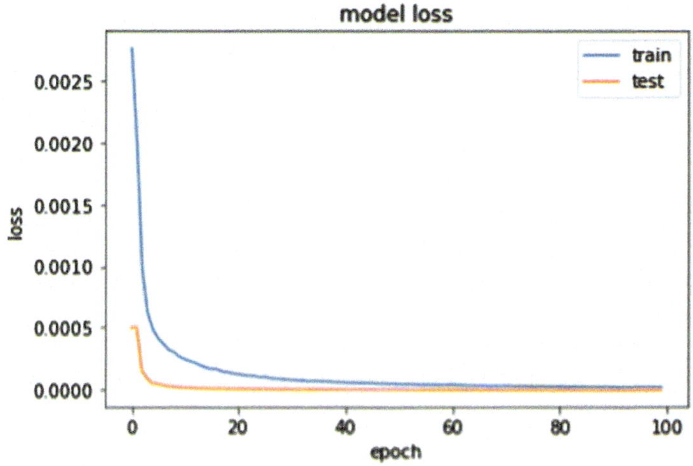

(a) Dataset 1

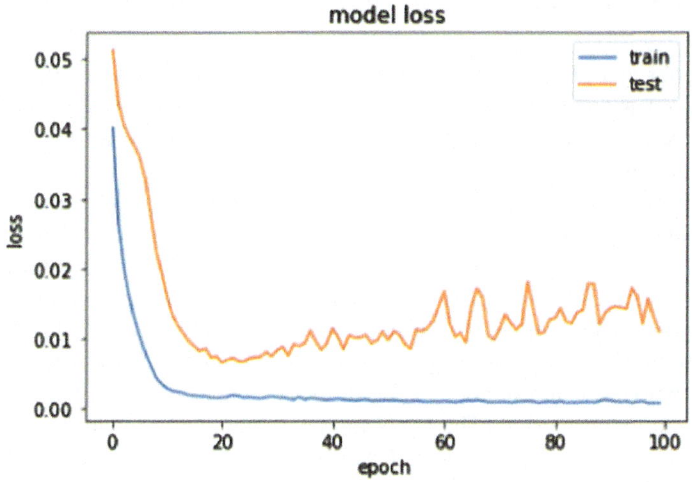

(b) Dataset 3

Fig. 10 Training performance of Keras-LSTM on Dataset 1 and Dataset 3

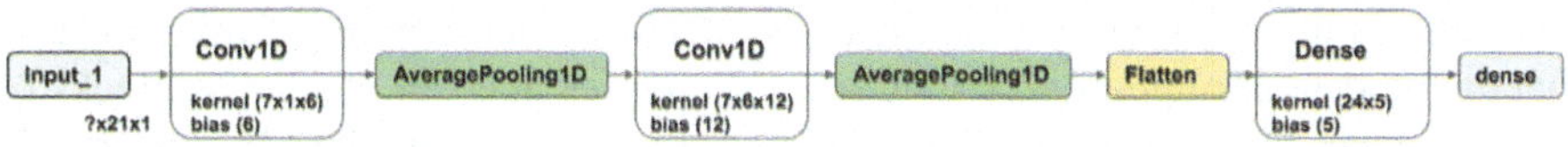

Fig. 11 CNN architecture

Table 5 Performance evaluation of CNN, CNN-LSTM, RNN-LSTM on the study Dataset 4

Parameters	CNN	CNN-LSTM	RNN-LSTM
Accuracy (%)	85.563	95.248	99.500
Value loss	0.043331	0.015	0.002

input volume. That is named as filter size or receptive field. Each filter learns a rule to be applied on random portions of the complete data set. CNN inspired ResNet [17] performed to achieve the highest accuracy score 99.9%. The architecture of ResNet is discussed in Sect. 4.2.

Table 5 displays the performance evaluation of the different deep learning models implemented to comprehend the performance on classification accuracy in the perspective of this study.

6.3 Dependency of Model Architecture on Nature of Data

To find better and effective deep learning models, we have trained the study data sets through various neural network architectures like: CNN, CNN-LSTM, RNN, ResNet. CNN requires huge data set but our HAR dataset is not that huge. As a result, the testing result didn't serve our expectations.

As RNN is used for sequential data, we have tried it to verify seasonality nature for our data set. Such kind of deep neural network is very prone to face vanishing gradient and gradient exploding issues. It also shows the lack of stability during training time when ReLU is applied as an activation function. Furthermore, compared to any other plain network, RNN takes significant time to be trained which makes it a time consuming process. While RNN is effective to maintain information over time, LSTM networks contain memory cells in every unit to cope up with information holding for a longer period. The CNN model contains more feature compatibility than RNN. With the advantages of CNN and LSTM, a combined model named CNN-LSTM has been trained.

When we have used ResNet as our training model, issues like vanishing gradient are resolved. The CNN works well in classifying the samples from the data set with an accuracy of 85.563%. It is a simple CNN with only one convolution filter layer and one dense layer. The convolution layer in combination with utilizing LSTM layer allows the creation of a hybrid CNN-LSTM model. Benefit of this model is that the model can support very long input sequences that can be read as blocks or sub-sequences by the CNN model, then pieced together by the LSTM model. The CNN-LSTM gives a slightly higher accuracy than using standard CNN models with no LSTM layers. This is encouraging as our CNN-LSTM uses far less parameters and layers compared to ResNet but the ResNet model is able to achieve 99.9.87% accuracy which surpasses all other models classification scores.

6.4 Resiliency of Deep Learning Model When Confronted by Non-targeted Adversarial Attack

The ResNet adopted in this study has asserted leading training performance in comparison to other benchmark deep learning models due to the specialty of architectural learning setting and hyperparameters. ResNet is an acclaimed deep learning model most known for its significant performance on computer vision based researches (object detection, image segmentation) [22, 33]. Exploration of ResNet performance on time-series data has demonstrated excellent prediction scores similar to our work. However, the encounter of ResNet with adversarial attack is still a relatively growing research field.

The findings from this paper demonstrated the shaky performance of pre-trained ResNet model when the data set is perturbed with FGSM and BIM methods through non-targeted adversarial attack. The accuracy has dropped from 99.9% to as low as 19% and 0.0031% by FGSM and BIM respectively at magnitude of perturbation value of $\varepsilon = 10$ on Dataset 4. ResNet has achieved slightly increased classification accuracy on Dataset 2 than Dataset 4 (by 0.003%), whereas in adversarial attack the lowest accuracy is obtained on Dataset 2 by BIM attack. At Lower values of $\varepsilon = 0.01, 0.025$, the accuracy decreases on both Dataset 2 and Dataset 4. The steep change occurred at $\varepsilon = 0.05$ on both the data set. ResNet has maintained prediction accuracy score per activity at around 30–50% at this value of ε, as presented in Fig. 12.

The drastic reduction of ResNet in face of non-targeted adversarial attack has since been proved to be a concern for future research and robust build of model architecture. In similar study by Ismail et al. [19] of adversarial attack on deep neural networks for 85 time-series data set from UCR archive, accuracy of ResNet has decreased by 43.2% and 56.89% by FGSM and BIM respectively. Some data set proved to be harder to fool for the synthetic data nature and small size of the data set. In a multivariate time-series regression based adversary example, FGSM and BIM has considerably decreased the accuracies by CNN, LSTM and GRU network trained on publicly available Google stock and household power consumption data sets [34]. Several experiments conducted by distinct research groups like Google Brain, Facebook AI run on deep learning models trained on time-series data set have consistently demonstrated the weak resiliency of deep learning models [7, 32]. The deep learning models are built to learn from specific input data pattern, thereby the adversary is able to penetrate into model strength just by perturbing the data set. The dependency of deep learning model on the size and nature of data influences the training performance, as well as the extended security threat posed by adversarial attack [4, 8]. The nature of the data set is critical in influencing the extent of adversarial domination over the deep learning model.

The perturbed data set after adversarial attack is able to fit into another deep learning architecture and reduce the training performance through black box attack [10, 35]. Hence, the security concern for adversarial attack on deployed deep learning model is increased to a greater extent. Two prevalent measures are currently being

Fig. 12 Prediction accuracy per activity of Dataset 4 following adversarial attack by BIM and FGSM in (**a**) and (**b**)

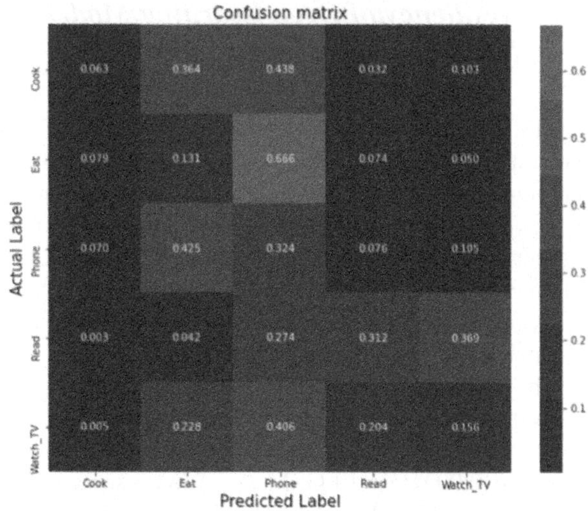

(a) BIM

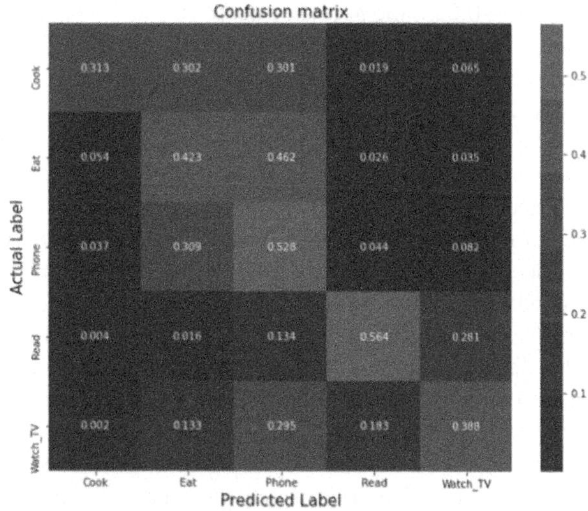

(b) FGSM

enacted to reduce risks of adversarial attack - reactive and proactive. Reactive measure pins down the noise injected instances and proactive measure increases the robustness of the model through adversarial training [40, 50]. These measures are still in growing phase and not established as proven steps to counter adversarial attack. The sensitivity of HAR techniques being intricate in daily lives and directly affecting individual lifestyle, usage of technology etc. sheds light on the necessity of building a secured learning model architecture and ensuring safe deployment against possible attacks. The research community can focus on improving upon current state-of-the-art deep learning models to deviate from dependency of model learning from specific input data fed into the system.

7 Future Work

There have been numerous significant developments made in the field of human activity recognition. But still this domain is entirely data-dependent. The number of available good public data sets are very few. Although our work in this paper has been based on the public data set published by the research team of Washington State University and made available through UCI Machine Learning repository [14]. The data set therefore is believed to be very properly managed. But for extensive research work we might need more general/robust data obtained from an available source. As raw data obtained from sensors do not help much to make a robust machine learning model, thus, the data collection process for human activity recognition-based research is pretty delicate. It needs costly-reliable hardware, time-wise monitoring, reliable network connection, and various data cleaning-noise reduction steps. The installed sensors need to be well-performing as well and needs to send data with a predefined sliding window. Most of the related work in this field is actually performed on available public data sets. Therefore, a good future goal should be set to introduce an easier yet effective data collection and data cleaning process, which could be followed by researchers in this domain to obtain their own data from uncomplicated and less costly processes. The processes should not be hardware dependent but algorithmically well established and robust. This can really help broaden the research domain so that researchers will not be dependent on particular public data sets only but can come up with their own reliable standard data. Acknowledging the need of data our research team built an android application to obtain a prototype data set from android smartphone sensors. We called volunteers to use the app and collect a data set for our research purpose. The application could label the selected activity upon button click, and the data was being dumped under a secured personal server obtained by our research team. The objective of this approach is to observe the training accuracy from the collected data set, without following any noise reduction and other state-of-the-art methodologies in order to compare it with a standard data set like the aforementioned one. Our intended aim is to prove that such noisy data obtained from the smartphone sensors are hard to train and adversarial attack upon such data set can actually be crucial as it can easily confuse any strong classifier even.

8 Conclusion

This paper has focused on extensive research on Human Activity Recognition Techniques and vulnerability testing of the existing benchmark deep learning model in counter with non-targeted adversarial attack. The data set is originally obtained from UCI machine learning repository, a filtered version of data set of five activities from the original source of 12 activities is selected for this study. The necessity of insightful feature engineering is highlighted in this study for feeding a meaningful version of the noisy time-series data for the classification model. We have explored four different deep learning models, assessed the training performance and the inherent structural dependency and advantages of respective models and finally chose ResNet as the stable model. Subsequent non-targeted adversarial attack by Fast Gradient Sign Method (FGSM) and Basic Iterative Method (BIM) on the data set of this study produced perturbed versions. The prediction accuracy obtained by pre-trained ResNet model achieved significantly lower scores. The findings from this study is presented in brief by the following:

- Data preprocessing steps achieved through normalization, per activity number of instance handling and meaningful feature engineering through Extra-Tree Classifier paved the way for smooth model training in a considerably short period of time.
- Various deep learning model exploration led to better understanding of relationship between the input data and the architecture, hyperparameters of the model in concern. Extensive feature-engineering on time-series data leads to distortion of ingrained time-series pattern and thereby, the state-of-the-art classifiers for TSC produced abrupt training performance. The deep learning models built from convolution neural network baseline achieved consistently excelling scores on this particular type of time-series classification. This finding is established by the ResNet training on the data which achieved 99.9% accuracy. The data set is over fitted due to the deep architectural configuration of the model, the classification performance is found to be similar in several studies of time-series classification by ResNet model.
- Non-targeted adversarial attack by two established methods, FGSM and BIM, has proved the frail resiliency of the pre-trained state-of-the-art ResNet model. The nature of the attacks by FGSM and BIM varies due to the differences in hyperparameters of each of the methods. FGSM has reduced the original accuracy score of ResNet by 72% where BIM has decreased it by 96%. The findings from non-targeted adversarial attack, where the adversary methods did not have prior access to the data set or the pre-trained model, has brought the importance of constructing robust deep learning models viable to handling such anticipated attacks in model deployment.

References

1. Abadi, M., Agarwal, A., Barham, P., Brevdo, E., Chen, Z., Citro, C., Corrado, G., Davis, A., Dean, J., Devin, M., Ghemawat, S., Goodfellow, I., Harp, A., Irving, G., Isard, M., Jia, Y., Kaiser, L., Kudlur, M., Levenberg, J., Zheng, X.: Tensorflow: Large-Scale Machine Learning on Heterogeneous Distributed Systems (2015)
2. Akhtar, N., Mian, A.: Threat of adversarial attacks on deep learning in computer vision: a survey. IEEE Access **6**, 14410–14430 (2018)
3. Aladag, C.H., Jin, J., Li, M., Li, M., Jin, L.: Data normalization to accelerate training for linear neural net to predict tropical cyclone tracks. Math. Prob. Eng. **2015** (2015)
4. Athalye, A., Carlini, N., Wagner, D.A.: Obfuscated gradients give a false sense of security: circumventing defenses to adversarial examples. In: ICML, pp. 274–283 (2018). http://proceedings.mlr.press/v80/athalye18a.html
5. Biggio, B., Corona, I., Maiorca, D., Nelson, B., Šrndić, N., Laskov, P., Giacinto, G., Roli, F.: Evasion attacks against machine learning at test time. In: Blockeel, H., Kersting, K., Nijssen, S., Železný, F. (eds.) Machine Learning and Knowledge Discovery in Databases, pp. 387–402. Springer, Berlin, Heidelberg (2013)
6. Breiman, L.: Random forests. Mach. Learn. **45**(1), 5–32 (2001)
7. Carlini, N., Wagner, D.: Adversarial examples are not easily detected: bypassing ten detection methods. In: Proceedings of the 10th ACM Workshop on Artificial Intelligence and Security, AISec '17, pp. 3–14. Association for Computing Machinery, New York, NY, USA (2017)
8. Carlini, N., Wagner, D.: Adversarial examples are not easily detected: bypassing ten detection methods. In: Proceedings of the 10th ACM Workshop on Artificial Intelligence and Security, pp. 3–14 (2017)
9. Carlini, N., Wagner, D.: Towards evaluating the robustness of neural networks. In: 2017 IEEE Symposium on Security and Privacy (SP), pp. 39–57. IEEE (2017)
10. Chen, S., Carlini, N., Wagner, D.: Stateful detection of black-box adversarial attacks. In: Proceedings of the 1st ACM Workshop on Security and Privacy on Artificial Intelligence, SPAI '20, pp. 30–39. Association for Computing Machinery, New York, NY, USA (2020)
11. Chollet, F., et al.: Keras. https://github.com/fchollet/keras (2015)
12. Chung, S., Lim, J., Noh, K.J., Kim, G., Jeong, H.: Sensor data acquisition and multimodal sensor fusion for human activity recognition using deep learning. Sensors **19**(7), 1716 (2019)
13. Cook, D.: Learning setting-generalized activity models for smart spaces. IEEE Intell. Syst. **27**(1), 32–38 (2012). https://doi.org/10.1109/MIS.2010.112
14. Dua, D., Graff, C.: UCI Machine Learning Repository (2017). http://archive.ics.uci.edu/ml
15. Geurts, P., Ernst, D., Wehenkel, L.: Extremely randomized trees. Mach. Learn. **63**(1), 3–42 (2006)
16. Goodfellow, I., Shlens, J., Szegedy, C.: Explaining and harnessing adversarial examples. In: International Conference on Learning Representations (2015). http://arxiv.org/abs/1412.6572
17. He, K., Zhang, X., Ren, S., Sun, J.: Deep residual learning for image recognition. In: 2016 IEEE Conference on Computer Vision and Pattern Recognition (CVPR), pp. 770–778 (2016). https://doi.org/10.1109/CVPR.2016.90
18. Ioffe, S., Szegedy, C.: Batch normalization: accelerating deep network training by reducing internal covariate shift. In: Proceedings of the 32nd International Conference on International Conference on Machine Learning, ICML'15, vol. 37, pp. 448–456. JMLR.org (2015)
19. Ismail Fawaz, H., Forestier, G., Weber, J., Idoumghar, L., Muller, P.A.: Adversarial attacks on deep neural networks for time series classification. In: IEEE International Joint Conference on Neural Networks (2019)
20. Ismail Fawaz, H., Forestier, G., Weber, J., Idoumghar, L., Muller, P.A.: Deep learning for time series classification: a review. Data Min. Knowl. Discov. **33**(4), 917–963 (2019)
21. Karegowda, A.G., Manjunath, A., Jayaram, M.: Comparative study of attribute selection using gain ratio and correlation based feature selection. Int. J. Inf. Technol. Knowl. Manag. **2**(2), 271–277 (2010)

22. Khan, R., Zhang, X., Kumar, R., Opoku Aboagye, E.: Evaluating the performance of ResNet model based on image recognition (2018). https://doi.org/10.1145/3194452.3194461
23. Krizhevsky, A., Sutskever, I., Hinton, G.E.: ImageNet classification with deep convolutional neural networks. Commun. ACM **60**(6), 84–90 (2017). https://doi.org/10.1145/3065386
24. Kurakin, A., Goodfellow, I., Bengio, S.: Adversarial examples in the physical world (2016)
25. Lei, Y., Wu, Z.: Time series classification based on statistical features. EURASIP J. Wirel. Commun. Netw. **2020**, 1–13 (2020)
26. Li, B., Vorobeychik, Y.: Feature cross-substitution in adversarial classification. In: Advances in Neural Information Processing Systems, pp. 2087–2095 (2014)
27. Liu, Y., Chen, X., Liu, C., Song, D.: Delving into transferable adversarial examples and black-box attacks. arXiv preprint arXiv:1611.02770 (2016)
28. Lowd, D., Meek, C.: Adversarial learning. In: Proceedings of the Eleventh ACM SIGKDD International Conference on Knowledge Discovery in Data Mining, pp. 641–647 (2005)
29. Lu, S., Hong, Q., Wang, B., Wang, H.: Efficient ResNet model to predict protein-protein interactions with GPU computing. IEEE Access **8**, 127834–127844 (2020). https://doi.org/10.1109/ACCESS.2020.3005444
30. van der Maaten, L., Hinton, G.: Visualizing data using t-SNE. J. Mach. Learn. Res. **9**(86), 2579–2605 (2008). http://jmlr.org/papers/v9/vandermaaten08a.html
31. Madry, A., Makelov, A., Schmidt, L., Tsipras, D., Vladu, A.: Towards deep learning models resistant to adversarial attacks (2017)
32. Madry, A., Makelov, A., Schmidt, L., Tsipras, D., Vladu, A.: Towards deep learning models resistant to adversarial attacks. arXiv preprint arXiv:1706.06083 (2017)
33. Mahmood, A., Bennamoun, M., An, S., Sohel, F.: Resfeats: residual network based features for image classification (2017). https://doi.org/10.1109/ICIP.2017.8296551
34. Mode, G., Hoque, K.: Adversarial examples in deep learning for multivariate time series regression (2020)
35. Narodytska, N., Kasiviswanathan, S.: Simple black-box adversarial attacks on deep neural networks. In: 2017 IEEE Conference on Computer Vision and Pattern Recognition Workshops (CVPRW), pp. 1310–1318 (2017). https://doi.org/10.1109/CVPRW.2017.172
36. Oguntala, G.A., Abd-Alhameed, R.A., Ali, N.T., Hu, Y.F., Noras, J.M., Eya, N.N., Elfergani, I., Rodriguez, J.: SmartWall: novel RFID-enabled ambient human activity recognition using machine learning for unobtrusive health monitoring. IEEE Access **7**, 68022–68033 (2019)
37. Papernot, N., Goodfellow, I., Sheatsley, R., Feinman, R., McDaniel, P.: cleverhans v1.0.0: an adversarial machine learning library. arXiv preprint arXiv:1610.00768 (2016)
38. Papernot, N., McDaniel, P., Jha, S., Fredrikson, M., Celik, Z.B., Swami, A.: The limitations of deep learning in adversarial settings. In: 2016 IEEE European Symposium on Security and Privacy (EuroS P), pp. 372–387 (2016). https://doi.org/10.1109/EuroSP.2016.36
39. Reyes-Ortiz, J.L., Oneto, L., Samà, A., Parra, X., Anguita, D.: Transition-aware human activity recognition using smartphones. Neurocomputing **171**, 754–767 (2016)
40. Shafahi, A., Najibi, M., Ghiasi, M.A., Xu, Z., Dickerson, J., Studer, C., Davis, L.S., Taylor, G., Goldstein, T.: Adversarial training for free. In: Advances in Neural Information Processing Systems, vol. 32, pp. 3358–3369. Curran Associates, Inc. (2019). https://proceedings.neurips.cc/paper/2019/file/7503cfacd12053d309b6bed5c89de212-Paper.pdf
41. Sherstinsky, A.: Fundamentals of recurrent neural network (RNN) and long short-term memory (LSTM) network. Phys. D: Nonlinear Phenom. **404**, 132306 (2020). https://doi.org/10.1016/j.physd.2019.132306
42. Siegel, S.: Nonparametric Statistics for the Behavioral Sciences. International student edition. McGraw-Hill (1956). https://books.google.com.bd/books?id=ebfRAAAAMAAJ
43. Sola, J., Sevilla, J.: Importance of input data normalization for the application of neural networks to complex industrial problems. IEEE Trans. Nucl. Sci. **44**, 1464–1468 (1997). https://doi.org/10.1109/23.589532
44. Szegedy, C., Zaremba, W., Sutskever, I., Bruna, J., Erhan, D., Goodfellow, I., Fergus, R.: Intriguing properties of neural networks. In: International Conference on Learning Representations (2014). http://arxiv.org/abs/1312.6199

45. Tama, B.A., Rhee, K.H.: A combination of PSO-based feature selection and tree-based classifiers ensemble for intrusion detection systems. In: Advances in Computer Science and Ubiquitous Computing, pp. 489–495. Springer, Singapore (2015)
46. Tasmin, M., Ishtiak, T., Uddin Ruman, S., Ur Rahaman Chowdhury Suhan, A., Shihab Islam, N.M., Jahan, S., Ahmed, S., Shahnawaz Zulminan, M., Raufus Saleheen, A., Rahman, R.M.: Comparative study of classifiers on human activity recognition by different feature engineering techniques. In: 2020 IEEE 10th International Conference on Intelligent Systems (IS), pp. 93–101 (2020). https://doi.org/10.1109/IS48319.2020.9199934
47. Pandas Development Team: pandas-dev/pandas: Pandas (2020). https://doi.org/10.5281/zenodo.3509134
48. Tramèr, F., Kurakin, A., Papernot, N., Boneh, D., McDaniel, P.: Ensemble adversarial training: attacks and defenses (2017)
49. Ukil, A., Bandyopadhyay, S., Pal, A.: Sig-R2ResNet: residual network with signal processing-refined residual mapping, auto-tuned L1-regularization with modified Adam optimizer for time series classification. In: 2020 International Joint Conference on Neural Networks (IJCNN), pp. 1–8 (2020). https://doi.org/10.1109/IJCNN48605.2020.9206827
50. Wang, X., Han, X., Liu, Z., Sun, M., Li, P.: Adversarial training for weakly supervised event detection. In: Proceedings of the 2019 Conference of the North American Chapter of the Association for Computational Linguistics: Human Language Technologies, vol. 1 (Long and Short Papers), pp. 998–1008. Association for Computational Linguistics, Minneapolis, Minnesota (2019). https://doi.org/10.18653/v1/N19-1105
51. Wang, Z., Yan, W., Oates, T.: Time series classification from scratch with deep neural networks: a strong baseline. In: 2017 International Joint Conference on Neural Networks (IJCNN), pp. 1578–1585 (2017). https://doi.org/10.1109/IJCNN.2017.7966039
52. Wilcoxon, F.: Individual comparisons by ranking methods. Biom. Bull. **1**(6), 80–83 (1945). http://www.jstor.org/stable/3001968
53. Yang, J., Nguyen, M.N., San, P.P., Li, X., Krishnaswamy, S.: Deep convolutional neural networks on multichannel time series for human activity recognition. In: IJCAI, vol. 15, pp. 3995–4001. Citeseer (2015)
54. Yuan, X., He, P., Zhu, Q., Li, X.: Adversarial examples: attacks and defenses for deep learning (2018)
55. Zeng, M., Nguyen, L.T., Yu, B., Mengshoel, O.J., Zhu, J., Wu, P., Zhang, J.: Convolutional neural networks for human activity recognition using mobile sensors. In: 6th International Conference on Mobile Computing, Applications and Services, pp. 197–205. IEEE (2014)

Complex Multivalued Hierarchical Logic (HS-Logic)

Vassil Sgurev

Abstract A complex multivalued propositional hierarchical logic is proposed, abbreviated as complex HS-logic. It combines two logics—real L_r with two states and an introduced new imaginary logic L_i with two other states $\{i, \neg i\}$ that are not equivalent to $\{T, F\}$. Through two additional requirements a complex logic is constructed with four unequal states $\{T, F, i, \neg i\}$. It has been proved that within the framework of the complex logic thus defined, the true and false states i and $\neg i$ of the imaginary logic have a greater force than those of the real logic $\neg T$ and F. This means that imaginary logic dominates over L_i-logic. This allows all four states of HS-logic to be sorted by degree of falsity and degree of truth. Rules have been derived that show the dependencies between the four states of complex logic in disjunction, equivalence and implication, conjunction. Some of these dependencies are shown graphically. Based on the obtained rules, four square Kelly tables were compiled—for disjunction, conjunction, implication and equivalence, respectively. Each of the tables is also four-element—for the four states of complex logic. It is a quaternion for each of the logical operations. Based on Modus Ponens and data from the Kelly table for implication, four inference rules were obtained. It is shown that the proposed complex hierarchical HS-logic meets all the requirements of Boolean algebra and lattice and this allows it to be used to solve logical equations in cases where this cannot be done within the classical propositional logic. Another example of possible applications of the proposed HS-logic is given.

Keywords Mathematical logic · Boolean algebra · Conjunction · Disjunction · Imaginary logic

---

V. Sgurev (✉)
Intelligent Systems Department, Institute of Information and Communication Technologies, Bulgarian Academy of Sciences, Acad. Georgi Bonchev Str, Sofia 1113, Bulgaria

V. Sgurev et al. (eds.), *Advances in Intelligent Systems Research and Innovation*, Studies in Systems, Decision and Control 379,
https://doi.org/10.1007/978-3-030-78124-8_7

1 Introduction

The difficulties in solving some specific classes of logical equations using the methods of classical propositional logic L_r can be overcome by introducing an auxiliary logical structure called imaginary logic L_i with two states $\{i, \neg i\}$.

In [3] the connection between the two logics is proposed—L_r and L_i to be realized through the dependence

$$F \wedge i = T. \tag{1}$$

This leads to a new complex logic L_s called S-logic with four mismatched states $\{T, F, i, \neg i\}$.

The performed researches show that in the requirement (1) this logic does not possess the property "associativity", i.e. it does not meet all the requirements of Boolean algebra. This creates inconveniences and difficulties in use and limits its application.

2 Complex Hierarchical Multivalued Logic

In the present work it is proposed on the basis of both logics L_r and L_i through appropriate connecting dependences to create a complex multivalued logic L_c that satisfies all axioms of Boolean algebra and lattice. Hereinafter it will be called hierarchical associative S-logic or HS-logic. Moreover, the additional requirements connecting the two logics L_r and L_i are selected in such a way that the imaginary logic dominates over the real one in terms of the properties "true" and "false".

It has been proved that in the disjunction and conjunction for an arbitrary idempotent semiring, a natural order can be introduced by the dependences

$$x \leq y \Leftrightarrow x \vee y = y; \tag{2}$$

$$y \leq x \Leftrightarrow x \wedge y = x, \tag{3}$$

where $\Leftrightarrow$ is a symbol of equivalence possessing the properties of reflexivity, symmetry and transitivity.

If $x = F$ and $y = T$ then the above equations and inequalities pass into

$$x \leq y \;\; \Leftrightarrow \;\; x \vee y = y; \;\; T \leq F \;\; \Leftrightarrow \;\; T \wedge F = F. \tag{4}$$

Boolean algebra (B, $\vee$, $\wedge$, $\neg$, 0, 1) is considered as an idempotent symmetric semi-ring in which for every x there exists a so-called complement $\neg x$, such that

$$x \vee \neg x = 1 \quad \text{and} \quad x \wedge \neg x = 0.$$

Table 1 Axioms in boolean algebra

№	Axiom	№	Axiom
1	$a \vee (b \vee c) \equiv (a \vee b) \vee c$	8	$a \wedge (b \wedge c) \equiv (a \wedge b) \wedge c$
2	$a \vee b \equiv b \vee a$	9	$a \wedge b \equiv b \wedge a$
3	$a \vee a \equiv a$	10	$a \wedge a \equiv a$
4	$a \vee 0 \equiv a$	11	$a \wedge 1 \equiv a$
5	$a \wedge (b \vee c) \equiv (a \wedge b) \vee (a \wedge c)$	12	$a \vee (b \wedge c) \equiv (a \vee b) \wedge (a \vee c)$
6	$a \wedge 0 \equiv 0$	13	$a \vee 1 \equiv 1$
7	$a \vee \neg a \equiv 1$	14	$a \wedge \neg a \equiv 0$

a, *b* and *c* are logical variables

In Boolean algebra, the disjunction $\vee$ is called a Boolean union, and the conjunction $\wedge$ is usually called a Boolean intersection.

Table 1 records the axioms of Boolean algebra, arranged in pairs [5]:

From the above axioms follow some important properties:

(a) Uniqueness of the complement: for each a its complement $\neg a$ is unique;
(b) Symmetricity of the complement: within Boolean algebra the identity is true:

$$\neg\neg\, a \equiv a; \tag{5}$$

(c) De Morgan's laws: in Boolean algebra the identities are true:

$$\neg(a \vee b) \equiv \neg a \wedge \neg b \quad \text{and} \quad \neg(a \wedge b) \equiv \neg a \vee \neg b; \tag{6}$$

(d) The property "absorption" leads to the algebraic structure "lattice" for which it is true:

$$x \vee (x \wedge y) \equiv x \quad \text{and} \quad x \wedge (x \vee y) \equiv x. \tag{7}$$

In the present work it is proposed to connect the two logical structures—the real L_r and the imaginary L_i—through the following two dependencies:

$$F \wedge \neg i = \neg i; \quad T \wedge i = T. \tag{8}$$

It is assumed that the logics L_r and L_i whose states are respectively $\{T, F\}$ and $\{i, \neg i\}$ satisfy separately the requirements of Table 4 and the properties from (a) to (d). The introduction of the two dependences on (8) leads to a new complex logic L_c with four states of $\{T, F, i, \neg i\}$. This makes it ambiguous. It will be shown that this logic also corresponds to the axioms of Boolean algebra and lattice, as well as to the statements (a) to (d). In that L_i dominates over L_r and has stronger properties of "true" and "false".

It can be proved that under assumptions (8) the following 12 dependences can be deduced, hereinafter referred to as rules:

1.

$$F \wedge \neg i = \neg i \Leftrightarrow \{F \le \neg i\}; \tag{9}$$

2.

$$T \wedge \neg i = \neg i \Leftrightarrow \{T \le \neg i\}; \tag{10}$$

3.

$$T \wedge F = F \Leftrightarrow \{T \le F\}; \tag{11}$$

4.

$$T \wedge i = T \Leftrightarrow \{i \le T\}; \tag{12}$$

5.

$$F \wedge i = F \Leftrightarrow \{i \le F\}; \tag{13}$$

6.

$$i \wedge \neg i = \neg i \Leftrightarrow \{i \le \neg i\}; \tag{14}$$

7.

$$T \vee i = i \Leftrightarrow \{T \le i\}; \tag{15}$$

8.

$$F \vee i = i \Leftrightarrow \{F \le i\}; \tag{16}$$

9.

$$F \vee T = T \Leftrightarrow \{F \le T\}; \tag{17}$$

10.

$$F \vee \neg i = F \Leftrightarrow \{\neg i \le F\}; \tag{18}$$

11.

$$T \vee \neg i = T \ \Leftrightarrow \ \{\, \neg i \leq T \,\}\,; \tag{19}$$

12.

$$\neg i \vee i = i \Leftrightarrow \{\neg i \leq i\}. \tag{20}$$

The dependences (11), (14), (17) and (20) follow directly from the algebraic structures of the two logics L_r and L_i.

The dependence (10) follows from (9) and (11), and from the chain:

$$T \wedge \neg i = T \wedge (F \wedge \neg i) = F \wedge \neg i = \neg i.$$

Equivalence (12) follows from the second equation in (8). The dependence (13) is a direct consequence of (11) and (12), namely,

$$F \wedge T = (T \wedge F) \wedge i = (T \wedge i) \wedge F = F \wedge F = F.$$

The dependence (15) arises from the addition of the first of the dependencies in (8):

$$\neg(F \wedge \neg i) = \neg\neg i; \ \ T \vee i = i.$$

The equivalence of (16) follows from (15) and (17):

$$T \vee i = (F \vee T) \vee i = F \vee i = i.$$

The dependence (18) follows from the complement of the second proposal in (8) and from (6):

$$\neg\,(\,T \wedge i\,) = \neg T; \ \ F \vee \neg\, i = F.$$

The equivalence in (19) is a consequence of (17) and (18):

$$T \vee \neg i = (F \vee T) \vee \neg i = (F \vee \neg i) \vee T = F \vee T = T.$$

The obtained 12 rules from (9) to (20) fully characterize the relations between the four states $\{T, F, I, \neg i\}$ of the complex multivalued logic L_c and show the established order between them in terms of true in disjunction and false in conjunction.

Conditionally graphically, these relations can be shown by the following two figures—Fig. 1 for the conjunction operation and Fig. 2 for the disjunction operation.

The symbol $\geq$ corresponds to the relations from (9) to (20), and the arrows show the directions from greater "false" to less In the next Fig. 2 the directions are reversed—from states with greater "true" to states with less:

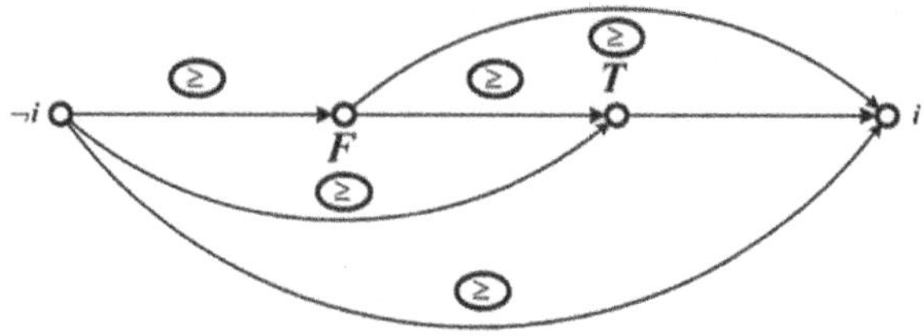

Fig. 1 Conjunction operation

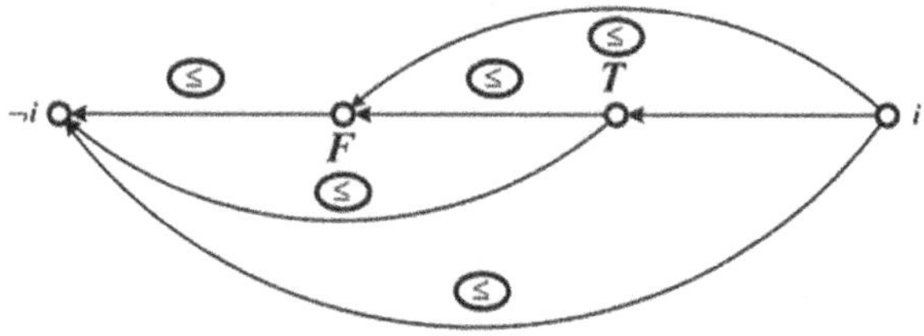

Fig. 2 Disjunction operation

The introduced order in the set of four states of complex logic can also be described by the following chain of inequalities:

(a) for the conjunction:

$$\neg i \geq F \geq T \geq i; \ \neg i \geq F \geq i; \tag{21}$$

$$\neg i \geq F \geq i; \ \neg i \geq i; \tag{22}$$

(b) for the disjunction:

$$\neg i \leq F \leq T \leq i; \ \neg i \leq F \leq i; \tag{23}$$

$$\neg i \leq T \leq i; \ \neg i \leq i. \tag{24}$$

Inequalities (21) to (24), respectively, rules 1–12 show that the axioms of Boolean algebra and lattice are observed for complex logic, i.e. the properties of associativity, commutativity, idempotence, distributivity, double negation and absorption are in force.

Rules 1 to 12 make it possible to construct Kelly tables for disjunction and conjunction, as well as for uncertainties (21) to (24): Tables 2 and 3 are respectively for disjunction and conjunction, and Tables 4 and 5—for their existing inequalities.

It follows from the dependences from (9) to (20) that in the algebraic structure of complex logic—analogously as in the imaginary logic—the role of 1 (unit) is played by the state i, and of zero (0)—the state $\neg i$.

The two logics—the real and the imaginary—are connected by the complex logic in such a way that due to the requirement (8) and the rule (15) the true state i of L_i

Table 2 Disjunction

	i	T	F	$\neg i$
i	i	i	i	i
T	i	T	T	T
F	i	T	F	F
$\neg i$	i	T	F	$\neg i$

Table 3 Conjunction

	i	T	F	$\neg i$
i	i	T	F	$\neg i$
T	T	T	F	$\neg i$
F	F	F	F	$\neg i$
$\neg i$	$\neg i$	$\neg i$	$\neg i$	$\neg i$

Table 4 Disjunction

	i	T	T	$\neg i$
i	$=$	$\geq$	$\geq$	$\geq$
T	$<$	$=$	$\geq$	$\geq$
T	$<$	$<$	$=$	$\geq$
$\neg i$	$<$	$<$	$<$	$=$

Table 5 Conjunction

	i	T	T	$\neg i$
i	$=$	$<$	$<$	$<$
T	$\geq$	$=$	$<$	$<$
T	$\geq$	$\geq$	$=$	$<$
	$\geq$	$\geq$	$=$	$<$

is more true than that of $L_r - T$. Similarly from the first assumption in (8) rule (9) follows that the false state $\neg I$ of logic is more false than that of $L_r - F$. And it follows that the statements in imaginary logic have greater force than the statements in real logic, i.e. L_i dominates over L_r.

Graphically, the three logics—L_i, L_r and L_c—in disjunction can be represented as follows (Fig. 3):

Since complex logic has four states—$\{T, F, i, \neg i\}$, the rules from (9) to (20) and Tables 2 and 3 allow these two Kelly tables to be considered as quaternions of disjunction and conjunction in complex logic, respectively. Similar Kelly tables can be constructed for the other two operations—implication and equivalence.

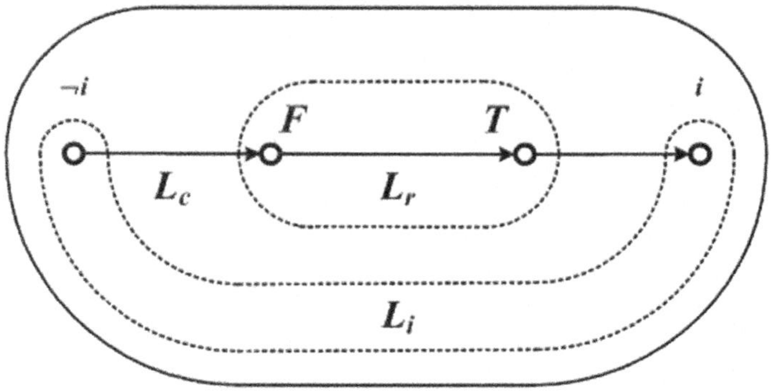

Fig. 3 The three logics L_i, L_r and L_c separated

The elements of the Kelly table of equivalence are determined by the disjunction and conjunction operations as follows:

$$A \sim B = (A \wedge B) \vee (\neg A \wedge \neg B). \tag{25}$$

The following four dependencies obviously follow from (25):

$$i \sim i = i;\ T \sim T = T;\ F \sim F = \text{И};\ \neg i \sim \neg i = i. \tag{26}$$

The other four elements are determined from Tables 2 and 3 by (25):

$$\begin{aligned} i \sim T &= (\, i \wedge T\,) \vee (\, \neg i \wedge F\,) = T \vee \neg i = T; \\ i \sim T &= (\, i \wedge T\,) \vee (\, \neg i \wedge F\,) = T \vee \neg i = T; \\ i \sim F &= (\, i \wedge F\,) \vee (\, \neg i \wedge T\,) = F \vee \neg i = F; \\ T \sim \neg i &= (\, T \wedge \neg i\,) \vee (\, F \wedge i\,) = \neg i \vee F = F; \\ F \sim \neg i &= (\, F \wedge \neg i\,) \vee (\, T \wedge \neg i\,) = \neg i \vee T = T. \end{aligned}$$

The following Kelly Table 6, shows the equivalent dependences between all pairs of the old states of complex logic L_c:

The equivalence table above shows that:

Table 6 Equivalence

	i	T	F	$\neg i$
i	i	T	F	$\neg i$
T	T	T	F	F
F	F	F	T	T
$\neg i$	$\neg i$	F	T	i

- The elements of the first and last row are respectively opposite to each other;
- The elements of the second and third rows are respectively opposite to each other;
- The equivalence between a pair only from the real L_r or respectively the imaginary logic L_i coincides with the analogous results from the classical propositional logic;
- Twelve elements (states) of the Table 6 are from the real logic L_r and only 4—from the imaginary logic L_i. This is due to the fact that in pairs in which there is at least one state of the real logic, the result is a state of the same logic that is dominated from imaginary logic;
- The elements in Table 6 are imaginary states only if the respective pairs are also imaginary states, and only then.

The elements of the Kelly table for the implication will be determined by the disjunction and conjunction operations according to the formula of the propositional logic:

$$A \to B = \neg A \vee B. \tag{27}$$

The results are as follows:

$$\begin{aligned}
&i \to i = \neg i \vee i = i;\ \ i \to T = \neg i \vee T = T;\ \ i \to F = \neg i \vee F = F;\\
&i \to \neg i = \neg i \vee \neg i = \neg i;\ \ T \to i = F \vee i = i;\ \ T \to T = F \vee T = T;\\
&T \to F = F \vee F = F;\ \ T \to \neg i = F \vee \neg i = F;\ \ F \to i = T \vee i = i;\\
&F \to T = T \vee T = T;\ \ F \to F = T \vee F = T;\ \ F \to \neg i = T \vee \neg i = T;\\
&\neg i \to i = i \vee i = i;\ \ \neg i \to T = i \vee T = i;\ \ \neg i \to F = i \vee F = i;\\
&\neg i \to \neg i = i \vee \neg i = 0.
\end{aligned}$$

Table 7 is a Kelly table for the implication in complex logic L_c:

The analysis of the data and dependencies from Table 7 allows draw the following general conclusions:

- The elements of the first column and the last row of the table coincide. They are all equal to the state i of imaginary logic, which is the truest of the four states of complex logic;
- The first three elements of the second column and the last three elements of the penultimate row coincide with each other and are equal to T;

Table 7 Implication

	i	T	F	$\neg i$
i	i	T	F	$\neg i$
T	i	T	F	F
F	F	T	T	T
$\neg i$	i	i	i	I

Table 8 Implication in imaginary logic

A	B	$A \to B$
i	i	i
i	$\neg i$	$\neg i$
$\neg i$	i	i
$\neg i$		i

Table 9 Implication in real logic

A	B	$A \to B$
T	T	T
T	F	F
F	T	T
F	F	T

- The first two elements of the third column and the last two rows of the second row coincide with each other and are equal to F.
- The Modus Ponens (MP) rule can be applied to Kelly's table, which is formally recorded as follows:

$$\frac{A \to B,\ \ A}{B}. \tag{28}$$

Similarly to the logics L_r an L_i from Table 7 and (28) the conclusions follow:

(a) From $(A \to B) = (i \to i) = i$ and $A = i$ follows $B = i$;
(b) From $(A \to B) = (T \to T) = T$ and $A = T$ follows $B = T$.

These two conclusions according to Modus Ponens are also confirmed by the following tables—respectively for the imaginary and for the real logic. The data in them are taken from the general table for the implication—Tables 7, 8 and 9:

Logical inferences can also be made through the implicative relationship $A \to B$ in which states A and B are of two different logics—real and imaginary. The following two conclusions are an example of such. They are derived by Modus Ponens based on (28) and the implication data from Table 7:

(c) From $(A \to B) = (T \to i) = i$ and $A = T$ follows $B = i$;
(d) From $(A \to B) = (i \to T) = T$ and $A = i$ follows $B = T$.

Tables 10 and 11 are part of Table 7 for the implication. They illustrate why these two conclusions are true:

The first row of both Tables 10 and 11, is crucial for the application of Modus Ponens.

In addition to logical equations, complex HS logic can have other applications. The following example shows this.

Table 10 Implication in complex logic

A	B	$A \to B$
T	i	i
T	$\neg i$	F
F	i	i
F	$\neg i$	T

Table 11 Implication in complex logic

A	B	$A \to B$
i	T	T
i	F	F
$\neg i$	T	i
$\neg i$	F	i

If there are two different sources of judgments on the same objects and one of these sources has a higher degree of confidence than the other, then its statements of true and false are accepted as states i and $\neg i$ by imaginary logic, respectively.

The statements of the less reliable source will be denoted accordingly by the states $\{T, F\}$ of real logic. Then, within the framework of complex logic, through the rules from (9) to (20) different logical operations can be performed, allowing the realization of more accurate and in-depth judgments, than individually through L_r and L_i. At the same time, within the complex logic, the imaginary logic L_i dominates over the real one L_r.

3 Conclusion

1 In the present work, on the basis of two logical propositional logical structures—real L_r with two states $\{T, F\}$ and imaginary L_i with two states $\{i, \neg i\}$, it is proposed to build a complex hierarchical logic L_c with four unequal states $\{T, F, i, \neg i\}$. It will be called complex hierarchical logic or HS-logic for short.

The unification of the two logics L_r and L_i into one complex logic L_c (or—which is the same—HS-logic) is carried out through appropriately selected requirements that ensure the dominance of L_i-logic in terms of the degree of true or false in compliance with all axioms of Boolean algebra and the lattice.

2. Twelve rules for disjunction and conjunction operations are derived and it is proved that the complex logic corresponds to all axioms of Boolean algebra and lattice. It has been proven that the truth and falsity of imaginary logic are "stronger" than those of real logic within complex logic. And this means that imaginary logic dominates over real logic.

3. Basic disjunctive and conjunctive dependences between the four unequal states of complex logic are obtained and are shown graphically and by Kelly's tables. They can be considered as logical quaternions of disjunction, conjunction, equivalence and implication in complex logic.
4. Dependencies are derived and Kelly tables for equivalence and implication are constructed. Based on these tables, through the Modus Ponens rule, four inference schemes are proposed.
5. The possibility for practical use of the offered complex logic is shown.

References

1. Kleene, S.C.: Mathematical Logic. Wiley, New York (1967)
2. Kowalski, R.: Logic for Problem Solving. Elsevier North Holland Inc., Amsterdam (1979)
3. Sgurev, V.: An essay on complex valued propositional logic. In: Artificial Intelligence and Decision Making, pp. 95–101 (2013)
4. Sgurev, V.: An approach to constructing a complex propositional logic, Comptes Rendus de l'Academie bulgare des Sciences **66**(11):1623–1632
5. Johnson, B.: Topics in Universal Algebra. Springer, New York (1972)

Design Considerations for a Multiple Smart-Device-Based Personalized Learning Environment

Masashi Katsumata

Abstract This paper proposes a personalized learning environment based on multiple smart devices. This learning environment comprises a learning content provider and analytical functions related to the learner's learning outcomes and biological data. The preferred learning style of the proposed system is assumed to use the mobile and wearable smart devices used in daily life. The goal of the proposed system is to provide educational services appropriate to a learner's personal situation by analyzing merged learning outcomes and biological data. To verify the feasibility of the proposed system, the design of the prototype system and an experimental scenario are presented. According to the experimental results, future system design considerations are presented for achieving a more realistic system.

1 Introduction

Online learning has become increasingly commonplace because of the growing popularity of portable smart devices that can access the Internet, such as smartphones and tablets. Currently, information and communication technology (ICT)-based teaching methods and classrooms are leveraged by educational institutions and families worldwide. In an ICT-based learning environment, a learning management system (LMS) is employed to manage learning content and test student progress. Learning-related data is stored in LMS are then utilized for learning support [1–3]. These data are called learning logs and used to perceive the learner's learning situation and progress.

In recent years, the practical use of learning logs has attracted enormous attention in the field of learning analytics. Moreover, many researchers have focused on intelligent tutoring systems (ITS), which aim to provide effective learning support to individual learners [4].

M. Katsumata (✉)
Department of Information Technology and Media Design, Nippon Institute of Technology, Saitama, Japan
e-mail: katumata@nit.ac.jp

V. Sgurev et al. (eds.), *Advances in Intelligent Systems Research and Innovation*, Studies in Systems, Decision and Control 379,
https://doi.org/10.1007/978-3-030-78124-8_8

Recent ITS research has focused on learners' emotional and psychological characteristics to provide more individualized support [5]. To identify these characteristics, researchers have introduced dedicated measurement devices that provide substantial insights into various learning-support approaches [6]. However, such devices are typically large, heavy, and invasive, and they cannot be used for daily learning [7].

It is important that learners maintain personal motivation in their online learning environments, because, unlike classroom learning, there are rarely proctors to prompt or support them. During classroom-based lessons, teachers have the advantage of observing students' facial expressions and energy levels. However, during personal online learning, the process is controlled by an LMS, which lacks such personal touches.

Various extant and developing *wearable* smart devices (e.g., eyeglasses and wristbands) are increasingly used in daily life to monitor and collect biological metrics (e.g., physical movements and heart rates). By utilizing this information, it is expected that personalized learning environments will become more reactive and tailorable to support students and intuitively manage their progress.

This study differs from previous studies, because we develop a personalized learning environment based on multiple smart devices. The objective is track and assess a learner's online learning situation using those devices and to subsequently provision learning-support services based on the learner's changing situation. A goal of this study is to identify the requirements needed to establish a realistic version of the proposed system. This is achieved through preliminary experimentation using a prototype implemented with a primal function that can measure the learner's situation during online learning.

In the Sect. 2, we describe related works, and the concept of a multiple smart-device-based personalized learning environment is presented, which provides learning services according to learners' real-time situations (Sect. 3). Then, we describe the prototype, the experimental scenario, and the results (Sect. 4). In Sect. 5, design considerations for realizing a realistic and usable system are presented. Finally, Sect. 6 provides the conclusion to this work.

2 Related Works

2.1 Estimating Individual Learner States During Learning

Alyuz et al. [7] assessed three emotional states (i.e., satisfied, bored, and confused) to represent learner situations in an ITS environment using a depth camera installed on a notebook computer. States were detected using information that integrated superficial features captured by the camera during learning. Other studies focused on learners' emotional states in ITSs [5, 8], aiming to improve instruction by sharing key learner metrics with their teachers.

Alyuz et al. [9] investigated learning engagement using three unobtrusive modalities (i.e., student appearance, interaction/performance data, and mouse movements). Traditional ITS research that captured learners' emotions employed single modalities for detection [10]. However, more recent research has used multiple factors, indicating that a single modality is insufficient. Therefore, machine-learning approaches have been applied to predict learner emotions by analyzing multimodal factors [7, 10].

Kapoor and Picard [10] described a system for classifying affective states related to the interest shown by children trying to solve a puzzle on a computer. The system used real-time tracking of facial features, head gestures, and chair postures. It focused on naturally occurring nonverbal behaviors so that it could learn to cope with the unconstrained nature of real data. Sensor data obtained by a camera and a pressure-sensitive chair were used to identify facial and posture features, respectively. The system successfully classified interest/disinterest with 87% accuracy, indicating good potential for detecting learners' affective states using multimodal data.

In contrast to these previous studies that used desktop computers, this study involves different learning environments, wherein learners wear multiple commonly used smart devices. Fortenbacher et al. [11] also proposed the concept of a sensor-based adaptive learning environment called "LISA" and developed a wearable prototype using electrodermal activity and electrocardiograph sensors. However, their objective was to investigate technical and ethical issues related to personal sensor data in the learning environment.

2.2 *Personalized or Adaptive Learning Services*

A personalized learning system provides learners with the most appropriate learning content according to their learning state. Cai [12] developed a web-based LMS using the Intellipath system currently employed at Colorado Technical University. Intellipath determines the learning path for each student based on various assessments of the student's current knowledge state or performance.

Raj and Renumol [13] proposed an adaptive personalized learning environment that assisted learners via content recommendations, adapting to their preferences and performance. Content recommendation was implemented using a rule-based prediction engine that employed learner attributes and learning-object data. The attribute data were obtained from questionnaires and tests. This system provided learners with content according to their states and preferences combined with dynamic Bayesian network model data and ontology-based knowledge. Ciolacu et al. [14] presented the Education 4.0 concept, which assists professors in identifying student performance and weakness factors using a didactic method based on learning analytics and machine learning with multimodal data obtained from sensor devices. Their focus was to advocate for professors to innovatively support learner success. However, the learner in this case did not directly affect learning content based on their learning state.

In contrast, this study determines students' learning states in relation to their learning outcomes and sets the pace by monitoring learner data using smartphones, tablets, and various wearable devices.

3 Multiple Smart-Device-Based Learning Environments

3.1 Overview

In this section, the design of a personalized learning environment based on multiple smart devices used in daily life is described. In such an environment, the learner predominantly uses a tablet or a smartphone as a learning terminal to browse and utilize content (e.g., documents, exercises, and videos). Additionally, the learner can use a wearable device (e.g., wristband or eyeglass) to help provide personalized learning experiences. The learning data acquired from this environment are collected in various ways (e.g., sensors, computers, and smart devices). Smartphones and tablets include webcam functions and various other sensors (e.g., accelerometers and gyroscopes) that can be combined to help determine the learner's situation. The proposed environment can analyze merged learning outcomes using sensor data divided into physical and biological characteristics.

3.2 System Components and Function

The proposed learning environment consists of a learner's learning environment and an LMS-based personalized online learning service (Fig. 1).

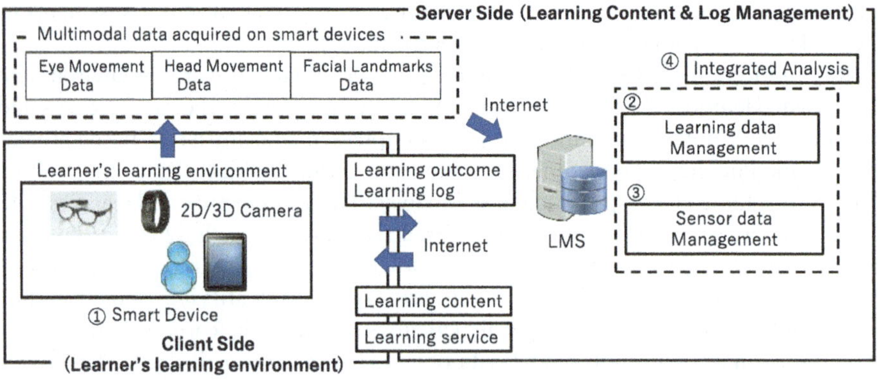

Fig. 1 Overview of multiple smart device based personalized learning environment

3.2.1 Smart Device

In proposed system, the smart device is a non-invasive and non-contact device used as the learner's learning environment. The learning environment estimates a learner's situation by combining multiple sensory data from different devices. Eyeglass- or wristband-type devices that operate with a tablet or a smartphone are preferred in our learning environment. It records the activity data of the learner during online learning using various smart devices.

3.2.2 Data Management

LMS is designed to manage a learning data and a sensor data. The function of learning data management manages the learner's learning data such as page browsing time of learning materials, the time spent of time on quiz, and the results of the quiz.

The function of sensor data management is to manage biomedical data and inertia data acquired from the wearable device. The facial expressions data analyzed through the camera (2D/3D) of the tablet used as the learning terminal is managed for estimating learner's situation.

3.2.3 Integrated Analysis

The function of integrated analysis is to integrate learning and sensor data to estimate the learner's situation during online learning. By combining the prediction functions of learning and sensor data, we can assess the learner's situation during online learning. Section IV describes how the prototype system is implemented to obtain learning and sensor data related to the proposed system.

4 Prototype

4.1 System Objective

The objective of the prototype system is to investigate which points should be considered in order to establish a more realistic system based on the proposed concept.

In proposed system concept, we are considering using the in-camera of a tablet to track the eye movements of a learner during online learning. However, the prototype system uses JINS MEME [15] as an eyeglass-type device. JINS MEME has a three-point electrography sensor, an accelerometer sensor, and a gyroscope sensor. In this prototype system, a learning terminal environment was set up with a 9.6-in iPad and eyeglass-type device.

4.2 System Implementation

In this prototype system, the LMS provided learning materials and quizzes related to an information technology (IT) passport examination: a national qualification for measuring basic IT knowledge in Japan. On the server-side, the LMS served two types of content: learning material and quizzes. The LMS also allowed learners to self-report their levels of understanding. The learning content appeared on the iPad's web browser. Page-browsing time, time spent on quizzes, and quiz scores were tracked by the LMS.

On the client side, we implemented a data logger application that ran on the iPad using the JINS MEME software development kit. The data measured were recorded on the iPad using the data logger. In this implementation, we manually transferred the recorded data to a desktop computer for further processing.

4.3 Experimental Scenario

The learner's situation was investigated using our experimental scenario. In this experiment, the examinee performed a learning session using the iPad and eyeglass-type device. After the session, the examinee completed a quiz to ascertain an understanding level. Each learning session was 10 min, and each quiz lasted 5 min. Afterwards, the eyeglass-type device and LMS data were collected and analyzed.

4.4 Experimental Result

We conducted the experiment with 30 university student volunteers. The following items were examined: browsing time for learning materials; time spent on the comprehension quiz; eye-movement data (i.e., up, down, right, and left); and strength and speed of eye blinks. Data are shown in Figs. 2, 3 and 4. Figure 2 shows browsing start and end times for each learning page. Each instruction section consisted of multiple pages. Figure 3 shows the quiz results. Figure 4 shows the measured values of eye movements and the strength and speed of eye blinks. The values measured by the eyeglass-type device were in the range 0–4 (0: "no detection;" 1: "small;" 2: "medium;" 3: "large;" and 4: "very large"). The strength of an eye blink was an integer value between 200 and 500. The speed of an eye blink was measured in milliseconds.

From the experiment, we obtained sensory data and learning logs. From these, we could interpret learners' behavioral characteristics from their online learning experiences. Alongside this experiment, we took video to provide ground truth. After the experiment, we interviewed the examinees to obtain more detailed learning anecdotes.

userid	section	page	start time	end time
2018032	5-3	1	2018/7/9 13:39:42.134	2018/7/9 13:41:29.642
2018032	5-3	2	2018/7/9 13:41:29.644	2018/7/9 13:43:47.479
2018032	5-3	3	2018/7/9 13:43:47.481	2018/7/9 13:45:38.918
2018032	5-3	4	2018/7/9 13:45:38.921	2018/7/9 13:47:46.520
2018032	5-3	5	2018/7/9 13:47:46.523	2018/7/9 13:49:48.761
2018032	5-3	6	2018/7/9 13:49:48.763	2018/7/9 13:50:07.477
2018032	5-3	5	2018/7/9 13:50:07.480	2018/7/9 13:50:14.556
2018032	5-3	6	2018/7/9 13:50:14.559	2018/7/9 13:51:38.416
2018032	5-5	1	2018/7/9 13:58:13.657	2018/7/9 13:59:36.944
2018032	5-5	2	2018/7/9 13:59:36.946	2018/7/9 14:00:25.561
2018032	5-5	3	2018/7/9 14:00:25.564	2018/7/9 14:01:24.756
2018032	5-5	4	2018/7/9 14:01:24.759	2018/7/9 14:02:00.293

Fig. 2 Data from the browsing time log stored in the LMS

userid	section	exercise no	start time	end time	answer
2018032	5-3	1	2018/7/9 13:54:43.044	2018/7/9 13:54:57.350	ウ
2018032	5-3	2	2018/7/9 13:54:59.277	2018/7/9 13:55:04.983	ウ
2018032	5-3	3	2018/7/9 13:55:07.219	2018/7/9 13:55:14.578	ウ
2018032	5-3	4	2018/7/9 13:55:17.459	2018/7/9 13:55:23.470	ウ
2018032	5-3	5	2018/7/9 13:55:26.403	2018/7/9 13:55:32.551	ア
2018032	5-3	6	2018/7/9 13:55:35.370	2018/7/9 13:55:55.637	イ
2018032	5-3	7	2018/7/9 13:55:57.460	2018/7/9 13:56:39.052	エ
2018032	5-3	8	2018/7/9 13:56:43.921	2018/7/9 13:57:36.778	ア
2018032	5-5	1	2018/7/9 14:02:12.765	2018/7/9 14:02:27.399	ウ
2018032	5-5	2	2018/7/9 14:02:31.065	2018/7/9 14:03:17.383	ウ
2018032	5-5	3	2018/7/9 14:03:19.706	2018/7/9 14:04:20.263	ウ
2018032	5-5	4	2018/7/9 14:04:25.603	2018/7/9 14:05:20.393	エ
2018032	5-5	5	2018/7/9 14:05:22.722	2018/7/9 14:05:46.157	イ
2018032	5-5	6	2018/7/9 14:05:49.616	2018/7/9 14:07:10.883	ウ

Fig. 3 Part of answer log stored in LMS

5 Considerations for System Design

Then, we provided design considerations for realizing a more realistic system from the clustering analysis, video monitoring, and interviews.

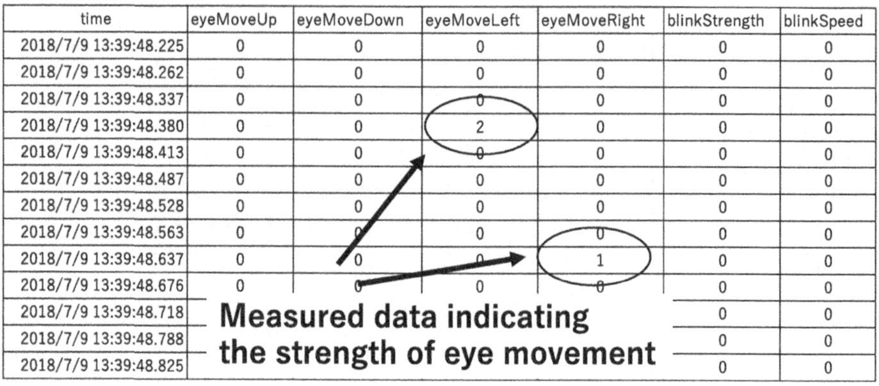

time	eyeMoveUp	eyeMoveDown	eyeMoveLeft	eyeMoveRight	blinkStrength	blinkSpeed
2018/7/9 13:39:48.225	0	0	0	0	0	0
2018/7/9 13:39:48.262	0	0	0	0	0	0
2018/7/9 13:39:48.337	0	0	0	0	0	0
2018/7/9 13:39:48.380	0	0	2	0	0	0
2018/7/9 13:39:48.413	0	0	0	0	0	0
2018/7/9 13:39:48.487	0	0	0	0	0	0
2018/7/9 13:39:48.528	0	0	0	0	0	0
2018/7/9 13:39:48.563	0	0	0	0	0	0
2018/7/9 13:39:48.637	0	0	0	1	0	0
2018/7/9 13:39:48.676	0	0	0	0	0	0
2018/7/9 13:39:48.718					0	0
2018/7/9 13:39:48.788					0	0
2018/7/9 13:39:48.825					0	0

Fig. 4 Sensor data from the eyeglass-type device

5.1 Hierarchical Clustering Analysis

We conducted two types of hierarchical clustering to explore the relationship between eye movements and learner understanding during online learning. These included eye-movement analysis related to each page of the learning material and that of eye movements vs. learning time.

5.1.1 Eye Movement per Page

Six metrics (i.e., up, down, left, right, strength of eye blink, and speed of eye blink) for the 21 examinees were analyzed using Ward's method of hierarchical clustering. Figure 5 shows the dendrogram of the hierarchical clustering result.

The eye movement data used for the clustering analysis represent the ratio of the eye movement data to the browsing time of the learning material page. From the results of hierarchical clustering analysis, the examinee groups can be roughly divided into two: C1 and C2. The number of learners belonging to each cluster is shown in the parentheses of Fig. 5.

Table 1 shows the ratios of eye movement data for each group to the browsing time of the learning material page and the correct answer rate for all eight quiz questions. The number of eye movements detected by the eyeglass-type device is represented as the detection number.

From the cluster classifications and correct answer rates for each learning material page, groups having a high number of eye movements and a high correct answer rate (clusters marked with * in Table 1) were classified as 1, 2, 3, and 4 of the learning material page. For learning materials having page numbers of 5 and 6, it was confirmed that the groups having a high number of eye movements and low correct answers (clusters marked with ** in Table 1) were classified. It is considered that these differences were influenced by the content structure that including a figure

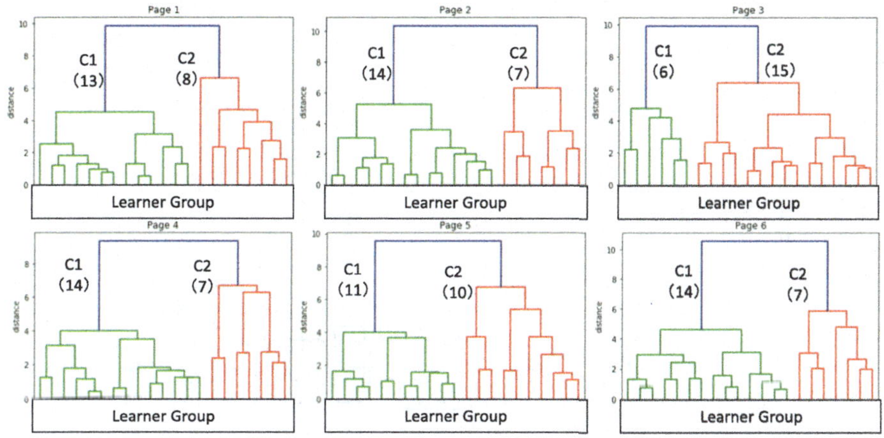

Fig. 5 Hierarchical clustering result for each page of learning material

Table 1 Correct rate of quiz and the detection number average of eye movement for each cluster

Learing material	Cluster	Direction of eye movement				Eye blinks		Correct rate of quiz (%)
		Up	Down	Left	Right	Strength	Speed	
Page	C1	0.1	0.2	0.4	0.3	9.2	17.4	69
	C2*	0.2	0.4	0.8	0.7	22.6	37.5	78
Page	C1	0.1	0.1	0.5	0.3	10.9	21.8	71
	C2*	0.2	0.3	0.7	0.6	26.4	42.6	77
Page	C1*	0.2	0.3	0.9	0.7	26.8	43.0	77
	C2	0.1	0.2	0.5	0.3	11.3	22.6	71
Page	C1	0.1	0.1	0.6	0.4	11.8	21.7	71
	C2*	0.2	0.4	0.8	0.7	24.3	40.1	71
Page	C1	0.1	0.1	0.5	0.3	9.9	21.1	75
	C2**	0.2	0.3	0.7	0.6	20.2	36.9	70
Page	C1	0.1	0.2	0.5	0.3	12.6	22.1	75
	C2**	0.3	0.4	0.8	0.7	25.6	45.5	68

to help with the understanding of the learning content on those pages. Figures were also used on other pages (1, 2, and 4), but they did not distract the reader in a similar manner. It is considered that the group having a high correct answer rate on pages 5 and 6 may have benefitted from the presence of the figure. Therefore, it is considered that the detection number of eye movements was low.

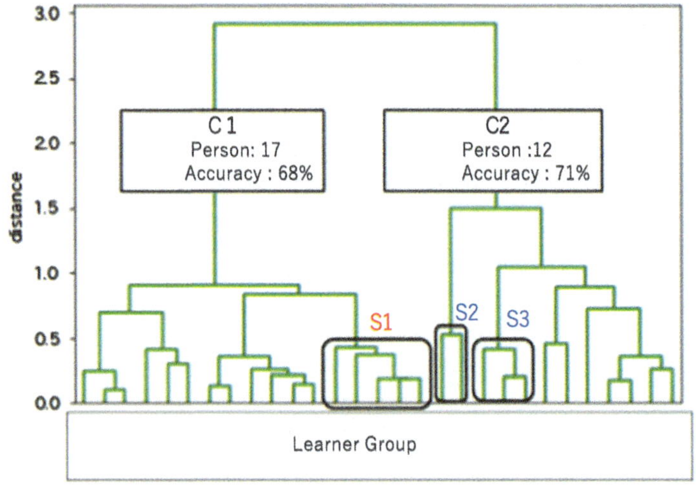

Fig. 6 Hierarchical clustering results for eye movement vs. learning time

5.1.2 Eye Movements per Learning Time

We conducted a clustering analysis of the eye-movement data vs. total learning time. From the results of hierarchical clustering analysis, the examinee's groups were divided into groups C1 and C2. The number of learners and the correct answer rates of the quiz for C1 and C2 learners are shown in Fig. 6. No substantial differences were observed in the correct answer rates. However, the S1, S2, and S3 subgroups (Fig. 6) exhibited a great deal of eye movement, with high correct answer rates of 90, 70, and 85%, respectively. Thus, the number of blinks and the number of eye movements are considered to be factors correlated to the degree of comprehension. To further analyze specific learning materials, a more detailed analysis should be conducted in the future by investigating different types of learning materials.

5.2 Effect of Learner's Reading Behavior

We analyzed the characteristics of reading behavior from the browsing time for learning materials in learning logs. We found that some examinees spent more time on particular pages. We also found a reading behavior in which one examinee browsed excessively while moving frequently between pages.

According to interviews, it was found that learners employed distinct reading behaviors for different types of content. Specifically, learners judged the parts highlighted in bold to be more important. Thus, they read those parts more carefully [16]. Additionally, learners focused on diagrams and illustrations, and they studied them with the written explanations. These types of learner reading behaviors were common, but they were difficult to capture from the log data. Thus, to determine a learner's learning situation more accurately, it will be necessary to consider the effect of different types of learning content in the future.

5.3 *Effect of Learner's Physical Behavior*

Here, we describe the characteristics of the examinee during online learning observed from video monitoring. We found that some examinees exhibited unique physical characteristics (e.g., head movements, shaking legs, thinking gestures, and chin-holding). Owing to these characteristics, we should design the future system to consider these individual behavioral characteristics. Furthermore, we found that it was difficult to capture the learner's situation using only smart devices. Thus, it may be necessary to consider equipping tables and chairs with sensor devices in the future.

6 Conclusions

In this study, a personalized learning environment based on multiple smart devices was proposed. A prototype experiment revealed points that should be considered to achieve a more realistic system of our system concept. The learning environment of this prototype comprised two parts: a client side, where data were collected from multiple smart devices, and a server side, where the learning content was managed and estimation functions and analyses were conducted. The estimation function was divided into two phases: an analysis phase, in which the acquired data were stored and analyzed, and a service phase, in which real-time learning support was provided. An aim of future research will be to include multimodal data analysis and personalization and to conduct actual scale experiments using multiple smart devices using additional metrics, such as heart rate, accelerations, and gyroscopic sensor data to represent the learner's physical situation and surrounding environment. Furthermore, it will be important to consider privacy issues related to the proposed system, because it utilizes human-generated data. Privacy concerns could be mitigated by conducting data analysis within the learner's device without sending private data to a server.

References

1. Alario-Hoyos, C., Estevez-Ayres, I., Perez Sanagustin, M., Leony, D., Delgado Kloos, C.D.: MyLearningMentor: a mobile app to support learners participating in MOOCs. J. Univ. Comput. Sci. **21**(5), 735–753 (2015)
2. Kavitha, M., Raj, D.: Educational data mining and learning analytics-educational assistance for teaching and learning. arXiv preprint arXiv:1706.03327 (2017)

3. Siemens, G., Long, P.: Penetrating the fog: analytics in learning and education. Educause Rev. **46**(5), 31–40 (2011)
4. Huerta-Pacheco, N.S., Rebolledo-Mendez, G., Hernandez-Gonzalez, S.: Cognitive-affective modelling approach in tutoring system. In: Proceedings of the 1st International Workshop on Emotion Awareness in Software Engineering, pp. 18–21 (2016)
5. Ammar, M.B., Neji, M., Alimi, A.M., Gouardères, G.: The affective tutoring system. Expert Syst. Appl. Expert Syst. Appl. **37**(4), 3013–3023 (2010)
6. Mills, C., Fridman, I., Soussou, W., Waghray, D., Olney, A.M., D'Mello, S.K.: Put your thinking cap on: detecting cognitive load using EEG during learning. In: Proceedings of the Seventh International Learning Analytics and Knowledge Conference, pp. 80–89 (2017)
7. Alyuz, N., Okur, E., Oktay, E., Genc, U., Aslan, S., Mete, S.E., Arnrich, B., Esme, A.A.: Semi-supervised model personalization for improved detection of learner's emotional engagement. In: Proceedings of the 18th ACM International Conference on Multimodal Interaction, pp. 100–107 (2016)
8. Charoenpit, S., Ohkura, M.: A proposal of new e-learning system design focusing on emotional aspects using biological signals. In: Proceedings of the International Conference on Biometrics and Kansei Engineering, pp. 178–182 (2013)
9. Alyuz, N., Okur, E., Genc, U., Aslan, S. Tanriover, C., Esme, A.A.: An unobtrusive and multimodal approach for behavioral engagement detection of students. In: Proceedings of the 1st ACM SIGCHI International Workshop on Multimodal Interaction for Education, pp. 26–32 (2017)
10. Kapoor, A., Picard, R.W.: Multimodal affect recognition in learning environments. In: Proceedings of the 13th Annual ACM International Conference on Multimedia, pp. 677–682 (2005)
11. Fortenbacher, A., Pinkwart, N., Yun, H.: [LISA] Learning analytics for sensor-based adaptive learning. In: Proceedings of the Seventh International Learning Analytics and Knowledge Conference, pp. 592–593 (2017)
12. Cai, R.: Adaptive learning practice for online learning and assessment. In: Proceedings of the 2018 International Conference on Distance Education and Learning, pp. 103–108 (2018)
13. Raj, N.S., Renumol, V.G.: Architecture of an adaptive personalized learning environment (APLE) for content recommendation. In: Proceedings of the 2nd International Conference on Digital Technology in Education, pp. 17–22 (2018)
14. Ciolacu, M., Tehrani, A.F., Binder, L., Svasta, P.M. Education 4.0—artificial intelligence assisted higher education: early recognition system with machine learning to support students' success. In: 2018 IEEE 24th International Symposium for Design and Technology in Electronic Packaging, pp. 23–30 (2018)
15. Kanoh, S., Ichi-Nohe, S., Shioya, S., Inoue, K., Kawashima, R.: Development of an eyewear to measure eye and body movements. In: Proceedings of the Annual International Conference of the IEEE Engineering in Medicine and Biology Society, pp. 2267–2270 (2015)
16. Larkin, J.H., Simon, H.A.: Why a diagram is (sometimes) worth ten thousand words. Cogn. Sci. **11**(1), 65–99 (1987)

Intelligent Assistive Sensors and Smart Systems for the Control and Analysis of Driver Reaction Times

David Sanders, Malik Haddad, Giles Tewkesbury, Tom Barker, Martin Langner, and Alex Gegov

Abstract Intelligent assistive sensors and smart systems for powered wheelchairs are presented and used to analyze ways to improve mobility and enhance the quality of life of powered wheelchair users. Time-delays are introduced to investigate their effects. Systems interpret joystick signals and sensor signals and then mix them together. Ultrasonics are used to identify hazards and the intelligent systems suggest safer directions and speeds. Wheelchair drivers were observed while they completed tasks. Time-delays were introduced, wheelchair users undertook tests with and without intelligent systems, as delays were extended then more failures were recorded. Systems were practically tested. Results from drivers completing a series of timed routes are presented Results are compared and contrasted with recent systems and systems presented in this paper outperformed recent systems. Also the level of support provided by the systems should be adjusted depending on conditions and environments. Clinical trials will be conducted at Chailey Heritage Foundation as part of broader research conducted by the authors.

Keywords Wheelchair · Disabled · Intelligent · Sensor · Time-delay · Ultrasonic

1 Introduction

A simple intelligent system [1, 2] is described that controls a powered-wheelchair [3–6]. Ultrasonics [7] were used to recognize obstacles and recommend appropriate speeds and safe directions. Systems described in the literature that are actually in use [8–11] tend to rely on the skill of wheelchair users. Methods are described here

D. Sanders (✉) · M. Haddad · G. Tewkesbury · A. Gegov
University of Portsmouth, Portsmouth, UK
e-mail: david.sanders@port.ac.uk

T. Barker
Gems Sensors, Basingstoke, UK

M. Langner
Chailey Heritage Foundation, North Chailey, Lawes, UK

V. Sgurev et al. (eds.), *Advances in Intelligent Systems Research and Innovation*, Studies in Systems, Decision and Control 379,
https://doi.org/10.1007/978-3-030-78124-8_9

that make the driving task easier by using intelligent systems to interpret sensor and joystick signals. Several different environments are considered and time-delays and their effect on performance were then explored.

Ways of improving the way that the wheelchairs interacted with their human drivers were explored. New algorithms mixed data from the ultrasonics with data from the joysticks controlling the wheelchairs [12–16]. Controllers have often been open-loop in the past. Wheelchair users indicated a bearing and their wheelchairs moved on those indicated bearings. Systems described here processed data from driver's joysticks and sensors and then employed the generated information to help drivers.

Reaction times can be influenced by a lot of things, for example distraction, gender, age, alcohol, tiredness, personality type, physical fitness, and if the stimuli are auditory or visual [13]. As an example, a patient with Parkinson's has a longer reaction time [17].

A joystick has usually been used to steer a powered-wheelchair [18–20] although other pieces of equipment can be manipulated: pointers [21, 22], switches [23], or virtual reality transducers [24]. Wheelchair systems have used a variety of sensors to avoid obstacles and assist with steering: ultrasonics [6, 7, 25], laser or light [26], and infrared [27]. Positioning has used gyroscopes, odometry, tilt and acoustics [28]. GPS [29] and assisted GPS [30] have sometimes been useful but they can be limited for use indoors. Vision [31–37] has become cheaper and more popular and available as computing power has increased [38, 39]. But despite the fact that sensors and sensor systems have become cleverer, a disabled human wheelchair user is still often the most useful and precise data source concerning the surroundings and task in hand. In this research work, ultrasonics were selected for use because the sensors tended to be cheaper, simpler, and robust [6, 7, 25, 40, 41].

Some key conclusions from the work described here showed that wheelchair drivers may perform more quickly in relatively straightforward and uncomplicated situations when they do not have any sensors to assist them, and failure rates increase if sensors are not used and time-delays increase.

2 Wheelchair Systems

A Bobcat II wheelchair with trailing casters and 2x big driving wheels was used. Cameras were attached between the big driving wheels. Three ultrasonic transducer pairs were secured at the front of the chair; one in the center and two above each driving wheel. Transmitter/receiver pairs transmitted and received 1 ms pulses of sound and the range to obstacles was established by using the time for an ultrasonic pulse to transmit and then return.

The envelope for the three sensors and the potential grid is shown in Fig. 1.

The joystick was disconnected from the wheelchair and a microcomputer put in between to handle the control. It could operate in one of three modes:

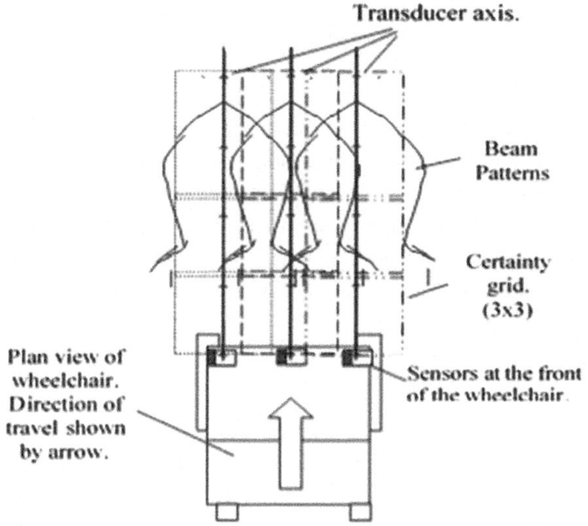

Fig. 1 Volume ahead of the wheelchair being represented by an array representing the working volume of the three sensors

- Wheelchair direction is modified using:
  - New algorithms described in this paper.
  - Some algorithms that were published recently.
- The direction of the wheelchair is modified using the joystick alone. The hierarchy and structure of the code is shown in Fig. 2.

The following rules were applied:

- Only modify trajectories when required.
- Use sensor data to create imaginary potential fields around objects [42].

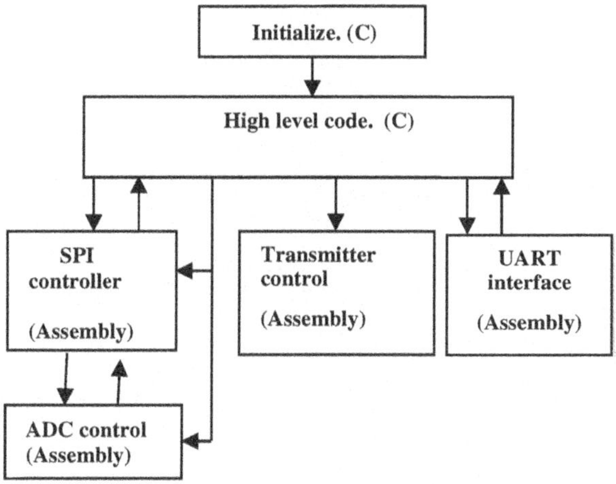

Fig. 2 Structure of the code, showing C Modules and Assembly functions

- User remains in overall control.
- Movements to be smooth and controlled.

Pulse lengths could be switched automatically by a "rangefinder". That meant that ranges could be shortened as a wheelchair moved closer to obstacles. If no obstacles were detected, the "rangefinder" slowly incremented the ultrasound pulse length to increase range.

Ultrasonics often gave misreadings as they were relatively noisy. Misreadings were filtered out with Histogramic In-Motion Mapping. Three volumes in front of the wheelchair were represented within a lattice that could be stored as an array:

- Nearby,
- Midway,
- and Distant.

Any detected object was classified as: nearby, midway or distant. Grid elements occupied by an object were incremented by 5 so that they increased in value relatively quickly. Other elements were reduced by 1 so that they reduced in value relatively slowly. Elements had a minimum value of zero and a maximum value of 15. Misreadings briefly increased the value in a grid, but then reduced in value quickly, when the systems updated. If obstacles moved into different elements then new elements increased in value and old grid elements reduced. Ranges were obtained within half a second. The object detection process is represented in Fig. 3.

Time-delays can come about because of slower reactions. These can add distortions to [43] and weaken performance [44]. To overcome this, wheelchair driving with time delays can be a compromise between transparency and stability [45]. Some approaches to control with fixed delay times [46] and varying times [47] have been presented.

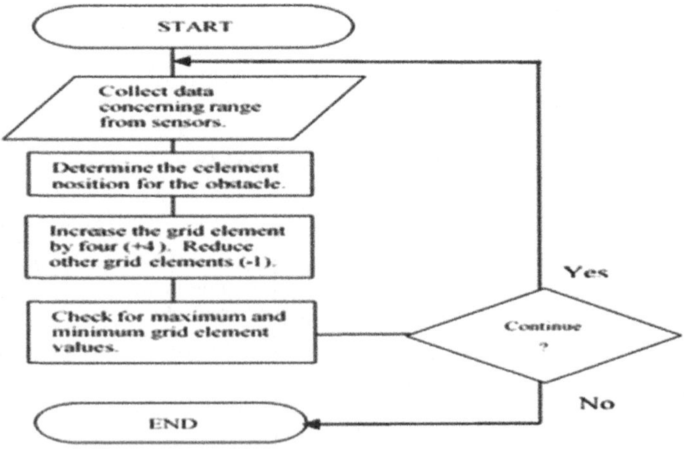

Fig. 3 Object detection

Reaction time is the time it takes for messages to travel to the brain from sensors and then to travel on to a muscle. Neurons transport messages to and from the spinal cord and brain. Many factors influence reaction times. For example, akinesia patients have significantly longer reaction times [48, 49] and any lesions in the right basal ganglia lead to extended reaction times [50].

A sensory neuron converts stimuli into an electro-chemical signal that travels along nerves to a motor neuron. A motor neuron causes glands to secrete or muscles to contract etc.

Reaction time can be affected by: Mental or Physical Fatigue, Learning Disorders, Distraction, Age, Alcohol, Warning of Imminent Stimulation, Threats, Personality Type, Stimulants, Stress, Right handed vs Left, Practice, Exercise, Errors, Stimulants, Illness and Brain Injury.

The work described in this paper was mostly interested in delays caused by brain injury, aging, illness and learning disorders. But the work also considered effects of practice, stimulants and distraction, making errors, and mental or physical fatigue on reaction times and therefore time delays.

2.1 Joystick

A Penny and Giles joystick was built-in to the wheelchairs. It used two potentiometers to produce voltages to represent joystick position. The voltages were converted by an analogue to digital converter. That produced values that represented the Cartesian co-ordinates of the joystick position. Cartesian co-ordinates were transformed into polar co-ordinates: $|\mathrm{Jo}|\ \angle\alpha$. Where $\angle\alpha$ was the angle of the joystick and represented the desired direction and $|\mathrm{Jo}|$ represented desired speed.

The angle and the magnitude values were determined f:

$$Argument = JoS0/JoS1;\ ATAN(opposite/adjacent) \tag{1}$$

$$Bearing = atan(argument);\ \angle\alpha \tag{2}$$

$$Magnitude = sqrt((JoS1 * JoS1) + (JoS0 * JoS0)) \tag{3}$$

where JoS0 and JoS1 were the Cartesian co-ordinates.

The system used $|\mathrm{Jo}|$ and $\angle\alpha$ to calculate the segment being occupied by the joystick. Position and confidence values were stored within arrays. Each segment held two quantities. “Magnitude” denoted joystick position for desired speed. “Angle Confidence” (0 to 15) denoted the likelihood that a joystick was within a sector representing a chosen direction.

The position of the joystick was symbolised by a histogram. The element in the histogram with the highest value represented the direction desired by the wheelchair user. A joystick histogram is represented in Fig. 4. A module called JoSArray tested

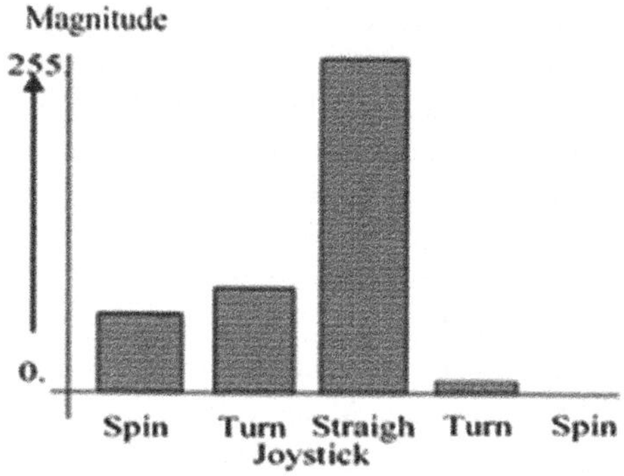

Fig. 4 Representation of joystick using histograms

the position and angle and determined which segment joystick was in. The equivalent element "angle confidence" (Anconf) increased by 60. Other Anconf elements reduced by 30. So histogram elements quickly decreased in value but took longer to grow larger.

3 Intelligent System

Experienced wheelchair drivers and rehabilitation engineers provided expert knowledge [51].

The systems had to be real time systems [52–57]. A wheelchair user provided desired velocities and the sensors gathered data about the surrounding area. A module named "Sensing-Expert" evaluated sensor data and recommended new trajectories to prevent collisions if that were necessary. Sometimes data disagreed. Then "Fuzz-Mix" wouold consider both of the inputs and decide on the output sent to the two motors. "JoMonitor" construed the intention of the driver.

The whole system was made up of: "Door", "Sensing-Expert", "JoMonitor", and "Fuzz-Mix". "Fuzz-Mix" distributed control between the joystick and the sensor system, matching the recommendation from the joystick with the recommendation from the sensor system. It arbitrated conflicts between the two and received advice from "Door", "JoMonitor" and "Sensing-Expert". "Stop" was a failsafe. It halted the wheelchair. "Fuzz-Mix" could override other inputs with "Stop". "Fuzz-Mix" mixed joystick confidence and sensor data values. If confidence was lower, then the wheelchair needed help to avoid things [58]. If confidence in the use of the joystick was higher, then the position of the joystick correctly replicated the wishes of the wheelchair user and the sensor systems had a smaller amount of influence.

"JoMonitor" monitored joystick consistency and position. Stable joystick positions confirmed that the driver was making a clear statement. If joysticks moved randomly then that implied uncertain or out of control drivers. Then the systems depended more on support from the sensors. "Sensing-Expert" applied knowledge about the sensors and sensor systems. "Sensing-Expert" placed the sensor data into a grid and then advocated promising actions to safely steer the wheelchair in order to avoid collisions. "Door" was the object avoidance program. "Door" used data from "Sensing-Expert". "Fuzz-Mix" could override "Door". Joystick data was merged with Sensor data so that:

$$V_{Out}(left) = V_{In}(left) - Dist(right) \tag{4}$$

$$V_{Out}(right) = V_{In}(right) - Dist(left) \tag{5}$$

where the vector quantities, V_{Out} was the voltage output to the controller, V_{In} was the joystick voltage, and Dist was the range to the closest obstacle. Each quantity had two values, one for each wheel (left and right).

"Door" turned the wheelchair away from the nearest obstacle, smoothly slowed the chair as the chair got close to obstacles and centered the chair between obstacles (e.g. door frames). "Fuzz-Mix" distributed control to the joystick or the sensor system depending on the environment, situation and wishes of the human wheelchair user. "Fuzz-Mix" assessed inputs and then control was distributed between the inputs:

$$\mathrm{Target_L} = (((\mathrm{JoS0} * \mathrm{Aconf[SectJo]}) + ((\mathrm{Target_L} - 125) * (255 - \mathrm{Aconf[SectJo]})))/255) + 125;$$
$$\mathrm{Target_R} = (((\mathrm{JoS1} * \mathrm{Aconf[SectJo]}) + ((\mathrm{Target_R} - 125) * (255 - \mathrm{Aconf[SectJo]})))/255) + 125;$$

where $\mathrm{Target_L}$ and $\mathrm{Target_R}$ were voltage values.

JoS0 and JoS1 = Values form the Joystick or simulated values.

Aconf [] = Confidence in values obtained form the joystick.

The algorithms used distance functions to generate values for right and left target voltages. The functions were:

$$Target_L = 2.5 * results[1] + 110; \tag{6}$$

$$Target_R = 2.5 * results[0] + 110; \tag{7}$$

where: results[] = real time ranges to objects.

Data from sensors were transformed into a type that was compatible with the Target analogue to digital converter data. The position of the joystick was required in order to make an assessment of the wheelchair user wishes. So a map of the joystick was separated into sectors: Right Spin, Left Spin, Right Turn, Left Turn, Reverse, Stop and Forward.

A microcomputer was fitted onto the wheelchair. Software was downloaded and then testing took place by driving through a variety of environments. Data was stored on a blackboard. Software was a mix of low and high level language and the whole code was compiled into a machine-level file. The file was uploaded to non-volatile memory. Algorithms were predictable and quick. If "JoMonitor" and

"Sensing-Expert" signified "forward", then a bearing straight ahead of the wheelchair was set. As the wheelchair approached obstacles then speed reduced.

Joystick data could be routed straight to the wheelchair controller without change. So the wheelchair then reacted directly to input from the joystick.

4 Testing

Wheelchairs were driven through a variety of environments. With the joystick held forward, the wheelchairs safely drove down corridors and aligned with doorways. "Sening-Expert" recommended satisfactory course adjustments and speed changes. The wheelchair successfully avoided objects and people.

The testing:

- Compared systems when they were being jointly controlled by the microcomputer and a human user, with when they were being controlled by the wheelchair user by themselves. Systems inserted varying time delays representing varying reaction times.
- Recorded the time it took to accomplish tasks using the sensor systems to help and without them helping. Delays were extended and spaces between objects were made smaller.
- Recorded width of the smallest gaps that wheelchair users could drive through. Time delays were increased and spaces between objects were made smaller. Tests were completed with sensors providing assistance and without them helping.

The sensors helped the human wheelchair users to steer their powered wheelchairs. Drivers quickly learnt how the wheelchair systems behaved. They moved their joysticks earlier when time delays were introduced and learnt to approximate stopping distances. Sets of test runs were undertaken to evaluate the speed of human drivers driving without any help and human drivers being assisted, using firstly an existing system [3, 9] and then human drivers using the new system described in this paper. Time taken to accomplish tasks was recorded when human users:

- Drove without any assistance from the sensor systems.
- Were helped by existing systems.
- Were helped by the new systems.

Human drivers drove their wheelchairs with sensor systems assisting them. Researchers recorded times taken to complete tasks using digital laboratory clocks. Additional scholars trailed the wheelchairs and counted collisions and recorded times using stop-watches. Tests:

- Captured suggestions for improvement and other comments.
- Observed tasks when wheelchairs were driven using a mix of human user and microcomputer controller control.
- Measured improvements.

- Measured time to complete tasks for human drivers without ant computer assistance and when driving with sensor systems providing assistance.

First, the tests were conducted without any sensors connected. Then the same tests were done again with sensors engaged and finally with intelligent systems providing assistance. The drivers used the following environments:

- LABORATORY.
- EMPTY CORRIDOR.
- COMPLICATED CORRIDOR.
- OUTSIDE.

Clear explanations about the study were provided that containing risks and benefits and ethics committee sanctioned tests. Tests were always completed more than once because of the variable performance of the human drivers. For each series of tests, drivers could make as many attempts as they liked (with or without the sensor assistance). So volunteers could ascertain how the computer and sensor systems behaved and then improve their performance as they learnt.

The testing led to some competition between drivers and they wanted to be better than other drivers and beat their own best times. If the drivers only finished a driving task with the sensors disconnected or only with the microcomputer and sensors, then results were rejected as a comparison couldn't be made.

The testing assessed the speed of the wheelchair users alone against their driving with help from the systems. If a quickest time was logged by a wheelchair user in a set of tests, then the user would make at least one attempt again at the other test to check that results were not just because the driver had learnt more about how the systems worked.

Numbers of successful tests and failed tests were noted as time delays were made longer. A successful run was defined as a run without any collisions. Failures were recorded if there was a collision.

The first sets of tests used routes with objects set 100 cm apart; 20 cm wider than the wheelchair (10 cm on either side). Then sets of tests were done again with narrower and narrower gaps. If a test was completed successfully with obstacles closer together then the volunteer attempted the other test (with or without the systems). Tests began from a stand still at a pre-determined start position and the distance between objects was confirmed by two or more researchers.

Delay was inserted before a movement instruction was sent to the wheelchair (h_2*in* Fig. 5). Velocity commands to the motors (v_1) could be delayed, that is signal (v_r) could be delayed. h was the whole delay. h consisted of h_2 (a delay in the forward path) and h_1 (a delay in the return path).

Figure 6 shows a Complicated Corridor. The wheelchair route is shown by arrows. Obstacles are represented by shaded blocks. The Complicated Corridor included a pair of double doorways. At these, one of the doors was open and the other door was closed. So a wheelchair would have to zig-zag to get through the doorways and around the obstacles.

Figure 7 shows the view ahead of the chair as it proceeded through the Complicated

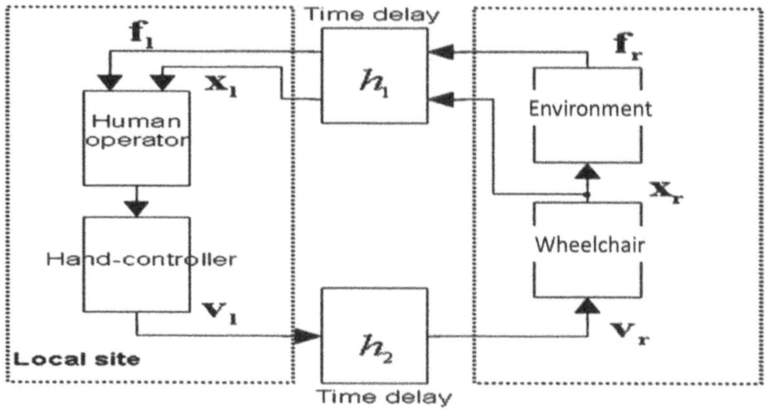

Fig. 5 Delays in the system

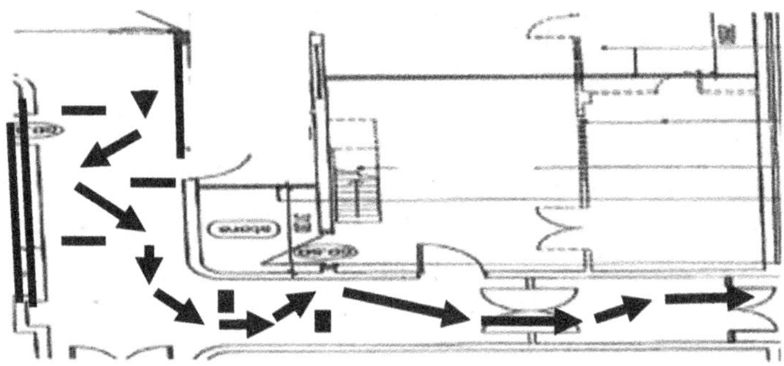

Fig. 6 Complicated Corridor

Corridor. A research student can be seen at the end of the corridor holding a digital clock. Another research student followed behind the chair with a stopwatch. The images show a test taking place with a delay of 1.5 s inserted.

5 Results

Typical results are shown in Fig. 8. The new systems described here were evaluated against the existing system and to a driver driving a chair without any assistance. The vertical scale shows the average best time to complete various routes for drivers not using any sensors (to the left in Fig. 8), employing the systems that were published recently as a comparison (in the centre) and the new systems (to the right). Different testing routes are shown along the horizontal axis.

Fig. 7 View ahead of a powered wheelchair moving along a complicated corridor

Fig. 8 Results

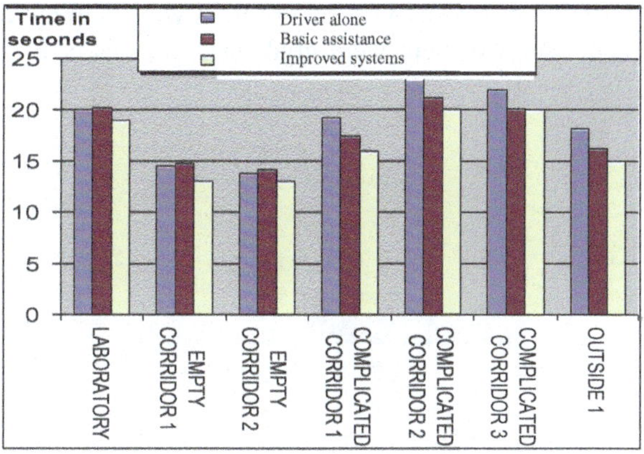

When sensor systems were connected then the wheelchairs automatically avoided obstacles.

5.1 Comparison When Using Sensors to Assist and Without Any Sensors

Figure 9 shows the average of best times to complete a number of different routes. The vertical axis shows the average of the times to accomplish successful test runs. Simpler situations are to the left in each graph. For example laboratory and empty corridors. The results demonstrate that wheelchair users could complete simpler routes more quickly without any assistance. More complicated routes are on the right, for example complicated corridors and routes outside.

Volunteers completed more complicated routes more quickly when the computer and sensor systems were working and providing assistance. The graph at the bottom of Fig. 9 shows the average of fastest times with a one second delay.

Every time a test was completed, the gaps were narrowed by half a centimetre. The narrowest gaps that a user successfully drove through were noted along with the number of successful and failed tests. Wheelchair users completed courses with narrower gaps when using the computer and sensor systems to assist them. Figure 10 shows the average improvement achieved in cm. The top graph is without any delay and the lower graph is for a one second delay. As environments became more complicated or the gaps narrowed then wheelchair users found it more and more difficult to estimate the gap widths so that it became harder for them to safely pass through the narrower gaps without colliding. Volunteers relied on their sensors and computer system to support and help them. Volunteers made it safely through narrower gaps when the microcomputer and sensors were helping them. The sensors also became more useful if there were gradients, surfaces or hills because they tended to turn

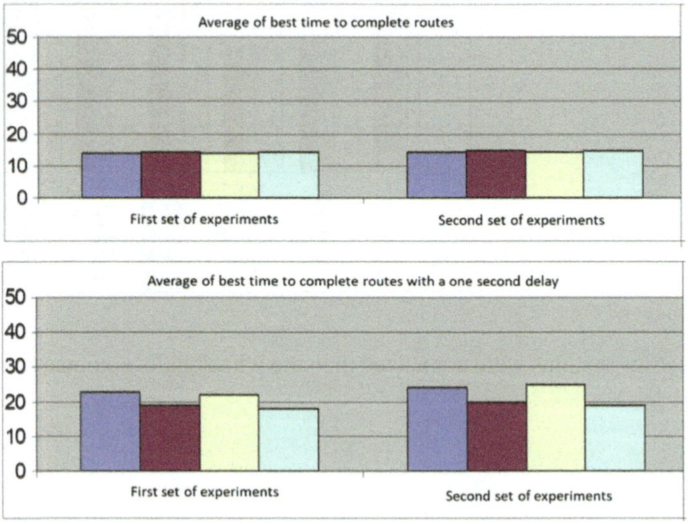

Fig. 9 Average of the fastest times to complete a Simple Corridor route

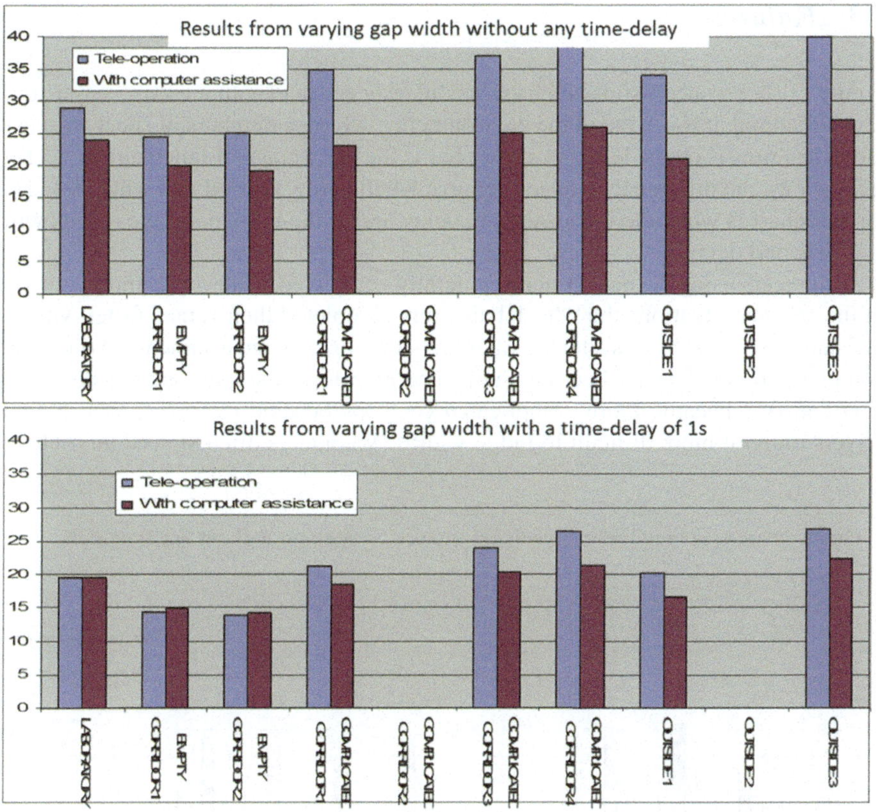

Fig. 10 Varying the distance between obstacles (with and without a time-delay of one second)

the wheelchairs. As time delays became longer then the improvement in the results became more obvious and the wheelchairs were driven faster and through narrower gaps with assistance from the sensors. Figure 10 shows the gaps that were safely accomplished with and without sensor systems engaged and without any time delay (top). The lower graph shows the gaps safely accomplished with and without the sensor systems helping but with a delay of one second.

5.2 Time Taken to Complete Test Runs

As the gaps were narrowed and the delay was increased, the quickest time to accomplish each route/task was recorded. When the distance between obstacles was larger, then the volunteers completed tests more quickly without any sensors helping them and when gaps were made narrower then volunteers completed tests/tasks more quickly with sensors and the microcomputer.

5.3 *Failures*

Figure 11 shows successful and unsuccessful tests with and without sensors assisting. The horizontal axis is a list of the environments. Average numbers of successful and failed test runs is on the left. The center bar is the percentage of failed attempts. The right shows the difference between failures when being assisted and when not. The top bar chart is when there was not any delay and at the bottom are the results with a one second delay.

Volunteers using the new systems generally completed routes/tasks more quickly. In the simpler situations the wheelchair users completed their routes faster without assistance from sensor systems. In more complicated environments, volunteers completed routes faster when they had some help form the intelligent systems. As environments became more complicated (or gaps became narrower) then human drivers found it more difficult to judge widths. Volunteers often slowed their chairs

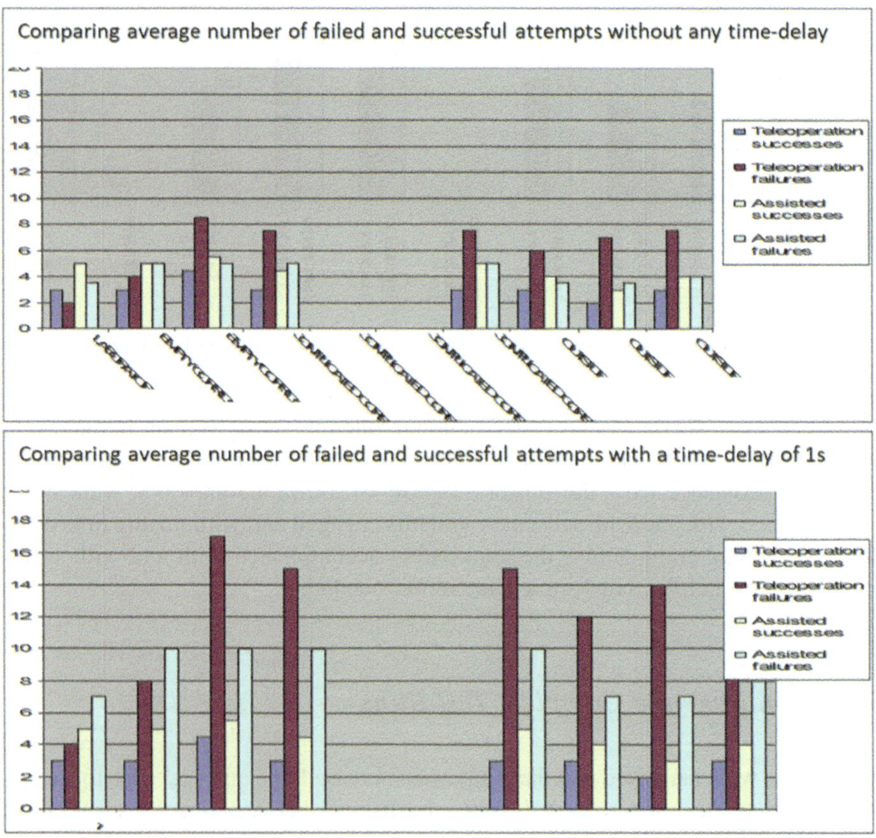

Fig. 11 Comparing average number of failed and successful attempts with 1 s delay

or stopped and had to reverse. In more complicated environments the volunteers performed better with some help.

The new systems described in this paper reliably and safely adjusted wheelchair direction and speeds and out-performed the recently published systems.

6 Discussion and Conclusions

Standard deviation was calculated and T-tests compared sample means. Paired-samples statistical tests were conducted results were in pairs. There were two sets of results: with sensors assisting and without any assistance. Pairing removed a lot of variability. Results were statistically significant. Volunteers driving without assistance and with assistance were significantly different. New systems and methods were significantly faster and safer than systems published recently.

Volunteers were quicker without sensors assisting them in simpler environments, but were quicker with sensors assisting them in more complicated situations or when there were longer delays. In simpler situations and/or with obstacles further apart, volunteers performed consistently faster without assistance. As obstacles were moved closer together or the environments were made more complex, or as delays were made longer, then volunteers found it increasingly demanding. In those situations, sensors and systems were more useful. Volunteers performed better with help in more complicated situations or when obstacles were closer together. Delays were only introduced in the forward path between the joystick and controller.

Results suggest that sensors may not be needed in freely navigable regions with good views but should be used in more complicated environments.

Current research is investigating decision making [52–57]. Intelligence [58–63], modelling and geometric representation of the environment [64] combining decision making methods [65–70] could be included.

Acknowledgements Research in this paper was supported by the University of Portsmouth and The Chailey Heritage Foundation and funded by EPSRC grant EP/S005927/1.

References

1. Sanders, D., Hudson, A.: A specific blackboard expert system to simulate and automate the design of high recirculation airlift reactors. Math. Comput. Simulat. **53**(1–2), 41–65 (2000)
2. Sanders, D., Okonor, O., Langner, M., Hassan, M., Khaustov, S., Omoarebun, P.: Using a simple expert system to assist a powered wheelchair user. In: Intelligent Systems and Applications. Advances in Intelligent Systems and Computing, vol. 1037, pp. 662–679. Springer (2019)
3. Sanders, D.: Comparing speed to complete progressively more difficult mobile robot paths between human tele-operators and humans with sensor-systems to assist. Assem. Auto. **29**(3), 230–248 (2009)

4. Haddad, M.J., Sanders, D.A.: Deep learning architecture to assist with steering a powered wheelchair. IEEE Trans. Neural Syst. Rehabil. Eng. (2020). https://doi.org/10.1109/TNSRE.2020.3031468
5. Sanders, D.A., Haddad, M., Tewkesbury, G.E., Thabet, M., Omoarebun P., Barker, T.: Simple expert system for intelligent control and HCI for a wheelchair fitted with ultrasonic sensors. In: 2020 IEEE 10th International Conference on Intelligent Systems (IS), pp. 211–216 (2020)
6. Sanders D., et al.: Intelligent control and HCI for a powered wheelchair using a simple expert system and ultrasonic sensors. In: Intelligent Systems and Applications. Advances in Intelligent Systems and Computing, vol. 1252, pp. 571–583. Springer (2021)
7. Rahiman, M., Zakaria, Z., Rahim, R., Ng, W.: Ultrasonic tomography imaging simulation of two-phase homogeneous flow. Sens. Rev. **29**(3), 266–276 (2009)
8. Sanders, D.: Controlling the direction of "walkie" type forklifts and pallet jacks on sloping ground. Assem. Auto. **28**(4), 317–324 (2008)
9. Sanders, D., Tewkesbury, G., Parchizadeh, H., Robertson, J., Omoarebun, P., Malik, M.: Learning to drive with and without intelligent computer systems and sensors to assist. In: Intelligent Systems and Applications. Advances in Intelligent Systems and Computing, vol. 868, pp. 1171–1181. Springer (2019)
10. Sanders, D., Gegov, A., Haddad, M., Ikwan, F., Wiltshire, D., Tan, Y.: A rule-based expert system to decide on direction and speed of a powered wheelchair. In: Intelligent Systems and Applications. Advances in Intelligent Systems and Computing, vol. 868, pp. 822–838. Springer, Cham (2019)
11. Sanders, D., Gegov, A., Tewkesbury, G., Khusainov, R., Sharing driving between a vehicle driver and a sensor system using trust-factors to set control gains. In: Intelligent Systems and Applications. Advances in Intelligent Systems and Computing, vol. 868, pp. 1182–1195. Springer, Cham (2019)
12. Sanders, D.A.: Comparing ability to complete simple tele-operated rescue or maintenance mobile robot tasks with and without a sensor system. Sens. Rev **30**(1), 40–50 (2010)
13. Sanders, D.A.: Analysis of the effects of time delay on the tele-operation of a mobile robot in various modes of operation. Ind. Rob. **36**(6), 570–584 (2009)
14. Haddad, M., Sanders, D., Langner, M., Omoarebun, P., Thabet M., Gegov, A.:Initial results from using an intelligent system to analyse powered wheelchair users' data. In: IEEE International Conference on Intelligent Systems, Bulgaria, pp. 241–245 (2020)
15. Haddad, M., Sanders, D., Ikwan, F., Thabet, M., Langner M., Gegov, A.: Intelligent HMI and control for steering a powered wheelchair using a raspberry pi microcomputer. In: IEEE International Conference on Intelligent Systems, Bulgaria, 2020, pp. 223–228 (2020)
16. M Haddad, M., Sanders, D., Langner, M., Ikwan, F., Tewkesbury G., Gegov, A.: Steering direction for a powered-wheelchair using the analytical hierarchy process. IEEE International Conference on Intelligent Systems, Bulgaria, pp. 229–234 (2020)
17. Draper, I.T., Johns, R.: The disordered movement in parkinsonism and the effect of drug treatment. Johns Hopkins Hosp. **115**, 465–480 (1964)
18. Stott, I., Sanders, D.: New powered wheelchair systems for the rehabilitation of some severely disabled users. J. Rehabil. Res. **23**(3), 149–153 (2000)
19. Sanders, D., Stott, I.: A new prototype intelligent mobility system to assist powered wheelchair users. Ind Rob **26**(6), 466–475 (1999)
20. Sanders, D.A., Baldwin, A.: X-by-wire technology. Total Vehicle Technology: Challenging Current Thinking, pp. 3–12 (2001)
21. Sanders, D.A., Tewkesbury, G.E.: A pointer device for TFT display screens that determines position by detecting colours on the display using a colour sensor and an ANN. Displays **30**(2), 84–96 (2009)
22. Sanders, D.A., Urwin-Wright, S.D., Tewkesbury, G.E. Gremont, B.: Pointer device for thin-film transistor and cathode ray tube computer screens. Electron. Lett. **4**(16), 894–896 (2005)
23. Sanders, D., Langner, M., Bausch, N., Huang, Y., Khaustov, S., Simandjunta, S.: Improving human-machine interaction for a powered wheelchair driver by using variable-switches and sensors that reduce wheelchair-veer. In: Intelligent Systems and Applications. Advances in Intelligent Systems and Computing, vol. 1038. pp. 1173–1191. Springer, Cham (2019)

24. Stott, I., Sanders, D.: The use of virtual reality to train powered wheelchair users and test new wheelchair systems. Int. J. Rehabil. Res. **2**(4), 321–326 (2000)
25. Goodwin, M.J., Sanders, D.A., Poland, G.A., Stott, I.: Navigational assistance for disabled wheelchair-users. J. Sys. Arch. **43**(1–5), 73–79 (1997)
26. Larsson, J., Broxvall, M., Saffiotti, A.: Laser-based corridor detection for reactive navigation. Ind. Rob. Int. J. **35**(1), 69–79 (2008)
27. Lee, S.: Use of infrared light reflecting landmarks for localization. Ind. Rob. **36**(2), 138–145 (2009)
28. Horn, O., Kreutner, M.: Smart wheelchair perception using odometry, ultrasound sensors, and camera. Robotica **27**, 303–310 (2009)
29. Milanes, V., Naranjo, J.E., Gonzalez, C., Alonso, J., De-Pedro, T.: Autonomous vehicle based in cooperative GPS and inertial systems. Robotica **26**, 627–633 (2008)
30. Lim, D., Lee, S., Cho, D.: Design of an assisted GPS receiver and its performance analysis. IEEE Symp. Circ. Syst. 1742–1745 (2007)
31. Bloss, R.: Vision and robotics team up at the 2007 Show. Ind. Rob. Int. J. **35**(1), 19–26 (2008)
32. Sanders, D., Tan, Y., Rogers, I., Tewkesbury, G.: An expert system for automatic design-for-assembly. Assem. Auto **29**(4), 378–388 (2009)
33. Sanders, D.: Environmental sensors and networks of sensors. Sens. Rev. **28**(4), 273–274 (2008)
34. Hopper, D.: The long perspective for robotic vision. Assem. Autom. **29**(2), 122–126 (2009)
35. Sanders, D., Lambert, G., Graham-Jones, J., Tewkesbury, G.E., Onuh, S., Ndzi, D., Ross, C.: A robotic welding system using image processing techniques and a CAD model to provide information to a multi-intelligent decision module. Assem. Auto. **30**(4), 323–332 (2010)
36. Sanders, D., Graham-Jones, J., Gegov, A.: Improving ability of tele-operators to complete progressively more difficult mobile robot paths using simple expert systems and ultrasonic sensors. Ind. Rob. An. Int. J. **37**(5), 431–440 (2010)
37. Sanders, D., Lambert G., Pevy, L.: Pre-locating corners in images in order to improve the extraction of Fourier descriptors and subsequent recognition of shipbuilding parts. Proc. Inst. Mech. Eng. Part B J. Eng. Manuf.**223**(9), 1217–1223 (2009)
38. Sanders, D.: Progress in machine intelligence. Ind. Rob. **35**(6), 485–487 (2008)
39. Sanders, D.: Recognizing shipbuilding parts using artificial neural networks and Fourier descriptors. Proc. Inst. Mech. Eng. Part B J. Eng. Manuf.**223**(3), 337–342 (2009)
40. Sanders, D., Stott, I.: Analysis of failure rates with a tele-operated mobile robot between a human tele-operator and a human with a sensor system to assist. Robotica **30**(6), 973–988 (2012)
41. Sanders, D., Langner, M., Tewkesbury, G.: Improving wheelchair-driving using a sensor system to control wheelchair-veer and variable-switches as an alternative to digital-switches or joysticks. Ind. Rob. Int. J. **37**(2), 157–167 (2010)
42. Khatib, O.: Real-time obstacle avoidance for manipulators and mobile robots. In: Proceedings of IEEE International Conference on Robotics and Automation, vol. 5, no. 1, pp. 90–98. Springer, New York, NY (1986)
43. Fiorini, P., Oboe, R.: Internet-based telerobotics: problems and approaches. In: Proceedings of the 8th ICAR'97, Monterey, CA. Teleoperation of robots with delay, vol. 681. pp. 765–770. IEEE (1997)
44. Richard, N.: Time delay systems: an overview of some recent advances and open problems. Automatica **39**, 1667–1694 (2003)
45. Kim, W., Hannaford, B., Bejczy, A.: Force reflection and shared compliant control in operating telemanipulators with time delay. IEEE Trans. Robot. Autom. **8**(2), 176–185 (1992)
46. Slawinski, E., Mut, V., Postigo, J.F.: Teleoperation of robots with time-varying delay. Robotica **24**, 673–681 (2006)
47. Brumlik, J., Boshes. B.: The mechanism of Bradykinesia in Parkinsonism. Neurology (Minneap) **16**, 337–44 (1966)
48. Joubert, M., Barbeau, A.: Akinesia in Parkinson's disease. Progress in Neurogenetics. Excerpta Med. Found. 366–376 (1969)

49. Wiesendanger, M., Schneider, P., Villoz, J.P.: Electromyographic analysis of a rapid volitional movement. Am. J. Phys. Med. **48**, 17–24 (1969)
50. Kosinski, R.J.: A literature review on RT kinds of RT experiments. **10** (1) https://www.semanticscholar.org/paper/A-Literature-Review-on-Reaction-Time-Kinds-of-Time-Kosinski (2012)
51. Hudson, A., Sanders, D., Golding, H., Tewkesbury, G.E., Cawte, H.: Aspects of an expert design system for the wastewater treatment industry. J. Syst. Architect. **43**(1–5), 59–65 (1997)
52. Sanders, D.: Introducing AI into MEMS can lead us to brain-computer interfaces and super-human intelligence (invited viewpoint review paper). Assembly Autom. **29**(4), 309–312 (2009)
53. Sanders, D.: Ambient-intelligence, rapid-prototyping and where real people might fit into factories of the future. Assembly Autom. **29**(3), 205–208 (2009)
54. Sanders, D., Haynes, B., Tewkesbury, G., Stott, I.: The addition of neural networks to the inner feedback path in order to improve on the use of pre-trained feed forward estimators. Math. Comput. Simulat. **41**(5–6), 461–472 (1996)
55. Sanders, D.: Perception in robotics. Ind. Rob. **26**(2), 90–92 (1999)
56. Sanders, D.: System specification 2. Microproc. Microprog. **38**(1–5), 833–833 (1993)
57. Sanders, D., Harris, P., Mazharsolook, E.: Image modelling in real-time using spheres and simple polyhedral. In: 4th International Conference on Image processing and its applications, vol. 354, pp. 433–436 (1992)
58. Sanders, D., Hudson, A., Tewkesbury, G.: Automating the design of high-recirculation airlift reactors using a blackboard framework. Expert Syst. Appl. **18** (3), 231–245. IET, London (2000)
59. Sanders, D.: Real time geometric modelling using models in an actuator space and Cartesain space. J. Robotic. Syst. **12**(1), 19–28 (1995)
60. Fahimi, F., Nataraj, C., Ashrafiuon, H.: Real-time obstacle avoidance for multiple mobile robots. Robotica **27**, 189–198 (2009)
61. Tewkesbury, G., Sanders, D.: The automatic programming of production machinery for de-flashing plastic parts. Adv. Manuf. Technol. **VIII**, 279–283 (1994)
62. Tewkesbury, G., Sanders, D.: The use of distributed intelligence within advanced production machinery for design applications. Total Veh. Technol. Challenging Curr. Thinking, 255–262 (2001)
63. Sanders, D., Robinson, D., Hassan Sayed, M., Haddad, M., Gegov, A., Ahmed, N.: Making decisions about saving energy in compressed air systems using ambient intelligence and artificial intelligence. In: Intelligent Systems and Applications. Advances in Intelligent Systems and Computing, vol. 869, pp. 1229–1236. Springer, Cham (2019)
64. Sanders, D., Wang, Q., Bausch, N., Huang, Y., Khaustov, S., Popov, I.A.: Method to produce minimal real time geometric representations of moving obstacles. In: Intelligent Systems and Applications. Advances in Intelligent Systems and Computing, vol. 868, pp. 881–892.Springer, Cham (2019)
65. Haddad, M., Sanders, D., Gegov, A., Hassan, M., Huang, Y., Al-Mosawi M.: Combining multiple criteria decision making with vector manipulation to decide on the direction for a powered wheelchair. In: Intelligent Systems and Applications. Advances in Intelligent Systems and Computing, vol. 1037, pp. 680–693. Springer, Cham (2019)
66. Haddad, M., Sanders, D., Bausch, N.: Selecting a robust decision making method to evaluate employee performance. Int. J. Manag. Decis. Mak. **18**(4), 333–351 (2019)
67. Haddad, M., Sanders, D., Tewkesbury, G.: Selecting a discrete multiple criteria decision making method to decide on a corporate relocation. Arch. Bus. Res. **7**(5), 48–67 (2019)
68. Haddad, M., Sanders, D.: Selecting a best compromise direction for a powered wheelchair using PROMETHEE. IEEE Trans. Neural Syst. Rehabil. Eng. **27** (2), 228–235 (2019)
69. Haddad, M., Sanders, D., Tewkesbury, G., Gegov, A., Hassan, M., Ikwan, F.: Initial results from using preference ranking organization methods for enrichment of evaluations to help steer a powered wheelchair. In: Intelligent Systems and Applications. Advances in Intelligent Systems and Computing, vol. 1037, pp. 648–661. Springer, Cham (2019)

70. Haddad, M., Sanders, D., Bausch, N., Tewkesbury, G., Gegov, A., Hassan, M.: Learning to make intelligent decisions using an Expert System for the intelligent selection of either PROMETHEE II or the analytical hierarchy process. In: Intelligent Systems and Applications. Advances in Intelligent Systems and Computing, vol. 868, pp. 1303–1316. Springer, Cham (2019)

Implementation and Experimental Validation of the IEC 63047 Standard for Data Transfer in Radiological and Nuclear Robotic Applications

Frank E. Schneider, Dennis Wildermuth, Francisco Garcia Rosas, and Jan Paepen

Abstract Conventionally, data from radiation detectors is integrated over time, delivering a simple dose rate or energy spectrum. The time integration over several seconds makes it hard to locate a radioactive source when either the detector or the source is moving. A novel approach is to use the detector's much higher internal data rate and combine it with precise time and position information. In principle, this so-called list-mode data acquisition allows innovative algorithms that can locate and identify a radioactive source much faster than in the conventional approach. IEC 63047 is a recently introduced binary standard for list-mode data collection, which is applicable to data files and streams. The standard is specified using Abstract Syntax Notation One (ASN.1). The format supports various types of time-stamped data and can be used in a wide range of applications involving radiation detection and measurement. It may also be used to represent data from other sensors than radiation detectors and supports positioning data from Global Navigation Satellite Systems (GNSS). The standard has a wide potential for CBRNE detection equipment, including robotics. To facilitate the use of the standard, an open-source solution for encoding and decoding IEC 63047 messages has been implemented and tested. In addition, a simple demonstration device was developed from off-the-shelf components. Extensive laboratory and field tests were conducted to show that the standard itself as well as the demonstration device are suitable for typical radiological and nuclear (RN) inspection and monitoring tasks. For a better use with unmanned robot systems, a software interface between IEC 63047 and the Robot Operating System (ROS) was implemented. Finally, using this ROS interface, the demonstration device

F. E. Schneider · D. Wildermuth (✉) · F. G. Rosas
Cognitive Mobile Systems (CMS), Fraunhofer Institute for Communication, Information Processing and Ergonomics FKIE, 53343 Wachtberg, Germany
e-mail: dennis.wildermuth@fkie.fraunhofer.de

F. E. Schneider
e-mail: frank.schneider@fkie.fraunhofer.de

J. Paepen
European Commission—Joint Research Centre (JRC), Geel, Belgium
e-mail: jan.paepen@ec.europa.eu

V. Sgurev et al. (eds.), *Advances in Intelligent Systems Research and Innovation*, Studies in Systems, Decision and Control 379,
https://doi.org/10.1007/978-3-030-78124-8_10

was integrated into an experimental robot system and successfully tested under RN field conditions at the European Robotics Hackathon (EnRicH).

Keywords Radiological and nuclear (RN) · Radiation detection · Standardization · Data transmission · List-mode data acquisition

1 Introduction

Natural disasters, industrial catastrophes and terror acts pose an unpredictable yet significant risk to everyone's live and prosperity. Properly assessing such situations, especially in combination with radiological and nuclear (RN) threats, is a significant challenge. Several incidents in the past decades (e.g. Chernobyl, Sellafield, Fukushima, Asse) have underlined the urgent need for robotic platforms in scenarios involving radiation. However, although robotics research has delivered impressive results in general, there is still significant room for improvement with respect to the use of robots in RN-related applications. Among others, one major drawback is the absence of standards, best practices and norms.

The International Electrotechnical Commission (IEC) published a new international standard, IEC 63047, specifying a data format for list-mode digital data acquisition used in radiation detection and measurement [1]. IEC 63047 was developed under the lead of the European Commission's Joint Research Centre (JRC) in Geel, Belgium. The pre-normative research for this standard was performed by the European Reference Network for Critical Infrastructure Protection (ERNCIP), in particular by its Thematic Group on Radiological and Nuclear threats to critical infrastructure (RN TG). This is a group of experts discussing various topics related to critical infrastructure, under the coordination of the JRC in Ispra, Italy.

The JRC and the ERNCIP RN TG promote the use of the standard in a wide range of applications involving radiation detection and measurement, novel detector technologies, nuclear security, reachback/expert support and robotics. The standard can be applied to chemical, biological, radiological, nuclear and explosives (CBRNE) detection equipment because it includes features to represent data from any kind of sensor, including geolocation.

The JRC developed a demonstration device from off-the-shelf components and proposed and tested an open-source solution for encoding and decoding binary IEC 63047 messages. The ERNCIP RN TG Robotics subgroup developed a software interface between IEC 63047 and the Robot Operating System (ROS) [2]. The demonstration device consists of a single-board computer, a spectrometric radiation detector and a Global Navigation Satellite System (GNSS) receiver. Tests performed at the JRC site in Geel assessed the performance of the demonstration device in laboratory and under field conditions. The Fraunhofer FKIE successfully integrated the demonstration device and the ROS software interface on one of their experimental land robot systems. The system took part in the European Robotics Hackathon (EnRicH) 2019 in Zwentendorf, Austria, and delivered impressive results, showing that the use of IEC 63047 for data transmission is a feasible approach for radiological and nuclear robotics applications.

2 Background

Since the late 1980s, robotic solutions have been utilised in several response efforts to catastrophes and natural disasters. Lessons learned from the Chernobyl and Fukushima disasters and from decommissioning of old nuclear power plants show that also in the RN domain robots have important advantages. They can operate in extreme surroundings with high radiation or danger of explosions, extreme heat or collapsing structures. They can manipulate the environment and take samples. Robots can also be used for long-time surveillance of contaminated areas and for monitoring the movements of a potential threat, using real-time data from mobile sensors. Possible scenarios can be roughly divided into two groups. First, there are prevention scenarios where unmanned systems can be used to prevent incidents involving radioactive material. Second, there are response scenarios where unmanned systems gather information after RN incidents have occurred. In both cases, three main tasks for robotic systems can be identified: spatial mapping of RN sensor data, active search for RN sources, and sampling (e.g. air sampling, sweep sampling or material sampling) which often requires mobile manipulation.

Only very few robotic systems have been designed for use in scenarios involving radiation, e.g. [3, 4], which present general purpose approaches for all the above mentioned tasks. Many systems, however, solely concentrate on finding, mapping and classifying radiation sources [5, 6]. Generally, most of these systems are research projects or technical demonstrators. Only on rare occasions, robots have been used in real RN disaster environments, and this usually caused huge adaption and hardening efforts, while at the same time producing only limited results [7, 8].

The European Reference Network for Critical Infrastructure Protection (ERNCIP) has therefore established a Thematic Group on the protection of critical infrastructure from Radiological and Nuclear threats (RN TG). The group looks at issues such as certification of radiation detectors, standardisation of deployment protocols, response procedures and the use of unmanned systems. Possible threats involve measurement and sampling scenarios that are too risky for humans to carry out, also in the event of criminal or unauthorised acts involving nuclear or other radioactive material.

In order to systematically analyse the field, the ERNCIP RN Thematic Group conducted a survey to assess the state of the art of unmanned systems with potential to be used for radiation measurements and sampling. One of the findings was that no standards, best practices or norms for sampling or taking measurements with unmanned systems have been introduced so far. Thus, a first set of basic standards for unmanned systems in RN measurement scenarios was developed. A widely accepted standard framework for robot software development is the Robot Operating System (ROS) [2]. Further important standards concerning communication with robots and control of unmanned systems are, for example, Battle Management Language (BML)

[9], Interoperability Profile (IOP) [10] and Joint Architecture for Unmanned Systems (JAUS) [11]. Furthermore, there have been efforts for standardisation in the International Electrotechnical Commission (IEC) regarding international standards for RN measurements with unmanned systems for several years already.

2.1 Benefits and Challenges of Robotics

Critical infrastructures like bridges, train tracks or nuclear power plants require regular inspections, maintenance, but also safe demolition and fast intervention in case of disasters. This kind of operations usually involves tasks of a great danger for the workers employed, but also for the nearby population in general. The United States Nuclear Regulatory Commission's regulation 10 CFR (Code of Federal Regulations) part 20—Standards for Protection Against Radiation—stipulates that a worker cannot receive a full body dose of more than 50 mSv per year. According to the Council Directive 2013/59/EURATOM, in European countries the yearly occupational dose is even limited to 20 mSv. Clearly, there is a need to replace the human worker on daily base maintenance work and especially in the case of emergency situations and nuclear decommissioning.

Autonomous or tele-operated robotic solutions could be of great help to this kind of operations, making them a lot safer and easier to process. Examples for such scenarios include decommissioning of nuclear power plants, clean-up after landslides, or crime scene interventions after terrorist attacks involving hazardous materials. One of the main advantages of robots in these scenarios is the protection of the personnel involved. Unmanned systems can operate in environments that under normal conditions are inaccessible to humans. Since the operating time of a robot is generally only limited by the battery charge condition it also facilitates 24/7 operations. Other benefits include

- operation in areas with high radiation as well as danger of explosives, collapsing structures, or heat;
- the ability to take samples and to manipulate;
- long-time measuring and surveying in contaminated areas;
- monitoring the movements of a (potential) threat;
- providing real-time data from multiple on-board sensor sources.

Unmanned systems also allow dirty, dull and dangerous repetitive work in pre-disaster or security-related operations. One application is the exploration and mapping of large areas. A typical example might be a container terminal at a seaport that has to be monitored for illicit trafficking of potentially dangerous radioactive material. Quite similar are scenarios which involve extensive patrolling, e.g. prior and during a major public event.

Due to its rich endowment with multiple sensors a robot is able to contribute enormously to the situation awareness of the involved personnel. It does not only deliver real time video but—equipped with 3D laser scanners—it is also possible to enable virtual reality output. This allows the operator to be in-situ at the place of the incident. The virtual reality environment can be attributed by real time RN measurements and accumulated measurements of a longer period.

There are challenges for utilizing robots on nuclear security. For example, major parts of reactors that are currently operational were commissioned before 1990. That time mobile robotic systems were still on their infancy and not spread at all. Thus, nuclear power plant infrastructure is not designed to support mobile robots. So far, successful deployments have been specially designed and tailored for specific use cases. In addition, existing robot technologies have only proven to be useful to assess a situation, but they are often too immature or fragile for the use in a real decommissioning, clean-up, or deconstruction scenario.

Fortunately, nuclear power plant accidents are very rare. Major accidents like Three Mile Island and Chernobyl happened before mobile robots became effectively usable. The last big accident in Fukushima confirmed drastically how important it is to assess the site conditions as soon as possible in order to provide reference information for emergency disposal. To achieve that, collaborative robotics approaches can be a feasible option. Additionally, there might be benefits by using teams of robots from different domains—land, air and water—to map the situation quickly and plan the recovery operations more precisely.

To enable experts and decision makers to build up real-time situation awareness, the data acquired by the robot has to be communicated seamlessly and loss-less. Clearly, the standardisation process with regard to communication, interfacing and interoperability needs a significant boost.

2.2 Benefits and Challenges of List-Mode Data Acquisition

In the context of RN measurements, list-mode data acquisition is the technique of recording data obtained from the individual pulses observed directly at the detector output. The technique has already been in use for a few decades, but its real advantages emerge when applied to novel detector systems using high-performance digital electronics, especially in applications where time-correlated data from one or several detectors is to be combined [12].

For example, in nuclear security the availability of time-stamped list-mode data enables automatic rejection of nuisance alarms caused by coincident neutrons generated by cosmic radiation and therefore improves the sensitivity of the detector system to nuclear material. More recently developed systems such as Compton imagers [13] and advanced muon tomography systems [14] rely on time stamped data to operate.

Because list-mode data acquisition with high-performance digital electronics observes the signal directly at the detector output and because all settings and parameters affecting the data acquisition process can be remotely set by software, the technique enables remote diagnosis of underperforming or malfunctioning detectors. Equipment manufacturers have the specialised knowledge in digital systems and can connect to the systems for remote diagnostics. For a number of cases, the technical expert no longer needs to go on site to adjust the system.

However, to realise such functionalities, comprehensive and detailed data needs to be shared. Moreover, to fully exploit novel detection instruments, the comprehensive data shall reach the reachback centre untampered over a reliable connection. Thus, it is clear that standards play an important role in enabling interoperability between these systems' hardware and software. Data format standards are an essential first step in improving interoperability. While the well-established data format standard IEC 62755 [15] (identical to ANSI N42.42) provides a solution at a higher level, it is not suitable to send low-level list-mode data. IEC 63047 is a binary data format standard for list-mode data and complements IEC 62755. The ERNCIP RN TG had a key role in developing the list-mode data format standard IEC 63047, which was published in October 2018. The standard provides the necessary features to represent data extracted from pulses observed at the detector output. In addition to this pure list-mode data, the standard IEC 63047 can be used to represent other types of data such as geolocation information and histograms such as energy spectra. The latter have been included deliberately to support the transition phase from conventional electronics to digital instrumentation. In addition to data from radiation detectors, the standard can also represent timestamped measurement results from any kind of sensor.

A challenge for integrated systems that need to share more comprehensive and detailed data is the availability of reliable and secure network connections with sufficient bandwidth. Inevitably, requirements for data storage will increase.

When timestamped list-mode data from spatially separated systems needs to be combined, the systems must be accurately synchronised. It is challenging yet feasible to realise such synchronisation for unwired (e.g. mobile) systems. For example, time synchronisation with an accuracy below 1 ns can be realised via GPS (or other GNSS) receivers that have multiple satellites in view. While this is generally sufficient for most applications, some applications using fast detector materials may require a timestamp resolution a few orders of magnitude higher.

Low-level data format standards such as IEC 63047 improve the interoperability between hardware and software. While this is expected to be a bless for system integrators and developers of software systems for data analysis, the success of the standard will depend on the willingness of manufacturers who offer complete systems to implement the standard as an alternative to the proprietary data format that they use between the hardware and software components.

3 IEC 63047 for List-Mode Data Acquisition

Depending on the application and the semantics of the information, different data formats are required. As described in the last section, very detailed timestamped raw data from CBRNE sensors may be represented in the list-mode data format IEC 63047. However, acquiring such detailed data may require a relatively high bandwidth, for example when RN sensors acquire data from individual interactions in the detector in a high radiation field. Apart from individual event data, the standard can also be used to send—amongst others—measurement results and positioning data.

3.1 IEC 63047 in Comparison to Other Data Formats

The IEC 62755 data format, which is better known but identical to ANSI N42.42, is commonly used by nuclear security equipment deployed against illicit trafficking of RN material. IEC 62755 can represent spectra, identification results and information on the calibration of a detector. The XML-based format has no tags to represent detailed list-mode event data. In addition, the size of IEC 62755 text files will become too large for list-mode data. However, IEC 62755 can be used to transfer a piece of binary IEC 63047 data. Vice versa, IEC 63047 may be used to transmit an IEC 62755 data file. The two standards are in fact complementary, with IEC 63047 leaning more toward applications where raw sensor data is required.

There are several other standards for RN-related data transfer. A complete discussion is far beyond the scope of this work, but some of them will be mentioned for completeness. The military standard NATO ATP-45 (Warning and Reporting and Hazard Prediction of Chemical, Biological, Radiological and Nuclear Incidents), for instance, was specifically developed for handling CBRN incidents, but is mainly used in the military domain. The International Radiological Information Exchange format (IRIX), as another example, is an XML data format to facilitate exchange of radiological information and data between organizations during a nuclear or radiological emergency. Starting in 2005, IRIX was developed by the International Atomic Energy Agency (IAEA) [16], and since 2013, for instance, it is fully implemented by the European Radiological Data Exchange Platform (EURDEP) [17]. IRIX data can be imported into the EURDEP platform for exchange between countries for expert support, regardless of the origin of the data: from fixed measurement stations or mobile radiation sensors deployed after emergency or mounted on robots. Obviously, IRIX is intended for high-level data exchange, comparable to the IEC 62755/ANSI N42.42 data format, and thus addresses different applications than the rather low-level IEC 63047.

In modern systems, digital data acquisition instruments are directly connected to the output of radiation detectors. Contrary to conventional data acquisition, the analogue-to-digital conversion occurs at the beginning of the electronics chain. The processing of the signals observed at the output of the radiation detector occurs on an event-per-event basis. Digital electronics is programmed to extract only the characteristics of interest, which is typically (but not limited to) the energy of the radiation observed in the detector and the timestamp at which the interaction took place.

List-mode data acquisition is of particular interest when data from more than one detector is combined. IEC 63047 provides the means to digitally represent data from a large number of detectors. Because of the high time resolution of some detector systems and large number of interactions that can be processed, the IEC 63047 is a binary data format. State-of-the-art electronics can sample at a rate up to several billion samples per second. Used with fast detectors, they are able to record time stamp and deposited energy for several million events per second, and do this continuously.

The timestamp and energy of a pulse are not the only properties of an interaction that are of interest. Sometimes the length of the pulse is required, given by the time that the signal exceeds a threshold. In addition, the rise or fall-time of the pulse can be of interest. In advanced cases, the individual samples that represent the pulse are needed for processing by software.

3.2 *ASN.1-Based Specification*

IEC 63047 describes a binary format using Abstract Syntax Notation One (ASN.1) (ISO/IEC 8824) [18]. ASN.1 is a standardized syntax used to describe data. The notation does not depend on the programming language implementation, nor the physical representation of the data, and consequently is vendor or platform independent.

The ASN.1 standards can be divided in two groups. The first group of standards, ISO/IEC 8824-1 (ITU-T X.680) to ISO/IEC 8824-4 (ITU-T X.683), specify the ASN.1 notation itself: the syntax used to define data types. The second group of standards, ISO/IEC 8825-1 (ITU-T X690) to ISO/IEC 8825-7 (ITU-T X.696), specify different encoding and decoding rules. These rules determine the precise bit-patterns of the data values. These encoding rules are completely independent of the ASN.1 notation itself, which means that, during the development process of a data format, the developers do not have to bother about how the data values will be stored on a medium or transferred to another computer system.

Besides the advantages originating from the complete separation of the data format syntax and its encoding rules, using ASN.1 has more benefits. The specification of the ASN.1 syntax and the encoding rules are platform and language independent. ASN.1 is extremely reliable, and has been implemented with success in many systems, since its first development in 1984 by the organization that is now known as the International Telecommunication Union (ITU), an Agency of the United Nations. ASN.1 can be used on embedded systems. It is extensible, which means that data can be exchanged between new and old systems: When a data format is updated, the old systems that were built on the previous version of the format can read the new data format. (Of course, the old system is unable to interpret the new elements that were added since it is unaware of how these new elements are formatted.) ASN.1 has several encoding rules. Data messages can be transmitted in XML as text, but also binary in a very compact way. The choice of the encoding rule depends mainly on the required throughput, available resources (both memory and speed) for the encoder/decoder and the need for the data to be human-readable or machine-readable.

Because the data format syntax and the encoding rules are well standardised, commercial and open-source tools exist to support the development and the use of data formats specified by means of ASN.1. One of the most powerful tools are the so-called ASN.1 compilers. These compilers are different from programming language compilers, which generate binary code that can be executed by a processor. ASN.1 compilers generate source code from the ASN.1 specification of a data format. An ASN.1 specification of a data format is a text file that precisely defines structured data types. The ASN.1 specification of IEC 63047 is included in Annex A of the standard, which is provided as a separate text file together with the PDF of the standard. From this piece of text and following the selected encoding rule, the ASN.1 compiler generates source code. The source code not only contains the type definitions, but also the routines for encoding and decoding data values.

ASN.1 compilers exist for a large number of programming languages and operating systems, supporting the various encoding rules. For a quick time-to-market, commercial solutions exist, but there are also free, open-source ASN.1 compilers. The use of these compilers is not obligatory. For simple data type definitions, it may be more efficient to develop encoding and decoding software for the specific application, based on the standard that precisely describes the encoding rule.

3.3 Exemplary Open Source Implementation

To lower the threshold and facilitate the use of the standard, JRC-Geel provides an open-source solution for writing (encoding) and reading (decoding) IEC 63047 data. The solution uses the open-source software `asn1c`, developed by Lev Walkin and contributors, which is available on GitHub [19]. This section provides guidance for implementing the standard in radiation detection systems and gives an overview how to use `asn1c` to generate an IEC 63047 application. For a detailed documentation of this open-source solution for encoding and decoding IEC 63047 data refer to Paepen and Lutter [20].

Figure 1 presents the ASN.1 workflow, applied to the standard IEC 63047. In general, four steps have to be followed to build a codec application. First, the `asn1c` source code has to be obtained. The open source software can be downloaded from GitHub and is free for private, educational and commercial use [19]. The application can be easily compiled an installed in all current major Linux distributions. In step 2, `asn1c` has to be used to generate C source code (`.h` and `.c` files) from the data format specification in Annex A of IEC 63047 [1]. (A text file with the ASN.1 specification is provided when obtaining the standard.) The C source code defines native C structures, which correspond to the IEC 63047 ASN.1 syntax, and provides encoding and decoding routines that convert the C values to the bit patterns and back, as defined by the Canonical Octet Encoding Rule (C-OER) [21]. This can be either done script-based and automatically or manually, following the step-by-step instructions given in [20].

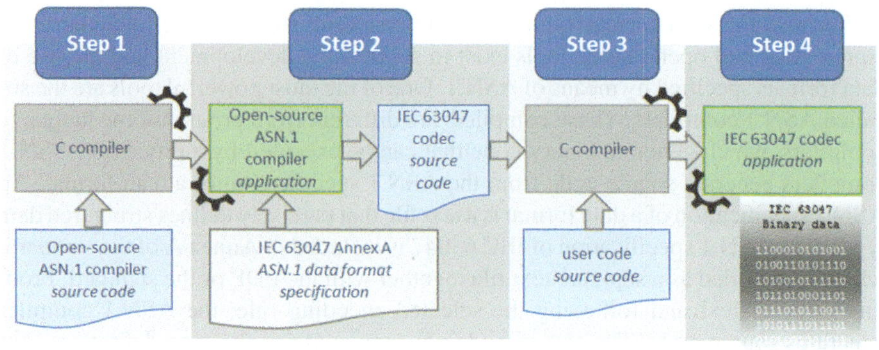

Fig. 1 ASN.1 workflow (after Paepen and Lutter)

After the codec source code has been generated it can be compiled and linked together with user code (or demo code provided by JRC-Geel) into an application that writes or reads IEC 63047 compliant binary data. The JRC demo code uses the C structures and encoding/decoding routines generated by `asn1c` to demonstrate how to encode and decode IEC 63047 data. Finally, as depicted in step 4 of Fig. 1, you can start using the IEC 63047 codec application. In case of the JRC demo application, this means the user can choose between two operating modes:

- Encoding: Randomly sampled data is encoded into the IEC 63047 format and saved to a file.
- Decoding: IEC 63047 formatted data is read from a file and printed on the screen.

In encoding mode, the code simulates the events recorded by a digital data acquisition instrument acquiring data from a NaI well-type detector. Every event contains a timestamp and a pulse height. The time between two successive events is randomly generated and determined by a parameter that represents the average count rate. The pulse height is randomly drawn from a histogram with 4096 channels. The demo code includes two histogram files of the radionuclides ^{111}In and ^{18}F, recorded on an $8'' \times 8''$ NaI well detector. The open source demo code is completely free for any kind of use and modification, and thus can be taken as an entry point for developing own IEC 63047 list-mode applications.

Generally, IEC 63047 can be applied to data files and streams. An IEC 63047 data message is a header, an event list or a footer. When stored to a file or sent to another computer system, the data shall be of the header type, followed by one or more data packages of the event list type and finally concluded with one encoding of the footer type. In addition to representing typical list-mode data from radiation detectors (e.g. timestamp, pulse height), the standard can be used to represent the following types of data:

- digital signals, allowing the use of the digital data acquisition device as a kind of oscilloscope, for example for troubleshooting or to process the digital signals by advanced analysis routines which cannot be implemented in the device itself;

- logic values and transitions of logic values, which can be used for example to represent status information or alarms;
- one and two-dimensional histograms, for example energy spectra;
- position, orientation, altitude and speed of objects, which is of particular interest in mobile applications;
- measurement results obtained by any kind of sensor;
- any kind of binary data or text message, for example NATO ATP-45, IEC 62755, images.

All these data types can be timestamped. Any mixture of them can be recorded in event lists. The standard specifies that IEC 63047 data shall be encoded using the Canonical Octet Encoding Rule (C-OER) [21]. Canonical encoding enables authentication and encryption, if this is required by the actual application. However, it is important to mention that the standard does not strictly impose the use of any authentication and encryption algorithm. There are software tools that can convert binary IEC 63047 data into the human-readable XML format.

4 Technical Approach

An IEC 63047 demonstration system was developed with components readily available on the market. Figure 2 presents the functional layout of the demonstration environment. One or more mobile demonstration devices operate in a wireless network and send data to a control centre.

The device can be built for less than 3700 €, making it ideal to be used in training and education, but also for developing sensor networks which could, for example, be deployed in case of CBRNE incidents. The relatively low cost and weight makes it also suitable for robotics applications. The device consists of a single-board computer extended with a GNSS receiver board and an antenna. A radiation detection module is connected to the single-board computer via USB. Power is delivered from a battery pack via a switching regulator. Alternatively, a power bank may be used. See Fig. 3 for a schematic drawing and Fig. 4 for the actual assembled device.

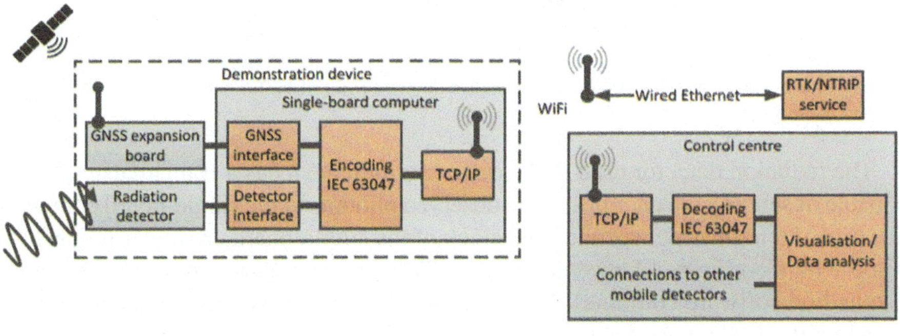

Fig. 2 Functional layout of the demonstration device

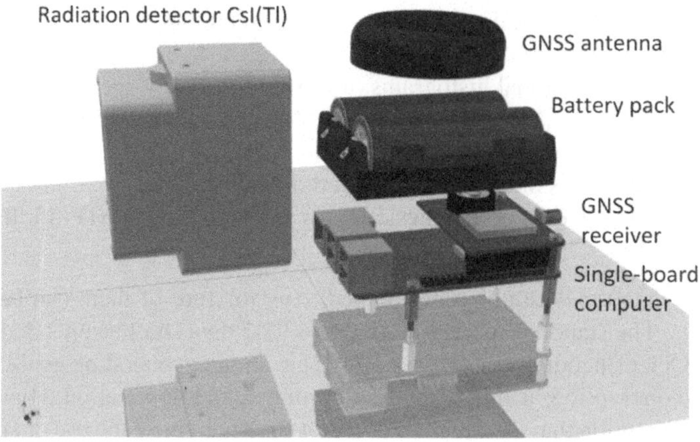

Fig. 3 Components of the IEC 63047 demonstration device as schematic drawing

Fig. 4 The assembled IEC 63047 demonstration device (without batteries)

The radiation detector used in the demonstration device is a Hamamatsu C12137 capable of sending the energies of the observed photons to the single-board computer at a rate of ten times per second. The clock of the single-board computer is synchronised with Coordinated Universal Time (UTC) via the GNSS connection. The data acquisition software combines the radiation data with the position and encodes the data into the IEC 63047 format, using the open-source software `asn1c` [19].

Encoded data can be stored locally on the single-board computer or transmitted to the control centre over a wireless ROS connection. Alternatively, systems that do not use ROS can receive IEC 63047 data packages over a TCP socket. Note that, in this demonstration environment, a Wi-Fi network was used but in real scenarios the use of a 3G/4G mobile communication network is more appropriate. This is however merely a small technological difference.

The Hamamatsu C12137 consists of a CsI(Tl) scintillator coupled to a multi-pixel avalanche photodiode. These photodiodes are operated in Geiger mode, which results in a high internal gain that enables single photon detection. The device also offers high photon detection efficiency, excellent timing resolution, low bias voltage operation, ruggedness, resistance to excess light, and immunity to magnetic fields.

4.1 Preliminary Tests

The demonstration device was tested at the JRC-Geel site. Special emphasis was put on testing the performance of the system from the point of view of fulfilling the requirements of the standard IEC 63047 and its applicability to field conditions for robotics.

4.1.1 Tests of the Radiation Detector in Laboratory Conditions

The basic performance of the Hamamatsu C12137 radiation detector was tested in laboratory conditions, without using GNSS positioning. Detector data was encoded in the IEC 63047 format and stored as a file on the memory card of the single-board computer. The files were then decoded and analysed to make sure that, on the one hand, the detector's basic parameters matched the specification and, on the other hand, all its readings were correctly encoded in the IEC 63047 standard. The tests were repeated with a second Hamamatsu detector of different type, the Hamamatsu C12137-01. Its scintillator is ten times larger compared to the C12137. Table 1 summarises the properties of the radiation detector modules tested, including the lot numbers and serial numbers of the detectors actually tested. It was not the aim of these tests to fully characterise the response of the detectors, only the basic parameters were assessed.

Background spectra (see Fig. 5) were acquired over a period of 65 h for both detector types. The logarithmic plot shows the expected ^{40}K peak at 1460 keV and the smaller peaks from the radionuclides in the ^{238}U decay series. The figure is expressed in units of count rate and confirms that the size of the scintillators differs by a factor of ten.

Measurements with a ^{137}Cs source (see Fig. 6) were performed in order to verify the specifications provided by the manufacturer with respect to dose rate, counting efficiency and energy resolution. In addition, the response time of the detector was tested. Note that the geometry is not completely symmetric along the horizontal axis

Table 1 Properties of the tested Hamamatsu radiation detection modules

Parameter	Hamamatsu C12137	Hamamatsu C12137-01
Scintillator	CsI(Tl)	
Scintillator size	13 mm × 13 mm × 20 mm = 3380 mm^3	38 mm × 38 mm × 25 mm = 36,100 mm^3
Mass	120 g	320 g
Counting efficiency (for ^{137}Cs, 662 keV)	> 40 cpm at 0.01 μSv/h (or > 66 cps at 1 μSv/h)	> 400 cpm at 0.01 μSv/h (or > 666 cps at 1 μSv/h)
Energy resolution at 662 keV (%)	< 8	< 8.5
Measurement range (dose rate)	0.01–100 μSv/h	0.001–10 μSv/h
Measurement error (dose rate)	< ± 20%	
Lot number and serial number tested	Lot 14L S/N 0015	Lot 17E S/N 0008

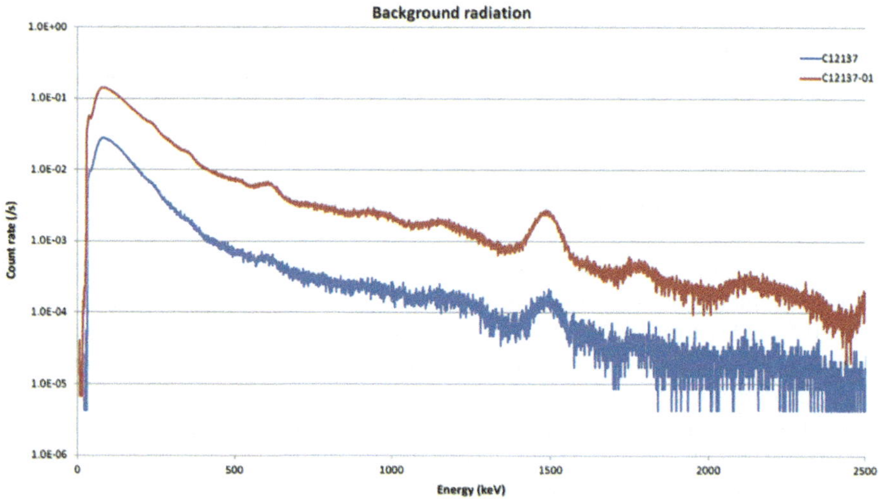

Fig. 5 Background spectra

through the middle of the scintillator, which may lead to errors for close source-detector distances. The ^{137}Cs source had an activity of about 91 kBq at the day of testing and was placed on top of the detector ('in contact') and at distances of 50, 100, 150 and 200 mm from the centre of the scintillator. Figure 7 shows the spectra recorded with the small detector (type C12137).

The correctness of the dose rate (from 662 keV gamma radiation emitted by ^{137}Cs) was verified by comparing the measured dose rate with a theoretical dose rate calculated using the Nucleonica 'Dosimetry and Shieding++' application [22]. Figure 8 shows that the dose rate is correct within the claimed ± 20%, except for the

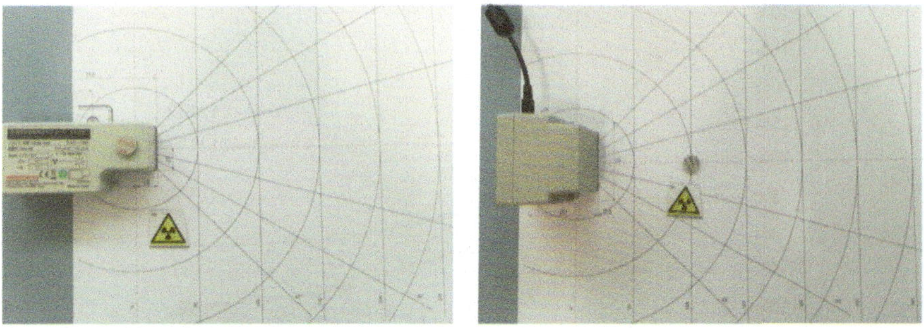

Fig. 6 Basic performance tests of the radiation detectors in laboratory conditions, Hamamatsu type C12137-01 (left) and type C12137 (right)

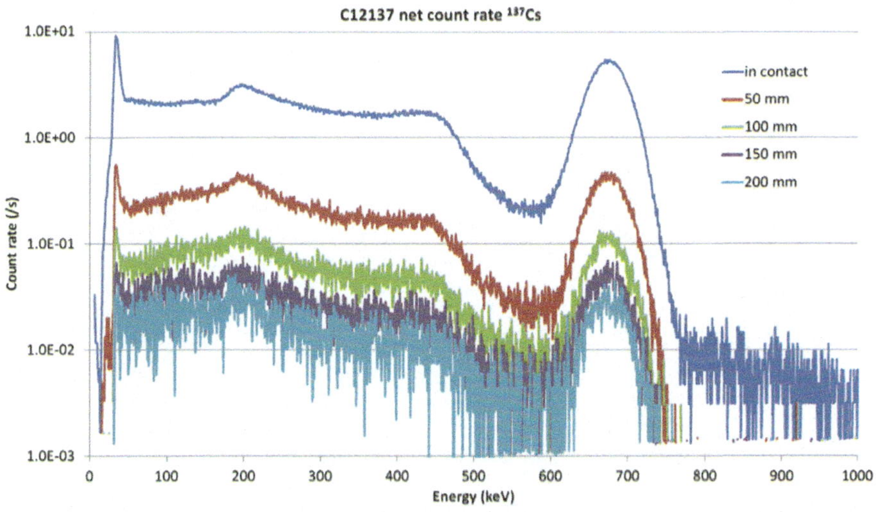

Fig. 7 Net (background corrected) energy spectra recorded with a ^{137}Cs source at different distances to the detector

larger detector with the source at 50 mm distance, which yielded an error of 24%.

The counting efficiency was determined by subtracting (in counts per second) the background spectrum from the energy spectrum obtained with the ^{137}Cs source, and then integrating the net spectrum for energies larger than 43 keV. The resulting count rate (attributed to ^{137}Cs only) is then divided by the measured dose rate, yielding a figure of efficiency (counts per second per 1 μSv/h). Figure 9 plots the ratio of this figure to the claimed (minimum) efficiency. The test was performed at different distances from the source to the detector. The figure shows that the efficiency is at least as good as claimed, except for the larger detector with the source in contact. In fact, for this particular measurement the net dose rate exceeded the maximum

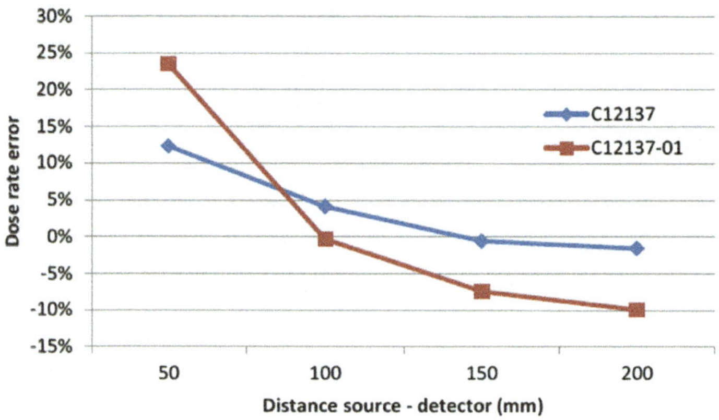

Fig. 8 Error on the dose rate

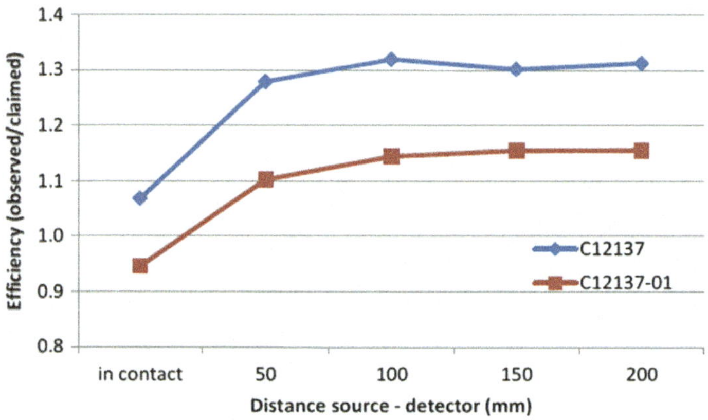

Fig. 9 Ratio of the observed counting efficiency versus the claimed efficiency

dose rate of 10 μSv/h for C12137-01. It is therefore justified to ignore this point and confirm the performance with respect to counting efficiency.

At count rates above 1500 per s, both detectors suffer from pile-up. This effect can be observed from the net (background corrected) spectra recorded with the ^{137}Cs source close to the detector. In principle, apart from statistical fluctuations in the background radiation, there should not be any counts in the spectra at energies beyond the full-energy Gaussian peak. Figure 10 was recorded with the small detector and shows an elevated number of counts at energies higher than about 750 keV when the source is in contact with the detector. As can be expected, the effect is larger for the larger detector, as demonstrated by Fig. 11 where pile-up is still slightly present at a distance of 50 mm.

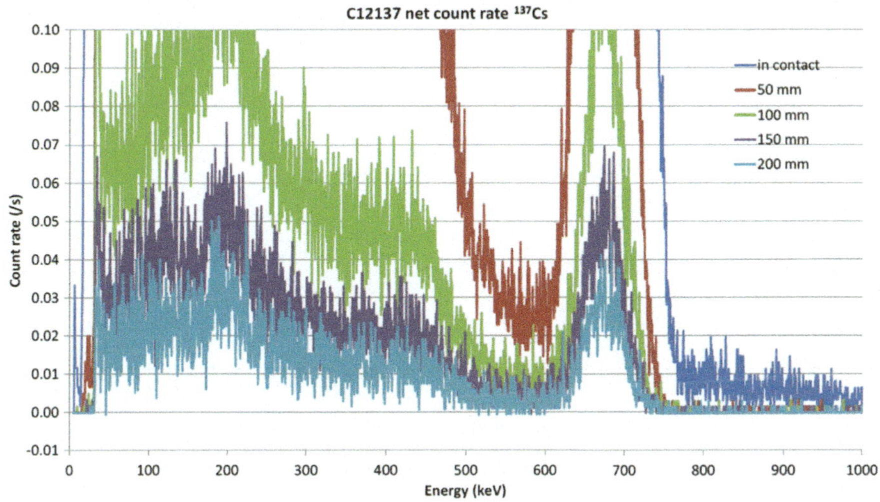

Fig. 10 Net count rate spectrum showing pile-up of events when the source is in contact with the small detector

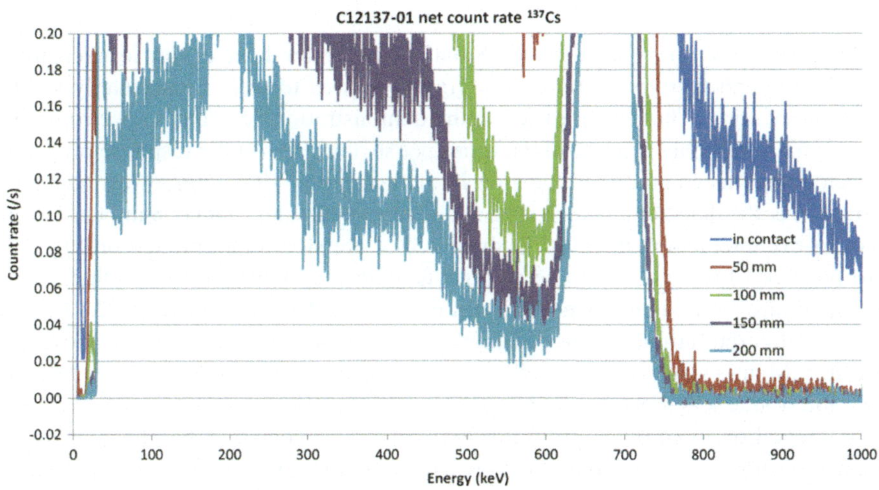

Fig. 11 Net count rate spectrum showing pile-up of events when the source is in contact with the larger detector. The pile-up effect is also present when the source is at a distance of 50 mm

The pile-up effect is caused by the detector electronics not being fast enough to remove all charges associated to the light observed from one gamma-ray interaction before the next gamma-ray interacts.

Earlier experience with robot-ported dose rate meters shows a delay in the response of the detector when exposed to a quickly changing radiation field, for example as observed when the robot moves quickly over a point source. The effect is caused due

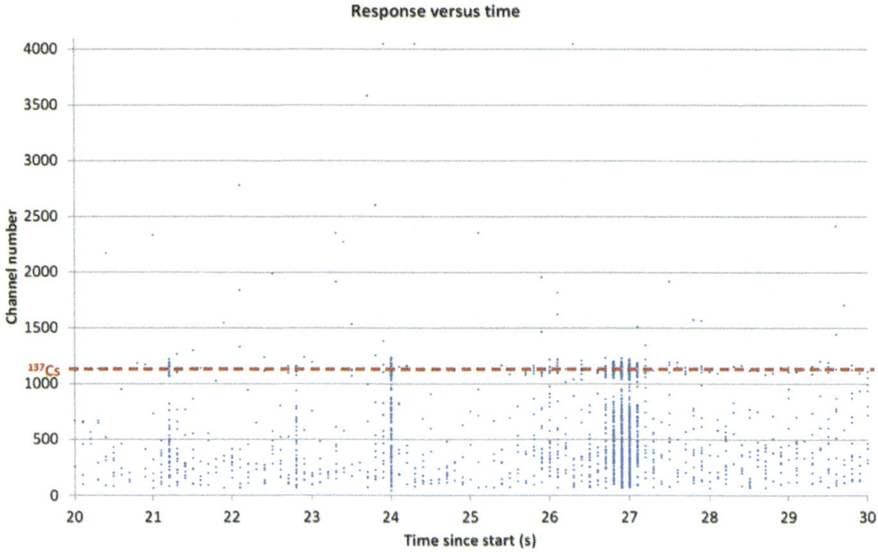

Fig. 12 Response versus time to a quickly moving ^{137}Cs source

to the integration time of the detector. Smaller and lighter detectors are less sensitive to radiation and therefore a time averaging is needed to yield a reasonably stable dose rate reading. The detector used in the demonstration device is a spectrometric device that sends data at a rate of ten times per second. To test the response time, the detector was placed at a fixed location in the lab and a ^{137}Cs source was manually moved over the detector at a high speed. The timestamp and channel number of each gamma-ray interaction observed in the detector are stored and plotted on Fig. 12. The position of the ^{137}Cs full-energy peak is indicated by the red dashed line.

The figure shows that indeed 10 times per second data is acquired. At 21.2, 22.8 and 24 s after the start of the measurement the source was passing a few cm above the detector. Many events are recorded within the small time interval of 0.1 s. Around 27 s, the source was moved slower, resulting in even more events, spread over a total time interval of about 0.5 s. From this experiment we conclude that the detector in the demonstration device does not show a delay in the response. Finally, the energy resolution measured at 662 keV was slightly better than the claimed values.

4.1.2 Basic Field Tests

As a next step, the complete demonstration device (including the radiation detector and GNSS receiver) was tested in outdoor conditions. Both detector and positioning data were encoded into the IEC 63047 data format and stored on the memory card of the single-board computer, at a rate of 10 times per s. Compliance of the generated data files with the data format standard was confirmed by importing the file into the software ASN1 Studio [23] to convert it to a human-readable format. Apart from the radiation data, which was already examined in the laboratory tests, this time also accurate GNSS positioning was checked.

To allow centimetre-accurate positioning, the GNSS expansion board needs an NTRIP connection to a RTK (Real-Time Kinematics) service. NTRIP is a protocol for standardised transmission of GNSS data over the internet. Real-time kinematic (RTK) was implemented with rtkexplorer [24], which is a modified version of RTKLib [25] and allows the acquisition of centimetre-accurate satellite positioning information. At the JRC site in Geel, where the demonstration device was tested, the RTK service is provided by the Flemish Government (FLEPOS). Once the position of a local base station is accurately established, the NTRIP connection can be replaced by the local RTK base station.

In a first stage, the centimetre-accurate GNSS positioning was tested without encoding data into the IEC 63047 format. A local outdoor Wi-Fi network was set up using a smartphone as mobile hotspot. The reference position of a 'base station' was first determined using the FLEPOS service. Using this base station instead of a virtual station determined by the FLEPOS service resulted in a very stable and accurate positioning signal.

The results of the positioning tests are shown in Fig. 13. The demonstration device was moved around by a person following the border of the road, in order to compare the position with satellite images. The position of the demonstration device was logged and plotted using Google Earth [26]. The RTKLib software assigns a quality flag to the calculated position. The figure only shows the points that are flagged as RTK 'Fixed'. These points are accurate up to a few centimetres. At the extremities of the track, the Wi-Fi signal was lost causing less accurate positioning data. High positioning accuracy could be sustained in open space at least 15 m away from buildings and trees and when moving at a maximum speed of 3 m/s. Faster speeds were not tested since not deemed necessary for Unmanned Ground Vehicles (UGVs). Experience from tests performed by University of Oulu shows that also for Unmanned Aerial Vehicles (UAVs), which may move faster, positioning accuracy up to a few centimetres can be achieved using the same technique: RTK referring to the known position of a base station.

Since all test results were satisfactory, the demonstration device appeared to be fit for its purpose. Of course, further systematic testing according to established standards, such as environmental, stability or sensitivity, has to be done and will be part of future work.

4.2 Integration into the Robot Operating System (ROS)

Robotic systems are generally composed by heterogeneous mobile robots and Operator Control Units (OCU) connected by wireless interfaces such as Wi-Fi or LTE. The communication between all involved parts is handled by the Robot Operating System ROS, enabling bi-directional data transfer using serialized messages through

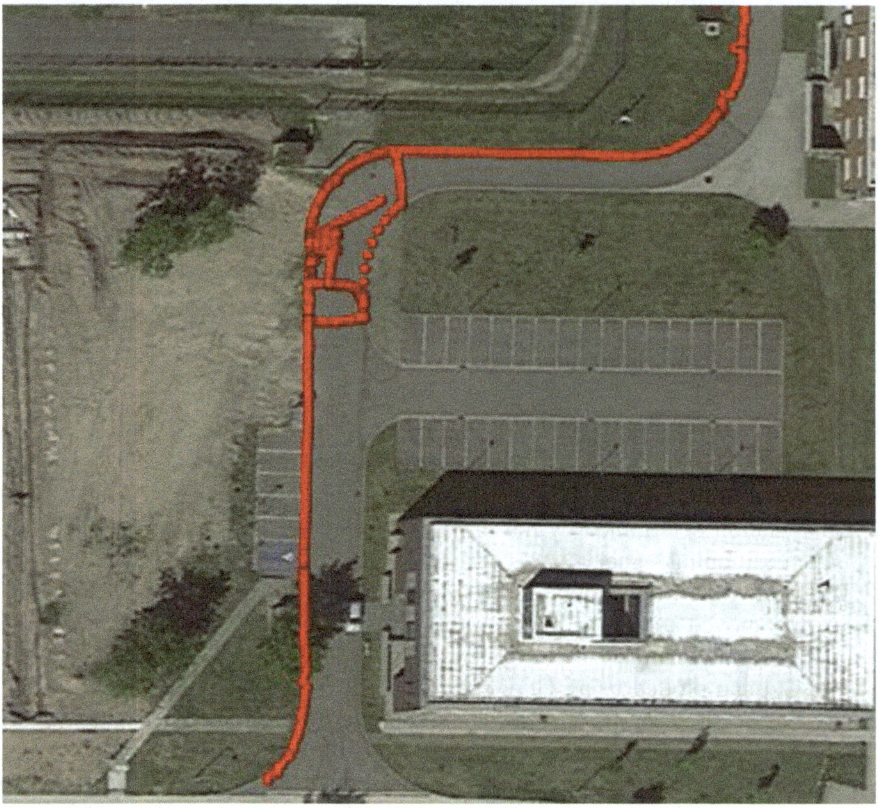

Fig. 13 Demonstrating centimetre-accurate positioning using real-time kinematic RTK. (Satellite image by Google)

TCP/UDP socket connections. Establishing a wireless communication network is not straightforward. Usually, wireless data communication is not very reliable, has limited bandwidth and therefore is often restricted to ROS messages. The robotic platform should locally process the acquired data and only use the ROS network to transmit data that is required for the mission.

In the ROS framework, all processes that perform computations are called nodes, and the software is organised using packages that contain source code, nodes and configuration files. ROS nodes communicate to each other by interchanging 'typed-fields messages', defined by a simplified message description language that enables automatic code generation. The fields of the messages are created using standard built-in types such as Boolean, 8-bit integer or 64-bit floating point numbers.

To integrate the demonstration device into robotic applications, Fraunhofer FKIE created a software interface using ROS packages based on the encoding/ decoding solution provided by the JRC. The developed software interface is divided into two main components: A multifunctional device driver node that is executed inside the

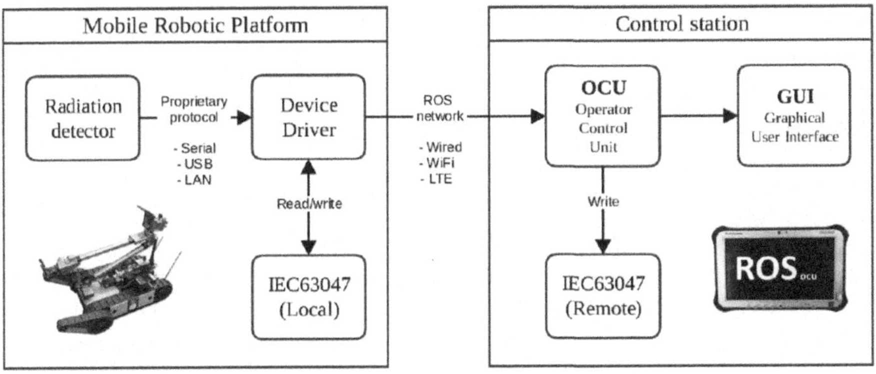

Fig. 14 Layout of the ROS integration

mobile robotic platform and an Operator Control Unit (OCU) driver node, running in an operator's control station. Figure 14 pictures this approach.

The radiation detector measures periodically (up to ten times per second) dose and gamma-energy count rate, which is used to construct the energy spectrum. The measurements are requested by the device driver using a vendor-specific protocol and interface. For the Hamamatsu detector used in the demonstration device, communication is performed over a USB connection. Simultaneously, the device driver receives geolocation data from the attached GNSS device using an emulated serial interface, containing information about current latitude, longitude, altitude and time.

4.2.1 The ROS Device Driver Node

The device driver node runs on the mobile robotic platform. The driver serializes the measurements and geolocation data into a measurement-generic ROS message named [MeasurementLocated], which only contains a basic subset of information, and publishes it on the ROS network. The definition of the generic message is shown in Table 2. It is worth noting that multiple nodes, for instance OCUs or even other robots, may subscribe simultaneously to receive the published data.

The device driver node also allows encoding the measurements and geolocation data into the IEC 63047 format. The binary IEC 63047 messages can then be stored in a local file or published on the ROS network using a built-in message of type [std_msgs/ByteMultiArray], which serves as a binary container. Local storage on the robot is particularly useful when the network connection is interrupted and/or the communication with the OCU is not available. Considering the design criteria in ROS, it is strongly advisable to avoid generic message types, because they do not convey any semantic meaning about their content. In principle, this design rule is violated when IEC 63047 encoded data is transmitted over the ROS network using the binary container. Therefore, the use of the built-in message of type

Table 2 ROS Message definition for collected measurements

Message name	Fields		Description
	Type	Name	
Measurement-located	Measurement	measurement	
	float64	latitude	Geo-location
	float64	longitude	
	float64	altitude	
Measurement	Header	header	Measurement time stamp
	string	device_name	Custom name of the device
	string	device_designation	Device manufacturing name
	string	classification	Classification of the device (if applicable)
	Value[]	values	Measurement values
Value	string	sensor	Name of the sensor
	string	source	Source of the sensor data, e.g. wind, rain or other
	string	type	Type of sensor data, e.g. speed, duration, intensity
	float64[]	value	Single measurement or a spectrum of values
	string	unit	Measurement units
	float64	min	Minimum value (if available)
	float64	avg	Average value (if available)
	float64	max	Maximum value (if available)
	string	info	Additional information

[std_msgs/ByteMultiArray] is only acceptable when the subscribers have the prior knowledge that the data is encoded into the IEC 63047 format.

The driver node supports both approaches simultaneously (typed-fields messages and the binary-container), the actual choice is application-dependent. For instance, if the application only requires plotting the measured data in the control station, the message type [MeasurementLocated] is more appropriate because it does not require any additional processing. However, if the application involves integration with additional systems, such as databases or external devices supporting the IEC 63047 format, the binary-container message is more convenient.

4.2.2 The ROS OCU Driver Node

The OCU driver node runs on the control station. ROS messages are streamed to the OCU subscriber node(s), which extract the corresponding measurements and geolocation data. Next, it is possible to visualize the data using a graphical user interface (GUI) or, if required, to encode the data into the IEC 63047 format and store in a local file or stream over a connection socket. Encoding to IEC 63047 is useful for integration with external systems or databases.

4.3 Encoding and Bandwidth Tests

Although both approaches (typed-fields messages and the binary-container) are technically feasible, some additional performance and design considerations need to be taken before selecting any of them.

The IEC 63047 standard requires the use of the C-OER encoding rule (Canonical Octet Encoding Rule [21]). Other data formats, for example the IEC 62755 format, use XML encoding, which is human readable. Because IEC 63047 is specified using ASN.1 (which decouples the specification of the data format from the encoding rule), it is easy to read C-OER encoded data and convert the data using another encoding rule. For comparison, tests were performed using XER (XML Encoding Rule) and BER (Basic Encoding Rule), in addition to C-OER. The encoding rule determines the size of the encoded data and therefore the required bandwidth for transmitting a measurement. It is important to realise that the requirements on resources (like memory and computing time) are also affected by the encoding rule. During the development of IEC 63047 it was found that from all the ASN.1 encoding rules, C-OER offered the best compromise between encoding size, speed and memory use in typical applications where the data format may be used.

As an example, data acquired and stored in the IEC 63047 format by the demonstration device during the previous preliminary tests can be used. The binary files were opened with the software package ASN1 Studio [23] and then saved as a text file in the ASN.1 value notation syntax. The most relevant parts of the text file are included. The comments (starting with `--`) have been added manually to explain the meaning of the data elements.

An IEC 63047 data file is a sequence of values of the `Listmodedata` data type, encoded using the C-OER encoding rule. The `Listmodedata` data type is a choice between the `Header` type, the `Eventlist` type and the `Footer` type. The first value in the file is always of the `Header` type, and is followed by values of the `Eventlist` type. Finally, the last value in the file is of the `Footer` type.

```
value1 Listmodedata ::= header : {
    standardID "IEC 63047 Ed. 1.0", -- Fixed element
    listModeDataID "17E0008", -- Identifies the data
    listModeDataPart 0, -- This is the first part of the data
    listModeDataNParts 1, -- There is only one part
    measSetupID "Hamamatsu C12137",
        -- Identification of the measurement setup
    measSetupDescription "Hamamatsu C12137 over USB",
        -- Description of the measurement setup
    iec62755 single : NULL,
        -- There is no associated file in the IEC 62755 format
    radSource "Background radiation",
        -- The source of radiation is natural background
    start {
        date-time "2019-05-14T13:23:32Z",
            -- The start of the measurement, in UTC
        fractional-seconds 6.41E-1
            -- Precise start was 0,641 seconds after
    },
    deviceList { -- The list of devices (here only one)
        {
            name "Radiation detector module",
            manuf "Hamamatsu",
            model "C12137",
            serial "17E0008"
                -- Combination of lot number and serial number
        }
    },
    channelList { -- The list of channels (here only one)
        {
            deviceID 0, -- Channel belongs to the first device
            kind physicalkind,
                -- A physical channel (real detector channel)
            physicalChannel 0, -- Number of the
            name "CH0",
            description "Multi-pixel photon counter coupled to a
            CsI(Tl) scintillator",
            parameters "[comparator threshold][30][keV];
            [lower limit energy range][30][keV];
            [higher limit energy range][ 2.000][MeV];
            [conversion factor][ 1.000][uSv/h];
            [ADC to keV][1000][keV];
            [energy threshold][60][keV]",
```

```
                -- Set of configuration parameters
            delay 0, -- No time delay applied
            adcSamplingRate int : 0, -- Not applicable here
            adcBitResolution 12,
                -- There are (2^12) bins in the energy histogram
            clockFrequency int : 10,
                -- Timestamp clock runs at 10Hz (one tick is 0,1 s)
            eventPropertyList { -- Event types to expect in the
                -- Eventlist, and their properties
                eventPulseProperty : {
                    description "CH0 records TimeStamp and
                    pulse height",
                    valueDescriptionList {
                        "Pulse height"
                    }
                },
                eventGeoProperty : {
                    description "RTKDEMO5 b31"
                }
            }
        }
    }
}

value2 Listmodedata ::= eventList : {
    listModeDataID "17E0008",
        -- Identifies the data (same as in Header)
    listModeDataPart 0, -- This is part 1 of the data
    id 0, -- This is the first Eventlist
    eventList {
        eventPulse : { -- A gamma ray in bin 1675 was detected
            -- 0,7 seconds after start
            channelID 0,
            timeStamp 7, -- To multiply with the inverse of
            -- clockFrequency from Header
            valueList {
                int : 1675
            }
        },
-- Other events of the eventPulse type have been omitted
        eventPulse : { -- A gamma ray in bin 163 was detected
            -- 2,9 seconds after start
            channelID 0,
            timeStamp 29,
            valueList {
                int : 163
            }
        },
        eventGeo : { -- The position at 3,2 seconds
            channelID 0,
            timeStamp 32,
            position geographicPoint : {
                latitude 5.1196566162E+1,
                longitude 5.041901482E+0,
                datum "WGS-84"
```

```
                }
            },
        }
}

value3 Listmodedata ::= eventList : {
        listModeDataID "17E0008",
            -- Identifies the data (same as in Header)
        listModeDataPart 0, -- This is part 1 of the data
        id 1, -- This is the second Eventlist (starts counting from 0)
-- Events are omitted here
}

-- Eventlists are omitted here

value1418 Listmodedata ::= eventList : {
        listModeDataID "17E0008",
            -- Identifies the data (same as in Header)
        listModeDataPart 0, -- This is part 1 of the data
        id 1416,
            -- This is the 1417th Eventlist (starts counting from 0)
-- Events are omitted here
}

value1419 Listmodedata ::= footer : { -- This is the end of the data
        listModeDataID "17E0008",
            -- Identifies the data (same as in Header)
        listModeDataPart 0, -- This is part 1 of the data
        lastEventListid 1416 -- The id of the last event list was 1416
}
```

Additionally, a performance test was executed to compare the required bandwidth between the different message definitions. The Hamamatsu radiation detector collected measurements of the background radiation. The ROS device driver node ran on the demonstration device and published the data while the ROS OCU driver node on the control station subscribed to read the data. The ROS [MeasurementLocated] message type needed a bandwidth of about 32 kB/s. The IEC 63047 data in the ROS binary container only took between 9 and 13 kB/s (depending on the encoding used)—although containing more information than the [MeasurementLocated] messages. The ASCII-based XML format XER, of course, requires a many times higher bandwidth and has been included only for comparison purposes. These results underline the efficiency of the IEC 63047 data format. Table 3 gives an overview.

Table 3 Bandwidth usage during the performance test

ROS message Type	Encoding rule	Average bandwidth used (1 kB = 1024 B) (kB/s)
Typed-fields message [MeasurementLocated]	ROS serialization	32
Binary container [std_msgs/ByteMultiArray] with IEC 63047 data	C-OER	9
	BER	13
	XER	235

Fig. 15 Demonstration device mounted on a mobile robotic platform

Finally, the developed ROS integration packages were successfully tested using the demonstration device mounted on an iRobot Packbot, showed in Fig. 15. The developed package can also be executed directly on the robot computer, without using explicitly the demonstration device. In this configuration, the radiation detector is connected to the robot computer using USB and the geolocation data is provided by the robot's integrated GPS.

The device driver developed by Fraunhofer FKIE supports IEC 63047 files as input, enabling future integration of radiation detectors that also support this standard. This is particularly important for improving and simplifying the integration of new radiation detection systems.

5 System Test at the European Robotics Hackathon 2019

After these preliminary tests Fraunhofer FKIE and JRC-Geel decided to take part in the European Robotics Hackathon (EnRicH) 2019 [27]. In 2017 the European Reference Network for Critical Infrastructure Protection (ERNCIP) task group on RN Threats planned to conduct a robotic trial with a strong focus on radiological and nuclear reconnoitring. As a result the so-called European Robotics Hackathon (EnRicH) started in 2017 as a biennial event, taking place in the inoperative nuclear power plant (NPP) in Zwentendorf, Austria (see Fig. 16). The boiling water reactor of the Zwentendorf NPP is one of the very few in the world which was completely

Fig. 16 The inoperative NPP in Zwentendorf, Austria, where the EnRicH 2019 took place

built but never put into operation. Instead, the plant was transferred into a training centre in which repairing and dismantling measures but also critical incidents can be trained under realistic conditions. Besides being a trial, the main goal of EnRicH is to foster the development of robotic solutions for the RN domain, bringing together roboticists and RN specialists.

The participating teams could choose among several tasks, all close to typical RN tasks [28]: map the scene, locate radiation sources inside a pipe system, and use a mobile manipulator to close the corresponding valves. The mapping task consisted of two sub-tasks: first, build a digital 3D map of the area and, second, create a radiation map of the scene in 2D or 3D. Figure 17 gives an overview of the area that the robots had to map, the ground floor of the reactor building. Additionally, the systems could search for "missing workers", dummies of human size and weight, and transport them out of the dangerous area. The teams could work on any combination of these tasks.

From these various RN-related tasks offered during the hackathon, two seemed most suitable for the FKIE/JRC system setup: build a 2D radiation map of the scene and locate the different radiation sources. Figure 18 shows the robot platform during a trial. Two sensors were mounted on the robot: a Hamamatsu C12137 and an FLIR Identifinder 2. Both sensors were connected to the robot using USB interfaces, and delivered measurements continuously to the ROS network. Generated sensor data

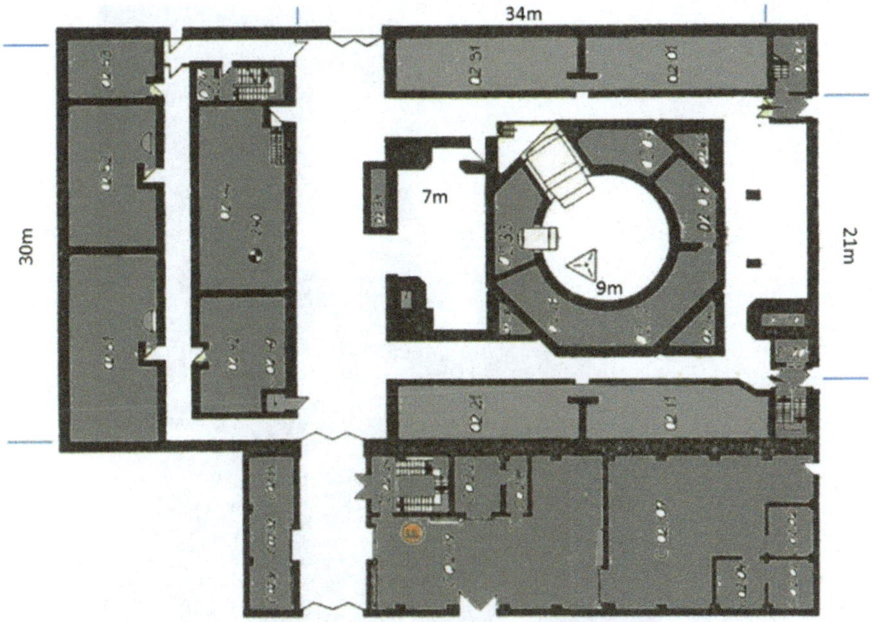

Fig. 17 Floor plan of the ground floor of the reactor building. The white areas could be entered and had to be mapped

included dose-rate (μSv/h), gamma-rate (cps) and energy spectra. The sensor drivers added positioning information to the radiation measurements, which is useful for post-processing and visualization purposes. In outdoor scenarios the exact robot position is usually computed by an external GPS module. Since the EnRicH provided an indoor scenario, a locally referenced position computed by the robot's Simultaneous Localisation and Mapping (SLAM) sub-system was used instead.

A special protocol converter node subscribed for the ROS radiation messages and generated the corresponding IEC 63047 data. This data either can be streamed using socket services or saved in local files. Due to communication constrains, we chose to save all converted measurements locally on the robot and, afterwards, transferred them to the control station once the exploration run was completed. Two formats were used in the competition, a compressed version of the binary Canonical Octet Encoding Rule (C-OER) for long-term storage, and a XER (XML-based format) version for debugging and testing.

During the exploration runs, the Hamamatsu sensor failed due to saturation because the maximum dose-rate of 10 μSv/h was unexpectedly surpassed. The source used in the scenario was quite active, and maximum values close to 320 μSv/h were measured during the runs. The sensor Identifinder 2, however, worked as expected, since it is able to handle a much higher dose-rate up to 10 mSv/h. As a result, the radiation in the whole scenario could be successfully measured, stored and transferred to the operator station using the IEC 63047 data format. Using this data together with

Fig. 18 The FKIE robot platform with the IEC 63047 demonstration device in the EnRicH 2019 scenario

the robot's 2D mapping sub-system, complete radiation maps of the scenario could be created; see Fig. 19 for an example.

6 Conclusion and Outlook

Conventionally, data from radiation detectors is integrated over time and presented in the form of a simple dose rate or energy spectrum. The time integration over several seconds makes it hard to locate a radioactive source when the detector is moving, for example on a land robot or drone. Novel in this approach is the use of precise time and position information. The energy of the individual photons is combined with centimetre-accurate position information of the detector at a rate of ten times per second. This way of collecting "list-mode data" allows the development of innovative algorithms that potentially can locate and identify a radioactive source much faster than in the conventional approach.

JRC-Geel developed a device from off-the-shelf components to demonstrate the use of the list-mode data format standard IEC 63047. Encoding and decoding

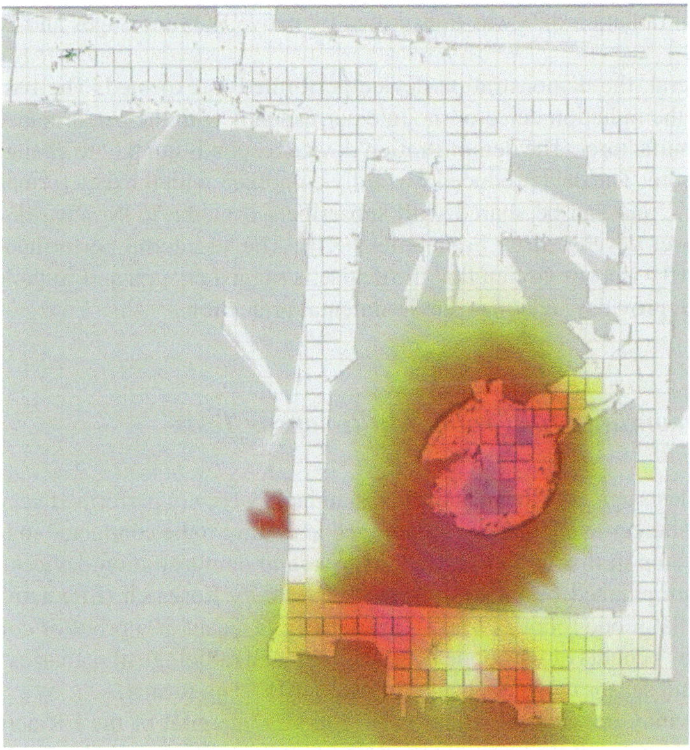

Fig. 19 An example radiation map created with the IEC 63047 test setup during EnRicH 2019

software for the binary IEC 63047 format has been developed using open-source code `asn1c` from GitHub. Demo software code is provided under the open-source European Union Public Licence (EUPL) [20].

The demonstration device implements a solution for encoding data into the IEC 63047 standard format using this open-source code. The demonstration device consists of a single-board computer, a spectrometric radiation detector and a GNSS receiver which enables real-time kinematics to achieve centimetre accurate positioning. The detector data is combined with the positioning information and encoded into the IEC 63047 format. A USB software interface between the radiation detector and the single-board computer has been written and all pieces of software have been integrated.

Fraunhofer FKIE extended the detector interface to a ROS device driver, capable of sending the data to the operator control unit over the ROS network. Tests performed at the JRC site in Geel assessed the performance of the demonstration device in laboratory and field conditions. The demonstration device and the ROS software were also successfully tested by Fraunhofer FKIE, who included the device into one of their experimental robot systems. With this robot system FKIE took part in the EnRicH 2019 hackathon and was able to fulfil the radiation mapping task. During the trial all radiation data was transmitted by native ROS messages and IEC 63047 in parallel.

In general, the demonstration device performed as expected. The basic performance of the radiation detector module corresponds with the performance claimed by the manufacturer. The demonstration device responds quickly to changing radiation exposure. The data produced was fully compliant with the data format standard IEC 63047, and can be shared with subscribers over the ROS network using the device driver developed by Fraunhofer FKIE. The successful performance during EnRicH 2019 is a strong hint that the device is rugged enough and suitable for field conditions, providing adequate environmental protection.

6.1 Upcoming Multi-robot Multi-sensor Tests

Profound tests against other established standards were not performed yet due to the limited resources available; of course, such tests have to be conducted in the future. As a next step in this direction, tests with several demonstration devices operating together are planned by the JRC in the Exploratory Research (ER) activity 2018-ER-A32 'list-mode robotics trial'. This activity is meant as a proof of concept that assesses the benefits of list-mode data acquisition in radiological scenarios involving robotics and sensor networks using the IEC 63047 data format.

Robots may operate individually, but the ultimate goal of the ER activity is to use several robots at the same time, in combination with fixed/relocatable sensors. The acquired data (time, position and radiation) from each sensor can then be used to investigate if the list-mode data protocol simplifies search operations when data from two or more sensors are combined (see Fig. 20).

It is expected that the high precision of the position of the radiation detector and its time resolution of 0.1 s allow a faster estimation of the position of a hidden radiation source, compared to conventional techniques where a spectrum is acquired over a certain time (several seconds). Over a period of several seconds, the robot may have travelled already a few meters and could have passed the source without noticing. The expected advantage of list-mode data acquisition is even larger when data from more than one moving detector is processed in real-time. Innovative algorithms could provide feedback to the robots in order to adjust their path, or to reposition a sensor.

During the ER activity 'list-mode robotics trial', the demonstration device will be tested in a real environment by mounting it on land robots (UGVs) provided by Fraunhofer FKIE and drones (UAVs) from University of Oulu. These robots will collect list-mode data acquired from several sensors operating at the same time and will perform a search operation to locate and identify a hidden radioactive source. The actual experiments will take place at the GELINA flight path area, which is an outdoor controlled area of the JRC-Geel site, allowing the use of active radiation sources and, at the same time, providing sufficient space for the parallel use of several UGVs and UAVs.

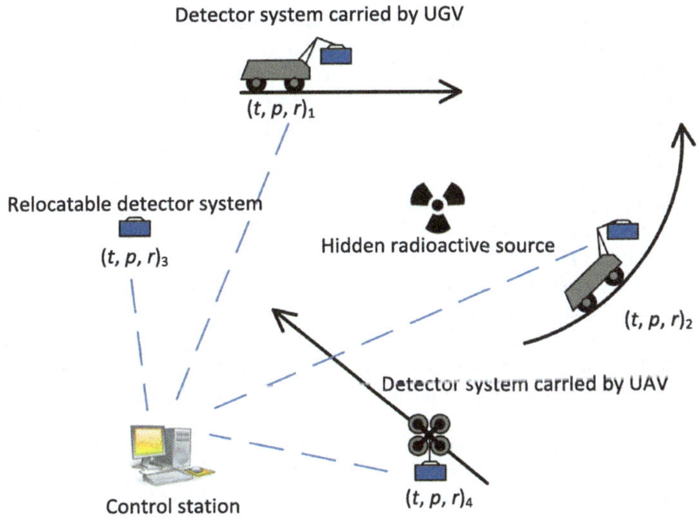

Fig. 20 Several sensors work together to locate and identify a radioactive source

After the field test, the acquired data will be analysed to investigate that the identification and localisation of a radioactive source is more effective when more than one moving and fast-responding radiation detector is used (with or without fixed detectors). The development of innovative algorithms to locate and identify a radioactive source using robots may be considered. Additionally, an integration of the proposed architecture with the European GRRASP (Geospatial Risk and Resilience Assessment Platform) project [29] is planned in order to enable data visualisation, interaction with drones and geo-analytics.

The technique of collecting timestamped list-mode data is not limited to RN sensors but applies to any kind of sensor. Moreover, the IEC 63047 data format supports any kind of sensor. Thus, in the future, the IEC 63047 format could be applied to CBRNE sensor networks, including data from other sensors than RN. This could support data collection and real-time data analysis in support of disaster risk management and critical infrastructure protection, for example by building an interface that allows IEC 63047 formatted data to be imported into modelling platforms (e.g. JRC GRRASP). List-mode data could also be used to develop new methods for source localisation and identification, for example via machine learning.

References

1. IEC 63047, Edition 1.0, 2018-10: Nuclear instrumentation–data format for list mode digital data acquisition used in radiation detection and measurement.
2. Robot Operating System (ROS), official web page, https://www.ros.org

3. Giesbrecht, J., Fairbrother, B., Collier, J., Beckman, B.: Integration of a high degree of freedom robotic manipulator on a large unmanned ground vehicle. In: Gerhart, G.R., et al. (eds.) Unmanned Systems Technology XII—Proceedings of SPIE, vol. 7692 (2010)
4. Schneider, F.E., Welle, J., Wildermuth, D., Ducke, M.: Unmanned multi-robot CBRNE reconnaissance with mobile manipulation. In: Proceedings of the 13th IEEE International Carpathian Control Conference (ICCC), pp. 637–642 (2012)
5. Christie, G., Shoemaker, A., Kochersberger, K., Tokekar, P., McLean, L., Leonessa, A.: Radiation search operations using scene understanding with autonomous UAV and UGV. J. Field Rob. **34**(3), 1450–1468 (2017)
6. Lazna, T., Gabrlik, P., Jilek, T., Zalud, L.: Cooperation between an unmanned aerial vehicle and an unmanned ground vehicle in highly accurate localization of gamma radiation hotspots. Int. J. Adv. Rob. Syst. 1–16 (2018)
7. Nagatani, K., Kiribayashi, S., Okada, Y., Otake, K., Yoshida, K., Tadokoro, S., Nishimura, T., Yoshida, T., Koyanagi, E., Fukushima, M., Kawatsuma, S.: Emergency response to the nuclear accident at the Fukushima Daiichi nuclear power plants using mobile rescue robots. J. Field Rob. **30**(1), 44–63 (2013)
8. Yoshida, T., Nagatani, K., Tadokoro, S., Nishimura, T., Koyanagi, E.: Improvements to the rescue robot quince toward future indoor surveillance missions in the Fukushima Daiichi nuclear power plant. In: Yoshida, K., Tadokoro, S. (eds.) Field and service robotics, Springer tracts in advanced robotics 92, pp. 19–32. Springer, Berlin/Heidelberg (2014)
9. Remmersmann, T., Tiderko, A., Schade, U., Schneider, F.E.: Standardised command and control of multi-robot systems using BML. In: Proceedings of the 11th International Conference Interactive Systems: Problems of Human-Computer Interactions, Ulyanovsk, pp. 49–56 (2015)
10. Pradhan, M., Tiderko, A., Ota, D.: Approach towards achieving interoperability between military land vehicle and robotic systems. In: Proceedings of the 2017 International Conference on Military Communications and Information Systems (ICMCIS), Oulu (2017)
11. Rowe, S., Wagner, C.R.: An Introduction to the Joint Architecture for Unmanned Systems (JAUS), vol. 1001, pp. 48108. Ann Arbor (2008)
12. Ihantola, S., Tengblad, O., Toivonen, H., Peräjärvi, K., Csome, C., Borg, J., Paepen, J., Tagziria, H., Gattinesi, P.: Novel Detection Technologies for Nuclear Security. EUR 29270, Publications Office of the European Union, Luxembourg (2018). https://doi.org/10.2760/703301
13. Turecek, D., Jakubek, J., Trojanova, E., Sefc, L.: Single layer Compton camera based on Timepix3 technology. J. Instrum. 15 (2020)
14. Wang, G., Schultz, L.J., Qi, J.: Bayesian image reconstruction for improving detection performance of Muon tomography. IEEE Trans. Image Process. **18**(5), 1080–1089 (2009)
15. IEC 62755, Edition 1.0, 2012-10: Radiation Protection Instrumentation—Data Format for Radiation Instruments used in the Detection of Illicit Trafficking of Radioactive Materials
16. International Atomic Energy Agency: International Radiological Information Exchange (IRIX) Format, IAEA, Vienna (2020). https://www.iaea.org/publications/12257
17. Official EURDEP Platform: https://remon.jrc.ec.europa.eu/About/Rad-Data-Exchange
18. ISO/IEC 8824-1:2015: ITU-T X.680, Information technology—Abstract Syntax Notation One (ASN.1): Specification of basic notation.
19. Walkin, L.: The ASN.1 Compiler asnlc. Available from GitHub, https://github.com/vlm/asn1c
20. Paepen, J., Lutter, G.: An open-source solution for encoding and decoding IEC 63047:2018 data. EUR 30409 EN, Publications Office of the European Union, Luxembourg (2020). ISBN 978-92-76-23523-1 (online), doi:https://doi.org/10.2760/644570 (online), JRC121882
21. ISO/IEC 8825-7:ITU-T X.696, Information Technology—ASN.1 encoding rules: Specification of Octet Encoding Rules (OER)
22. Nucleonica Dosimetry and Shieding++ web application. https://www.nucleonica.com/Application/ShieldingPlus.aspx
23. ASN1 Studio®: OSS Nokalva. https://www.oss.com
24. rtkexplorer: Exploring High Precision GPS/GNSS with Low-Cost Hardware and Software Solutions. http://rtkexplorer.com
25. RTKLib: An Open Source Program Package for GNSS Positioning. http://www.rtklib.com

26. Google Earth: https://earth.google.com
27. The European Robotics Hackathon (EnRicH): official web site, https://enrich.european-robotics.eu
28. Schneider, F.E., Wildermuth, D.: Real-world robotic competitions for radiological and nuclear inspection tasks. In: Proceedings of the 20th International Carpathian Control Conference (ICCC), Wieliczka (2019)
29. Azzini, I., Dido, M., Giannopoulos, G.: GRRASP: Geospatial Risk and Resilience Analysis Platform—Version 2.0. Report JRC90531, European Commission, UR Scientific and Technical Research Series, Ispra (2014)

Data Science Modeling and Constraint-Based Data Selection for EEG Signals Denoising Using Wavelet Transforms

Magdalena Garvanova, Ivan Garvanov, and Vladimir Jotsov

Abstract This work presents basic information about Electroencephalogram (EEG) signals, their processing and application in practice. Modeling and constraint satisfaction cases have been considered aiming at diminishing the manual labor during the wavelet signal filtering and fitting to medical applications. The EEG signals are easily affected by various noise sources. The noise can be electrode noise or can be generated from the body itself. The noises in the EEG signals are called artifacts and these artifacts are needed to be removed from the original EEG signals for the proper analysis of the signals. This work presents denoising algorithm based on the combination of wavelet transform (WT), threshold processing and inverse wavelet transform. The proposed algorithm is tested using real EEG signals. To improve its efficiency, different modeling and data preprocessing methods have been applied. In case when there is a need of constraint shift/modification/elimination, new types of constraints are considered and applied.

Keywords Wavelet transform · Denoising of EEG signals · Intelligent system · Data science · Data selection · Constraint satisfaction · Puzzle methods

M. Garvanova · I. Garvanov · V. Jotsov (✉)
University of Library Studies and Information Technologies, Sofia, Bulgaria
e-mail: v.jotsov@unibit.bg

M. Garvanova
e-mail: m.garvanova@unibit.bg

I. Garvanov
e-mail: i.garvanov@unibit.bg

I. Garvanov · V. Jotsov
Al-Farabi Kazakh National University, Almati, Kazakhstan

V. Sgurev et al. (eds.), *Advances in Intelligent Systems Research and Innovation*, Studies in Systems, Decision and Control 379,
https://doi.org/10.1007/978-3-030-78124-8_11

1 Introduction

The considered research aims to unify and improve the medical applications of signal filtering using wavelet transforms. The denoising stage could be implemented by using artificial neural networks and deep learning methods (ANNs) but even in this case the process is time and cost consuming and it does not adapt to unknown denoising problems like dynamic body and external noise occurrence, the change of the reasons for the noise, and so on. In these conditions original deep modeling applications have been elaborated and considered, they are naturally combinable with classical constraint satisfaction cases. Deep modeling methods are fit to describe and gradually sustain different thinking and reasoning processes. They are effective in conditions when the logical reasoning is rather constrained and the machine learning using ANN couldn't be widely applied because of the lack of proof explanations. It is considered that both classical and non-classical logical operators may be applied in one scheme. On the other hand, the effectiveness of data selection and data processing significantly increases.

Electroencephalogram (EEG) signal is often contaminated by electrocardiogram (ECG), electromyography (EMG) and eye blinking, which are physiologic sources of noise [1]. Also, EEG signal includes line interference and electrode noise. Analyzing depth of anesthesia using corrupted EEG signals may result in an incorrect result. Most of the contaminated signal comes from the eye ball movement and blinking which is known as electrooculogram (EOG) and electromyogram (EMG) signals from the muscles [2]. Therefore, it is necessary to filter the noise from EEG signal. The major noise source of EEG signal is EOG. It is because of the movement of the eye ball causes an electric field around the eye and affected the electric field of the scalp. This electric field could contaminate neurons potential of the brain and as a result EEG signal is contaminated. EOG is considered as a signal with high amplitude and low frequency. This signal is usually affected the lower band of EEG signal. Also, the EOG signal could increase the power of low frequency band [3].

There are many techniques to remove the artifacts from EEG signal, such as adaptive filter, frequency domain regression technique, Wiener filter technique, FIR filter, independent component analysis, and wavelet transform. However, each technique has its own ability to remove one particular artifact. For instance, the popular technique to remove the EOG artifact is adaptive filter. Kumar has used the adaptive filter technique combined with wavelet to remove the EOG [4]. On the other hand, the most popular technique to filter EMG artifact from the EEG signal is discrete wavelet transform [5]. Furthermore, in order to remove the EMG signal from the EEG signal, Lanlan used the db4 wavelet and decomposed it using 8 layers [6]. The result shows that the wavelet is more effective to remove the noise from the EEG signal. However, this technique works only in removing the EMG signal from the EEG signal. Different to the method used by [6], Araghi used the bior3.3 discrete wavelet transform and decomposed it using six layers to remove the artifact from the EEG signal [7]. In contrast, Palendeng et al. explain that the wavelet transform technique which is combined with adaptive least mean square able to remove the

EMG artifact as well as other artifact in low frequency [8]. Filtering the EEG signal is essential before analyzing it. Removing artifact from the EEG signal would reduce the error in calculations. In this paper, wavelet transform is proposed to remove the artifact from the EEG signal. The method is using WT because of its time invariance and it has better sampling rates in the low frequency bands. Also, WT could improve power of the wavelet transform and effectively eliminates noise in signal denoising. It has good ability in filtering the noise and retains the information. The main challenge in using wavelet transform is to select the most optimum mother wavelet for the given tasks, as different mother wavelet applied on to the same signal may produce different results. The selection of a mother wavelet function is an important step because studies have yet to provide specific mother wavelet basis functions that cater to all EEG channels. Four mother wavelets basis functions from families were investigated. This paper reviews on the mother wavelet selection methods and EEG signal denoising by using artificial neural networks and deep learning methods.

The remaining of the current chapter is organized as follows: In Sect. 2, a deep knowledge modeling and its puzzle methods applications had been considered. In Sect. 3 we comment a Electroencephalography applications. In Sect. 4, we discuss EEG signal denoising algorithm using wavelet transform. In Sect. 5, we present analysis of variance (ANOVA). In Sect. 6, we discuss experimental results. Finally, in Sect. 7, we present the conclusions.

2 Deep Knowledge Modeling and Its Applications

Wavelet transform is a powerful method to analyze different signals. This type of research presents the time-scale information of the signal. Deep modeling and data preprocessing may significantly improve its versatility and applications, and diminish the manual work much of which is hidden behind the perfect mathematically described apparatus. One of the problems in Data Science concerning deep modeling problems is related to the selection of data in order to process them more efficiently. Such methods could be used for data analysis and correction. These issues are directly related to the methods for deep modeling of data and knowledge without which the application cycle for logical and statistical data processing in order to extract hidden patterns cannot be built. Different types of non-classical logics, data-driven approaches and/or machine learning methods are used for this purpose [9–11]. Classical selection tools include both various statistical applications and applications of non-classical logics (descriptive logics etc.). There are many studies for solving crossword puzzles [12–16], but in most of them the problems are solved in a probabilistic way, by chance, by using random numbers and combinations, which is not effective. This paper proposes to increase efficiency by using logically oriented modeling tools. To solve the problem, it is proposed to use three groups of non-classical constraints. They are considered in connection with the research of the Puzzle method using the system of classical constraints [17]. The goal is to form a

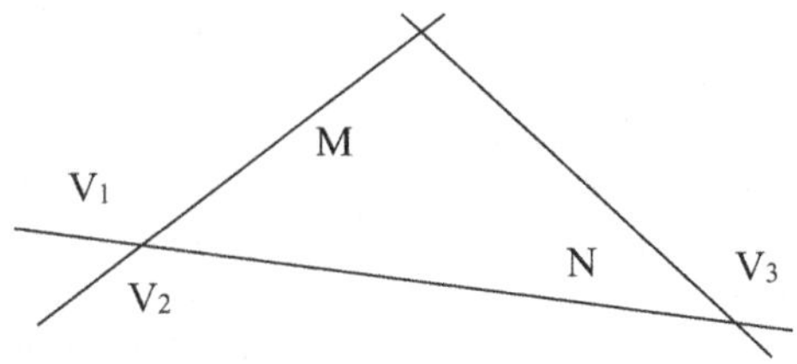

Fig. 1 A system of linear constraints with focus on objects M and N

closed area is formed aiming at focus attention of a software agent to certain data including two objects M and N (Fig. 1).

The specified area contains the required data for objects denoted by M and N. The closed area contains much less data than in the standard case but still it may be too large to explore. If the data selection process is enough effective, the focusing process/area closure helps to reveal a new information concerning data, metadata, knowledge and/or meta knowledge for objects M and N, for example: 'it follows from M that N is true' or 'there is a significant relation between M and N'. Some disadvantages: the constraint satisfaction is a rather static process. The set of constraints cannot be modified during the solving process. On the other hand, what is the constraint in the application sense? If this is a line/curve drawn on a surface, it should be enough to notice it to prevent crossing it. And what is its functionality? When the crossing itself is harmful not just to the intruder but only to the others, in many cases the violation will be a fact. Many other application problems arise. Below are the proposed solutions to the problems.

The constraint satisfaction solution should take in the account *why* does the constraint appear, *when* it could disappear, *how* does it apply, etc. Aiming at the data selection flexibility, new meta knowledge forms have been elaborated concerning logical-based control of the constraint application process.

First of all, the number of the linear constraints may be less than 3 depending on the selected set of goals. Then, one and the same constraint may be treated differently depending on the current conditions.

Denote the object M is a city museum area rounded by ancient city walls with a visitor entrance, service entrance and a hidden tunnel beneath the walls. Let the population P of individuals A, B, C, D, I $\in$ P consists of individuals A and B with some ordinary and many hidden secret intelligence functions, a museum guardian C, a set of ordinary visitors D and an intruder I disguised as a tourist. Every individual can reveal a place on the wall where one can jump out outside the constraints (museum walls) but cannot return. Hence the constraint role could be broken only in one direction and this could be done only in an illegal manner. On the other hand, the individuals A, B, C could penetrate through the set of constraints legally through the service entrance, and all considered individuals could go through the constraints via the official entrance. The intruder appeared at the site M because of a gathered information that something peculiar happens. Could the drawing attention event be described using the standard set of objective or fitness functions? In general case

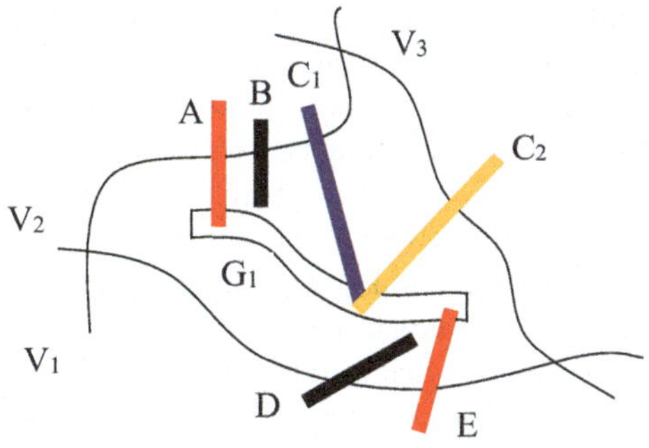

Fig. 2 A system of nonlinear constraints and the three groups of logically-based constraints

this is possible. A generalized case is described below, it supports a more deep and effective modeling.

Let only A has means to use the hidden tunnel. Every such usage could attract the attention of I. Also I could conclude its existence analyzing the distribution of external guarding devices, and other collateral information. As a result I will reveal the most possible location of the tunnel entrance. This is modeled as a binding region with the maximum probability value in the centre. The usage of visual surveillance/drones/satellite information constantly or temporary extends the guarded region outside the walls. The described toy problem case aims to reveal the need of an introduction of new, logical-based types of constraints.

Aiming at better deep modeling, it is proposed to use three non-classical groups of constraints: binding, pointing and crossword constraints. Binding constraints model situations where the solution is/was not far or close to the center of the binding area. A case is best investigated wren the farther from the center the less likely it is to find the desired solution. There are several types of binding constraints in this group. It is explored that the binding process may be unconditional or conditional, it may invalidate the linearity/range/the region of the usage and/or other properties of other constraints intersecting the binding domain. In Fig. 2 the binding constraints are represented by the lines {B, D} intersecting the surface containing the required/searched solution G_1.

Together with the binding constraints of Fig. 2, it is convenient to use the pointing (indicating) constraints {A, E} in order to determine not only the area, but also the direction of the search. The group of pointing constraints can be considered as a generalization of the classical systems of goal, target or fitness functions. In contrast, pointing constraints can change the direction of the search depending on the accumulated data or the knowledge (circumstances), in other words, the logic of data and based on them events can be followed. For example, if there is information that there is a pain, the data on his coordinates are probably no longer up to date. In this case, the direction has prognostic significance and the exact result is in doubt until the proof is found.

The last group of constraints is the crossword one depicted by $\{C_1, C_2\}$ from Fig. 2. Unlike {A, E}, they reveal not only the direction, but also parts of the desired solution (of the goal). The purpose of their application is to solve the problem G_1 when only some parts of the solution $\{C_1, C_2\}$ are known. For this purpose, combinations of other types of constraints are most often used, the relationships between $\{C_1, C_2\}$, etc. are studied. $\{C_1, C_2\}$ belongs to G1 but is only the subset of the desired solution. Many algorithms may be investigated how to calculate the other, unknown parts of the goal, and how to estimate their fitness to the solution of the problem. As a whole, the quoted problem is how to make links from the known set $\{C_1, C_2\}$ from the example to the unknown knowledge, how to use the pointing and other constraints aiming to diminish the set of possible solutions. No matter what is the application algorithm, its goal is to produce a set of knowledge which is the most appropriate to $\{C_1, C_2\}$. The considered descriptions do not have a purpose to comprise all possible methods and applications, but to consider a synthetic view of the field as a whole.

In EEG applications the pointing examples frequently concern the points in the centers of the binding areas where the signals were successfully obtained. The binding examples also concern best signal processing practice, good medical practice and so on. In the latter case little changes are possible: some binding models naturally support fuzzy and neuro-fuzzy applications. The pointing constraints are used for description and maximization of the effectiveness of specific brain-computer interface features and communications to other medical devices. Not only the good practice but also bad practice examples could be described using pointing and/or binding constraints. The proposed novel groups of constraints are not purposed to change the signal processing schemes.

3 Electroencephalography (EEG) Applications

3.1 Electroencephalography

Electroencephalography (EEG) measures the electrical activity generated by the brain using electrodes placed on the scalp [18]. EEG (popularly known as *brain waves*) measures the sum of the post-synaptic potentials generated by thousands of neurons having the same radial orientation with respect to the scalp (Fig. 3).

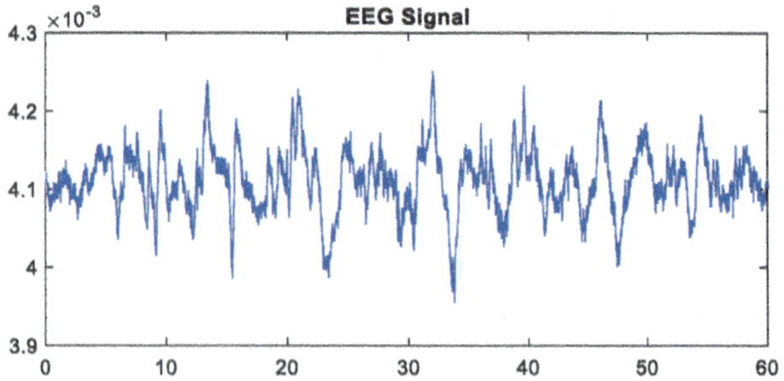

Fig. 3 Example of EEG signal for a healthy subject (time is displayed in seconds)

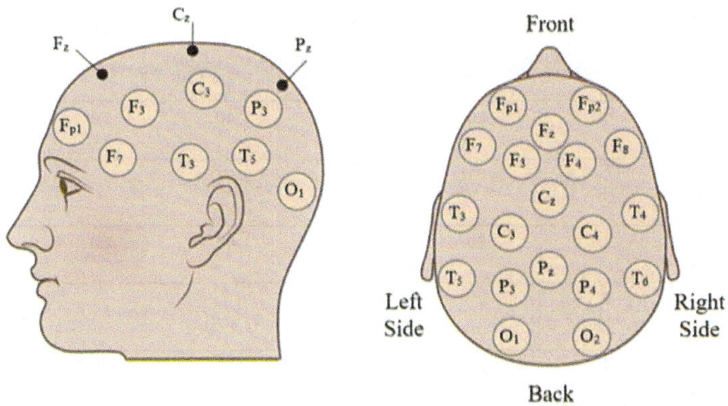

Fig. 4 Positions and names of the 10-20 international system electrodes

Signals recorded by EEG have a very weak amplitude, in the order of some microvolts. It is thus necessary to strongly amplify these signals before digitizing and processing them. Typically, EEG signals measurements are performed using a number of electrodes which varies from 1 to about 256, these electrodes being generally attached using an elastic cap. The contact between the electrodes and the skin is generally enhanced by the use of a conductive gel or paste. This makes the electrode montage procedure a generally tedious and lengthy operation. Electrodes are generally placed and named according to a standard model, namely, the 10-20 international system (Fig. 4). This system has been initially designed for 19 electrodes, however, extended versions have been proposed in order to deal with a larger number of electrodes. EEG signals are composed of different oscillations named "rhythms". These rhythms have distinct properties in terms of spatial and spectral localization. The normally used terms for EEG frequency bands whose sample is shown in Fig. 5.

- **Delta rhythm**: This is a slow rhythm (0.5–4 Hz), with a relatively large amplitude, which is mainly found in adults during a deep sleep. Delta waves occur during meditation in a state of deep sleep or coma. Abnormal delta activity may occur with the person, has learning disabilities or have difficulties maintaining conscious awareness (such as in cases of brain injuries).
- **Theta rhythm**: This a slightly faster rhythm (4–7 Hz), observed mainly during drowsiness and in young children. Theta waves are involved in sleep or daydreaming. This brain wave can indicate intuition or automatic tasks.
- **Alpha rhythm**: These are oscillations, located in the 7–13 Hz frequency band, which appear mainly in the posterior regions of the head (occipital lobe) when the subject has closed eyes or is in a relaxation state. Alpha waves connect the gap between our conscious thinking and subconscious mind. It helps us to calm down or it promotes a feeling of relaxation.
- **Beta rhythm**: This is a relatively fast rhythm, belonging approximately to the 13–30 Hz frequency band. It is a rhythm which is observed in awaken and conscious

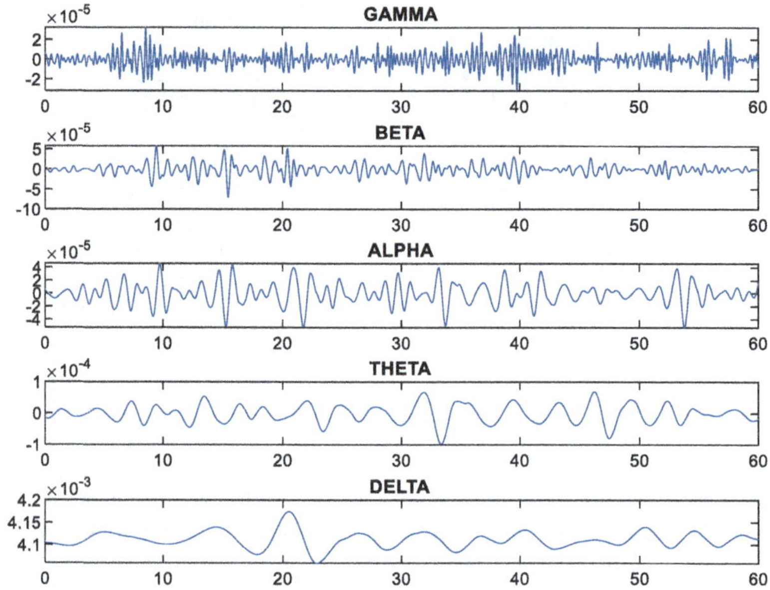

Fig. 5 Five major brain waves distinguished by their different frequency ranges

persons. This rhythm is also affected by the performance of movements, in the motor areas. Beta waves are active in a waking state. This frequency is visible in logical-analytical reasoning. In their activity, we focus on a problem solving.

- **Gamma rhythm**: This rhythm concerns mainly frequencies above 30 Hz. This rhythm is sometimes defined has having a maximal frequency around 80 Hz or 100 Hz. It is associated to various cognitive and motor functions. Gamma waves are important for learning, memory and information processing.

3.2 Signal and Data Processing of EEG

Due to very low in amplitude, EEG signals are prone to artifacts and noise [19]. The noise can be electrode noise or can be generated from the body itself. These various types of noises that can contaminate the signals during recordings are the electrode noise, baseline movement, EMG disturbance, eye movements, eye blinks and sometimes ECG disturbance. The noises in the EEG signals are called the artifacts and these artifacts are needed to be removed from the original signal as noise/artifacts makes analysis and further processing of the EEG signals difficult. The signal and data processing of EEG could be gathered into a single and more general, higher level component, which could be denoted as "EEG processing". This component is a key element in the design of a processing algorithm as it aims at transforming the input brain signals into a command for a given application. As such, the "EEG processing"

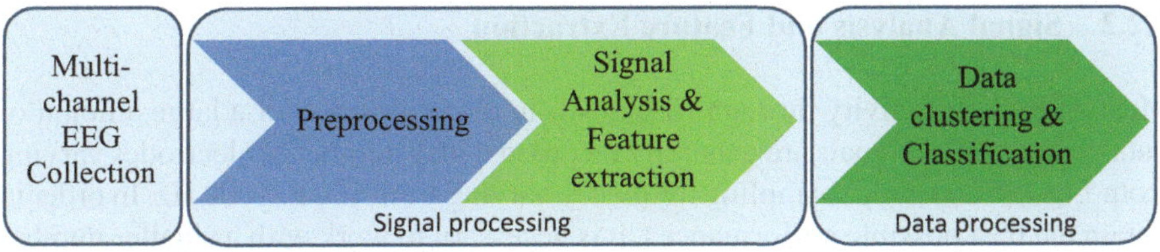

Fig. 6 Signal and data processing of EEG signals

component can be seen as the preprocessing, feature extraction and classification of EEG signals (Fig. 6).

It is important to note that the boundaries between the "preprocessing", "Signal analysis and feature extraction" and "Data clustering and classification" components are not hard boundaries, and these boundaries may even appear as fuzzy. Thus, the preprocessing and feature extraction components are sometimes merged into a single algorithm, whereas the classification algorithm can be missing or reduced to its simplest form, i.e., a decision threshold on the feature values. However, it is interesting to distinguish these components, as they have different inputs and outputs as well as different goals.

3.2.1 Preprocessing

Once the data have been acquired, they are generally preprocessed in order to clean (denoise) the signals and/or to enhance relevant information embedded in these signals [20]. Indeed, EEG signals are known to be very noisy, as they can be easily affected by the electrical activity of the eyes (EOG: ElectroOcculoGram) or of the muscles (EMG: ElectroMyoGram), e.g., face or jaw muscles. These muscle artifacts are especially annoying as they have an amplitude which is much larger than the one of EEG signals. As such, it appears as difficult to remove these artifacts without accidentally removing relevant information embedded in these EEG signals. Moreover, it is interesting to remove the background brain activity which is not related to the neurophysiological signals of interest. Overall, the preprocessing step can be defined as a method which transforms a set of signals into a new set of signals which are supposedly denoised. In other words, the preprocessing step aims at increasing the signal-to-noise ratio of the input signals.

In order to perform this preprocessing, various spatio-spectro-temporal filters are used [21, 22]. These filters can be simple frequency filters or more advanced filters such as independent component analysis [23] or common spatial patterns [24].

3.2.2 Signal Analysis and Feature Extraction

Measuring brain activity through EEG leads to the acquisition of a large amount of data. Indeed, EEG signals are generally recorded with a number of electrodes varying from 1 to 256 and with a sampling frequency varying from 100 to 1000 Hz. In order to obtain the best possible performances, it is necessary to work with a smaller number of values which describe some relevant properties of the signals. These values are known as "features". Such features can be, for instance, the power of the EEG signals in different frequency bands. Features are generally aggregated into a vector known as "feature vector". Thus, feature extraction can be defined as an operation which transforms one or several signals into a feature vector. Identifying and extracting good features from signals is a crucial step in Signal analysis and feature extraction. Indeed, if the features extracted from EEG are not relevant and do not describe well the neurophysiological signals employed, the classification algorithm which will use such features will have trouble identifying the class of these features, i.e., the mental state of the user. Consequently, the correct recognition rates of mental states will be very low, which will make the use of the interface not convenient or even impossible for the user. Thus, even if it is sometimes possible to use raw signals as the input of the classification algorithm, it is recommended to select and extract good features in order to maximize the performances of the system by making easier the task of the subsequent classification algorithm. According to some researchers, it seems that the choice of a good preprocessing and feature extraction method have more impact on the final performances than the choice of a good classification algorithm [25].

In the following we focus on different features extracted from EEG signals for emotion detection. These techniques can be divided in three main groups, which are: (1) the methods that exploit the temporal information embedded in the signals [26], (2) the methods that exploit the frequential information [22] and (3) hybrid methods, based on time-frequency representations, which exploit both the temporal and frequential information [27].

1. *Time domain features*
   These features are extracted from the signal in its original domain. First of all, the statistical properties of the signal such as mean, standard deviation, maximum, minimum, difference of maximum and minimum, median, mode and power are mostly used. Another feature is entropy, which is the symbol of scattering data [28]. Also, Hjorth [29], introduced some features called activity, mobility and complexity, which are then used by researchers for EEG emotion detection [30]. Other common features in time domain include fractal dimension [31], non-stationary index [32], Higher Order Crossings [33].
2. *Frequency domain features*
   Presentation of a signal in frequency domain is possible by using Fourier transform. The mostly used algorithm to compute Discrete Fourier Transform (DFT) is Fast Fourier Transform (FFT). The most popular feature in the context of emotion recognition from EEG is power in different frequency bands [34]. It is assumed that the signal is stationary for the duration of a trial. Different

frequency bands are found by wavelet transform. The frequency bands of EEG signals slightly vary in different studies. The extracted features from the resulting representation of the signal in frequency domain are: average power (mean) of frequency bands [35], and relative minimum, maximum, and variance. Additionally, the ratio of mean band powers is calculated for each channel. A set of frequency domain features has been introduced by Hosseini and et al. in [36] which is called higher order spectra.

3. *Time frequency domain features*
   If a signal is non-stationary, time-frequency methods can provide additional information by considering dynamical changes. In this category we can mention the features extracted by applying the wavelet transform to the signal [37]. Also, Hilbert-Huang spectrum of intrinsic mode functions is used [38].

3.2.3 Classification

After feature extraction, the extracted features should be classified for identifying neurophysiological signals by using an appropriate classification algorithm. Classification is achieved using algorithms known as "classifiers". Classifiers are able to learn how to identify the class of a feature vector, thanks to training sets. These sets are composed of feature vectors labeled with their class of belonging.

These classifiers can be divided into several categories which are: linear analysis, nonlinear analysis, adaptive algorithms, clustering and fuzzy techniques, and neural networks [18, 20, 39].

3.3 Applications of EEG

Applying appropriate signal processing of brain signals and evaluation of some of the parameters of the signals, it is possible to solve a wide range of tasks. Some applications of EEG signals are used to assess various emotional and physiological states of a person such as whether he is healthy or sick, happy or unhappy, calm or anxious and many others. It is possible to assess many other conditions, such as whether he has consumed alcohol or not, whether he has used GSM or not, and others. EEG devices can provide valuable information about human mental health states, thoughts, and imagination [20]. Electroencephalography can be used for various applications. Below is a list of the three most common applications of EEG technology:

3.3.1 Neuroscience and Clinical Applications

Neuroscience attempts to understand the functionality of the nervous system. It allows clinical or non-clinical researchers to get an idea about how the brain acts when humans experience different emotional states and how the brain works in various mental states. Researchers have applied EEG devices in their studies in the below fields. Most generally, psychological studies utilize EEG to study the brain processes underlying attention, learning, and memory. How do we perceive the world? How do our expectations shape the way we see our surroundings?

Whenever brain processes are impaired, deficits in behavioral, attentional and cognitive processing can be observed. Clinical and psychiatric fields use EEG to evaluate the patient's cognitive states, determine lesion sites, and classify symptoms.

Also, EEG is heavily used to evaluate the effect of medical and psychological treatment. More and more therapies utilize virtual reality technology and record EEG data to monitor how the patient's brains improve over time.

Real-time EEG signals can be used to provide immediate information about brain-wave activities. EEG data have been applied for diagnosing and predicting many abnormal brain diseases and cognitive impairments, as Epilepsy; Parkinson's Disease; Memory problems like Alzheimer's; Language impairments such as Dyslexia; Attention Deficit Hyperactivity Disorder; Seizures; Schizophrenia; Autism in adults and children; Sleep disorders and insomnia; Anxiety; Post-traumatic stress disorder; Huntington's disease; Multiple sclerosis diagnosis; Amyotrophic lateral sclerosis; Traumatic brain injury; Coma; Level of consciousness and Neurosurgery [40–44] (Fig. 7).

Neuroscience can also be applied to understanding human emotion in VR with/without the ability to touch the environment by displaying various types of media, such as: Real world or VR pictures; Images of nature and city environments; TV advertisements; Auditory stimuli; Multimedia, along with memory recall and dreams [45–47]. Cognitive neuroscience is studding Measuring cognitive load; Detecting differences between brain wave activity during suicidal and non-suicidal states. Understanding brain activity during insight (insight is a moment where a human understands how to solve a puzzle or gains knowledge); Analyzing brain workload during decision making or learning a new task; Studying sleep pattern [48, 49].

Behavioral neuroscience is studding Changing the workplace light and measuring brain alertness status; Measuring drowsiness or sleep detection for drivers and pilots; Measuring mental workload of deaf children exposed to a noisy environment during a word recognition task; Determining surgeon stress level while performing surgery; Identifying and reducing stress level; Environmental Psychology. Neurophysiology is studding Measuring changes in brain after drinking alcohol; Detecting fatigue, GSM talking [50, 51].

3.3.2 Brain-Computer Interfaces (BCI)

BCIs, sometimes called brain-machine interfaces (BMIs), are one of the most common applications of EEG. BCIs use real-time EEG data to control and direct mechanical and electronic devices.

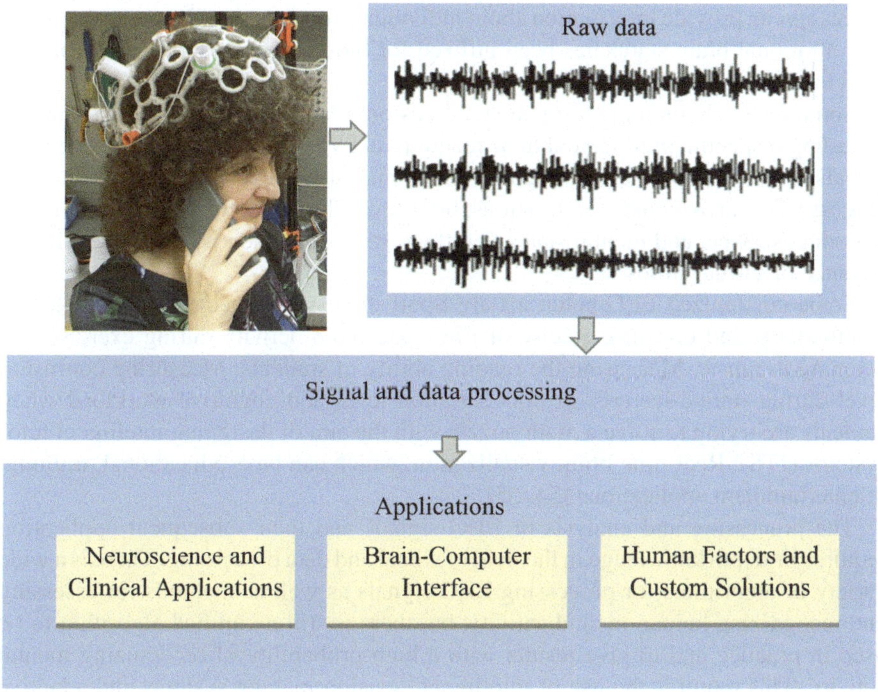

Fig. 7 Applications of EEG signals

BCI devices are commonly used as a human-machine interface to help individuals with mild to severe motor disabilities, including those who are not able to communicate with others. This can help paralyzed patients steer their wheelchairs or move a cursor on a screen. The most common BCI applications are aimed at people with disabilities or motor activity impairment also control of mechatronic devices by control of mobile phone apps using eyewinks; directing electrical wheelchair movement; control of artificial body part such as prosthetic hand or arm; controlling a robot using body gestures; speech recognition system for people with speech disability; mouse cursor control using imagined hand movement; controlling a video game or virtual reality (VR) environment using body gesture and eye movement; controlling of robotic body parts, robot, drone and vehicle. Monitoring and controlling sensors inside of smart houses [52, 53].

3.3.3 Human Factors and Custom Solutions

Originating from Psychology, the field of Human Factors focuses on workplace optimization; both with respect to tools and interfaces as well as social interaction. In this area, EEG research is used to identify brain processes related to specific personality traits such as intro-/extroversion or social anxiety. Additionally, brain processes can provide information about individuals' differences. Recently, cognitive and emotional brain status has been utilized for biometrics, meaning EEG data are used to identify people.

Some research attempts to understand customer's preferences and expectations regarding a specific product and their reaction to TV advertising by analyzing EEG signals. In the field of neuro marketing, economists use EEG research to detect brain processes that drive consumer decisions, brain areas that are active when we purchase a product/service, and mental states that the respective person is in when exploring physical or virtual stores.

Some customized EEG solutions are Sport, fitness and meditation: Monitoring health status and boosting quality of life using brain activity during exercise and listening to music. Measuring the reading ability of students; Measuring confusion level during online lectures or concentration level and cognitive workload when students are trying to solve a math puzzle with the aim of designing intelligent tutor systems (ITS); Real-time brain visualization, which can have educational, training, or entertainment applications [54, 55].

The processing and analysis of EEG signals and their subsequent application requires in-depth knowledge in the field of signal and data processing. There is a wide variety of algorithms for processing EEG signals as well as methods for assessing and recognizing human mental activity, but there is still no unified algorithm to be used in practice and to give results with a high probability of recognizing mental activity. This requires the use of intelligent decision-making systems and adaptive replacement and tuning of the algorithms used. In this article, we consider only one exemplary task in EEG signal processing. Our goal is to offer an intelligent solution to the problem of denoising of EEG signal by using Wavelet transform. The problems in solving this task are, choice of mother wavelet, decomposition level and threshold. The selection of these parameters will be made through the application of statistical analysis using the ANOVA method. To automate this process, the article proposes the use of deep modeling and the use of intelligent decision-making systems to configure the best algorithm.

4 EEG Signal Denoising Using Wavelets

4.1 Wavelet Transform

Wavelet transform is a powerful method to analyze the EEG signals. This method could present the time-scale information of the signal. Wavelet is able to perform a different time and scale resolution, so the user could choose which particular signal they want. The other advantage of the wavelet is capable to localize the area of larger signals. Recently, WT has been extensively used with non-stationary signals because WT shown powerful in removal several EEG artifact noises, which can be corrupted the original EEG signal during recording time, such as eye blinking noise,

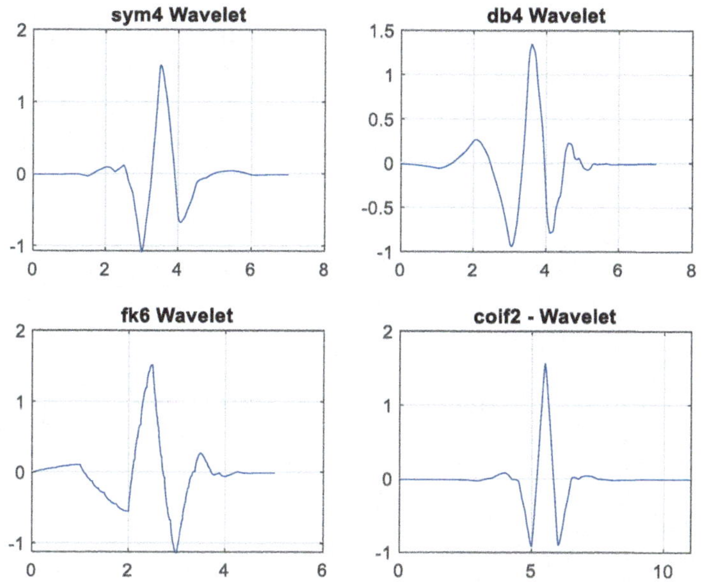

Fig. 8 Mother wavelet functions used for denoising of EEG signal

eye movement noise, muscles activity noise, power line noise, and electromyogram noise [56]. The benefit of wavelet analysis is capable to disclose information contain of the signals. The disclose information of the signal are trends, breakdown points, discontinuities and self-similarity [57]. In addition, wavelet is able to filter the signal as well as classify the signal. In the wavelet analysis, it is essential to pay attention to the scaling function and the type of mother wave. There is a wide variety of family waves that have proven to be particularly useful in signal processing. In practice, various wavelets are used in the decomposition of signals such as waves of Symlets, Daubechies, Fejer-Korovkin, Coiflets, and others (Fig. 8) [56, 58]. WT can be classified into two categories: continuous wavelet transform (CWT) and discrete wavelet transform (DWT).

The CWT performs a multiresolution analysis by contraction and dilatation of the wavelet functions. The discrete wavelet transform (DWT) uses filter banks for the construction of the multiresolution time-frequency plane. The time-frequency resolution problem is caused by the Heisenberg uncertainty principle and exists regardless of the used analysis technique. By using an approach called multiresolution analysis it is possible to analyze a signal at different frequencies with different resolutions. A filter bank consists of filters which separate a signal into frequency bands. An example of a three channel filter bank is shown in Fig. 9. A discrete time signal enters the analysis bank and is filtered by a low-pass filter LPF and a high-pass filter HPF which separate the frequency content of the input signal in frequency bands of equal width. The output of the filters each contain half the frequency content, but an equal amount of samples as the input signal. The two outputs together contain the same frequency

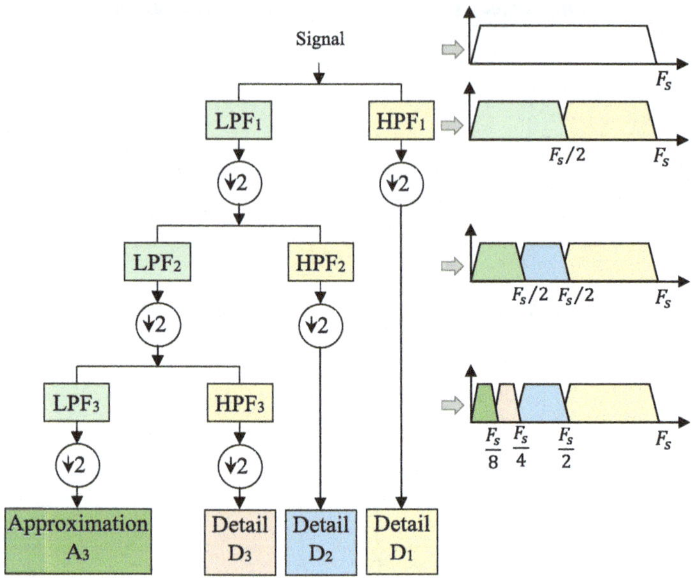

Fig. 9 Discrete wavelet decomposition tree

content as the input signal, however the amount of data is doubled. Therefore down sampling by a factor two, denoted by ($\downarrow$ 2), is applied to the outputs of the filters in the analysis bank. The DWT decomposes a signal into different frequency bands by passing it through two quadrature mirror filters via a finite impulse response, where g is a high-pass filter and h is a low-pass filter. h is related to the scaling function, whereas g is related to the mother WT.

Using Mallat's fast algorithms [59] the gradual decomposition of the original signal become:

$$A_k(n) = \sum_{l=-\infty}^{\infty} h(l-2n)A_{(k-1)}(l) \tag{1}$$

$$D_k(n) = \sum_{l=-\infty}^{\infty} g(l-2n)A_{(k-1)}(l) \tag{2}$$

The signal $A_{(k-1)}(l)$ convolves with $h(l-2n)$ when this signal acts as an LPF; otherwise, this signal acts as an HPF and convolves with $g(l-2n)$. The result transforms the original signal into two sub-bands $\left[0 \div \frac{F_S}{2}\right]$ and $\left[\frac{F_S}{2} \div F_S\right]$. $A_k(n)$ is the approximation component A that represents low-resolution components; $D_k(n)$ is the detail decomposition component D that describes high-resolution components [57, 58].

Reconstruction of the original signal (Inverse Wavelet Transform) is possible using the synthesis filter bank, where the signals are up sampled and passed through a low-pass and a high-pass filters. The filters in the synthesis bank are based on the filters in the analysis bank. The outputs of the filters in the synthesis bank are summed, leading to the reconstructed signal.

4.2 *Wavelet Denoising Algorithm*

The wavelet denoising algorithm will be applied to filter the signal. The denoising algorithm includes three major stages:

1. Decompose the signal from the original signal to different level composition by using a wavelet transform: In this phase, the original EEG signal will be divided into several levels based on the decomposition level value. At each level, the EEG signal will decompose into two parts namely Approximation coefficients (cA), and Detail coefficients (cD). The detail coefficients will process using high-pass filter and approximation coefficients will continue decompose for next level. Figure 10 shows the decomposition process using DWT for three decomposition level. The first two parameters of WT which are wavelet function name or mother wavelet function (Φ) and Decomposition level (L) must be selected in this phase.

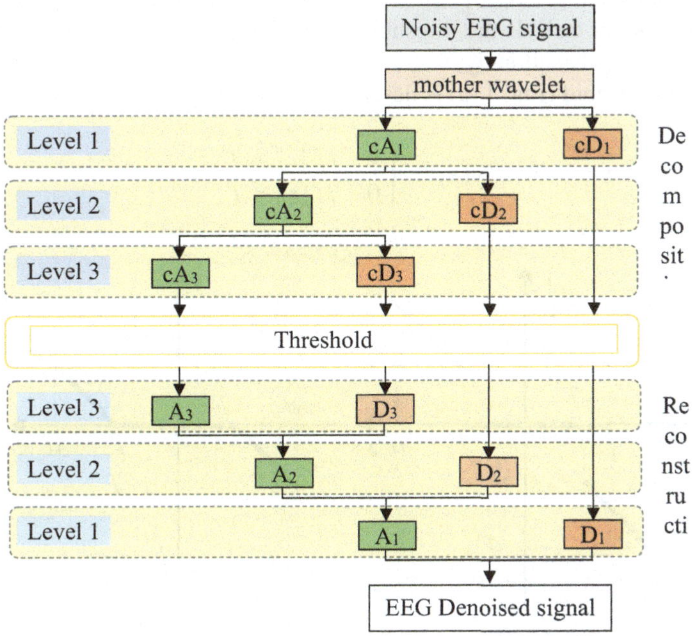

Fig. 10 Wavelet denoising algorithm

2. Threshold the signal based on the boundary of the noise: In this phase, the thresholding type (β), thresholding selection method (λ), and wavelet rescaling approach (ρ) must be determined for each level according to the coefficients noise level in the corrupted EEG signal.
3. Reconstruct the signal by using the reverse wavelet transform. In this phase, the EEG denoised signal is reconstructed using Inverse Discrete Wavelet Transform (IDWT). Figure 10 shows the signal reconstruction process for three decomposition level.

The denoising algorithm is applied to eliminate the noise from the useful signal [56] if the received signal consists of a useful and noisy constituent, as given in the equation:

$$y(t) = x(t) + \varepsilon(t) \tag{3}$$

The analysis of $y(t)$ is equal to the sum of the analyses of signal $x(t)$ and the noise $\varepsilon(t)$.

The basis of this technique is the use of threshold functions with different forms, on the basis of which the limitation of the detailing coefficients is carried out. By setting a threshold value it is possible to "cut" the signal value below this threshold, it can significantly reduce the noise and shrink the signal. The threshold functions discussed in this paper and most commonly used in modern filtering algorithms have been represented in Fig. 11 [60]. The choice of threshold values for wavelet transform is a very important task. If the threshold value is too large or too small, the signal may not be detected accurately.

The hard threshold (Fig. 11a) is described by equation [59]:

$$\mathrm{y(x)} = \begin{cases} x & if\,|x| \geq T \\ 0 & if|x| < T \end{cases} \tag{4}$$

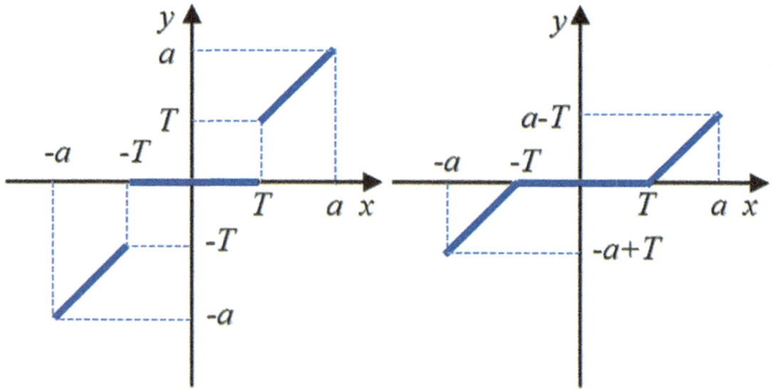

Fig. 11 Threshold functions for processing the conversion coefficients: **a** a hard threshold function; **b** a soft threshold function

The T-dimension can occupy values discussed below in the paper as x and y: the input and output coefficients of the wave transform.

Figure 11b shows the soft threshold function (soft threshold estimate), which is given by the expression [60]:

$$y(x) = \begin{cases} sgn(x)(|x| - T) & if\ |x| \geq T \\ 0 & if\ |x| < T \end{cases} \tag{5}$$

The function $sgn(x)$ specifies the sign of the x coefficient.

$$sgn(x) = \begin{cases} 1, & if\ x > 0 \\ 0, & if\ x - 0 \\ -1, & if\ x < 0 \end{cases} \tag{6}$$

The main difference between the soft threshold and the hard threshold is that the first (the soft threshold function) does not contain a jump at the point determined by the threshold value T. In other words, the soft function of the threshold, unlike the hard threshold, is continuous. Due to this circumstance, in the case of soft threshold processing, better signal processing is obtained near the break point. It should be noted that the reduction of the values of decomposition coefficients in the threshold value, in the case of a mute threshold processing, as a whole, for a large number of signals has a negative impact on the final evaluation of the quality of the recovered signal. Therefore, in general, as shown experimentally, better in terms of the numerical assessment of the quality of the recovered signal is the hard threshold estimate. In the case of soft threshold processing, the digital evaluation of the quality of the recovered signal is close to the latter in the case of hard threshold processing and it is necessary to select an appropriate threshold value T.

Wavelet selection, as well as wavelet decomposition level, and selected threshold can be made by applying statistical analysis using the ANOVA method.

5 Analysis of Variance (ANOVA)

There are two types of variables in the ANOVA (Analysis Of Variance): independent and dependent variables. Independent variables or factors are used to group the data. They have two, three or any finite number of levels. One-Way ANOVA assumes the presence of one independent variable (factor) with two or more levels. The dependent variable sets the differentiating feature by which the groups (samples) are compared. Changes in the dependent variable are, or are assumed to be, a consequence of changes in the independent variable, which is why it is often called the result variable. In the univariate analysis, the simultaneous independent effects (so-called "main effects") and interactions of two or more independent variables on the dependent variable are evaluated by a generalized statistical procedure.

ANOVA is used to test hypotheses, whether several means are equal. This technique is an extension of the t-test for independent samples, making it possible to determine not only the differences between the mean values, but also which exact mean values differ from the others. There are two types of tests to compare means: a priori contrasts and post hoc tests. Contrasts are tests performed before the experiment, and post hoc tests are performed after the experiment. The trends between categories can also be tested.

The values of the factor (discrete independent variable) should be integers and the dependent variable must be quantitative (interval level of measurement). Each group is an independent random sample of the normal population. The analysis of variance is resistant to deviations from normality, although the data should be symmetrical. The groups should come from populations with equal variances. To test this hypothesis, the Levene test for dispersion homogeneity is applied.

Therefore, the logic of the analysis of variance is based on the decomposition of the total variance of the variable (deviation of all values from their total mean) into two important components: intergroup variance—the deviations of group means from the total mean and intragroup variance—the individual deviations of the values from the mean within a given category (group). ANOVA uses the statistic F as the ratio of between and within group variances. The stronger the effect of the factor value, the greater the differences between the intergroup variance and the intragroup variance (the value of F will be great).

6 Results

To investigate the proposed denoising algorithm in this paper, the real EEG signal from healthy person was used. The EEG signal includes physiologic signals from the person and electrode noise. Part of the recording EEG signal is shown in Fig. 12. Wavelet transform is proposed to remove the artifact from the EEG signal. In most cases, optimal mother wavelet functions are selected on the basis of the compatibility

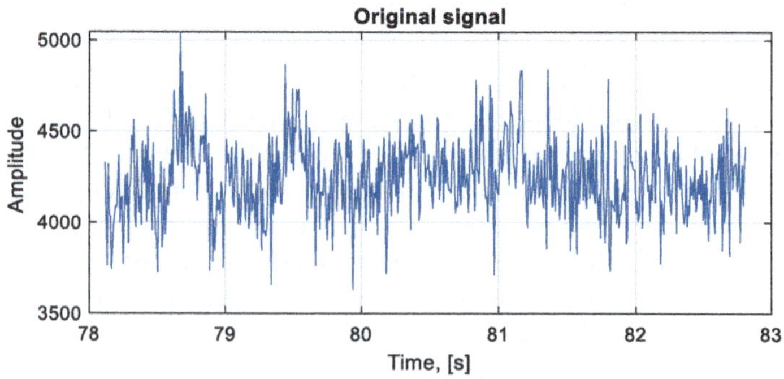

Fig. 12 Original and denoised EEG signals

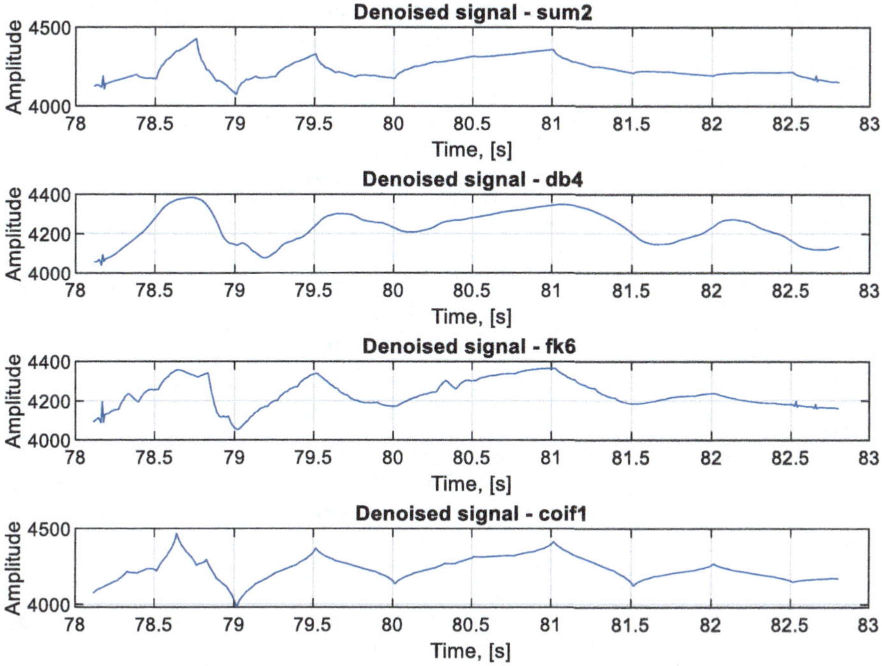

Fig. 13 Denoised EEG signal by soft threshold

with the EEG signal characteristics to be analyzed. Accurate mother wavelet selection not only helps retain the original signal, but also enhances the frequency spectrum of the denoised signal.

By using proposed algorithm and using wavelets Symlets (sim4), Daubechies (db4), Fejer-Korovkin (kf6), Coiflets (coif2), we obtained the result in Figs. 13 and 14. The result shows that this algorithm is able to remove noise from the EEG signal effectively. The applied wavelet denoising algorithms are with threshold processing using thresholds, soft (Fig. 13) and hard (Fig. 14).

An important point in EEG signal processing via wavelet is the selection of a suitable mother wavelet and decomposition level to reduce the artifacts that contaminate EEG signals. The best selection of the wavelet function from the wavelet families helps conserve the decomposed EEG signal and obtain optimal re-constructed signals.

Using ANOVA, we determined the mother wavelet functions with the most significant differences to maximize their cross-correlation with the EEG signals. Statistical analysis was performed through ANOVAs in SPSS 22.

The significant differences among the four types of filtering and both types of threshold processing as dependent variable were evaluated. Post-hoc comparison was performed through Scheffe test. The significance was set at $p < 0.05$ [61].

In our case, the best results were obtained using mother wavelet "sym4" and hart threshold. Therefore, the most compatible mother wavelet with the EEG signals

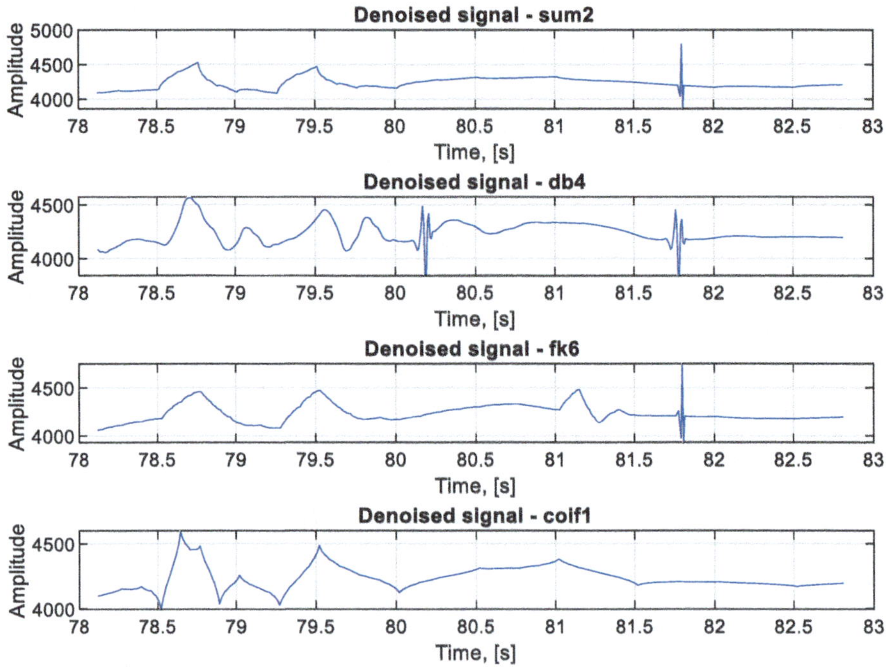

Fig. 14 Denoised EEG signal by hard threshold

should be selected to achieve wavelet denoising, decomposition, reconstruction, and good feature extraction. The cross-correlation coefficients of EEG signals are shown in Table 1. The highest cross-correlation coefficient was obtained at mother wavelet “sym4” and hart threshold. The results show that the choice of the wavelet function, as well as the type of threshold are essential for the denoising of the EEG signal. To optimize the selection process, this article proposes an intelligent upgrade for modeling and preprocessing EEG signals using a library of wavelets and a plurality of thresholds to denoise the EEG signal.

For this purpose, Binding and Pointing are used as constraints for post-processing of the obtained wavelet results.

Table 1 Cross-correlation of EEG signals

Name of wavelet	Type of threshold	
	Soft	Hard
sym4	0.868	0.885
db4	0.869	0.882
kf6	0.869	0.883
coif2	0.868	0.882

The default threshold value from the control system is pointed to hard, and the set of wavelets is pointed to the considered four cases from Fig. 14. The usage of the pointing constraints aims to avoid the expensive and time consuming deep learning or analogical procedures. This does not mean that the pointed values will be set in all cases. The values are shifted via system inquiries on different fault events or manually. The considered set of faults includes grip liquid contact problems, necessary adjustments and so on. The events including skin tremors, external noise, and work in rapidly changing environments will influence the defeat or adjustment of the pointed values. Hence, the control of wavelet/threshold values is realized in a data-driven manner without the usage of ANNs or probabilistic means. The system gradual improvement procedures are based on the analysis of the answers to typical questions, among them: WHY the outcome is so different when using wavelets from the library in the region 81–82 s in Fig. 14, WHEN the reason leading to error occurrence is planned to be revealed and eliminated, HOW the wavelet application influences the analysis of weakest brain signals, and so on. In this case, the research goes into ontological direction described in the book chapter [17].

The automatic answers to the quoted questions are still out of scope of the considered research, but their appearance in the system log files makes the system more versatile and data-driven. It gives opportunity to make more effective the existing manual work.

The binding process aims to eliminate the repeating error patterns caused by eyeball movements, and different non expected processes. The rare/irregular error events are more difficult to be revealed. In this case the above mentioned manual signal processing preparation procedures had been applied. This is avoided in the considered constraint modeling application. The binding of repeating error signal events helps to reveal analogical errors like double eye blinking etc. The binding constraint analysis emphasizes repeating patterns in the outcome EEG signals produced by pulse or other reasons. The analysis of the reasons raises new questions WHY, WHAT, WHEN earlier described in this section and points to the new unknown knowledge groups which should be linked to the existing knowledge base and experience.

The binding and pointing constraints allow us to model and process both interconnected EEG and error signals in a more universal way where the connections and larger parameter changes stay in the focus. For example, we may point to some knowledge or dataset as peculiar/odd that later return to analyze it. Also, binding constraints are frequently applied to defeat different features, among them: how to make linear constraints slightly nonlinear in binding or certain pointed regions. A research on the transition of classical constraint sets into a system of ontologies is in progress. The mutual application of binding and pointing constraints helps us to reveal different hidden interconnections between signals or parts of a signal. Their analysis with all possible means leads to significant improvements.

7 Conclusions

Different types of noise and artifacts contaminate EEG signals. In this paper, wavelet transform is proposed to remove the artifact from the EEG signal. The method is using wavelet transform because effectively eliminates noise and retains the information. In this study, the compatibility of four mother wavelet basis functions from the Daubechies, Symlets, Fejer-Korovkin, and Coiflets families were selected and subjected to analysis. Using ANOVA, we determined the mother wavelet and the type of threshold which maximize the cross-correlation with the EEG signals. From the obtained results we can conclude that the choice of mother wavelet is a complex process and the denoising stage could be implemented by using artificial neural networks and deep learning methods (ANNs). Examples are considered of new constraint applications from binding and/or pointing groups. Their usage doesn't change the considered signal processing schemes but makes them more effective and the modeling tools mode deep and universal.

Acknowledgements This work is supported by the Bulgarian National Science Fund, Project title "Synthesis of a dynamic model for assessing the psychological and physical impacts of excessive use of smart technologies", KP-06-N 32/4/07.12.2019 and by Project No. AP09259370 "Development of a technological platform for virtual learning based on artificial intelligence approaches" due to grant funding from the Ministry of Education and Science of the Republic of Kazakhstan.

References

1. Subha, D., Joseph, P., Acharya, R., Lim, C.: EEG signal analysis: a survey. J. Med. Syst. **34**(2), 195–212 (2010)
2. Wang, S., Liu, X., Yianni, J., Aziz, T., Stein, J.: Extracting burst and tonic components from surface electromyograms in dystonia using adaptive wavelet shrinkage. J. Neurosci. Methods **139**(2), 177–184 (2004)
3. Sakkalis, V.: Review of advanced techniques for the estimation of brain connectivity measured with EEG/MEG. Comput. Biol. Med. **41**(12), 1110–1117 (2011)
4. Kumar, P., Arumuganathan, R., Sivakumar, K., Vimal, C.: An adaptive method to remove ocular artifact from EEG signal using wavelet transform. J. Appl. Sci. Res. **5**(7), 741–745 (2009)
5. Singh, V., Sharma, R.: Wavelet based method for denoising of electroencephalogram. Int. J. Adv. Res. Comput. Sci. Softw. Eng. **5**(4), 1113–1117 (2015)
6. Lanlan, Y.: EEG denoising based on wavelet transformation. In: 3rd International Conference on Bioinformatics and Biomedical Engineering, Beijing, China, pp. 1–4 (2009). http://doi.org/10.1109/ICBBE.2009.5162680
7. Araghi, L.: A new method for artifact removing in EEG signals. In: International Multi-Conference of Engineers and Computer Scientists, Hong Kong, vol. 1, pp. 420–423 (2010)
8. Palendeng, M., Wen, P., Goh, S.: Investigation of Bispectral Index (BIS) filtering and improvement using wavelet transform adaptive filter. In: IEEE International Conference on Nano/Molecular Medicine and Engineering, Hung Hom, China, pp. 11–15 (2010)
9. Makridis, M., Papamarkos, N.: A new technique for solving puzzles. IEEE Trans. Syst. Man Cybern. Part B Cybern. 1–10 (2009) (A Publication of the IEEE Systems, Man, and Cybernetics Society)

10. Kochan, O., et al.: Methods of reducing the effect of the acquired thermoelectric in homogeneity of thermocouples on temperature measurement error. J. Meas. Tech. **58**, 327–331 (2015)
11. Levitin, A.: Algorithmic puzzles: history, taxonomies, and applications in human problem solving. J. Probl. Solving **10**, 1–15 (2017)
12. Alajlan, N.: Solving square jigsaw puzzles using dynamic programming and the Hungarian procedure. Am. J. Appl. Sci. **6**(11), 1941–1947 (2009)
13. Jotsov, V., Sgurev, V.: Applications in intelligent systems of knowledge discovery methods based on human-machine interaction. Int. J. Intell. Syst. (IJIS) **23**(5), 588–606 (2008)
14. Jotsov, V.: Machine self-learning applications in security systems. In: 6th IEEE International Conference on Intelligent Data Acquisition and Advanced Computing Systems, Prague, Czech Republic, Sept 15–17, pp. 727–732 (2011)
15. Afanasyev, I., et al.: Blockchain solutions for multi-agent robotic systems: related work and open questions. In: Balandin, S., Deart, V., Tyutina, T. (eds.) Proceedings FRUCT'24 Proceedings of the 24th Conference of Open Innovations Association FRUCT, Article No. 76 (2019)
16. Jotsov, V.: Evolutionary parallels. In: 1st International Symposium on Intelligent Systems, Varna, Bulgaria, 10–12.09.2002 (2002). ISBN: 0-7803-7134-8
17. Jotsov, V.: New proposals for knowledge driven and data driven applications in security systems, innovative issues in intelligent systems. In: Sgurev, V., Yager, R., Kacprzyk, J., Jotsov, V. (eds.) Studies in Computational Intelligence, vol. 623, pp. 231–294. Springer, Berlin (2016)
18. Dimitrov, G., Garvanova, M., Kovatcheva, E., Aleksiev, K., Dimitrova, I.: Identification of EEG brain waves obtained by emotive device. In: 9th International Conference on Advanced Computer Information Technologies, Ceske Budejovice, Czech Republic, pp. 244–247 (2019)
19. Padiri, G.R.: Using EEG to assess programming expertise against self-reported data. Iowa State University Capstones, Theses and Dissertations (2018)
20. Lotte, F.: Study of electroencephalographic signal processing and classification techniques towards the use of brain-computer interfaces in virtual reality applications. Human-Computer Interaction. INSA de Rennes (2008)
21. McFarland, D., McCane, L., David, S., Wolpaw, J.: Spatial filter selection for EEG-based communication. Electroencephalographic Clin. Neurophysiol. **103**(3), 386–394 (1997)
22. Besserve, M., Garnero, L., Martinerie, J.: Cross-spectral discriminant analysis (CSDA) for the classification of brain computer interfaces. In: 3rd International IEEE/EMBS Conference on Neural Engineering, pp. 375–378 (2007)
23. Kachenoura, A., Albera, L., Senhadji, L., Comon, P.: ICA: a potential tool for BCI systems. IEEE Sig. Process. Mag. **25**(1), 57–68 (2008)
24. Congedo, M., Lotte, F., Lécuyer, A.: Classification of movement intention by spatially filtered electromagnetic inverse solutions. Phys. Med. Biol. **51**(8), 1971–1989 (2006)
25. Hammon, P., de Sa, V.: Preprocessing and meta-classification for brain-computer interfaces. IEEE Trans. Biomed. Eng. **54**(3), 518–525 (2007)
26. Rakotomamonjy, A., Guigue, V., Mallet, G., Alvarado, V.: Ensemble of SVMs for improving brain computer interface P300 speller performances. In: International Conference on Artificial Neural Networks (2005)
27. Fatourechi, M.A., Bashashati, R., Ward, G.B.: A hybrid genetic algorithm approach for improving the performance of the LF-ASD brain computer interface. In: IEEE International Conference on Acoustics, Speech, and Signal Processing, Philadelphia, PA, vol. 5, pp. 345–348 (2005)
28. Zamanian, H., Farsi, H.: A new feature extraction method to Improve emotion detection using EEG signals. Electron. Lett. Comput. Vision Image Anal. **17**(1), 29–44 (2018)
29. Hjorth, B.: EEG analysis based on time domain properties. Electroencephalogr. Clin. Neurophysiol. **29**(3), 306–310 (1970)
30. Horlings, R., Datcu, D., Rothkrantz, L.: Emotion recognition using brain activity. In: International Conference on Computer Systems and Technologies (Comp Sys Tech), pp. 1–6 (2008)

31. Liu, Y., Sourina, O.: EEG-based subject-dependent emotion recognition algorithm using fractal dimension. In: IEEE International Conference on Systems, Man and Cybernetics, pp. 3166–3171 (2014)
32. Kroupi, E., Yazdani, A., Ebrahimi, T.: EEG correlates of different emotional states elicited during watching music videos. In: International Conference on Affective Computing and Intelligent Interaction, vol. 6975, pp. 457–466. Springer, Berlin (2011)
33. Petrantonakis, P., Hadjileontiadis, L.: Emotion recognition from EEG using higher order crossings. IEEE Trans. Inf. Technol. Biomed. **14**(2), 186–197 (2012)
34. Nie, D., Wang, X., Shi, L., Lu, B.: EEG-based emotion recognition during watching movies. In: IEEE International Conference on Neural Engineering, pp. 667–670 (2011)
35. Reuderink, B., Muh, C., Poel, M.: Valence, arousal and dominance in the EEG during game play. Int. J. Auton. Adapt. Commun. Syst. **6**(1), 45–62 (2013)
36. Hosseini, S., Khalilzadeh, M., Naghibi-Sistani, M., Niazmand, V.: Higher order spectra analysis of EEG signals in emotional stress states. In: IEEE International Conference on Information Technology and Computer Science, pp. 60–63 (2010)
37. Murugappan, M., Nagarajan, R., Yaacob, S.: Classification of human emotion from EEG using discrete wavelet transform. J. Biomed. Sci. Eng. **3**(4), 390–396 (2010)
38. Hadjidimitriou, S., Hadjileontiadis, L.: Toward an EEG-based recognition of music liking using time-frequency analysis. IEEE Trans. Biomed. Eng. **59**(12), 3498–3510 (2012)
39. Poorna, S., Baba, P., Ramya, G., Poreddy, P., Aashritha, L., Nair, G., Renjith, S.: Classification of EEG based control using ANN and KNN-A comparison. In: IEEE International Conference on Computational Intelligence and Computing Research, Chennai, India, pp. 1–6 (2016)
40. Acharya, U., Subbhuraam, V., Goutham, S., Martis, R., Suri, J.: Automated EEG analysis of epilepsy: a review. Knowl. Based Syst. **45**, 147–165 (2013)
41. Klassen, B., Hentz, J., Shill, H., Driver-Dunckley, E., Evidente, V., Sabbagh, M., Adler, C., Caviness, J.: Quantitative EEG as a predictive biomarker for Parkinson disease dementia. Neurology **77**, 118–124 (2011)
42. Melissant, C., Ypma, A., Frietman, E., Stam, C.: A method for detection of Alzheimer's disease using ICA-enhanced EEG measurements. Artif. Intell. Med. **33**, 209–222 (2005)
43. Rippon, G., Brunswick, N.: Trait and state EEG indices of information processing in developmental dyslexia. Int. J. Psychophysiol. **36**, 251–265 (2000)
44. Lansbergen, M., van Dongen-Boomsma, M., Buitelaar, J., Slaats-Willemse, D.: ADHD and EEG-neuro feedback: a double-blind randomized placebo-controlled feasibility study. J. Neural Transm. **118**, 275–284 (2011)
45. Campbell, A., Choudhury, T., Hu, S., Lu, H., Mukerjee, M., Rabbi, M., Raizada, R.: Neurophone: brain-mobile phone interface using a wireless EEG headset. In: 2nd ACM SIGCOMM Workshop on Networking, Systems and Applications on Mobile Handhelds, New Delhi, India, pp. 3–8 (2010)
46. Mirza, I., Tripathy, A., Chopra, S., D'Sa, M., Rajagopalan, K., D'Souza, A., Sharma, N.: Mind-controlled wheelchair using an EEG headset and Arduino microcontroller. In: International Conference on Technologies for Sustainable Development, Mumbai, India, pp. 1–5 (2015)
47. Petukhov, I., Glazyrin, A., Gorokhov, A., Steshina, L., Tanryverdiev, I.: Being present in a real or virtual world: a EEG study. Int. J. Med. Inform. **136**, 103977 (2020)
48. Cernea, D., Kerren, A., Ebert, A.: Detecting insight and emotion in visualization applications with a commercial EEG headset. In: SIGRAD 2011, Evaluations of Graphics and Visualization-Efficiency, Usefulness, Accessibility, Usability, Stockholm, Sweden (2011)
49. Sun, S.: Multitask learning for EEG-based biometrics. In: 19th International Conference on Pattern Recognition, Tampa, FL, USA, pp. 1–4 (2008)
50. Garvanova, M., Garvanov, I., Borissova, D.: The influence of electromagnetic fields on human brain. In: 21st International Symposium on Electrical Apparatus and Technologies, Bourgas, Bulgaria (2020)
51. Garvanova, M., Garvanov, I., Kashukeev, I.: Business processes and the safety of stakeholders: Considering the electromagnetic pollution. In: Shishkov, B. (ed.) Business Modeling and Software Design. BMSD 2020. Lecture Notes in Business Information Processing, vol. 391, pp. 386–393 (2020)

52. Stoyanov, S., Zhelezov, S.: New functionalities of a virtual computer model design and construction. Math. Softw. Eng. **5**(2), 23–33 (2019)
53. Hawsawi, O., Semwal, S.: EEG headset supporting mobility impaired gamers with game accessibility. In: IEEE International Conference on Systems, Man, and Cybernetics, San Diego, CA, USA, pp. 837–841 (2014)
54. Frey, J., Gervais, R., Lainé, T., Duluc, M., Germain, H., Fleck, S., Lotte, F., Hachet, M.: Scientific Outreach with Teegi, a Tangible EEG Interface to Talk About Neuro Technologies. Association for Computing Machinery, New York (2017)
55. Boryana, U.-D., Stanimir, Z., Hristo, P.: Intelligent methods for evaluation of student written works. J. Eng. Appl. Sci. **12**(Specialissue10), 8780–8784 (2017)
56. Garvanov, I., Jotsov, V., Garvanova, M.: Data science modeling for EEG signal filtering using wavelet transforms. In: IEEE 10th International Conference on Intelligent Systems, Varna, Bulgaria, pp. 352–357 (2020)
57. Croft, R., Barry, R.: Removal of ocular artifact from the EEG: a review. Neurophysiol. Clin./Clin. Neurophysiol. **30**(1), 5–19 (2000)
58. Kavitha, P., Lau, C.T., Premkumar, A.: Modified ocular artifact removal technique from EEG by adaptive filtering. In: 6th International Conference Information, Communications and Signal Processing, Singapore, pp. 10–13 (2007)
59. Mallat, S.: A theory for multi-resolution signal decomposition: the wavelet representation. IEEE Trans. Biomed. Eng. Pattern Anal. Mach. Intell. **11**, 674–693 (1989)
60. Garvanov, I., Iyinbor, R., Garvanova, M., Geshev, N.: Denoising of pulsar signal using wavelet transform. In: 16th International Conference on Electrical Machines, Drives and Power Systems, Varna, Bulgaria, pp. 637–640 (2019)
61. Гарванова, М.: Статистическа обработка и анализ на данни със SPSS. С., Издателство “За буквите – О писменехь”, 292 с (2014). ISBN 978-619-185-046-4

Intuitionistic Fuzziness, Standard and Extended Modality

Krassimir Atanassov

In memory of my mother Vassia Atanassova
(1 Feb. 1928–19 Nov. 2020)

Abstract Ever since the first publication over intuitionistic fuzziness, the two standard modal operators "necessity" ($\Box$) and "possibility" ($\Diamond$) are introduced over intuitionistic fuzzy sets. In the years, these two operators are object of a lot of extensions and modifications. In the present paper, a new direction of two modifications and two extensions of the two standard modal operators is discussed. We introduce six new operators and study their properties.

1 Introduction

In his "$M\varepsilon\tau\alpha\varphi\upsilon\sigma\iota\kappa\alpha$" and "$A\nu\alpha\lambda\upsilon\tau\iota\kappa\alpha$", 23 centuries ago, Aristotle introduced the notions of "sentence" and "predicate", with the proposal to evaluate them as "true" or "false", and the logical operations "negation", "conjunction", "disjunction" and "implication". Also, he gave the idea for quantifiers "*for all*" and "*there exists*" and the modal operators. In general, his latest idea was understood and realized for the first time in 1910 by Clarence Irving Lewis in his Harvard thesis [9] and in a series of articles beginning in 1912. In it, the two modal operators "necessity" ($\Box$) and "possibility" ($\Diamond$) are defined and for each variable x, they are connected by equalities

K. Atanassov (✉)
Institute of Biophysics and Biomedical Engineering, Bulgarian Academy of Sciences, Acad. G. Bonchev Str., bl. 105, 1113 Sofia, Bulgaria

Intelligent Systems Laboratory, Prof. Asen Zlatarov University, 8010 Bourgas, Bulgaria
e-mail: krat@bas.bg

V. Sgurev et al. (eds.), *Advances in Intelligent Systems Research and Innovation*,
Studies in Systems, Decision and Control 379,
https://doi.org/10.1007/978-3-030-78124-8_12

$$\Box x = \neg\Diamond\neg x, \tag{1}$$

$$\Diamond x = \neg\Box\neg x. \tag{2}$$

Therefore, they are conjugated and in some sense symmetric (see, e.g., [6–8, 10, 11]).

The first research over intuitionistic fuzzy sets (IFSs [1–3]) started with definition of the two modal operators over proper IFSs and for them it was shown that they lose sense in the case of fuzzy, but not in the case of intuitionistic fuzzy sets. These two intuitionistic fuzzy operators, have similar properties like the standard modal operators. In particular, they satisfy the above two equalities (1) and (2).

The present paper is an extension and continuation of my plenary talk [4]. Here and in [4], we discuss new forms of both standard intuitionistic fuzzy modal operators and study some of their properties. Only here, we introduce two extended intuitionistic fuzzy modal operators and study some of their properties, too. It is important to mention that these operators are not symmetric.

2 Preliminaries

Let us have a fixed universe E and its subset A. The set

$$A^* = \{\langle x, \mu_A(x), \nu_A(x)\rangle \mid x \in E\},$$

where

$$0 \le \mu_A(x) + \nu_A(x) \le 1 \tag{3}$$

is called IFS and functions $\mu_A : E \to [0, 1]$ and $\nu_A : E \to [0, 1]$ represent the *degree of membership (validity, etc.)* and *non-membership (non-validity, etc.)*. Now, we can define also the function $\pi_A : E \to [0, 1]$ by means of

$$\pi(x) = 1 - \mu(x) - \nu(x)$$

and it corresponds to *degree of indeterminacy (uncertainty, etc.)*.

For brevity, we shall write below A instead of A^*, whenever this is possible.

The most important geometrical interpretation of the IFSs is shown on Fig. 1.

The intuitionistic fuzzy forms of the two simplest modal operators are

$$\Box A = \{\langle x, \mu_A(x), 1 - \mu_A(x)\rangle | x \in E\};$$
$$\Diamond A = \{\langle x, 1 - \nu_A(x), \nu_A(x)\rangle | x \in E\}.$$

They are analogous to the modal logic operators "*necessity*" and "*possibility*" and their geometrical interpretation is shown on Fig. 2.

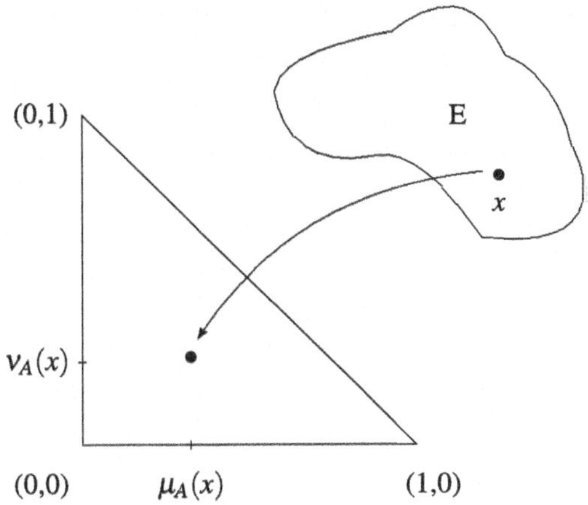

Fig. 1 Geometrical interpretation of element $x \in E$

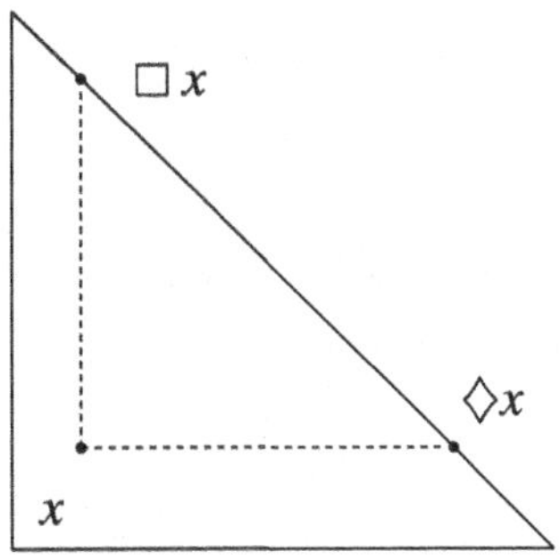

Fig. 2 Geometrical interpretation of elements $\Box x$ and $\Diamond x$

As it is mentioned in [2, 3], for a fuzzy set A the modal operators $\Box$ and $\Diamond$ would satisfy

$$\Box A = A = \Diamond A,$$

i.e., both operators lose their sense, while for a proper IFS A inclusions:

$$\Box A \subset A \subset \Diamond A$$

hold and

$$\Box A \neq A \neq \Diamond A.$$

This example shows that IFSs are essential extensions of fuzzy sets.

3 Definitions of Four New Standard Intuitionistic Fuzzy Modal Operators

We will call that the IFS A is *modally-full* if and only if (iff)

$$(\forall x \in E)(\Box x, \Diamond x \in E).$$

Therefore, if the triple $\langle x, \mu_A(x), \nu_A(x)\rangle \in A$, then

$$\langle \Box x, \mu_A(x), 1 - \mu_A(x)\rangle \in A$$

and

$$\langle \Diamond x, 1 - \nu_A(x), \nu_A(x)\rangle \in A.$$

We will say that the IFS A is *classical negation-full* (or $\neg_1$*-full*) iff

$$(\forall x \in E)(\neg_1 x \in E).$$

Therefore, if the triple $\langle x, \mu_A(x), \nu_A(x)\rangle \in A$, then $\langle \neg_1 x, \nu_A(x), \mu_A(x)\rangle \in A$. We define

$$\boxdot_A x = \{\langle y, \mu_A(y), \nu_A(y)\rangle | y \in E \ \& \ \mu_A(y)$$

$$= \mu_A(x) \ \& \ \nu_A(y) \geq \nu_A(x)\},$$

$$\diamond_A x = \{\langle y, \nu_A(y), \nu_A(y)\rangle | y \in E \ \& \ \nu_A(y)$$

$$= \nu_A(x) \ \& \ \mu_A(y) \geq \mu_A(x)\}.$$

Therefore,

$$\boxdot_A x \sqsubseteq A$$

and

$$\diamond_A x \sqsubseteq A,$$

where $\sqsubseteq$ is the relation "inclusion" in set-theoretical sense.

The IFS A is an *IF tautological set (IFTS)* if

$$(\forall x \in E)(\mu_A(x) \geq \nu_A(x)).$$

The element $x \in E$ is an *IF Tautological Element (IFTE)* in the IFS A, if for it $\mu_A(x) \geq \nu_A(x)$. Therefore, all elements of an IFTS are IFTEs. Opposite, the element $x \in E$ is an *IF False Element (IFFE)* in the IFS A, if for it $\mu_A(x) < \nu_A(x)$.

Therefore, the IFS A is an IFTS iff all its elements are IFTEs.

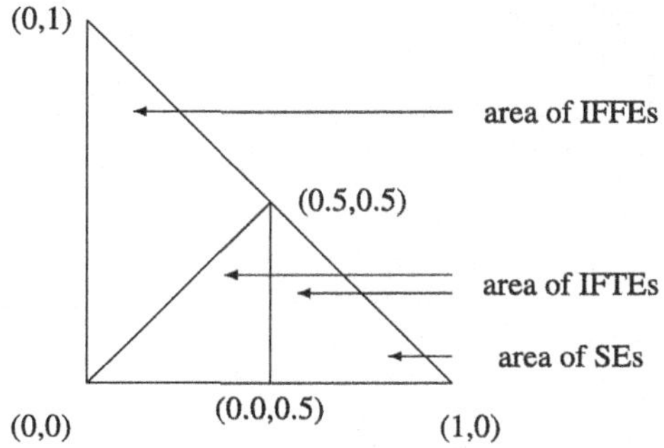

Fig. 3 The areas of the IFFEs, IFTEs and SEs

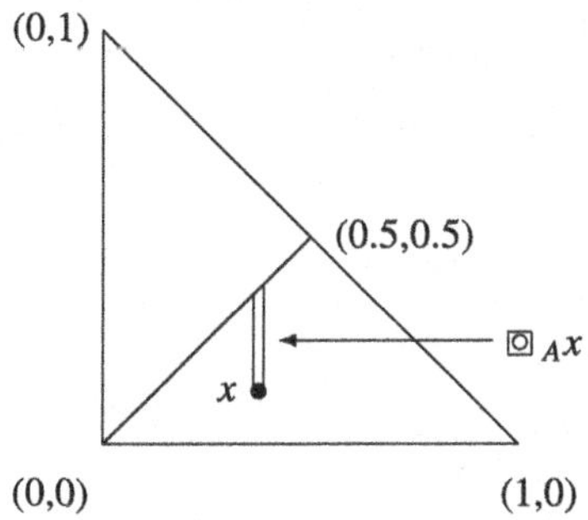

Fig. 4 Geometrical interpretation of $\boxdot_A x$—first case

The element $x \in E$ is *sure (SE)* in the IFS A, if for it $\mu_A(x) \geq 0.5 \geq \nu_A(x)$. The IFS A is *sure* if each its element is sure.

Therefore, the IFS A is a sure iff all its elements are SEs.

Moreover, each SE of an IFS A is an IFTE in the same set, but the opposite is not always true.

The regions of the IFTEs, IFFEs and SEs are shown in Fig. 3.

Proposition 1 *If x is an IFTE of the IFS A, then $\langle x, \mu_A(x), \nu_A(x)\rangle \in \boxdot_A x$ (see* Fig. 4*).*

Proof Let x be an IFTE. Therefore, $\mu_A(x) \geq \nu_A(x)$. Hence, from the definition of $\boxdot_A$ it follows that $\langle x, \mu_A(x), \nu_A(x)\rangle \in \boxdot_A x$.

Corollary 1 *If $\mu_A(x) = \nu_A(x)$ for $x \in E$ that is an element of the IFS A, then $\boxdot_A x = \{\langle x, \mu_A(x), \mu_A(x)\rangle\}$ (see* Fig. 5*).*

Corollary 2 *If $\mu_A(x) < \nu_A(x)$ for $x \in E$ that is an element of the IFS A, then $\boxdot_A x = \emptyset$ (see* Fig. 6*).*

Corollary 3 *If x is a sure element of the IFS A, then $\boxdot_A x$ contains all elements of E whose geometrical interpretations are points that lie on the section between points x and $\square x$, i.e., $\square x \in \boxdot_A x$ (see* Fig. 7*).*

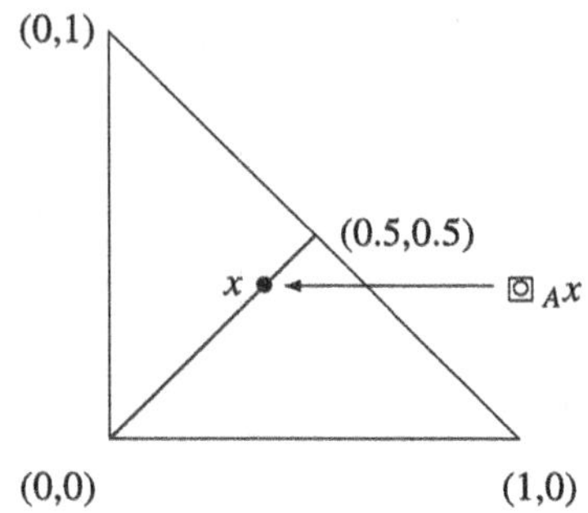

Fig. 5 Geometrical interpretation of ⧇ $_A x$—second case

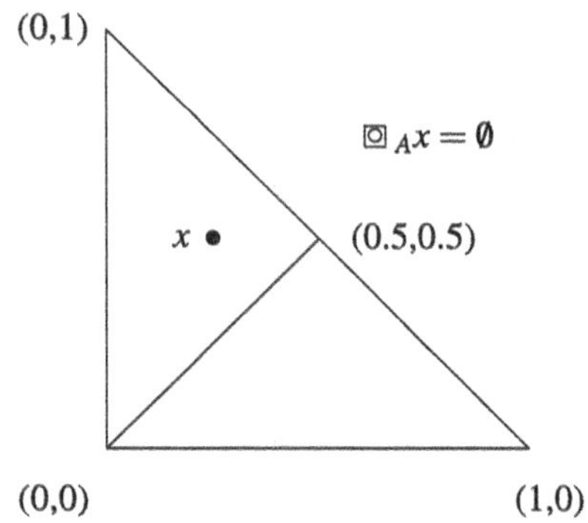

Fig. 6 Geometrical interpretation of ⧇ $_A x$—third case

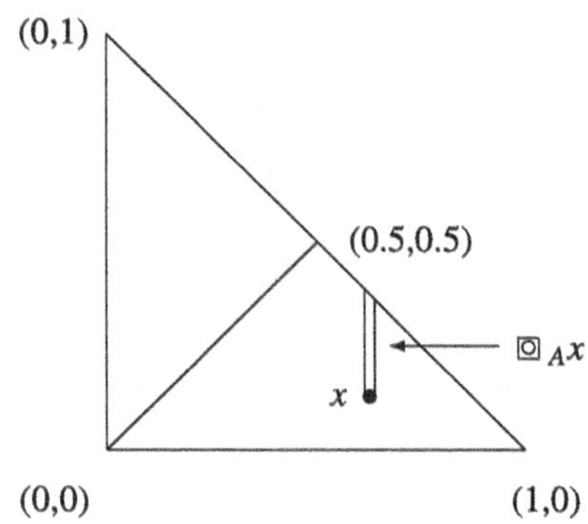

Fig. 7 Geometrical interpretation of ⧇ $_A x$—fourth case

More generally:

$$\Box x \in ⧇_A x \text{ iff } x \text{ is a sure element of } E.$$

Proposition 2 *If x is an IFTE of the IFS A, then* $\langle x, \mu_A(x), \nu_A(x)\rangle \in$ ◈ $_A x$ *(see* Fig. 8*).*

The proof is similar to the above one.

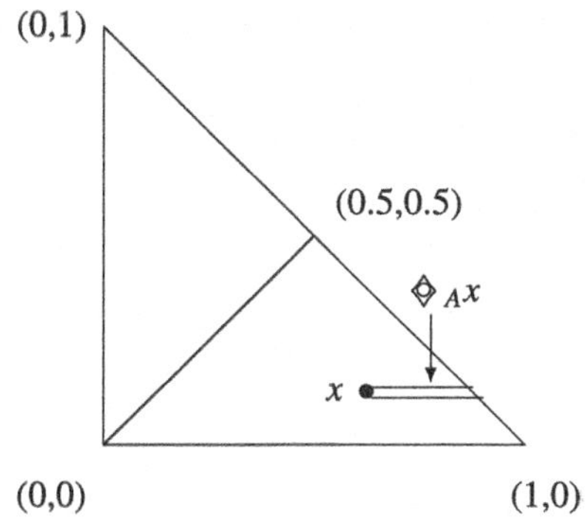

Fig. 8 Geometrical interpretation of $\diamondplus_A x$—first case

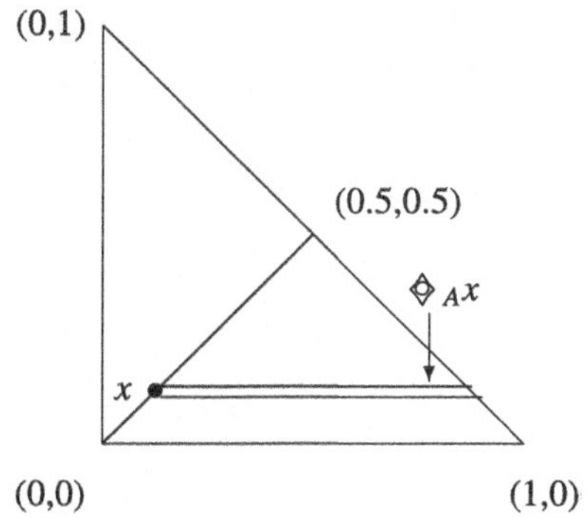

Fig. 9 Geometrical interpretation of $\diamondplus_A x$—second case

Corollary 4 *If $\mu_A(x) = \nu_A(x)$ for $x \in E$ that is an element of the IFS A, then $\diamondplus_A x$ contains all elements of E whose geometrical interpretations are points that lie on section between points x and $\Diamond x$ (see* Fig. 9*).*

Corollary 5 *If $\mu_A(x) < \nu_A(x)$ for $x \in E$ that is an element of the IFS A, then $\diamondplus_A x$ contains all elements of E whose geometrical interpretations are points that lie on the section between points x and $\Diamond x$ (see* Fig. 10*).*

Corollary 6 *If $\mu_A(x) < 0.5 < \nu_A(x)$ for $x \in E$ that is an element of the IFS A, then $\diamondplus_A x = \emptyset$ (see* Fig. 11*).*

Let for $x, y \in E$ and for the IFS A over E: $\mu_A(x) = \mu_A(y)$ and $\nu_A(x) = \nu_A(y)$. We will call x and y *equipollent elements* of E.

Now, we can extend the regions of the objects related to modal operators "necessity" and "possibility" by the forms:

We define

$$\begin{aligned}\boxast_A x = \{\langle y, \mu_A(y), \nu_A(y)\rangle | y \in E \\ \& \nu_A(x) \le \mu_A(y) \le \mu_A(x) \\ \& \nu_A(x) \le \nu_A(y) \le \mu_A(x)\},\end{aligned}$$

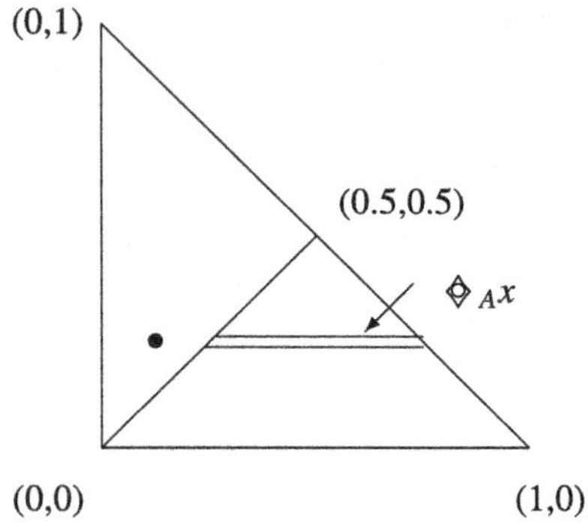

Fig. 10 Geometrical interpretation of $\Diamond_A x$—third case

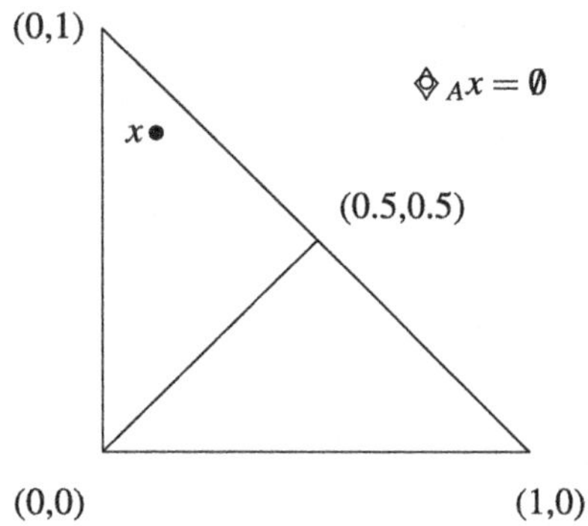

Fig. 11 Geometrical interpretation of $\Diamond_A x$—fourth case

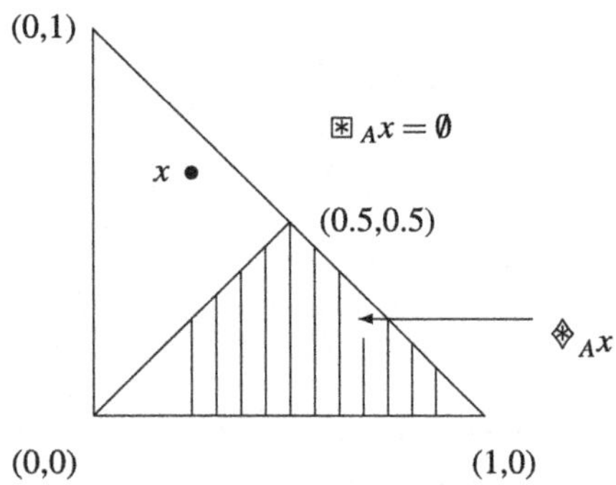

Fig. 12 Geometrical interpretation of $\boxast_A x$ and $\circledast_A x$—first case

$$\circledast_A x = \{\langle y, \mu_A(y), \nu_A(y)\rangle | y \in E \& \ \mu_A(y) \geq \mu_A(x) \ \& \ \nu_A(y) \leq \nu_A(x)\}.$$

The four basic places for point x are shown on Figs. 12, 13, 14 and 15.

Following [2, 3], we introduce the two extended modal operators

$$H^*_{\alpha,\beta}(A) = \{\langle x, \alpha.\mu_A(x), \nu_A(x) + \beta.(1 - \alpha.\mu_A(x) - \nu_A(x))\rangle$$

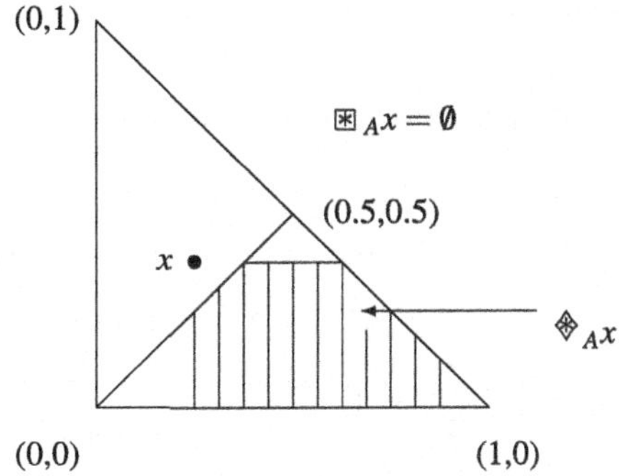

Fig. 13 Geometrical interpretation of $\boxast_A x$ and $\diamondast_A x$—second case

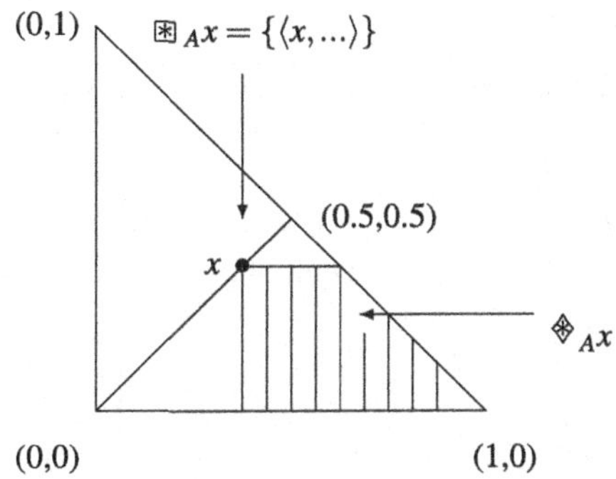

Fig. 14 Geometrical interpretation of $\boxast_A x$ and $\diamondast_A x$—third case

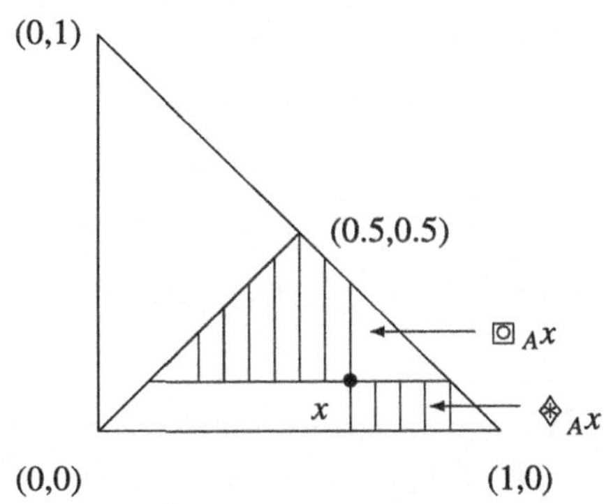

Fig. 15 Geometrical interpretation of $\boxast_A x$ and $\diamondast_A x$—fourth case

$$|x \in E\},$$

$$\overline{H}_{\alpha,\beta}(A) = \{\langle x, \alpha.\mu_A(x), \nu_A(x) + \beta - \beta\nu_A(x)\rangle | x \in E\},$$

$$J^*_{\alpha,\beta}(A) = \{\langle x, \mu_A(x) + \alpha.(1 - \mu_A(x) - \beta.\nu_A(x)), \beta.\nu_A(x)\rangle$$

$$|x \in E\},$$

$$\overline{J}_{\alpha,\beta}(A) = \{\langle x, \mu_A(x) + \alpha - \alpha\mu_A(x), \beta.\nu_A(x)\rangle | x \in E\},$$

where $\alpha, \beta \in [0, 1]$.

By analogy with the Intuitionistic Fuzy Pairs (IFPs, see [5]) we define for the same α, β:

$$H^*_{A,\alpha,\beta}(x) = \langle x, \alpha.\mu_A(x), \nu_A(x) + \beta.(1 - \alpha.\mu_A(x) - \nu_A(x))\rangle,$$
$$\overline{H}_{A,\alpha,\beta}(x) = \langle x, \alpha.\mu_A(x), \nu_A(x) + \beta - \beta\nu_A(x)\rangle,$$
$$J^*_{A,\alpha,\beta}(x) = \langle x, \mu_A(x) + \alpha.(1 - \mu_A(x) - \beta.\nu_A(x)), \beta.\nu_A(x)\rangle,$$
$$\overline{J}_{A,\alpha,\beta}(x) = \langle x, \mu_A(x) + \alpha - \alpha\mu_A(x), \beta.\nu_A(x)\rangle.$$

Theorem 1 *For each $\langle y, \mu_A(y), \nu_A(y)\rangle \in \boxdot_A x$ there are $\alpha, \beta \in [0, 1]$ such that $\langle y, \mu_A(y), \nu_A(y)\rangle$ and $H^*_{A,\alpha,\beta}\langle x, \mu_A(x), \nu_A(x)\rangle$ are equipollent.*

Proof Let elements x and y be fixed and $\langle y, \mu_A(y), \nu_A(y)\rangle \in \boxdot_A x$. Therefore, $\mu_A(x) \geq \nu_A(x)$. Then, we put

$$\alpha = \begin{cases} \frac{\mu_A(y)}{\mu_A(x)}, & \text{if } \mu_A(x) > 0 \\ 0, & \text{if} \mu_A(x) = 0 \end{cases},$$

$$\beta = \begin{cases} \frac{\nu_A(y) - \nu_A(x)}{1 - \alpha.\mu_A(x) - \nu_A(x)}, & \text{if } \alpha.\mu_A(x) - \nu_A(x) < 1 \\ 1, & \text{if } \alpha.\mu_A(x) - \nu_A(x) = 1 \end{cases}.$$

Then,

$$H^*_{A,\alpha,\beta}(\langle x, \mu_A(x), \nu_A(x)\rangle)$$
$$= \langle x, \alpha\mu_A(x), \nu_A(x) + \beta(1 - \alpha\mu_A(x) - \nu_A(x))\rangle.$$

We will discuss the following three cases.
Case 1. Let $\mu_A(x) > 0$ and $\alpha\mu_A(x) + \nu_A(x) < 1$. Then,

$$\begin{aligned} &H^*_{\alpha,\beta}(\langle x, \mu_A(x), \nu_A(x)\rangle) \\ &= \langle x, \frac{\mu_A(y)}{\mu_A(x)}.\mu_A(x), \nu_A(x) + \frac{\nu_A(y) - \nu_A(x)}{1 - \alpha\mu_A(x) - \nu_A(x)} \\ &\ .(1 - \alpha\mu_A(x) - \nu_A(x))\rangle \\ &= \langle x, \mu_A(y), \nu_A(x) + \nu_A(y) - \nu_A(x)\rangle \\ &= \langle x, \mu_A(y), \nu_A(y)\rangle. \end{aligned}$$

Case 2. Let $\mu_A(x) > 0$ and $\alpha\mu_A(x) + \nu_A(x) = 1$. Therefore, $\beta = 1$. If $\alpha < 1$, then

$$1 = \alpha\mu_A(x) + \nu_A(x) < \mu_A(x) + \nu_A(x) \leq 1,$$

which is impossible. Therefore, $\alpha = 1$, i.e., $\mu_A(y) = \mu_A(x)$. Then,

$$1 - \mu_A(y) \geq \nu_A(y) \geq \nu_A(x) = 1 - \mu_A(x) = 1 - \mu_A(y),$$

i.e., $1 - \mu_A(y) = \nu_A(y) = \nu_A(x)$. Therefore,

$$\begin{aligned} H^*_{\alpha,\beta}(\langle x, \mu_A(x), \nu_A(x)\rangle) &= H^*_{1,1}(\langle x, \mu_A(x), \nu_A(x)\rangle) \\ &= \langle x, \mu_A(x), \nu_A(x)\rangle = \langle x, \mu_A(y), \nu_A(y)\rangle. \end{aligned}$$

Case 3. Let $\mu_A(x) = 0$. Then, $\alpha = 0$ and from

$$\nu_A(x) \leq \mu_A(y) \leq \mu_A(x) = 0$$

it follows that $\nu_A(x) = \mu_A(y) = 0$. From

$$\nu_A(x) \leq \nu_A(y) \leq \mu_A(x) = 0$$

it follows that $\nu_A(y) = 0$. Therefore,

$$H^*_{\alpha,\beta}(\langle x, \mu_A(x), \nu_A(x)\rangle) = \langle x, 0, 0\rangle = \langle y, 0, 0\rangle.$$

Hence, in all (the three) cases x and y are equipollent elements of E.

Theorem 2 *For each* $\langle y, \mu_A(y), \nu_A(y)\rangle \in \diamondsuit_A x$ *there are* $\alpha, \beta \in [0, 1]$ *such that* $\langle y, \mu_A(y), \nu_A(y)\rangle$ *and* $J^*_{\alpha,\beta}\langle x, \mu_A(x), \nu_A(x)\rangle$ *are equipollent.*

Proof Let elements x and y be fixed and $\langle y, \mu_A(y), \nu_A(y)\rangle \in \diamondsuit_A x$. Therefore, $\mu_A(x) \leq \mu_A(y)$. Then we put

$$\alpha = \begin{cases} \frac{\mu_A(y)-\mu_A(x)}{1-\mu_A(x)-\beta\nu_A(x)}, & \text{if } \mu_A(x) - \beta\nu_A(x) < 1 \\ 1, & \text{if } \mu_A(x) - \beta\nu_A(x) = 1 \end{cases},$$

$$\beta = \begin{cases} \frac{\nu_A(y)}{\nu_A(x)}, & \text{if } \nu_A(x) > 0 \\ 0, & \text{if } \nu_A(x) = 0 \end{cases}.$$

After this, the proof is similar to the above one.

The next three theorems are proved analogously.

Theorem 3 *For each* $\langle y, \mu_A(y), \nu_A(y)\rangle \in \boxast_A x$ *there are* $\alpha, \beta \in [0, 1]$, *such that* $\langle y, \mu_A(y), \nu_A(y)\rangle$ *and* $\overline{H}_{\alpha,\beta}\langle x, \mu_A(x), \nu_A(x)\rangle$ *are equipollent.*

Theorem 4 *For each* $\langle y, \mu_A(y), \nu_A(y)\rangle \in \diamondplus_A x$ *there are* $\alpha, \beta \in [0, 1]$, *such that* $\langle y, \mu_A(y), \nu_A(y)\rangle$ *and* $\overline{J}_{\alpha,\beta}\langle x, \mu_A(x), \nu_A(x)\rangle$ *are equipollent.*

Theorem 5 *Let A be a modally-full IFS. For each* $x \in E$*:*

$$\Box x \in \boxast_A x \text{ iff } x \text{ is SE}.$$

4 Definitions of Two New Extended Intuitionistic Fuzzy Modal Operators

In [2, 3], the following two extended modal operators are introduced and their basic properties are discussed:

$$D_\alpha(A) = \{\langle x, \mu_A(x) + \alpha\pi_A(x), \nu_A(x) + (1-\alpha)\pi_A(x)\rangle | x \in E\},$$

$$F_{\alpha,\beta}(A) = \{\langle x, \mu_A(x) + \alpha\pi_A(x), \nu_A(x) + \beta\pi_A(x)\rangle | x \in E\}.$$

Obviously, for each IFS A:

$$\Box A = D_0(A),$$

$$\Diamond A = D_1(A),$$

$$D_\alpha(A) = F_{\alpha,1-\alpha}(A).$$

Similarly to above, we define:

$$D_{A,\alpha}(x) = \langle x, \mu_A(x) + \alpha\pi_A(x), \nu_A(x) + (1-\alpha)\pi_A(x)\rangle,$$

$$F_{A,\alpha,\beta}(x) = \langle x, \mu_A(x) + \alpha\pi_A(x), \nu_A(x) + \beta\pi_A(x)\rangle.$$

The two new extended intuitionistic fuzzy modal operators, that correspond to operators $\boxbox_A$ and $\diamondplus_A$ have the following forms:

$$\mathscr{D}_{A,\alpha}(x) = \{\langle y, \mu_A(y) + \alpha\pi_A(y), \nu_A(y) + (1-\alpha)\pi_A(y)\rangle | y \in E \ \& \mu_A(y) + \alpha\pi_A(y) \geq 0.5\},$$

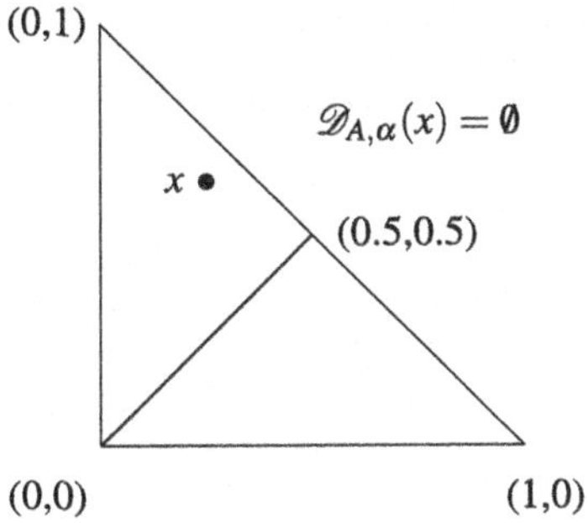

Fig. 16 Geometrical interpretation of $\mathscr{D}_{A,\alpha}(x)$—first case

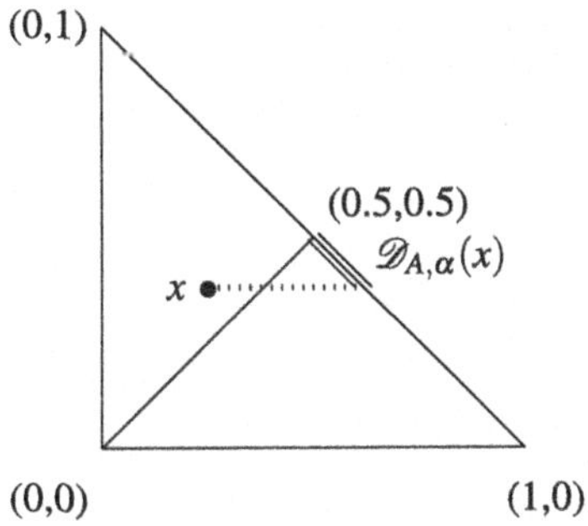

Fig. 17 Geometrical interpretation of $\mathscr{D}_{A,\alpha}(x)$—second case

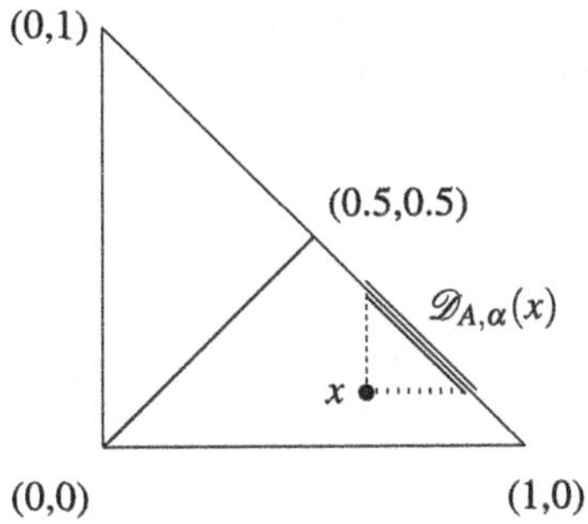

Fig. 18 Geometrical interpretation of $\mathscr{D}_{A,\alpha}(x)$—third case

$$\mathscr{F}_{A,\alpha,\beta}(x) = \{\langle y, \mu_A(y) + \alpha\pi_A(y), \nu_A(y) + \beta\pi_A(y)\rangle | y \in E \ \& \mu_A(y) + \alpha\pi_A(y) \geq 0.5\}.$$

If x is an IFFE and $\nu_A(x) > 0.5$, then $\mathscr{D}_{A,\alpha}(x) = \emptyset$ (see, Fig. 16).
If x is an IFFE and $\nu_A(x) \leq 0.5$, then $\mathscr{D}_{A,\alpha}(x)$ is the section shown on Fig. 17.
If x is an IFTE and $\mu_A(x) \geq 0.5$, then $\mathscr{D}_{A,\alpha}(x)$ is the section shown on Fig. 18.
If x is an IFFE and $\nu_A(x) > 0.5$, then $\mathscr{D}_{A,\alpha,\beta}(x) = \emptyset$ (see, Fig. 19).
If x is an IFFE and $\nu_A(x) \leq 0.5$, then $\mathscr{F}_{A,\alpha,\beta}(x)$ is the section shown on Fig. 20.
If x is an IFTE and $\mu_A(x) \geq 0.5$, then $\mathscr{F}_{A,\alpha,\beta}(x)$ is the section shown on Fig. 21.

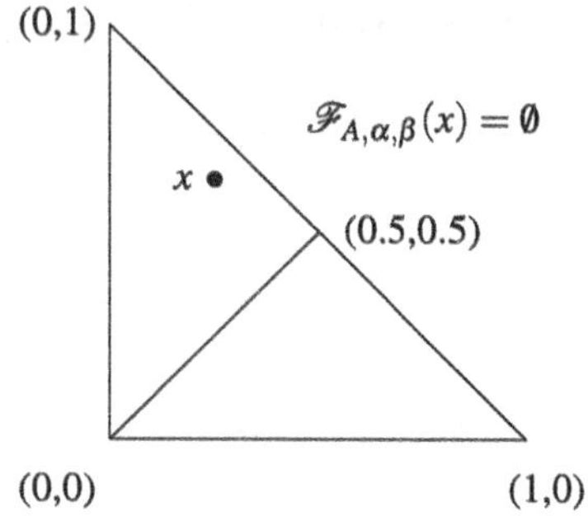

Fig. 19 Geometrical interpretation of $\mathscr{F}_{A,\alpha,\beta}(x)$—first case

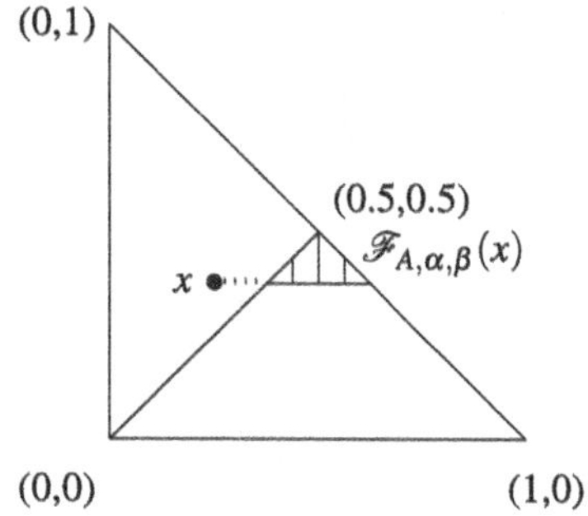

Fig. 20 Geometrical interpretation of $\mathscr{F}_{A,\alpha,\beta}(x)$—second case

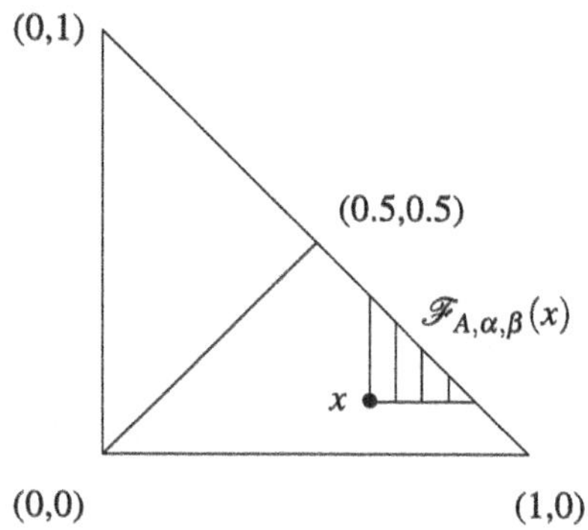

Fig. 21 Geometrical interpretation of $\mathscr{F}_{A,\alpha,\beta}(x)$—third case

Theorem 6 *For each $\langle y, \mu_A(y), \nu_A(y)\rangle \in \mathscr{D}_\alpha(x)$ there is $\alpha \in [0, 1]$, such that $\langle y, \mu_A(y), \nu_A(y)\rangle$ and $D_\alpha\langle x, \mu_A(x), \nu_A(x)\rangle$ are equipollent.*

Theorem 7 *For each $\langle y, \mu_A(y), \nu_A(y)\rangle \in \mathscr{F}_{\alpha,\beta}(x)$ there are α, $\beta \in [0, 1]$, such that $\langle y, \mu_A(y), \nu_A(y)\rangle$ and $F_{\alpha,\beta}\langle x, \mu_A(x), \nu_A(x)\rangle$ are equipollent.*

5 Conclusion, or Discussion on the Basic Idea of the Modality

Aristotle's idea for modality is that for each sentence x we can construct the set of sentences

$$Y(x) = \{y_1, y_2, \ldots\}$$

so that if some of them are valid (here—IFTE), then it follows that x is also IFTE. The set Y can be interpreted as $\boxdot_A x$, or in the second form—as $\boxast_A x$. On the other hand, for each sentence x we can construct the set of sentences

$$Z(x) = \{z_1, z_2, \ldots\}$$

so that if x is an IFTE, then it follows that elements of Z are also IFTEs. The set Z can be interpreted as $\Diamond_A x$, or in the second form—as $\Diamond\!\!\!\!\ast\;_A x$.

As it was mentioned in the Introduction, in standard modal logic, the equalities (1) and (2) are valid. In the intuitionistic fuzzy case, the situation is the same for the initial form of modal operators and this fact is discussed in the first paper over IFSs [1]. But for the four new operators the situation is different and we will illustrate it with the following example.

Let $x \in E$ be an arbitrary element of the universe and let $y \in \boxdot_A x$ (see Fig. 22). Then we construct $\neg x$ and $\neg y$ (see Fig. 23). For both them there is a unique set $\Diamond_A \neg x$ (that coincides with $\Diamond_A \neg y$) and let $z \in \Diamond_A \neg x$ be its arbitrary element (see Fig. 24). Then $\neg z$ is the negation of z (see Fig. 25), but it is not an element of $\boxdot_A x$, i.e., the equality

$$\boxdot_A x = \neg \Diamond_A \neg x$$

does not hold.

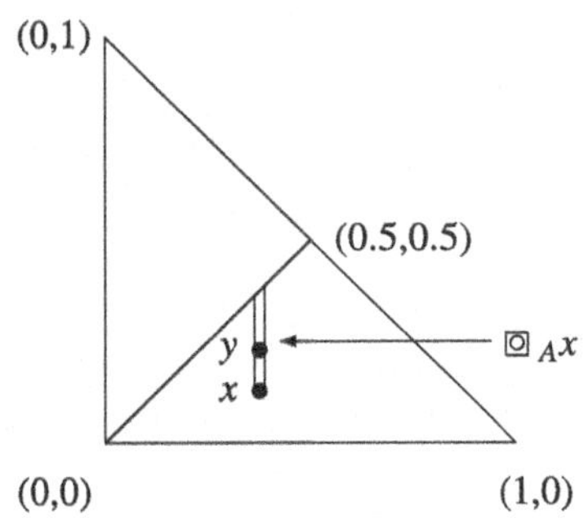

Fig. 22 An example—first step

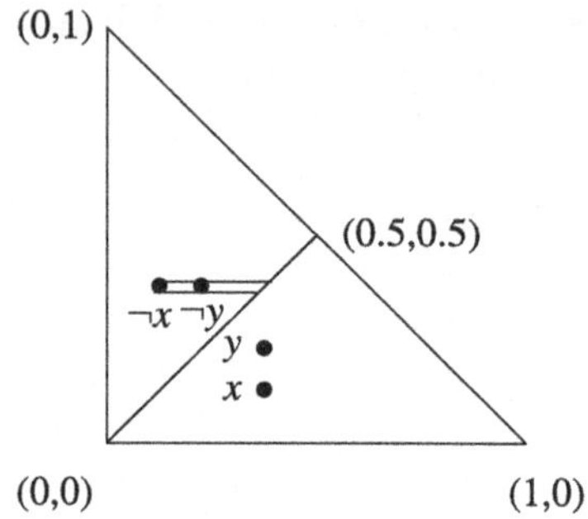

Fig. 23 An example—second step

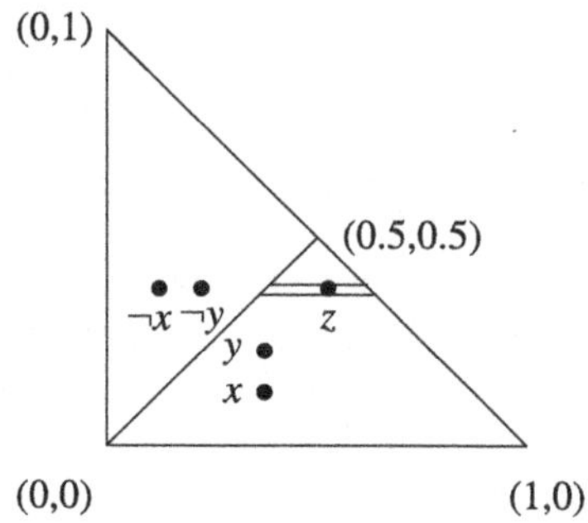

Fig. 24 An example—third step

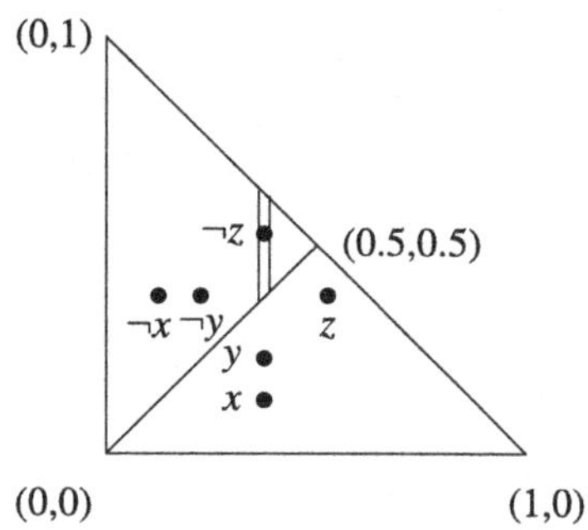

Fig. 25 An example—fourth step

In the same manner we see that the equality

$$\Diamond_A x = \neg \boxdot_A \neg x$$

does not hold.

The situation regarding the two other operations, introduced in the paper is similar.

In a future research some other intuitionistic fuzzy operators will be extended to new ones and their properties will be studied.

Acknowledgements This research was funded by Bulgarian National Science Fund, grant number KP-06-N22/1/2018 "Theoretical research and applications of InterCriteria Analysis".

References

1. Atanassov, K.: Intuitionistic fuzzy sets. VII ITKR's Session, Sofia, 1983 (Deposed in Central Sci.—Techn. Library of Bulg. Acad. of Sci., 1697/84) (in Bulg.)
2. Atanassov, K.: Intuitionistic Fuzzy Sets. Springer, Heidelberg (1999)
3. Atanassov, K.: On Intuitionistic Fuzzy Sets. Springer, Berlin (2012)
4. Atanassov, K.: Intuitionistic fuzziness and modality. In: Proceedings of 2020 IEEE 10th International Conference on Intelligent Systems, Varna, 28–30, 2020, pp. 19–25
5. Atanassov, K., Szmidt, E., Kacprzyk, J.: On intuitionistic fuzzy pairs. Notes Intuit. Fuzzy Sets **19**(3), 1–13 (2013)
6. Hughes, G., Cresswell, M.: A New Introduction to Modal Logic. Routledge, London (1996)
7. Konyndik, K.: Introductory Modal Logic. University of Notre Dame Press, Notre Dame (1986)
8. Lemmon, E., Scott, D.: An Introduction to Modal Logic. Blackwell, Oxford (1977)
9. Lewis, C.I.: The Place of Intuition in Knowledge. Ph.D. thesis, Harvard University (1910)
10. Mints, G.: A Short Introduction to Modal Logic. University of Chicago Press, Chicago (1992)
11. https://plato.stanford.edu/entries/logic-modal/

Introduction to Octopus-Inspired Soft Robots: Pipe-Climbing Robot TAOYAKA-S II and Ladder-Climbing Robot MAMEYAKA

Kazuyuki Ito, Yoshihiro Homma, Hiroaki Shimizu, and Yusei Sakuhara

Abstract In this study, we focus on the intelligent behavior of a soft body and propose a climbing robot inspired by an octopus. The robot can grasp various objects using its flexible arm and can climb ladders and rugged walls. We developed a prototype made of silicone and conducted experiments on it. The results confirmed that the proposed soft arm is effective, and the robot can climb various types of ladders and walls.

Keywords Soft robot · Octopus-like robot · Climbing robot

1 Introduction

In a conventional intelligence framework, the brain is considered a source of intelligence, and many researchers are attempting to develop intelligent systems similar to the brain using computers with high specifications. Currently, the calculation speed of a high spec computer is considered superior to the brain. However, autonomously controlling a robot with varying degrees of freedom in the complex real world remains difficult.

Alternately, when we focus on the real world, we find much rebutting evidence exists. For example, some smaller creatures lack brains yet exhibit intelligent adaptive behavior. Typically, creatures have varied degrees of freedom, and their bodies realize complex intelligent behavior. How they control their varying degrees of freedom is a pertinent question.

K. Ito (✉) · Y. Sakuhara
Department of Electronic Engineering, Hosei University, 3-7-2 Kajino-cho, Koganei, Tokyo 184-8584, Japan
e-mail: ito@hosei.ac.jp

Y. Homma
Mitsubishi Chemical Corporation, 1-1 Marunouchi 1-chome, Chiyoda-ku, Tokyo 100-8251, Japan

H. Shimizu
Toyota Motor Corporation, 1 Toyota-cho, Toyota 471-8571, Aichi, Japan

V. Sgurev et al. (eds.), *Advances in Intelligent Systems Research and Innovation*, Studies in Systems, Decision and Control 379,
https://doi.org/10.1007/978-3-030-78124-8_13

In all our previous works, we focused on the interaction between the soft body and the complex environment and developed various bio-inspired robots [1–6]. From these robots, we confirmed that the key is to utilize the dynamics of the body to realize intelligent behavior.

In this paper, we summarize our previous works [7, 8] and introduce soft climbing robots inspired by an octopus. The legs of the robots are soft and can grasp various objects through their octopus-like actions. We developed two types of robots; one can climb columnar objects such as a pipe, a pillar, and a branch of a tree, and the other can climb ladders and rugged walls.

The most significant advantage of the proposed robots is that they have lesser number of motors than the degrees of freedom of the body, and they can be controlled by fewer dimensional control inputs despite this. To demonstrate the effectiveness of the proposed robots, we conducted experiments and confirmed that they could grasp various unknown objects and climb in complex settings, such as columnar objects, ladders, and rugged walls.

2 Problem of a Conventional Framework and the Goal of This Study

In general, soft robots have significant number of degrees of freedom, and they can adapt to any complex environment using their flexible bodies. However, there are serious problems, such as the significant cost for control [9, 10]. This can be explained by an example as shown in Fig. 1.

Fig. 1 Grasping an object

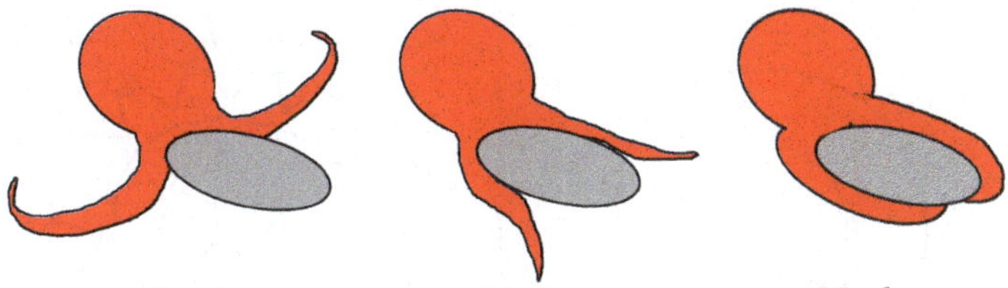

Fig. 2 Behavior of an octopus

Here, the task is to grasp an unknown object. In this case, the robot-hand is not soft and has only six joints. Nevertheless, the required calculation cost is high as it increases exponentially with increase in the degrees of freedom. Specifically, if we employ one motor for each joint, the dimension of the control input become six. Additionally, to control the joints, the robot has to measure the shape of the object.

Alternately, lower creatures have soft bodies, and they can act appropriately in real time despite their small brains. Conventional studies consider that their bodies play important roles in realizing intelligent and adaptive behaviors [11, 12].

In this paper, we propose to utilize the dynamics of the soft body to realize adaptive behavior, and we develop climbing robots inspired by an octopus.

3 Behavior of an Octopus

An octopus can wrap and grasp unknown objects of various shapes without seeing them. Figure 2 shows an example of an octopus's behavior. It touches its legs from the root to the tip owing to which it can wrap various shaped objects.

In our previous works, we proposed a manipulator inspired by an octopus and developed a soft finger that reproduces octopus-like behavior utilizing the dynamical properties of the soft finger [1, 3].

In this paper, we apply this mechanism to the leg of a multilegged-robot and develop a pipe-climbing as well as a ladder climbing robot.

4 Proposed Leg Mechanism

Figure 3 shows the basic concept of the leg mechanism. The leg is made of silicone rubber and its natural shape is a spiral.

Due to the spiral shape and the dynamics of the rubber, the leg comes in contact with an object from the root to the tip by just pulling the string as shown in Fig. 4. This action of "contacting from the root to the tip" is like that of an octopus [13–15].

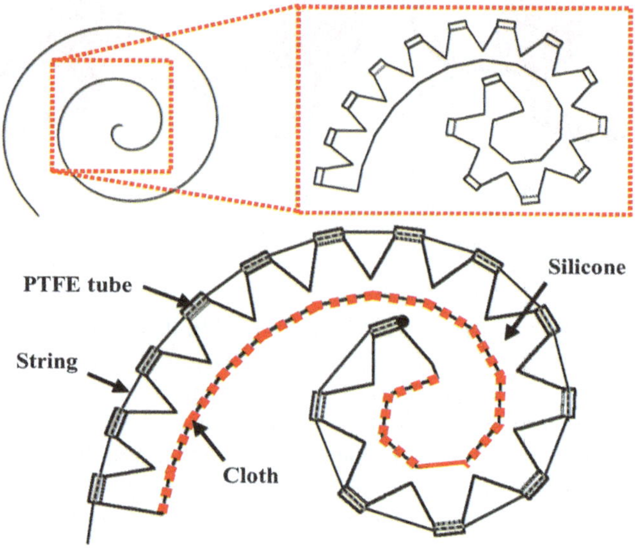

Fig. 3 Basic concept of the leg mechanism

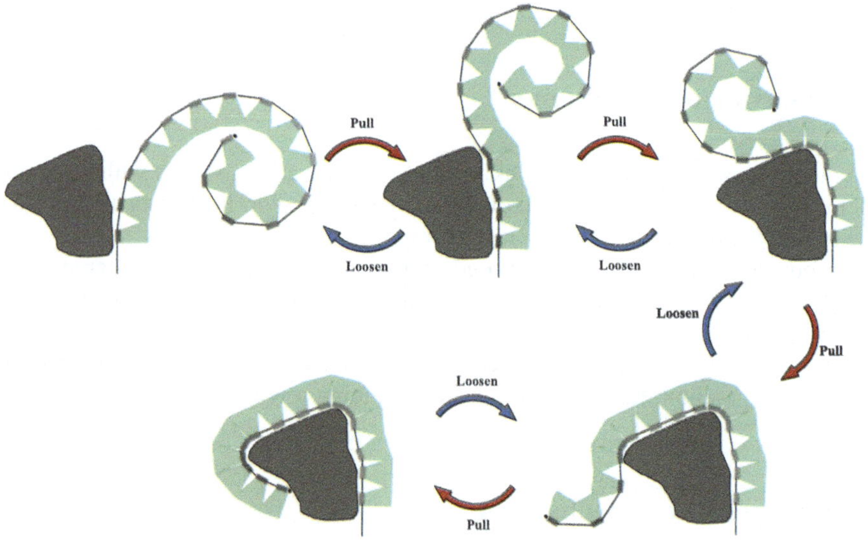

Fig. 4 Movement of the leg

5 Pipe Climbing Robot

Figure 5 shows the developed robot: TAOYAKA-S II [7].

The robot was designed to realize adaptive behavior using the interaction between the soft body and the environment. It had eight flexible legs and a trunk made of three flexible silicone links. To contract the legs and the trunk, plastic strings were connected to the motors through Polytetrafluoroethylene (PTFE) tubes. The motors were controlled by the PC, and the robot was driven by pulling the strings using the motors, as shown in Fig. 6.

5.1 Flexible Leg

Figure 7 shows the TAOYAKA-S II's flexible leg. It is composed of six segments, which are made of silicon 60A and a plastic string to close the leg, which is installed through the PTFE tubes. To realize enough friction, soft silicone blocks (30A) are attached to the inside of the leg. To prevent the leg from stretching, a ribbon is attached to its outside.

The leg's joint angle between segments increases gradually from the root to the tip, as shown in Fig. 7. Thus, the necessary force needed to close the joint also increases. Owing to this physical property, the leg gradually closes from the root to the tip just by pulling the string (Fig. 8). This is similar to the octopus's grasping behavior. It is reported that an octopus usually touches to an object with its tentacles from the root

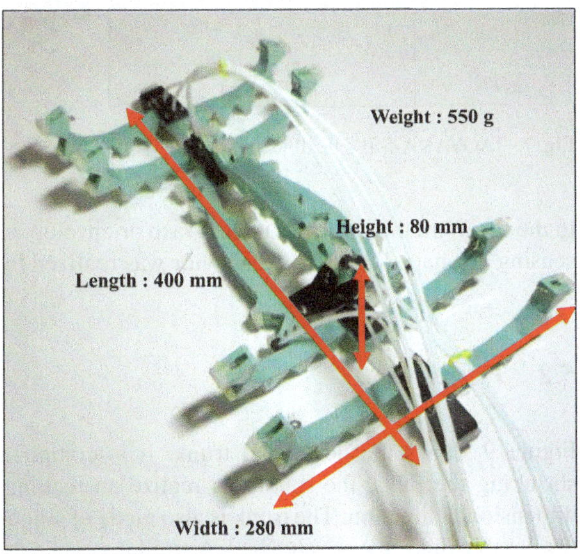

Fig. 5 TAOYAKA-S II

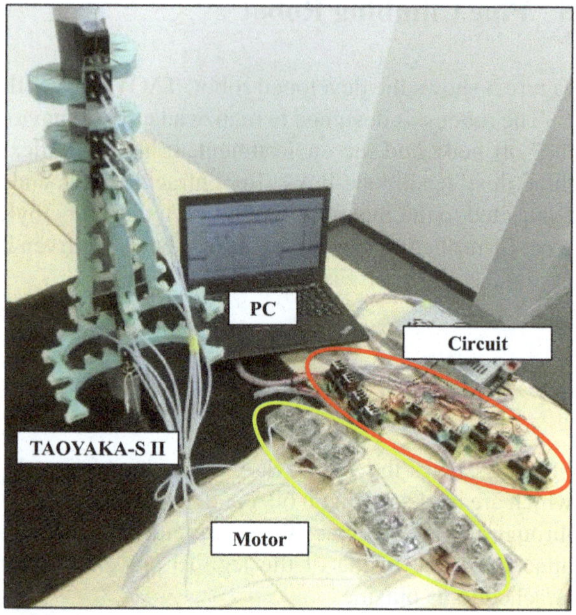

Fig. 6 Whole system developed for TAOYAKA-S II

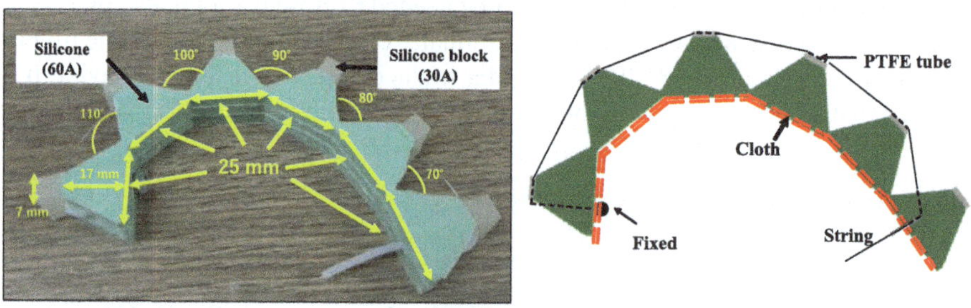

Fig. 7 TAOYAKA-S II's flexible leg

to the tip. With this strategy, it can grasp or envelop an unknown object without first sensing its shape [1, 3]. This behavior was realized by a simple mechanism.

5.2 *Trunk*

Figure 9 shows the developed trunk. It is composed of three flexible links. By shrinking the links, the robot can realize an inching motion in the desired three-dimensional direction. The trunk is also made of silicon 60A and ribbons to reinforce the flexible links that are attached. A similar string mechanism is employed, to shrink the links, as shown in Fig. 10.

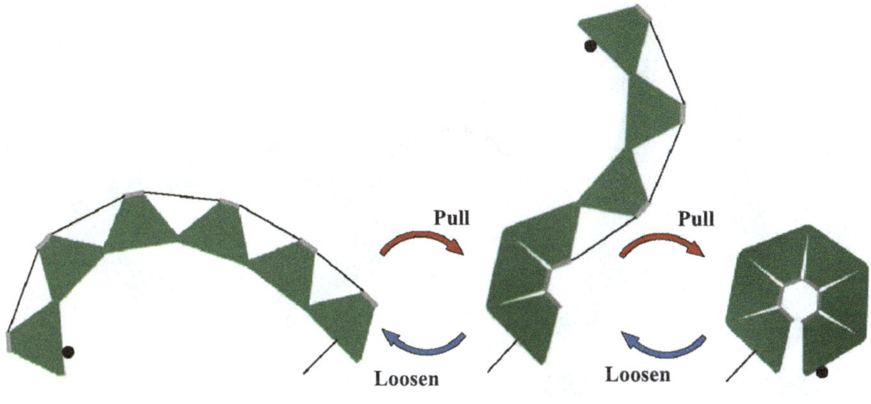

Fig. 8 Movement of the flexible leg

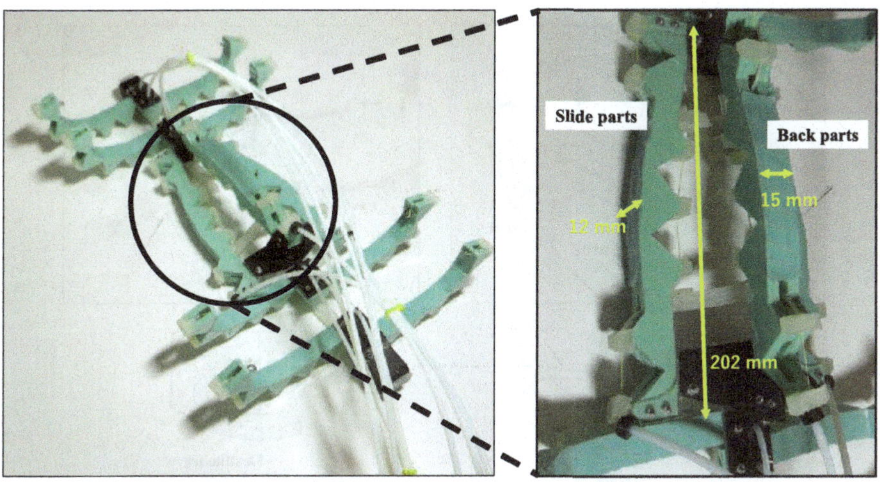

Fig. 9 Developed trunk

5.3 Actuation Mechanism and Control

Eleven motors are employed to drive the strings which pull the legs and the trunk. To reduce the weight of the robot, we placed the motor outside the robot. Figure 11 shows the pulling mechanism of the string which is connected to the motor through the PTFE tube. By rotating the motor, the pulley on the motor pulls the string. The motor is controlled by a PC via a motor driver circuit. To limit the maximum pulling force, we set a limit on the electrical current for the motor driver.

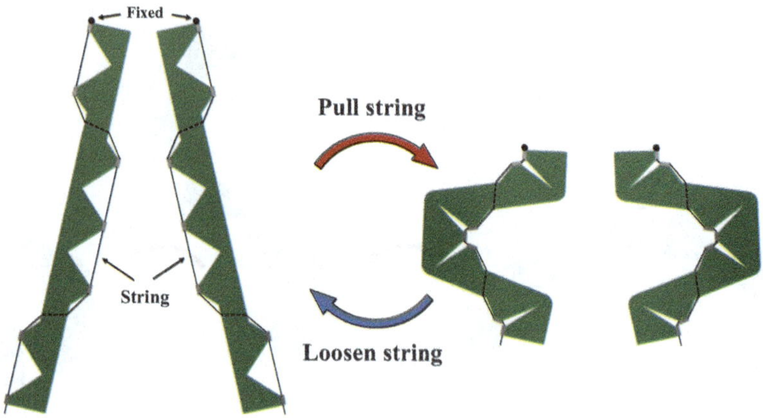

Fig. 10 Movement of the trunk

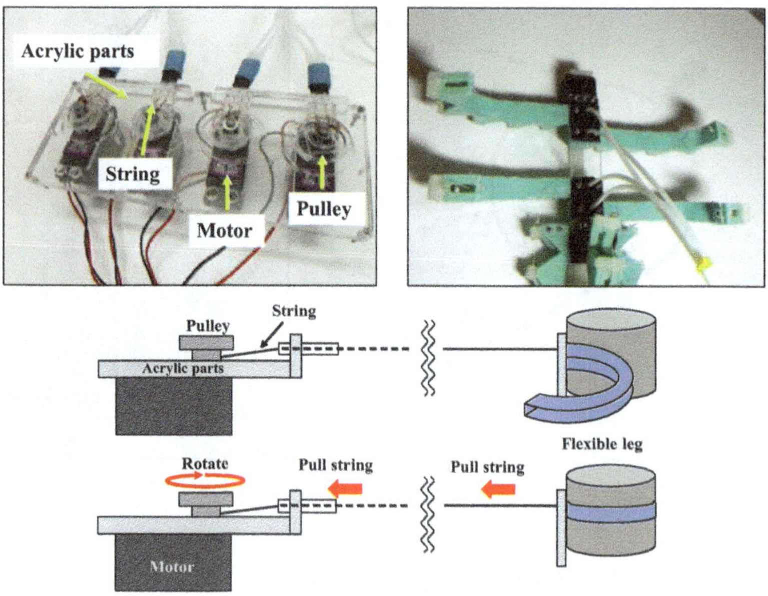

Fig. 11 Actuation mechanism for TAOYAKA-S II

The main advantage of the proposed robot is its high adaptability in unknown environments, utilizing its octopus-like behavior. This is realized passively by utilizing the dynamics of the soft body. Thus, there is no need to measure the environment or obtain complex feedback; alternately, we employ a preprogramed simple pattern for climbing. Figure 12 shows the timing chart. Using this pattern, the robot repeats its inching motion by changing the grasping point.

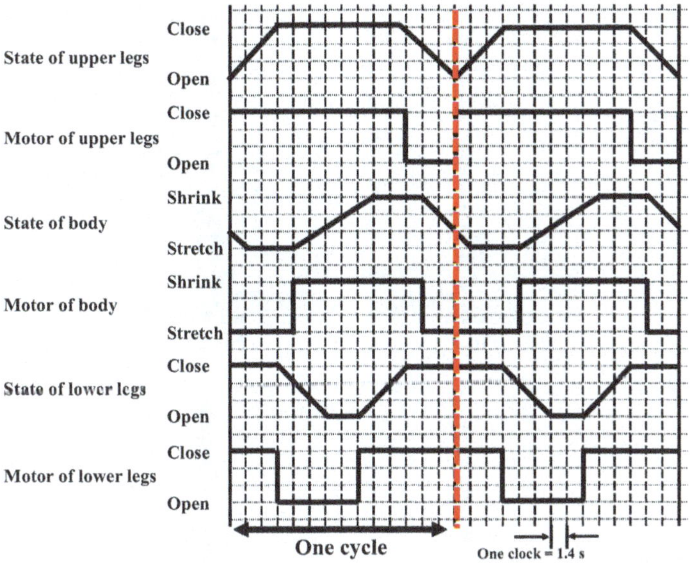

Fig. 12 Timing chart

5.4 Experimental Verification

To confirm its adaptability to various objects, we employ connected columnar pipes, square pipes, long balloons, and natural trees [7].

Figure 13 shows the result of a columnar pipe, and Fig. 14 shows the result of connected columnar pipes (18 and 22 mm). These pipes are not parallel and the distance between them changes while climbing. Nevertheless, the robot could climb using the same preprogramed pattern.

Figure 15 shows the result of using a square pipe (30 mm × 30 mm). Here, the robot could climb the square pipe without changing its locomotion pattern; the legs passively adapted to the square shape.

Figure 16 shows the result of using a long balloon connected to a columnar pipe. In this case, the right legs must grasp the rigid pipe, and left legs must grasp the soft balloon. Nevertheless, the robot could climb it using the same locomotion pattern.

Figure 17 shows the result of using a natural tree. In this case, the robot had to be controlled by a human operator. We confirmed that, by controlling the trunk in the desired three-dimensional direction, the robot could climb complex unknown objects such as a tree.

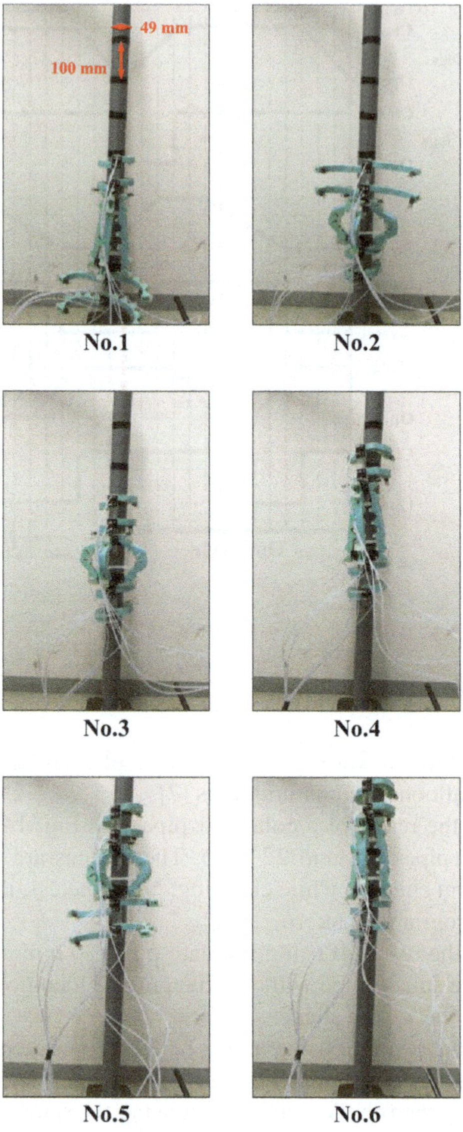

Fig. 13 Experimental result (column pipe d = 49 mm)

6 Ladder Climbing Robot

Figure 18 shows the proposed robot: MAMEYAKA, and Fig. 19 shows the whole system. The robot has two flexible arms, and each arm is driven by only one motor. The motors are connected to a PC using a motor driver circuit, and move based on

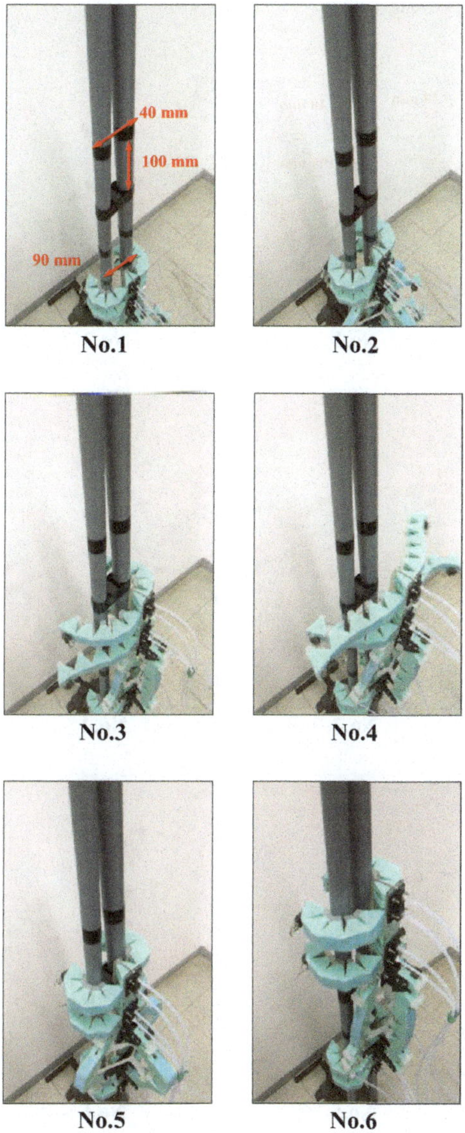

Fig. 14 Experimental result (two pipes)

the control inputs from the PC. The robot can autonomously move by imputing a simple cyclic pattern from the PC and can be manually controlled using a GUI in the PC.

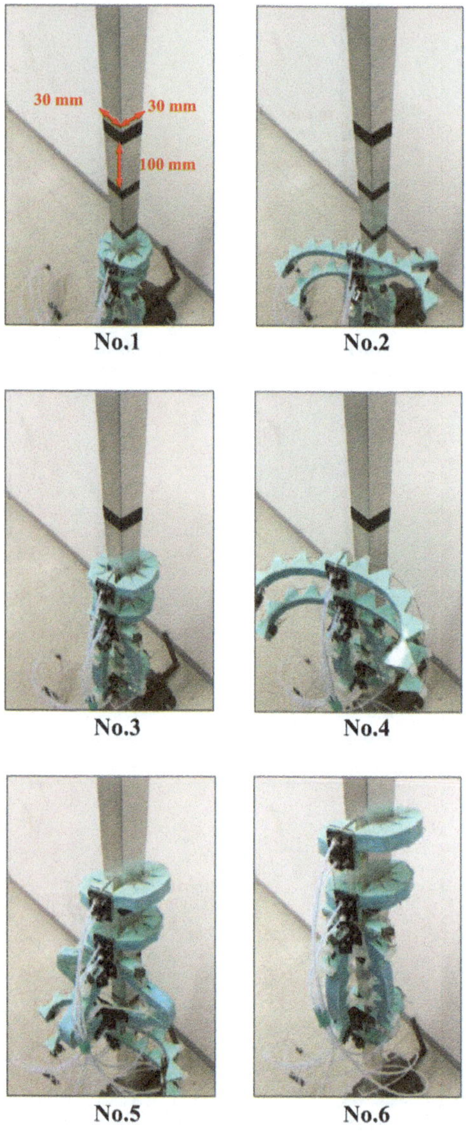

Fig. 15 Experimental result (square pipe)

6.1 Grasping Part

The arm is composed of a grasping and a lifting part. Figure 20 shows the grasping part which is made of silicone rubber 60A, and a cloth which is embedded on both sides of the arm for reinforcement. Soft blocks of silicone rubber 30A are attached

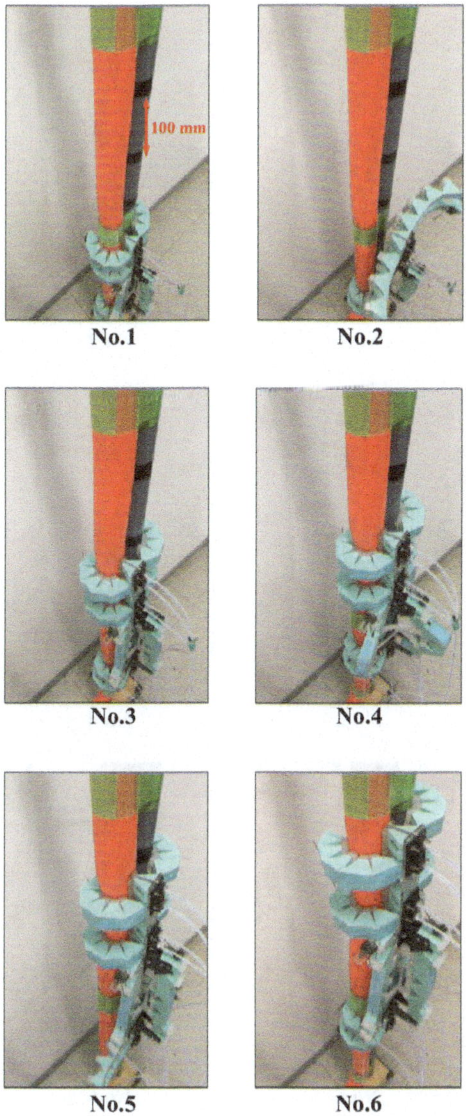

Fig. 16 Experimental result (balloon and pipe)

to produce enough friction to grasp objects. The grasping part is moved by pulling the string, which is installed through the frictionless PTFE tubes in the soft blocks.

The initial shape of the grasping part is a portion of a spiral curve. Owing to the dynamics of the rubber, the arm comes into contact with an object from the root to the tip by pulling the string, similar to the pipe climbing robot (Fig. 4), and through this action, the arm can cover and grasp an unknown object without measuring its

Fig. 17 Experimental result (natural tree)

shape. This complex grasping behavior is realized by the interaction between the arm and the object using the dynamics of the soft arm.

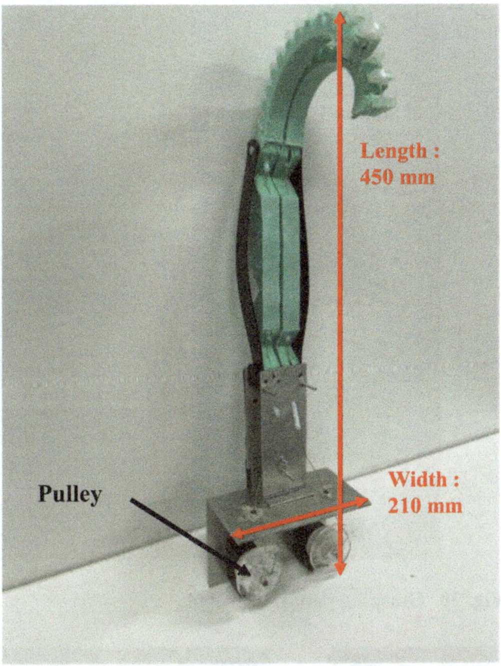

Fig. 18 MAMEYAKA—ladder climbing robot

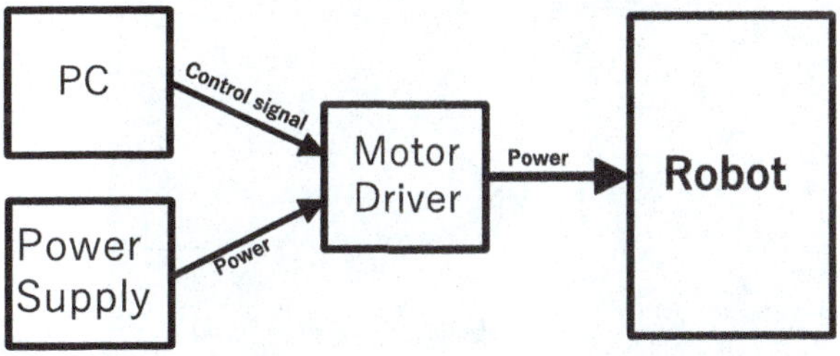

Fig. 19 Outline of the system for MAMEYAKA

6.2 *Lifting Part*

Figure 21 shows the lifting part, which is also made of silicone rubber 60A, and a cloth, which is embedded on both sides of the arm for reinforcement. By pulling the string, the lifting part shrinks (Fig. 22), and the other part of the arm is lifted.

6.3 *Actuation Mechanism and Control*

Figure 23 shows the whole arm. The string is connected to the pulley and is pulled and loosened by the motor. Since the lifting part has a higher elasticity than the

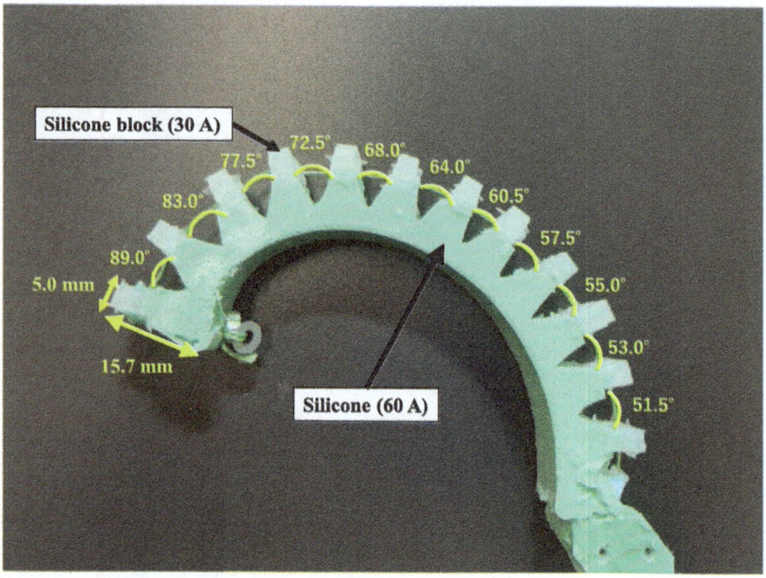

Fig. 20 Grasping part of the robot

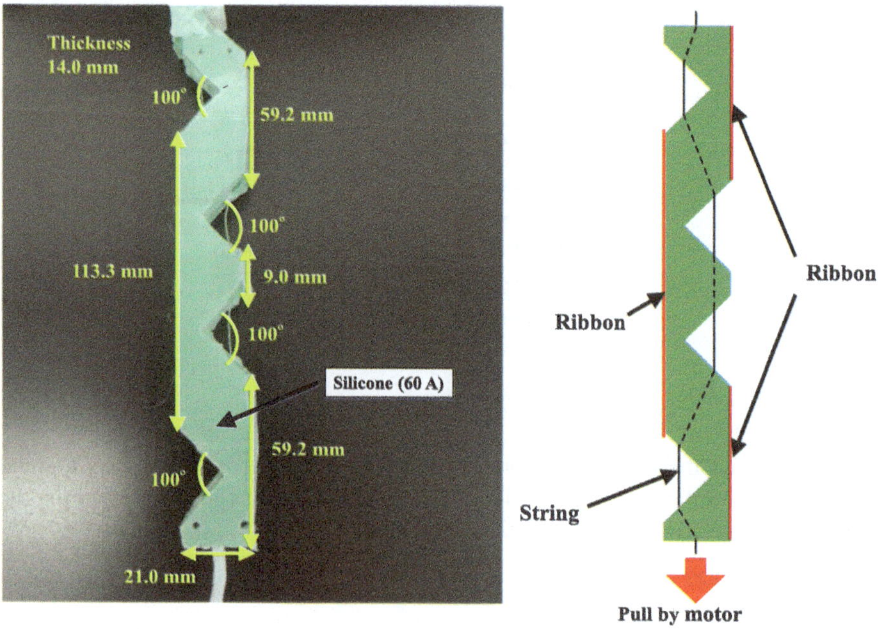

Fig. 21 Lifting part of the robot

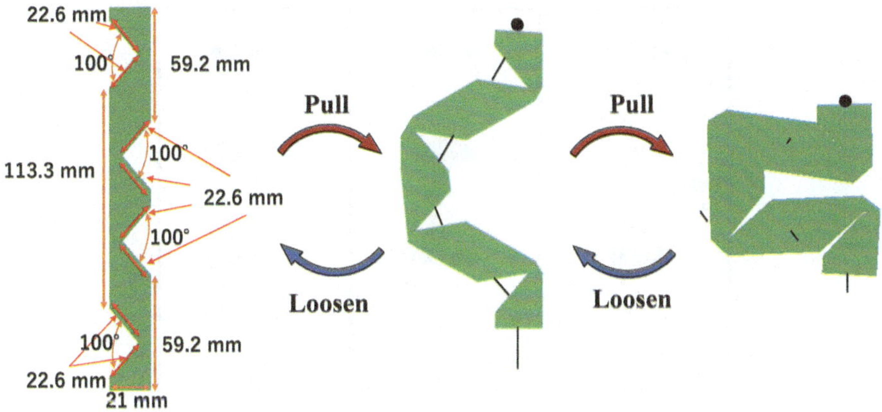

Fig. 22 Movement of the arm's lifting part

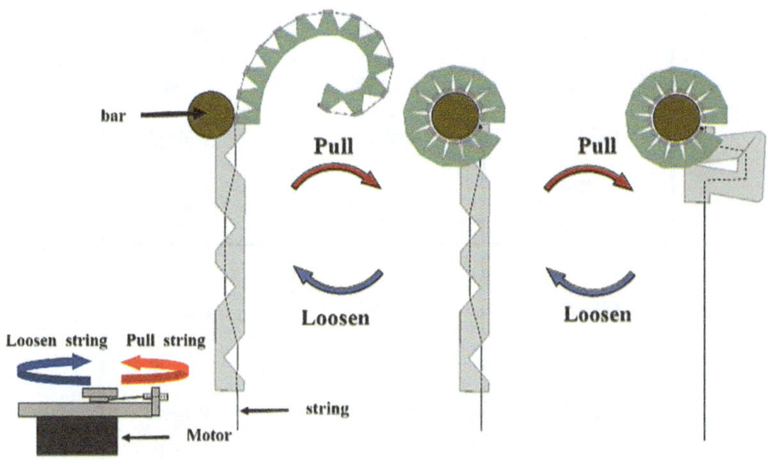

Fig. 23 Mechanism of the proposed robot

grasping part, the grasping part holds the object first; then, the lifting part lifts the object.

Figures 24 and 25 show the climbing pattern. By repeating the pattern of pulling and loosening the strings, the robot can climb a ladder. Complex maneuvers to grasp bars with different cross-sectional shapes are realized by the dynamics of the soft body.

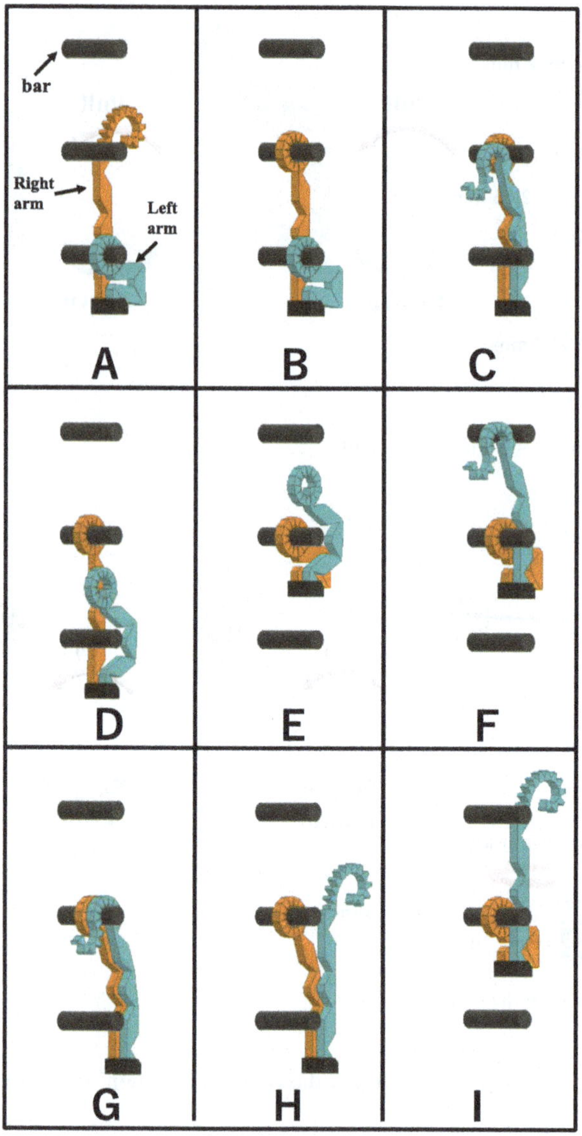

Fig. 24 Climbing pattern of the robot

6.4 Experimental Verification

6.4.1 Ladder

Figures 26, 27, 28, 29 and 30 show the movements used to grasp various bars,

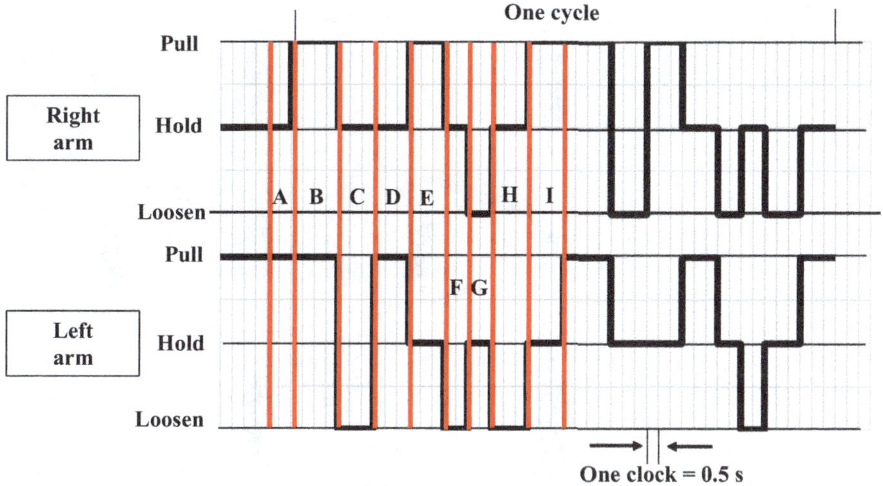

Fig. 25 Timing chart of climbing

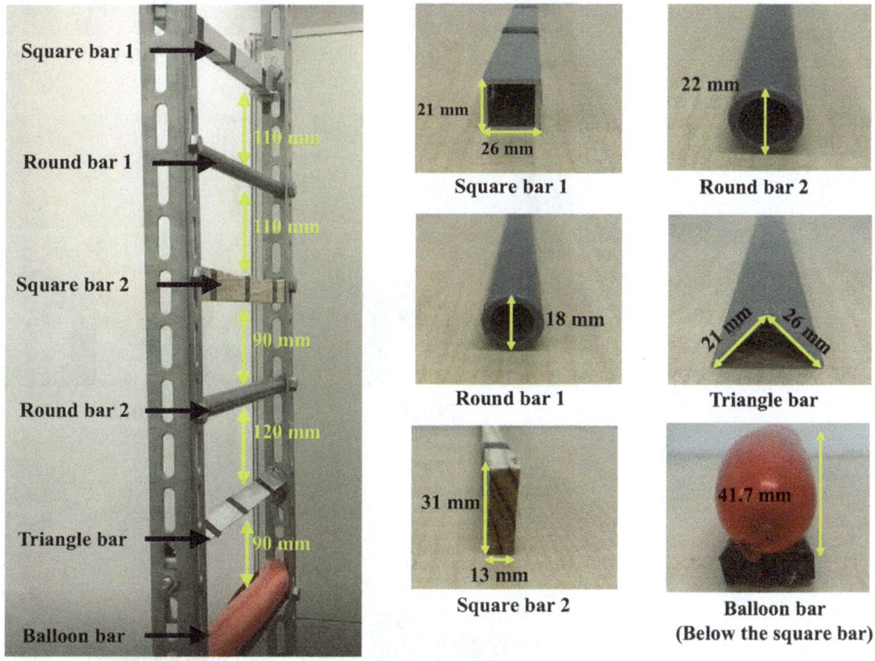

Fig. 26 Experimental environment (ladder)

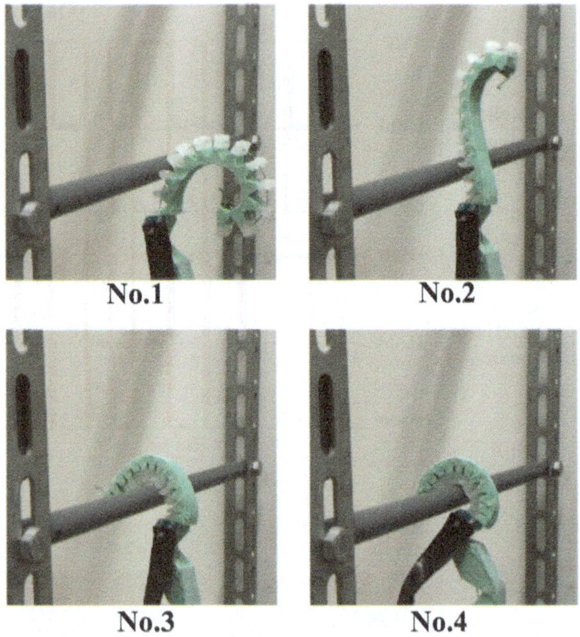

Fig. 27 Grasping behavior of the arm (round bar 2)

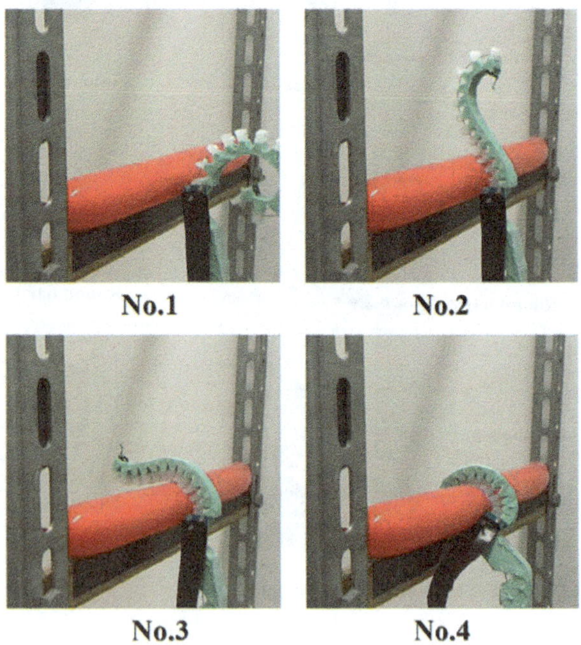

Fig. 28 Grasping behavior of the arm (balloon bar)

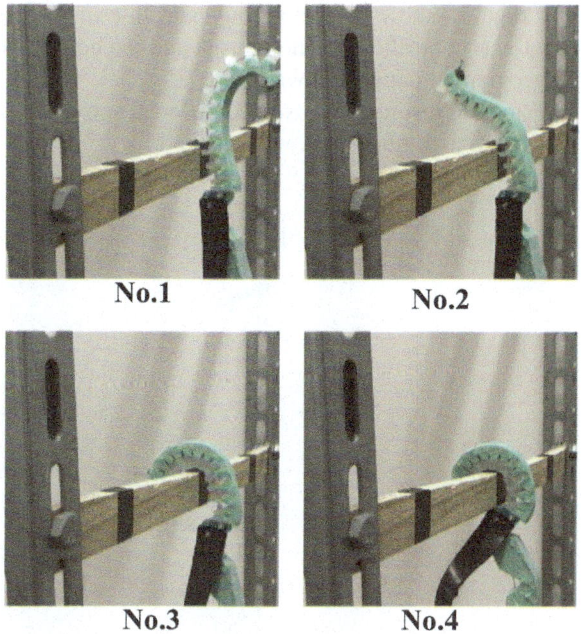

Fig. 29 Grasping behavior of the arm (square bar 2)

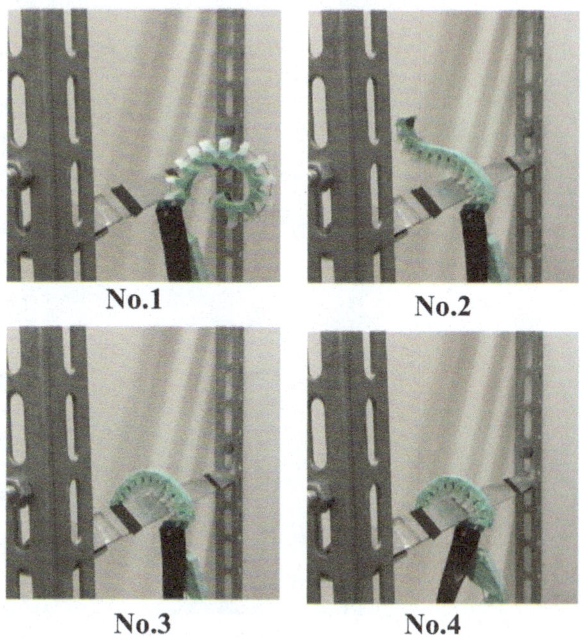

Fig. 30 Grasping behavior of the arm (Triangle bar)

and Fig. 31 shows the climbing behavior of the robot. In this experiment, the robot was controlled autonomously, and the control signal was an emulation of the preprogramed pattern shown in Fig. 25.

From the results, we observed different movements of the arm to grasp different bars and the climbing task was realized.

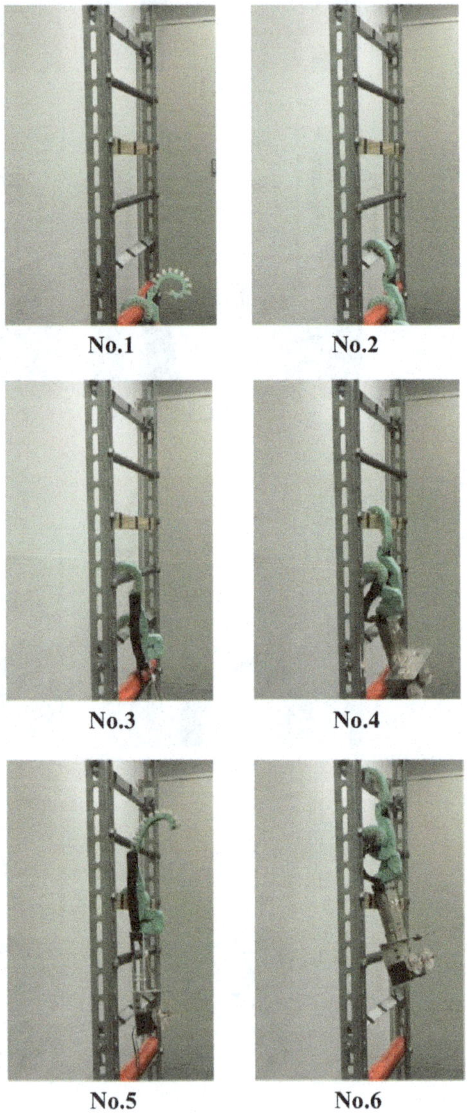

Fig. 31 Realized behavior (ladder)

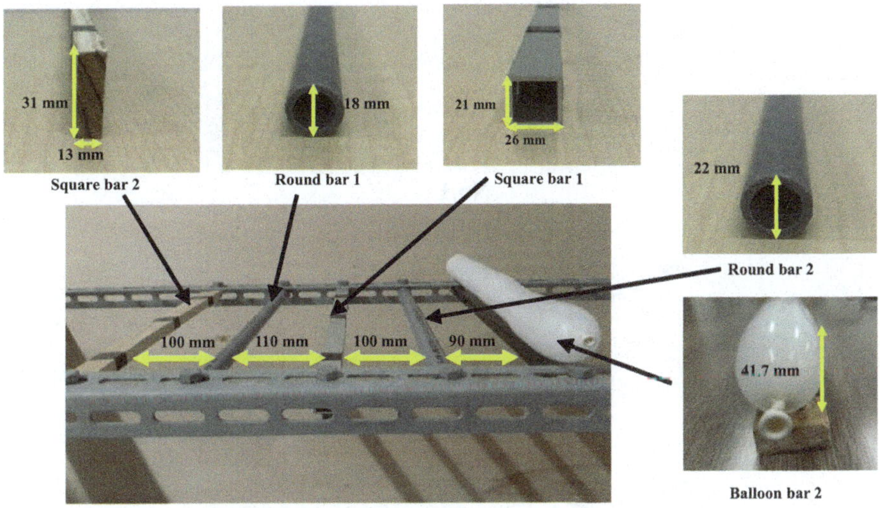

Fig. 32 Experimental environment (aerial ladder)

6.4.2 Application in Other Tasks

We placed the proposed robot on an aerial ladder and a bumpy wall and the robot was manually controlled.

Figures 32, 33, 34, 35, 36 and 37 show the different environments of the robot and the results. The results confirm that the complex behavior to complete these tasks can be realized by pulling and loosening the strings by the two motors. We conclude that we can control various degrees of freedom of the soft body and produce intelligent behavior through the robot's dynamics.

7 Conclusion

In this paper, we focused on the physical properties of the soft body and developed a pipe-climbing robot and a ladder-climbing robot inspired by an octopus. Further, we conducted experiments to demonstrate the effectiveness of the developed robots and confirmed that adaptive intelligent behavior could be realized by fewer dimensional control inputs compared to the large degrees of freedom of the soft body. Complex movements with many degrees of freedom were produced by the interaction between the soft body and the environment. In our future work, we will apply machine learning to the soft robot to acquire the control input and demonstrate the effectiveness of the soft body with machine learning.

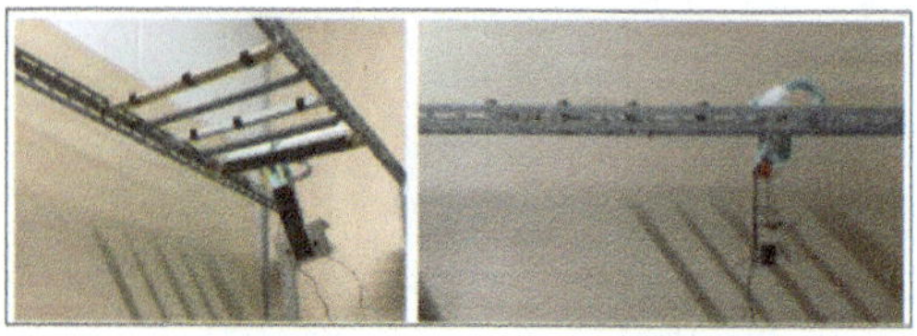

No.1

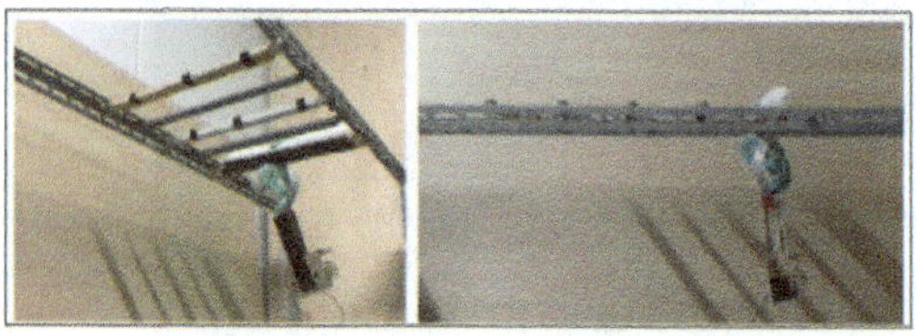

No.2

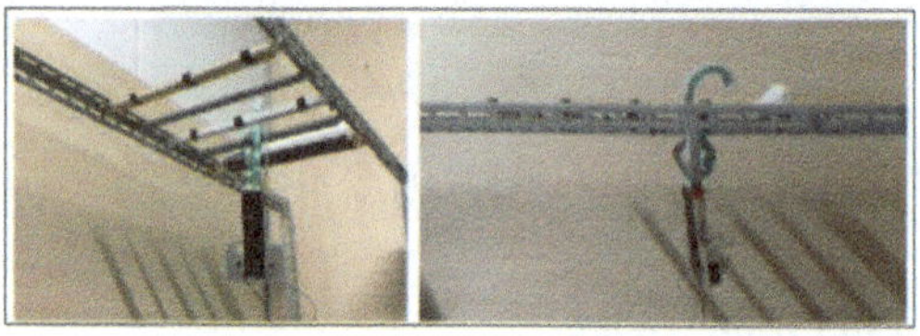

No.3

No.4

No.5

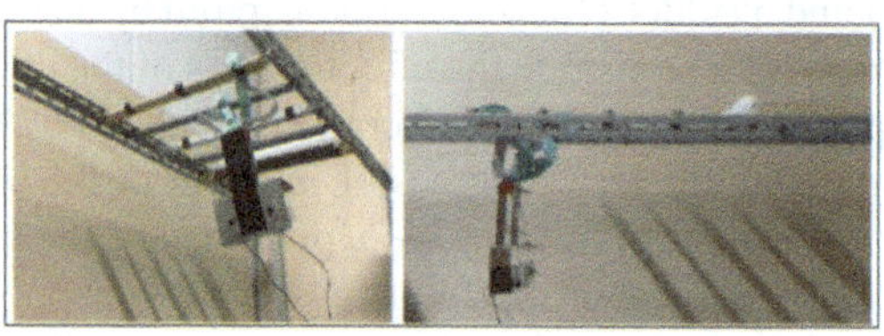

No.6

Fig. 33 Realized behavior (aerial ladder)

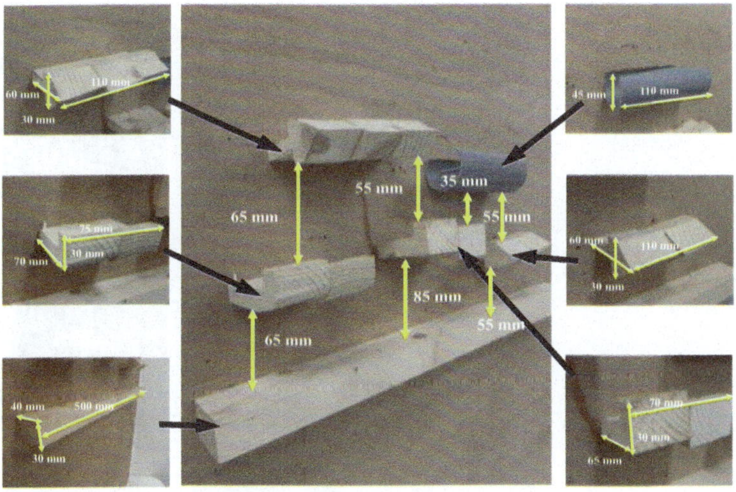

Fig. 34 Experimental environment (bumpy walls)

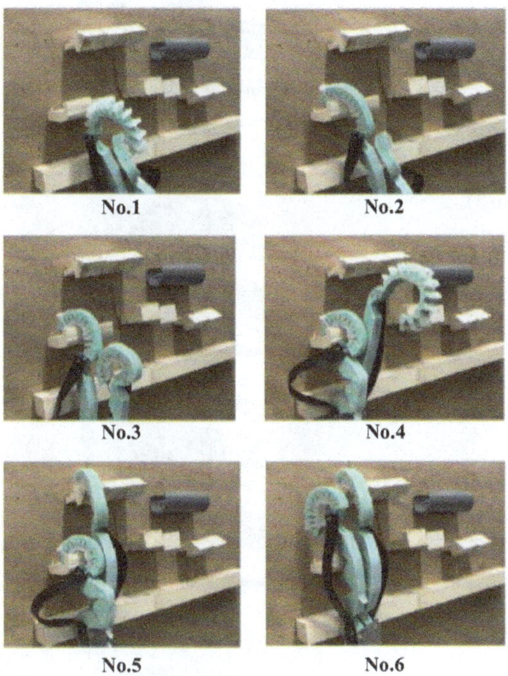

Fig. 35 Realized behavior (bumpy walls—pattern 1)

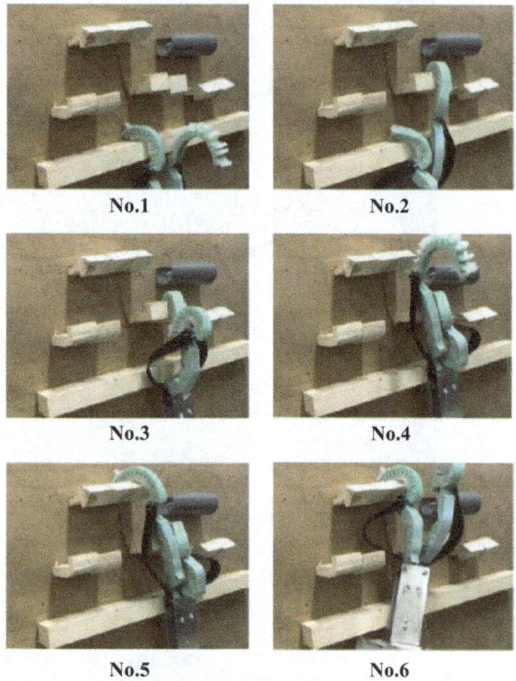

Fig. 36 Realized behavior (bumpy walls—pattern 2)

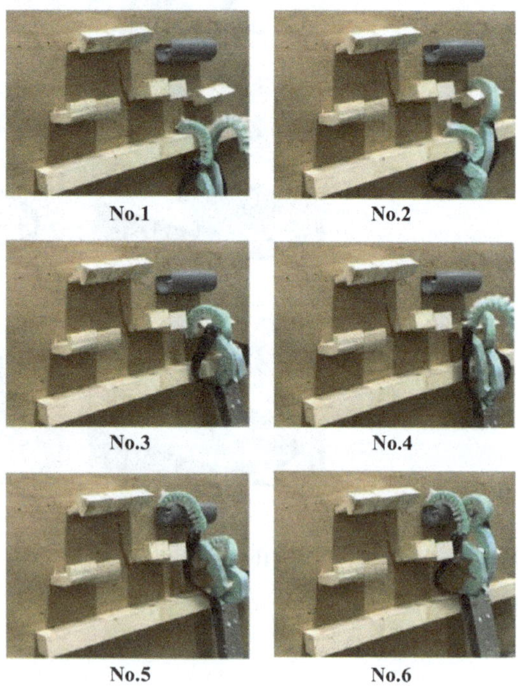

Fig. 37 Realized behavior (bumpy walls—pattern 3)

Acknowledgements This research was partially supported by the Japan Society for the Promotion of Science through the Grant-in-Aid for Scientific Research (18K11445).

References

1. Ito, K., Hagimori, S.: Flexible manipulator inspired by octopus: development of soft arms using sponge and experiment for grasping various objects. Artif. Life Rob. **22**(3), 283–288 (2017)
2. Ito, K., Aoyagi, R., Homma, Y.: TAOYAKA-III: a six-legged robot capable of climbing various columnar objects. J. Rob. Mech. **31**(1), 78–87 (2019)
3. Ito, K., Mukai, T.: Flexible manipulator inspired by octopi: advantages of the pulling Mechanism. Artif. Life Rob. (AROB) **25**(1), 167–172 (2020). https://doi.org/10.1007/s10015-019-00573-7
4. Ito, K., Aso, Y., Aihara, K.: Multi-legged robot for rough terrain: SHINAYAKA-L VI. In: Proceedings of the International Conference on Advanced Mechatronic Systems (ICAMech 2019), vol. 32 (2019)
5. Ito, K., Maruyama, H.: Semi-autonomous serially connected multi-crawler robot for search and rescue. Adv. Robot. **30**(7), 489–503 (2016)
6. Saito, A., Nagayama, K., Ito, K., Oomichi, T., Ashizawa, S., Matsuno, F.: Semi-autonomous multi-legged robot with suckers to climb a wall. J. Rob. Mech. **30**(1), 24–32 (2018)
7. Ito, K., Homma, Y., Rossiter, J.: The soft multi-legged robot inspired by octopus: climbing various columnar objects. Adv. Robot. **34**, 1096–1109 (2020)
8. Sakuhara, Y., Ito, K.: Climbing soft robot inspired by octopus. In: Proceedings of 2020 IEEE 10th International Conference on Intelligent Systems, pp. 463–468 (2020)
9. Ito, K., Fukumori, Y.: Autonomous control of a snake-like robot utilizing passive mechanism. In: Proceedings of IEEE International Conference on Robotics and Automation, pp. 381–386 (2006)
10. Ponsen, M., Taylor, M., Tuyls, K. E.: Abstraction and generalization in reinforcement learning: a summary and framework. In: International Workshop on Adaptive and Learning Agents, pp. 1–32 (2009)
11. Lee, C., Kim, M., Kim, J.Y., Hong, N., Ryu, S., Kim, J.H., Kim, H.: Soft robot review. Int. J. Control Autom. Syst. **15**(1), 3–15 (2017)
12. Rus, D., Tolley, M.: Design, fabrication and control of soft robots. Nature **521**, 467–475 (2015)
13. Sumbre, G., Gutfreund, Y., Fiorito, G., Flash, T., Hochner, B.: Control of octopus arm extension by a peripheral motor program. Science **293**(5536), 1845–1848 (2001)
14. Gutfreund, Y., Flash, T., Fiorito, G., Hochner, B.: Patterns of arm muscle activation involved in octopus reaching movements. J. Neurosci. **18**(15), 5976–5987 (1998)
15. Laschi, C., Cianchetti, M., Mazzolai, B., Margheri, L., Follador, M., Dario, P.: Soft robot arm inspired by the octopus. Adv. Robot. **26**(7), 709–729 (2012)

Automatic Log Analysis to Prevent Cyber Attacks

Andre Brandao and Petia Georgieva

Abstract Organizations nowadays are more dependent than ever on the internet, becoming necessary to provide better security to their networks. However, the rise of automated cyber attacks and their complexity is making this task harder for static approaches and manual examination of the logs, leading to a necessity of studying automated ways for identifying these cyber attacks. In this paper we propose an effective Log-based Intrusion Detection System (LIDS), to predict an attack or not, based on carefully selected features. The logs from various sources are aggregated into one dashboard and the most discriminative features are first determined. For the attack prediction, a few machine learning techniques were comparatively tested, with the Decision tree being the winner. The proposed system is illustrated with the largest publicly available labelled log file dataset *KDD Cup 1999*.

1 Introduction

Cyber attacks are growing due the appearance of more and more sophisticated toolkits and hacking as a service platforms, which facilitate attacks and allow people, even with low technical knowledge, to make a lot of damage.

The majority of the current intrusion detection systems (IDS) focus in anomaly-based and misuse-based detection techniques, being the last one favoured in commercial products given its high accuracy. However anomaly-based is more interesting given its potential to detect new forms of attacks. The IDSs generate a high number of events and many false positives, making it difficult for humans to determine the actual attack.

A. Brandao · P. Georgieva (✉)
Department of Electronics, Telecommunications and Informatics/IEETA,
University of Aveiro, Aveiro 3810-193, Portugal
e-mail: petia@ua.pt

A. Brandao
e-mail: andrebrandao@ua.pt

V. Sgurev et al. (eds.), *Advances in Intelligent Systems Research and Innovation*,
Studies in Systems, Decision and Control 379,
https://doi.org/10.1007/978-3-030-78124-8_14

In this paper we propose a Log-based Intrusion Detection System (LIDS) to predict if the network log is an attack or not. Log Analysis For Intrusion Detection is the process used to detect attacks on a specific environment using log files as the primary source of information.

2 Related Work

In order to avoid network attacks, the companies normally install several security devices such as Firewalls, IPS, anti-virus systems, etc., which are monitored by the Security Operation Center (SOC). These devices generate lots of logs. Additionally, there must be added the logs from Internet browsing, mail system, database, user authentication, Wireless Access, etc. On top of it, the logs from the networking devices have different structures, because they use different protocols or they are from different vendors (Cisco, Fortinet, SNMP, NetFlow, Linux or Windows platforms). All these problems leveraged the development of complex feature engineering approaches, for each individual client, such as log standardization, prioritization and normalization (building word dictionaries, identify keywords of event description, removing weak words, etc.). As a consequence, the commercial SOCs become very complex, with expensive and time-consuming maintenance. While larger businesses such as banks, insurances companies, service providers, etc. can afford such costly SOCs, small or medium companies are in a less favourable situation.

In order to address these issues some previous works focused into exploring the use of machine learning techniques with security logs. The ones that provided numerical results for the performance of their approaches are listed in Table 1. Most of these works are based on the data from the KDD Cup 99 competition, which is the only sufficiently large and publicly available labelled dataset based on log files.

Portnoy and co-authors, [1] applied a variant of single-linkage clustering to the KDD Cup 99 with the assumption that the bigger clusters contain the normal data and the smaller ones the attacks. Based on unbalanced training set of 1–1.5% attack and 98.5–99% normal instances, they have achieved 65.7% detection rate. The ROC curve was used to visualize the trade-off between true positive (TP) and false positive (FP) rates.

Latter on, [2] applied both supervised and unsupervised algorithms taking into account all features as inputs and achieved the best prediction rates with the C4.5 algorithm.

The authors of [3] used private logs from a number of services to detect infected machines and suspicious activities applying PCA (Principal Component Analysis) and k-means.

Different approach was followed in [4], where instead of using the logs to detect an attack, the alerts generated by sensors (IDSs such as for example Snort) were considered. These events are used to create meta-alerts that are then used to feed the classification models (Support Vector Machine, Bayesian Network and C4.5.

Table 1 Related work (data, methods, performance)

Reference	Data (labelled (yes/no)	Methods	Best performance
2000 [1]	KDD Cup 99 (yes)	Single-linkage clustering	Detection rate: 65.7%
2005 [2]	KDD Cup 99 (yes)	C4.5 algorithm	TP: 95% FP: 1%
2013 [5]	RBS (radio base station) from Ericsson (No)	K-means, DBSCAN, SOFM/SOM	F-score: 95% (SOFM-DBSCAN)
2018 [6]	Honeynet Project (no)	Decision tree	Acc: 98.2%
2018 [7]	KDD Cup 99 (yes)	Deep learning and others	Acc: 92.9% (DNN)
2019 [8]	Twitter (yes)	DNN	F1-score: 92%
This work	KDD Cup 99 (yes)	Feature selection & decision tree	Acc: 99.97%, FP = 0.05%

In [5], logs from the Ericsson Radio Base Station (RBS) were used, applying Self-organizing Map (SOM) and Density-based spatial clustering of applications with noise(DBSCAN). This dataset is not publicly available and according to the paper the labels are also not sufficiently reliable.

The work in [6] is based on the publicly available unlabelled Honeynet project dataset. After data normalization through the linearised Intrusion Detection Message Exchange Format (IDMEF) structure, the authors achieved 98.2% accuracy using a Decision Tree classifier.

Deep Neural Network (DNN) with 3 fully connected hidden layers and dropout layers was applied in [7] to KDD Cup 99 data. The network was trained with Backprop, ReLU functions in the hidden layers and sigmoid function in the last layer to predict attack or benign traffic.

Publicly available data from Twitter streaming API was considered recently in [8] to detect cyber threats. Data was labelled manually, corresponding to the activity of selected subset of accounts over four months. End-to-end threat intelligence tool was proposed using a sequence of two DNNs: a binary classifier based on a Convolutional Neural Network (CNN) architecture inspired by Natural Language Processing (NLP) and a Bidirectional Long Short-Term Memory (BiLSTM) network borrowed from Named Entity Recognition (NER). Data-driven approach was proposed also in [9] based on reinforcement learning for Intrusion detection system with log files. A common pain point of both deep and reinforcement learning is the need of a big amount of labelled training data to generalize, which is a labour intensive problem. Data visualization methods, as discussed in [10], are also intensively explored to facilitate the root cause analyses of network intrusion detection.

3 Data

Choosing the dataset was a difficult task as there are not many labelled datasets with complete and comprehensive descriptions, and those that exist are old. We decided to use the DARPA Intrusion Detection Evaluation 1999, despite its age because it is a common benchmark for evaluation of intrusion detection techniques. The DARPA Intrusion Detection Evaluation 1999 dataset was acquired by Lincoln Labs over nine weeks in a local-area network (LAN) simulating a typical U.S. Air Force LAN. They operated the LAN as if it were a true Air Force environment, but peppered it with multiple attacks. The data (available at: https://tinyurl.com/y2l64r8y) has about five million connection records. A connection is a sequence of TCP packets starting and ending at some well-defined times, between which data flows to and from a source IP address to a target IP address under some well-defined protocol. Each connection is labeled as either normal, or as an attack, with exactly one specific attack type. Each connection record consists of about 100 bytes.

The logs come from multiple services (e.g. http, smtp, telnet, etc.). Typical log structure is depicted in Appendix 1, see Fig. 6. The format, content and examples of the log items such as Access, Error, IPtables, Mail, Messages, etc. are also detailed in Appendix 1. This rich source of information was merged in one table with 41 features (given in Table 2) and a total of 22 attack types (Fig. 1). To create a balanced dataset, we selected about 400,000 records with 50% normal records and 50% attack records (Fig. 2). In this work we focus into a binary classification, e.g. if a record is an attack or not.

The selected categorical features were encoded applying one hot encoding. In order to create smoothness and rotational invariance, data were also normalized:

$$x_{normalized} = \frac{x}{\sqrt{\sum_{i=0}^{N} X_i^2}} \tag{1}$$

4 Feature Selection

Selecting relevant features in network intrusion detection is an open research question. Finding the important features not only speed up the data manipulation but holds the promise to improve the detection rate. Most of the references discussed in Sect. 2 use some kind of feature filtering as a preprocessing step without explicit feature selection. In [11], a feature selection based on the rough set theory is proposed, however without applying it for intrusion detection.

Here we propose to apply two feature selection approaches, namely Optimal Feature Selection (OFS) and Factor Analysis (FA).

Table 2 KDD Cup 1999 data features

Feature	Description	Type
Basic features		
Duration	Connection length (measured in sec.)	cont.
protocol_type	Protocol type (udp, tcp, etc.)	Categorical
service	Service on the destination (telnet, http, etc.)	Categorical
src_bytes	Quantity of data bytes from source to destination	cont.
DST_bytes	Quantity of data bytes from destination to source	cont.
flag	Connection status (normal or error)	Categorical
land	1 if connection is from/to the same host/port; 0 otherwise	Binary
wrong_fragment / urgent	# of "wrong" fragments/# of urgent packets	cont.
Content features		
hot	# of "hot" indicators	cont.
num_failed_logins	# of failed login attempts	cont.
logged_in	Successfully login (1); otherwise (0)	Binary
num_compromised	# of "compromised" conditions	cont.
root_shell	root shell is obtained (1); Otherwise (0)	Binary
su_attempted	1 if "su root" command attempted; 0 otherwise	Binary
num_root	# of "root" accesses	cont.
num_file_creations	# of file creation operations	cont.
num_shells	# of shell prompts	cont.
num_access_files	# of operations on access control files	cont.
num outbound cmds	# of outbound commands in an ftp session	cont.
hot_login	1 if the login belongs to the "hot" list; 0 otherwise	Binary
guest_login	1 if the login is a "guest" login; 0 otherwise	Binary
Time-based traffic features of the connection records (2 sec. window)		
count	# of connections to the same host as the current connections in the last 2 s	cont.
serror_rate	% of connections that have "SYN" errors	cont.
rerror_rate	% of connections that have "REJ" errors	cont.
same_srv_rate	% of connections to the same service	cont.
diff_srv_rate	% of connections to different services	cont.
srv_count	# of connections to the same service as the current connection in the last 2 s	cont.
srv_serror_rate	% of connections that have "SYN" errors	cont.
srv_rerror_rate	% of connections that have "REJ" errors	cont.
srv_diff_host_rate	% of connections to different hosts	cont.
Host-based traffic features		
DST_host_count	# of connections to the same destination host as the current connection	cont.
DST_host_srv_count	# of connections to the same destination service as the current connection	cont.
DST_host_same_srv_rate	% of connections to the same destination service	cont.
DST_host_diff_srv_rate	% of connections to different destination services	cont.
DST_host_same_src_port_rate	% of connections to the same source port	cont.
DST_host_srv_diff_host_rate	% of connections to different hosts	cont.
DST_host_serror_rate	% of connections that have "SYN" errors	cont.
DST_host_srv_serror_rate	% of connections that have "SYN" errors	cont.
DST_host_rerror_rate	% of connections that have "REJ" errors	cont.
DST_host_srv_rerror_rate	% of connections that have "REJ" errors	cont.

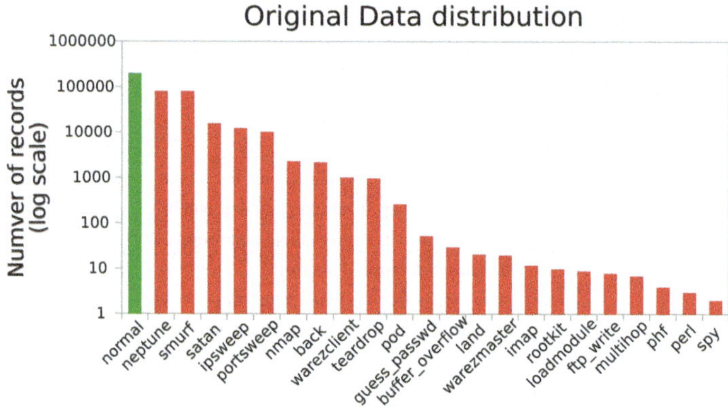

Fig. 1 KDD Cup 1999 data (red attack), green (normal)

Fig. 2 Data distribution

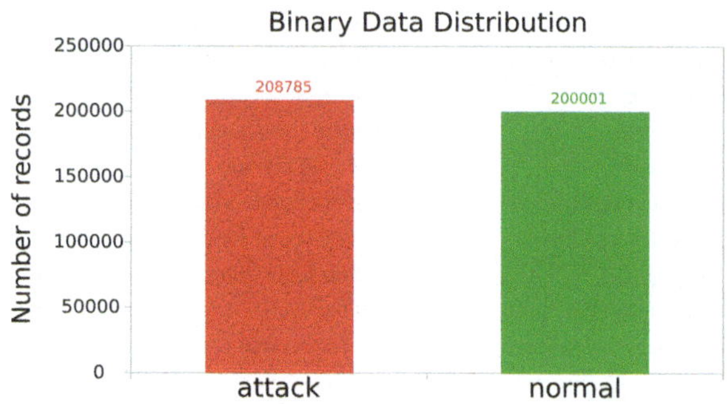

4.1 *Optimal Feature Selection*

OFS is a deterministic greedy algorithm that takes the locally optimal choice at each stage, [12]. OFS can be implemented in two ways: forward selection or backward elimination.

Forward Selection. Build data models based on individual features. Choose the best *k1* features. Make all two by two combinations of the *k1* selected features and build new data models. Choose the next group of best *k2* features and make all three by three combinations of them. Repeat the process until no more improvement in the model performance. The choice of subsequent feature subset combinations may vary in order to reflect the experience gained during the extraction process and the application in hand.

Backward Elimination. The model is trained on the complete set of features and weights are assigned to each of them. The features with the smallest weights are then pruned from the set. This process is repeated until the model performance deteriorates below a certain threshold.

The OFS, applied to the KDD Cup99 data, ranked the features in three categories: green (most important), yellow (neutral) and red (least important). 23 features were ranked as green: *protocol_type, service, src_bytes, DST_bytes, flag, hot, num_failed_logins, logged_in, hot_login, count, serror_rate, same_srv_rate, diff_srv_rate, srv_count, srv_serror_rate, DST_host_count, DST_host_srv_count, DST_host_same_srv_rate, DST_host_diff_srv_rate, DST_host_same_src_port_rate, DST_host_srv_serror_rate, DST_host_srv_diff_host_rate, DST_host_serror_rate.*

4.2 Factor Analysis

Factor Analysis (FA) is an explorative method similar to Principal Components Analysis (PCA) [13]. Both methods try to reduce the dimensionality of the dataset based on the concept of common variance. Common variance is the amount of variance that is shared among a set of features. Features that are highly correlated will share a lot of variance. PCA assumes that the total variance is build on all common variances, while the FA assumes that total variance can be partitioned into common and unique variance. Unique variance is any portion of variance that is not common. Thus, in the FA framework, the total variance is sum of the common variance and unique variance.

FA identifies the latent (unobserved) factors which represent the hidden organization of the observable measures. Factor scores (aka factor loadings) indicate how each latent factor is associated with the observable variables used in the analysis. If the goal is to simply reduce the variable list down into a linear combination of smaller number of components then PCA is the way to go. However, if the goal is to determine what are the features that describe a latent factor then the FA is more appropriate.

In Table 3 are presented the FA results of a 4-factor model performed with software R libraries. In the first column are represented the dominant features that identify the four hidden factors. The other columns are the factor loadings on these factors. Based on FA, a reduced feature set was constructed accounting for features with factor loadings with absolute values greater that 0.75:

$$\textbf{Reduced feature vector} = [service,\ protocol\_type,\ flag,\ count,\ logged\_in,\ DST\_host\_count].$$

Note that, according to both approaches, OFS and FA, the most important features are the categorical *service*, *protocol_type*, and *flag*. Their histograms, shown in Figs. 3 and 4, confirm that their class distributions are different, though some overlapping is also observed.

Most of the other selected features (*DST_xxx*) belong to the so-called host-based traffic features. These are connection records sorted by the destination host. They are constructed using a window of 100 connections to the same host instead of a time window. The reason for adding these features, according to [14], is that some probing

Table 3 Factor analysis loadings for *varimax* rotated matrix of 4-factor model (abs. values > 0.75) in bold

	F1	F2	F3	F4
service	**0.91**	− 0.13	0.07	0.02
protocol_type	**0.78**	0.10	− 0.44	0.01
flag	**0.77**	0.04	0.44	0.23
count	**− 0.79**	0.01	− 0.01	− 0.25
logged_in	− 0.14	**0.88**	0.03	0.07
DST_host_count	− 0.02	**0.88**	− 0.11	0.09
DST_host_diff_srv_rate	− 0.23	− 0.11	0.75	− 0.13
DST_host_same_srv_port_rate	− 0.34	− 0.02	− 0.72	− 0.13
DST_host_serror_rate	0.13	0.00	− 0.27	0.73

attacks scan the hosts (or ports) using a much larger time interval than two seconds, for example once per minute. Hence, their discrimination capacity is reflected by the feature selection mechanisms.

The visual assessment further confirmed the consensus between the OFS and FA regarding the feature relevance to the model construction and the class discriminability. Data class distribution of all features are provided in Appendix 2.

5 Cyber Attacks Predictive Models

The Decision Tree (DT), K-Nearest Neighbours (kNN), Neural Network (NN) and Naive Bayes were selected as predictive models. If the feature selection is well done (based on OFS or FA), the classifiers are expected to demonstrate similar performance. Data was split into training and cross validation subset (80%) and testing subset (20%). The model selection was done after K-fold cross validation over a grid of possible values for relevant hyper-parameters of the classifiers. The models were trained and validated with the 23 features, selected by the OFS method.

5.1 Decision Tree

The DT model was evaluated varying the maximum depth of the tree in the range [3, 30]. The optimal DT structure was found with maximum depth of 22 (Fig. 5).

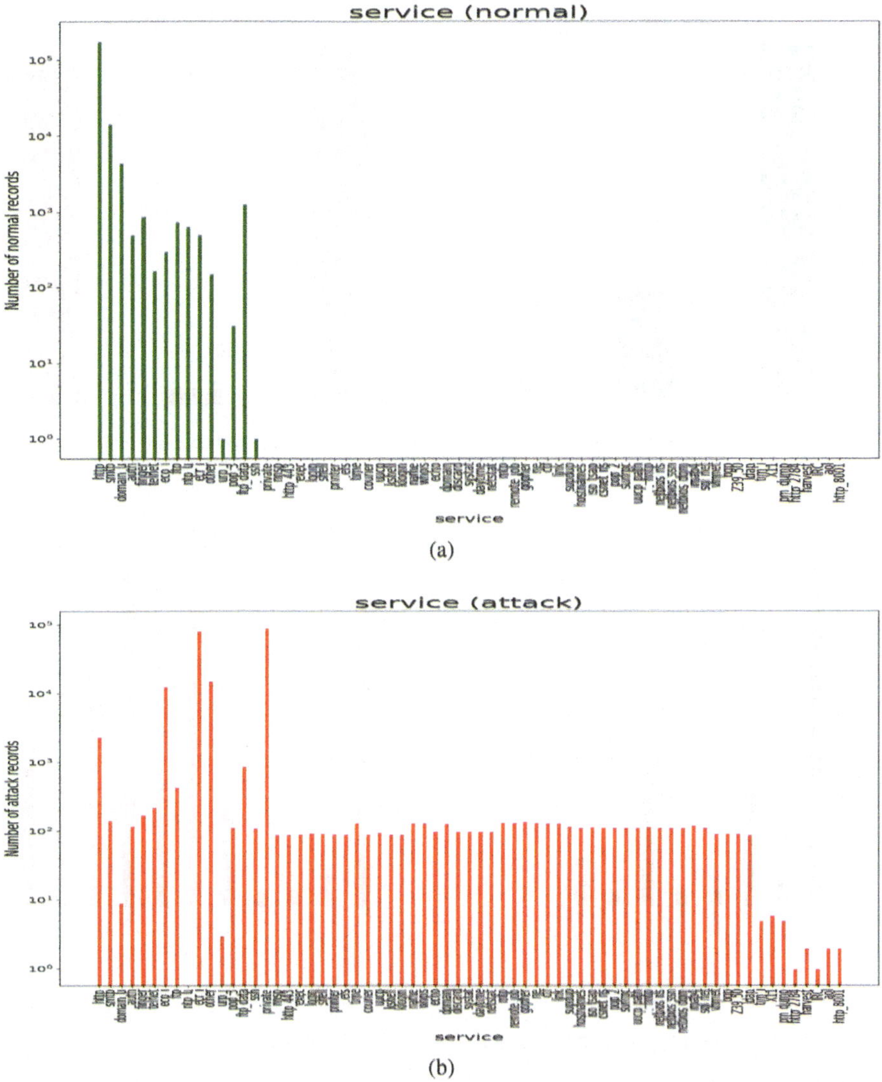

Fig. 3 Distribution of feature *service* **a** normal; **b** attack

5.2 *K-Nearest Neighbours*

The kNN classifier was optimized varying the number of neighbours k in the range [3, 15], with the optimal value equal to 3 (Table 4). As distance measure for computing the k-nearest neighbours was chosen cosine similarity. Cosine similarity is a measure of similarity between two non-zero vectors expressed by the cosine of the angle between them.

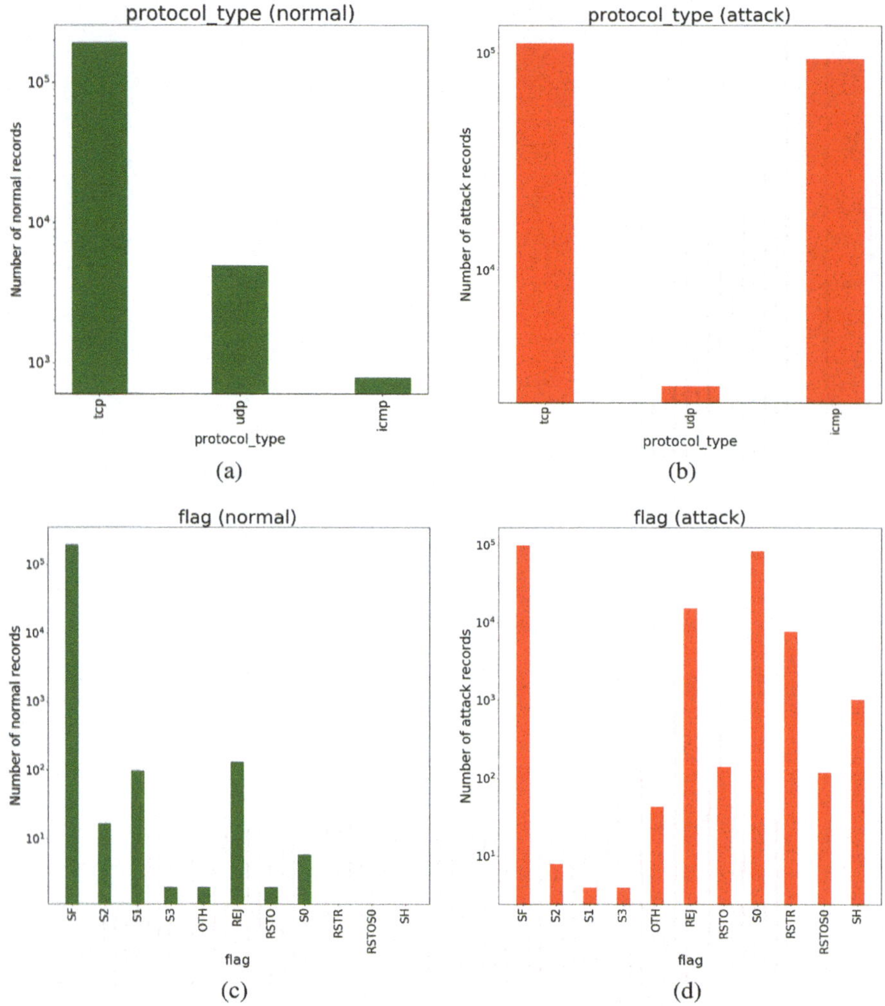

Fig. 4 Distribution of categorical features: *protocol* (**a**, **b**); *flag* (**c**, **d**)

5.3 Neural Network

A fully connected NN with one hidden layer (HL) and ReLu activation functions was chosen. The learning rate was picked from the range [0.001, 0.1, 05]. The number of HL units was optimized in the range [3, 5, 10]. Although the best performance was achieved for HL with 80 units (Table 5), simpler model with 5 HL units was chosen as the optimal structure due to the negligible difference in performance.

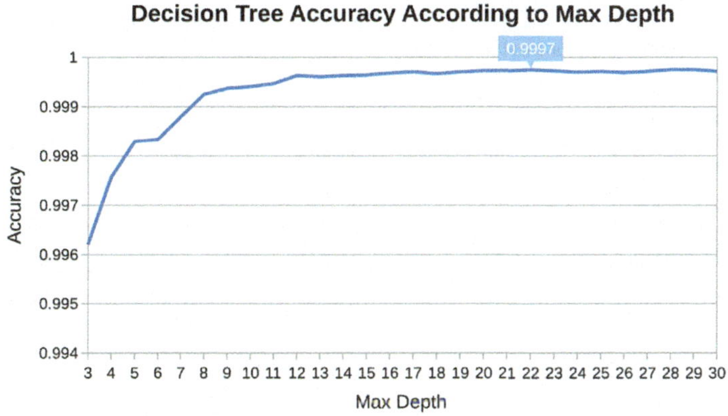

Fig. 5 DT model (variation of maximum depth)

Table 4 kNN model (variation of number of neighbours)

	Number of neighbours k				
	3	5	7	11	15
Accuracy	**99.90**	99.87	99.86	99.84	99.83

Table 5 NN model (variation of HL size and learning rate)

		Learning rate		
		0.001	0.1	0.5
Hidden layer size	3	98.97	98.86	98.73
	5	99.63	99.45	98.86
	10	99.65	99.36	98.87
	20	99.62	99.42	98.92
	50	99.66	99.34	98.80
	80	**99.89**	99.08	98.70

5.4 *Naive Bayes*

The probability of the features for the Naive Bayes is computed assuming Gaussian distribution:

$$P(x_i|y) = \frac{1}{\sqrt{2\pi\sigma_y^2}} \exp(-\frac{(x_i - \mu_y)^2}{2\sigma_y^2}) \tag{2}$$

The parameters standard deviation (σ_y) and the mean (μ_y) are estimated using maximum likelihood. With this model and the 23 features selected by the OFS method we have obtained the worse results of 97.21% accuracy on test data.

Table 6 Test performance of 3 classifiers (DT, kNN, KNN) with feature selection methods: optimal feature selection (OFS), factor analysis (FA)

	TP	FP	TN	FN	Acc (%)
OFS&DT	39,922	17	41,809	10	99.97
OFS&kNN	39,828	39	41,848	43	99.90
OFS&NN	39,835	196	41,620	107	99.63
FA&DT	39,901	28	41,798	31	99.93
FA&kNN	39,806	39	41,848	65	99.87
FA&NN	39,795	246	41,570	147	98.52

5.5 *Model Evaluation*

The classifier with the optimal hyper-parameters where the ones that achieved the maximum accuracy as shown in the previous section. Now, these models were retrained with the whole set of train and validation data, provided either with the OFT or FA selected features, respectively. Naive Bayes was excluded due to its relatively low performance. The performance of the models on test data is summarized in Table 6. The three classifiers demonstrate overall similar performance with the DT slightly outperforming the other models. These results are in accordance to the reviewed literature, where DT-based classifiers are frequently the suggested models.

6 Conclusions

Selecting relevant feature is an important problem in learning systems. KDD Cup'99 dataset was used for benchmarking intrusion detection problem based on network traffic logs and a suitable combination of features. The elimination of the insignificant features simplified the problem and did not hurt the detection rate. Our findings show that if the most relevant features are determined, the choice of the classifier is less critical. It does become more important to focus into data mining to attain best results.

With regards to future work, there is most certainly room for improvement. Alternative avenues could be explored either in the form of new algorithms or in the form of new features. In many attacks nowadays the attack vector used is a phishing email, and to detect the phishing emails it is necessary to make a higher level analysis, content and context analysis, using for example natural language processing. As such, a complete Intrusion Detection System would benefit from being composed of different models for different tasks, that combined will be more effective for detecting various forms of attack.

Acknowledgements This work was funded by National Funds through the FCT—Foundation for Science and Technology, in the context of the project UID/CEC/00127/2019.

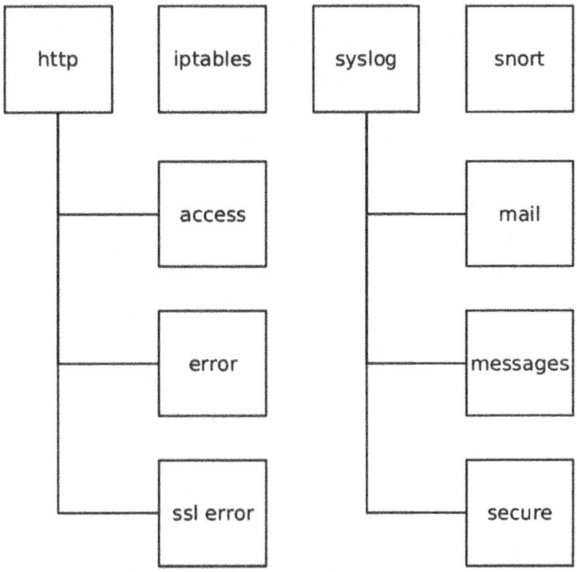

Fig. 6 Log structure

Appendix 1: Logs Structure and Formats

Before formulating attack hypotheses, it's important to learn as much about the logs structure and involved systems as possible. A better understanding of the environment and settings will allow to make an educated judgement on which attacks actually had a chance of succeeding and which had no chance at all. A typical logs structure is depicted in Fig. 6.

Access

The http access log records all requests processed by the server. Storing the information in the access log is the start of log management. The next step is to analyze this information to produce useful statistics. The format of the access log is highly configurable.

Typical Access Log Format: <ClientIP> - >ClientID> >Timestamp> >Method> >RequestResource> >Protocol> >StatusCode> >RetObjSize> >Referer> >UserAgent>

- ClientIP: This is the IP address of the client (remote host) which made the request to the server.
- ClientID: This is the userid of the person requesting the document as determined by HTTP authentication.
- Timestamp: The time that the request was received.

- Method: HTTP request method (e.g. GET, POST, PUT, DELETE, etc.).
- RequestResource: Path to the resource requested by the client.
- Protocol: HTTP protocol used by the client.
- StatusCode: This is the status code that the server sends back to the client.
- RetObjSize: The size of the object returned to the client, not including the response headers.
- Referer: This gives the site that the client reports having been referred from.
- UserAgent: The User-Agent HTTP request header. This is the identifying information that the client browser reports about itself.

Log Example: 222.95.39.192—[07/Feb/2005:19:52:35-0500] "GET http://ad.trafficmp.com/tmpad/banner/ad/tmp.asp?poID=el0w HTTP/1.0" 404 1187 "http://www.besteach.com/" "Mozilla/4.0 (compatible; MSIE 5.5; Windows 98)"

- ClientIP: 222.95.39.192.
- ClientID: - 1.
- Timestamp: [07/Feb/2005:19:52:35-0500].
- Method: GET.
- RequestResource: http://ad.trafficmp.com/tmpad/banner/ad/tmp.asp?poID=el0w.
- Protocol: HTTP/1.0.
- StatusCode: 404.
- RetObjSize: 1187.
- Referer: http://www.besteach.com/.
- UserAgent: Mozilla/4.0 (compatible; MSIE 5.5; Windows 98).

Error

Contains information about errors that the web server encountered when processing requests, such as when files are missing. It looks something like this:
Log Format: <Timestamp< <LogLevel< <ClientIPAndPort< <LogMessage<

- imestamp: The current time.
- LogLevel: Loglevel of the message (e.g. error, notice, warn, etc.).
- ClientIPAndPort: Client IP address and port of the request.
- LogMessage: The actual log message.

Log Example: [Tue Feb 22 11:04:40 2005] [error] [client 211.59.0.40] File does not exist: /var/www/html/scripts

- Timestamp: [Tue Feb 22 11:04:40 2005].
- LogLevel: error.
- ClientIPAndPort: 211.59.0.40.
- LogMessage: File does not exist: /var/www/html/scripts.

The Secure Sockets Layer (SSL) is used to create a safe connection between the client and the server which transmits data. This information is encrypted using two keys a private one and a public one. The SSL error format is similar as the error log format above.

IPtables

Log Format: <Timestamp< <MachineName< kernel: <TrafficDirection< < Protocol<: IN=<IN< PHYSIN=<PHYSIN< OUT=<<UT< PHYSOUT=<PHY-SOUT< SRC=<SRC< DST=<DST< LEN=<LEN< TOS=<TOS< PREC= <PREC< TTL=<TTL< ID=<ID< PROTO=<PROTO< SPT=<SPT< DPT= <DPT<

- Timestamp: The current time.
- MachineName: Name of the machine.
- TrafficDirection: Traffic direction.
- Protocol: Protocol.
- IN: This indicates the interface that was used for this incoming packets.
- PHYSIN: This indicates the physical interface that was used for this incoming packets.
- OUT: This indicates the interface that was used for outgoing packets.
- PHYSOUT: This indicates the physical interface that was used for outgoing packets.
- SRC: The source ip-address from where the packet originated.
- DST: The destination ip-address where the packets was sent to.
- LEN: Length of the packet.
- TOS: Type of Service.
- PREC: Precedence bits.
- TTL: Time To Live.
- ID: Packet identifier.
- PROTO: Indicates the protocol (e.g. ICMP, TCP, etc.).
- SPT: Indicates the source port.
- DPT: Indicates the destination port.

Log Example: Feb 25 12:11:24 bridge kernel: INBOUND TCP: IN=br0 PHYSIN =eth0 OUT=br0 PHYSOUT=eth1 SRC=220.228.136.38 DST=11.11.79.83 LEN=64 TOS=0x00 PREC=0x00 TTL=47 ID=17159 PROTO=TCP SPT=1629 DPT=139

- Timestamp: Feb 25 12:11:24
- MachineName: bridge
- TrafficDirection: INBOUND
- Protocol: TCP
- IN: br0
- PHYSIN: eth0

- OUT: br0
- PHYSOUT: eth1
- SRC: 220.228.136.38
- DST: 11.11.79.83
- LEN: 64
- TOS: 0x00
- PREC: 0x00
- TTL: 47
- ID: 17159
- PROTO: TCP
- SPT: 1629
- DPT: 139

Mail

Log Format: <Date> <Host> sendmail[Pid]: <Qid>: <What>=<Value>

- Date: Month, day and time that the line was logged.
- Host: The name of the host that produced this information (may dier from the logging host).
- sendmail: Literal, even if sendmail is invoked as mailq or newaliases, 'sendmail' is printed here.
- Pid: The process id of the sendmail invocation that produced this log line.
- Qid: The queue id, a message identifier unique on the host producing the log lines.
- What=Value: A comma-separated list of equates. Which equate appears in which line depends on whether the line documents the sender or the recipient and whether delivery succeeded, failed, or was deferred.
  - lass: The queue class: the numeric value defined in the sendmail configuration le for the keyword given in the Precedence: header of the processed message.
  - Ctladdr: The "controlling user", that is, the name of the user whose credentials we use for delivery.
  - Delay: The total message delay: the time difference between reception and final delivery or bounce). Format is delay=HH:MM::SS for a delay of less than one day and delay=days+HH:MM::SS otherwise.
  - From: The envelope sender. Format is from=addr, with addr defined in [2] by the "address" keyword. This can be an actual person, or a postmaster.
  - Mailer: The symbolic name (defined in the sendmail configuration file) for the program (known as delivery agent) that performed the message delivery.
  - Msgid: A world-unique message identifier. The msgid = equate is omitted if it (incorrectly) is not defined in the configuration file.
  - Nrcpts: The number of recipients for the message, after all aliasing has taken place.

- pri: The initial priority assigned to the message. The priority changes each time the queued message is tried, but this equate only shows the initial value.
- Proto: The protocol that was used when the message was received; this is either SMTP, ESMTP, or internal, or assigned with the -p command-line switch.
- Relay: Shows which user or system sent/received the message; the format is one of relay=user(a)domain [IP], relay=user(a)localhost, or relay=fqdn host.
- Size: The size of the incoming message in bytes during the DATA phase, including end-of-line characters. For messages received via sendmails' standard input, it is the count of the bytes received, including the newline characters.
- Stat: The delivery status of the message. For successful delivery, stat=Sent (text) is printed, where text is the actual text that the other host printed when it accepted the message, transmitted via SMTP. For local delivery, stat=Sent is printed. Other possibilities are stat=Deferred: reason, stat=queued, or stat=User unknown. [complete list of possible values to be made]
- to: Address of the final recipient, after all aliasing has taken place. The format is defined in [2] by the "address" keyword.
- Xdelay: The total time the message took to be transmitted during final delivery. This ders from the delay= equate, in that the xdelay= equate only counts the time in the actual final delivery.
- dsn: Delivery Status Notifications.

Log Example: Mar 15 04:04:36 combo sendmail[13337]: j2F94C6S013336: to=<root@combo.honeypotbox.com>, ctladdr=<root@combo.honeypotbox.com> (0/0), delay=00:00:00, xdelay=00:00:00, mailer=local, pri=31702, dsn=2.0.0, stat= Sent

- Date: Mar 15 04:04:36.
- Host: combo.
- sendmail: 'sendmail'.
- Pid: 13337.
- Qid: j2F94C6S013336.
- What=Value:
- Ctladdr: root@combo.honeypotbox.com.
- Delay: 00:00:00.
- Mailer: local.
- pri: 31702.
- Stat: Sent.
- to: root@combo.honeypotbox.com.
- Xdelay: 00:00:00.
- dsn: 2.0.0.

1.1 Messages

Log Format: <Date> <Host> <Program> [Pid]: <Action>

- Date: Date and time that the line was logged (no year is present, which is a syslog peculiarity).
- Host: Executing program's hostname.
- Program: Name of the utility, program or daemon that caused the message.
- Pid: The process id of the program that produced this log line.
- Action: The action that occurred.

Log Example:

- Date: Feb 1 11:50:36
- Host: combo.
- Program: sshd(pam unix).
- Pid: 32603.
- Action: authentication failure; logname= uid=0 euid=0 tty=NODEVssh ruser= rhost=208.29.180.3 user=nobody.

1.2 Secure

Authentication messages, xinetd services, etc. are logged here. Log Format: <Date> <Host> <Program> [Pid]: <Action>

- Date: Date and time that the line was logged (no year is present, which is a syslog peculiarity).
- Host: Executing program's hostname.
- Program: Name of the utility, program or daemon that caused the message.
- Pid: The process id of the program that produced this log line.
- Action: The action that occurred.

Log Example: Mar 13 22:50:55 combo sshd[9356]: Failed password for root from 67.103.15.70 port 55639 ssh2

- Date: Mar 13 22:50:55.
- Host: combo.
- Program: sshd.
- Pid: 9356.
- Action: Failed password for root from 67.103.15.70 port 55639 ssh2.

1.3 Snort

Log Format:

- Date: Date and time that the line was logged (no year is present, which is a syslog peculiarity).
- Host: Executing program's hostname.
- GID: Generator ID, the component of Snort generated this alert.
- SID: Snort ID (sometimes referred to as Signature ID),
- RID: Revision ID.
- Msg: Message.
- Class: Classification.
- Priority: Priority.
- PROTO: Protocol.
- SOURCE IP PORT: Source IP and port.
- DEST IP PORT: Destination IP and port.

Log Example: Feb 25 12:23:54 bastion snort: [1:2003:8] MS-SQL Worm propagation attempt [Classification: Misc Attack] [Priority: 2]: UDP 61.185.28.41:1067 -> 11.11.79.89:1434

- Date: Feb 25 12:23:54.
- Host: bastion.
- GID: 1.
- SID: 2003.

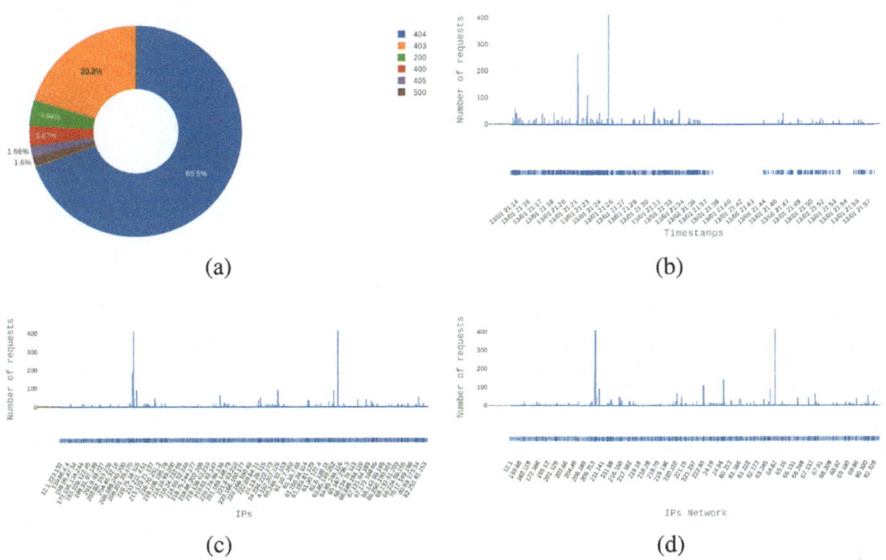

Fig. 7 Data visualization: **a** number of requests by status code: **b** distribution of requests by time; **c** distribution of requests by client IP; **d** distribution of requests by IP network

- RID: 8.
- Msg: MS-SQL Worm propagation attempt.
- Class: Misc Attack.
- Priority: 2.
- PROTO: UDP.
- SOURCE IP PORT: 61.185.28.41:1067.
- DEST IP PORT: 11.11.79.89:1434 (Fig. 7).

Appendix 2: Data Class Distribution

Fig. 8 Class distribution of features—part 1

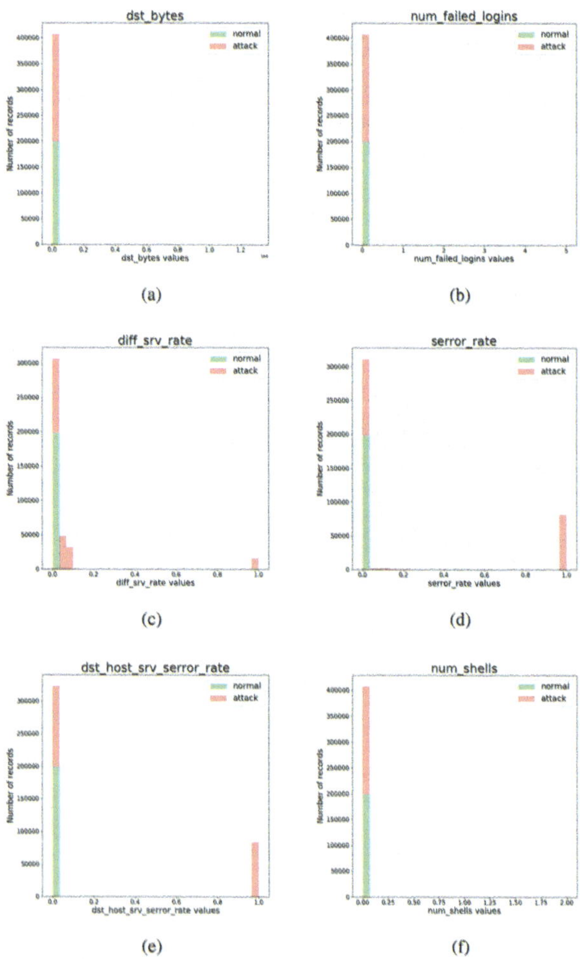

Fig. 9 Class distribution of features—part 2

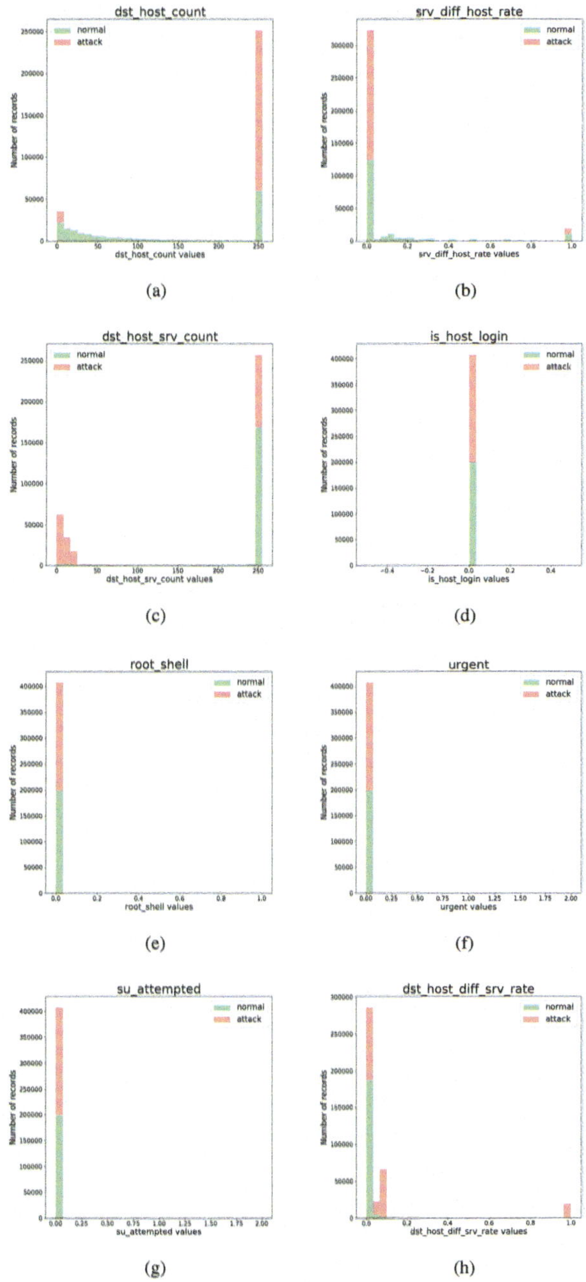

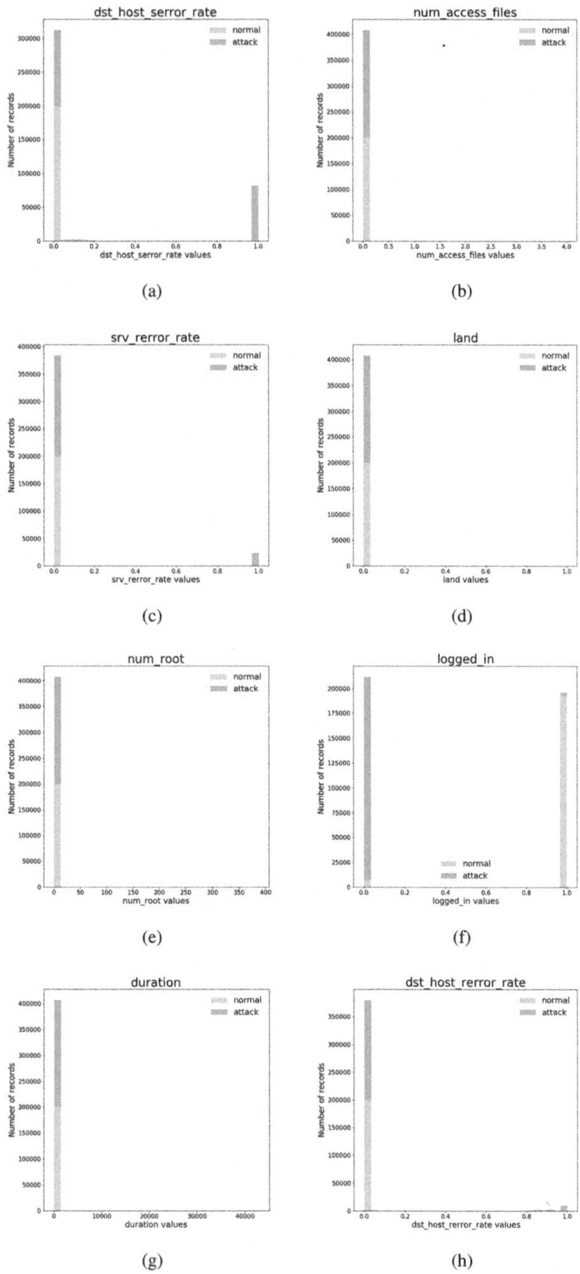

Fig. 10 Class distribution of features—part 3

Fig. 11 Class distribution of features—part 4

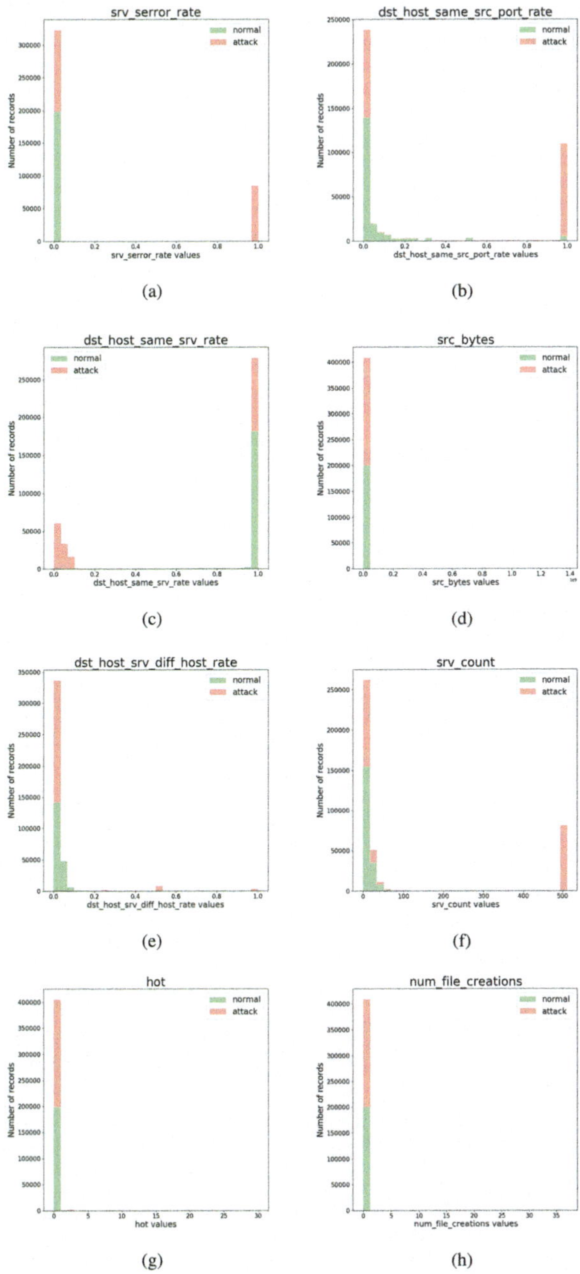

Fig. 12 Class distribution of features—part 5

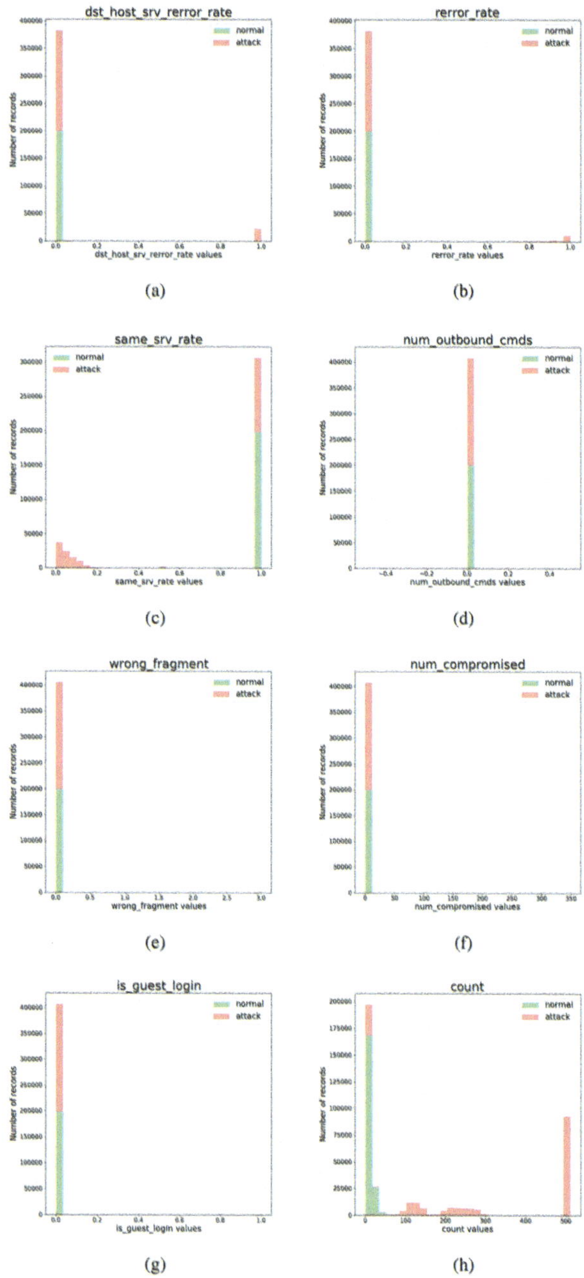

References

1. Portnoy, L.: Intrusion detection with unlabeled data using clustering. Ph.D. dissertation, Columbia University (2000)
2. Laskov, P., Düssel, P., Schäfer, C., Rieck K.: Learning intrusion detection: supervised or unsupervised? In: International Conference on Image Analysis and Processing, pp. 50–57. Springer (2005)
3. Yen, T.-F., Oprea, A., Onarlioglu, K., Leetham, T., Robertson, W., Juels, A., Kirda, E.L Beehive: large-scale log analysis for detecting suspicious activity in enterprise networks. In: Proceedings of the 29th Annual Computer Security Applications Conference, pp. 199–208. ACM (2013)
4. Stroeh, K., Madeira, E.R.M., Goldenstein, S.K.: An approach to the correlation of security events based on machine learning techniques. J. Internet Serv. Appl. **4**(1), 7 (2013)
5. Li, W.: Automatic log analysis using machine learning: awesome automatic log analysis version 2.0 (2013)
6. Vasquez Villano, E.G.: Classification of logs using machine learning technique. Master's thesis, NTNU (2018)
7. Vigneswaran, K.R., Vinayakumar, R., Soman, K., Poornachandran, P.: Evaluating shallow and deep neural networks for network intrusion detection systems in cyber security. In: 2018 9th International Conference on Computing, Communication and Networking Technologies (ICCCNT), pp. 1–6. IEEE (2018)
8. Dionísio, N., Alves, F., Ferreira, P.M., Bessani, A.: Cyberthreat detection from twitter using deep neural networks. arXiv preprint arXiv:1904.01127 (2019)
9. Deokar, B., Hazarnis, A.: Intrusion detection system using log files and reinforcement learning. Int. J. Comput. Appl. **45**(19), 28–35 (2012)
10. Bulavas, V.: Investigation of network intrusion detection using data visualization methods. In: 2018 59th International Scientific Conference on Information Technology and Management Science of Riga Technical University (ITMS), pp. 1–6. IEEE (2018)
11. Rampure, V., Tiwari, A.: A rough set based feature selection on KDD Cup 99 data set. Int. J. Database Theory Appl. **8**(1), 149–156 (2015)
12. Bozhkov, L., Georgieva, P.: Brain neural data analysis with feature space defined by descriptive statistics. In: Iberian Conference on Pattern Recognition and Image Analysis, pp. 415–422. Springer (2015)
13. Tucker, L.R., MacCallum, R.C.: Exploratory factor analysis. Unpublished manuscript, Ohio State University, Columbus (1997)
14. Stolfo, J., Fan, W., Lee, W., Prodromidis, A., Chan, P.K.: Cost-based modeling and evaluation for data mining with application to fraud and intrusion detection. Results from the JAM Project by Salvatore, pp. 1–15 (2000)

Generalized Net Model for Collecting, Evaluating and Including of Facts in the Educational Content

Krassimir Atanassov, Anthony Shannon, Evdokia Sotirova, Valentin Vasilev, and Sotir Sotirov

Abstract The research expounded in this paper is a continuation of previous investigations into the modelling of information flow with a typical university. A generalized net model which describes the process of creating and evaluation of educational content using a set of criteria is constructed. The example for education of medical students in diabetes mellitus and possibilities for venous thrombotic disease is given.

Keywords Generalized nets · University · Educational content

1 Introduction

In a series of papers, the authors have described some of the most important processes of functioning of universities and educational processes [11–20]. For modelling the theory of Generalized Nets (GNs, see [1, 2, 4]) is used. For assessment the modelling processes the intuitionistic fuzzy evaluations according theory of Intuitionistic Fuzzy Sets (IFSs, [3]) are proposed.

K. Atanassov
Department of Bioinformatics and Mathematical Modelling, Institute of Biophysics and Biomedical Engineering, Bulgarian Academy of Sciences, Sofia, Bulgaria

K. Atanassov · S. Sotirov (✉)
Intelligent System Laboratory, Prof. Assen Zlatarov University, 1 Prof. Yakimov str., 8010 Burgas, Bulgaria
e-mail: ssotirov@btu.bg

A. Shannon
Warrane College, The University of New South Wale, Kensington, NSW, Australia
e-mail: t.shannon@warrane.unsw.edu.au

E. Sotirova · V. Vasilev
Faculty of Public Health and Health Care, Prof. Assen Zlatarov University, 1 Prof. Yakimov str., 8010 Burgas, Bulgaria
e-mail: esotirova@btu.bg

V. Vasilev
e-mail: vailvasilev@mail.bg

V. Sgurev et al. (eds.), *Advances in Intelligent Systems Research and Innovation*, Studies in Systems, Decision and Control 379,
https://doi.org/10.1007/978-3-030-78124-8_15

The purpose of the present paper is to offer GN model of the process of creating, maintaining and updating of the educational content within university courses. A different GN models of the of learning and e-learning processes within university educational environment were constructed: GN model for evaluation of the problems solved by students [9, 14], GN model of the process of evaluation by lecturers [16], GN model of the evaluation of the lecturers [17], and educational course [18], etc. The educational content contains of different training materials and information resources [6, 10, 11]. In the presented here GN model there are several databases (DB) with facts: DB of sure facts; DB of possible facts; DB of useful facts, that can be used in the educational content; DB of current educational content; and DB of current concepts for the educational content. The new facts that enter the model, extend the respectively database. The fact can be from reliable or from not reliable source as well as the facts can be sure fact or not sure fact. It can coincide with an existing fact, or the fact can be in contradiction with an existing fact, or the fact can be unknown.

The paper is based on our lecture [19], but here, we modified and corrected it. The paper is organized as follows. Section 2 presents the constructed GN model. The example for education of medical students in diabetes mellitus and possibilities for venous thrombotic disease is given. Section 3 presents our conclusions and future work.

2 A Generalized Net Model

This GN model for the creation and evaluation of educational content contains 7 transitions and 19 places (see Fig. 1).

Initially the following tokens remain, respectively, in places l_1, l_{10}, l_{13}, l_{16} and l_{19} with initial characteristics:

- sequence of tokens (φ_1, φ_2,...,) that for brevity will denote only as an φ-token which enters place l_1 with initial characteristic

  "a new fact",

- δ_s-token that stays permanently in place l_{10} with initial and current characteristic

  "DB of sure facts",

- δ_p-token that stays permanently in place l_{12} with initial and current characteristic

  "DB of possible facts",

- δ_u a token that stays permanently in place l_{13} with initial and current characteristic

  "DB of useful facts that can be used in the educational content",

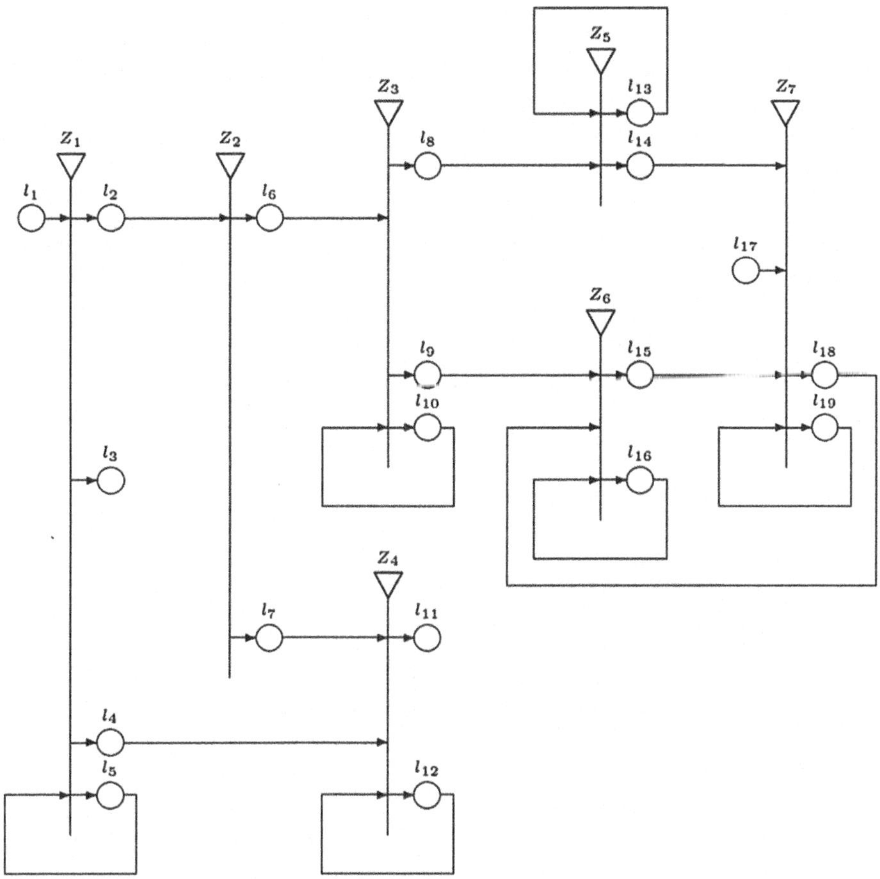

Fig. 1 GN model of a creation and evaluation of educational content

- δ_l a token that stays permanently in place l_{16} with initial and current characteristic

  "DB of current educational content",

- δ_c a token that stays permanently in place l_{19} with initial and current characteristic

  "DB of current concepts for the educational content".

$$Z_1 = \langle \{l_1, l_5\}, \{l_2, l_3, l_4, l_5\}, \begin{array}{c|cccc} & l_2 & l_3 & l_4 & l_5 \\ \hline l_1 & false & false & false & true \\ l_5 & W_{5,2} & W_{5,3} & W_{5,4} & false \end{array} \rangle,$$

where,

- $W_{5,2}$ = "the new fact is in a contradiction with an existing fact in the DB of confirmed facts, DB of possible facts or DB of useful facts",
- $W_{5,3}$ = "the new fact coincides with an existing fact in the DB of confirmed facts, DB of possible facts or DB of useful facts",
- $W_{5,4}$ = "the new fact is unknown".

Token φ enters place l_2, l_3 or l_4 without a new characteristic and it enters place l_5 with the characteristic

"a status of the new fact (it is in a contradiction with another fact/it coincides with another fact/it is unknown").

$$Z_2 = \langle \{l_2\}, \{l_6, l_7\}, \begin{array}{c|cc} & l_6 & l_7 \\ \hline l_2 & W_{2,6} & W_{2,7} \end{array} \rangle,$$

where,

- $W_{2,6}$ = "the source of the fact is reliable",
- $W_{2,7} = \neg W_{2,6}$.

in which ¬P is the negation of predicate P.

Token φ enters place l_6 or l_7 without a new characteristic.

$$Z_3 = \langle \{l_6, l_{10}\}, \{l_8, l_9, l_{10}\}, \begin{array}{c|ccc} & l_8 & l_9 & l_{10} \\ \hline l_6 & false & false & true \\ l_{10} & W_{10\,8} & W_{10\,9} & true \end{array} \rangle,$$

where,

- $W_{10,8}$ = "the new fact is suitable for the current educational content",
- $W_{10,9}$ = "the new fact is suitable for generating of concepts for an educational content".

Token φ from place l_6 enters place l_{10} where it is merged with token δ_s and token δ_s extends its previous characteristic with the new fact.

When some among the predicates $W_{10,8}$ or $W_{10,9}$ is valid, token δ_s splits to the same token δ_s which continues to stay in place l_{10} without a new characteristic and to one or two φ-tokens with respect to the number of valid predicates. The respective φ-token enters place l_9 (if predicate $W_{10,8}$ is true) and/or place l_9 (if predicate $W_{10,9}$ is true) without a new characteristic.

$$Z_4 = \langle \{l_4, l_7, l_{12}\}, \{l_{11}, l_{12}\}, \begin{array}{c|cc} & l_{11} & l_{12} \\ \hline l_4 & W_{4,11} & W_{4,12} \\ l_7 & W_{7,11} & W_{7,12} \\ l_{12} & false & true \end{array} \rangle,$$

where,

- $W_{4,11} = W_{7,11} =$ "in δ_p there is an older version of the new fact that already was rejected",
- $W_{4,12} = W_{7,12} = \neg W_{4,11}$.

When some one of the predicates $W_{4,11}$ or $W_{7,11}$ is valid, token φ enters place l_{11} without a new characteristic.

When some one of the predicates $W_{4,12}$ or $W_{7,12}$ is valid, token φ enters place l_{12} where it is merged with token δ_p and token δ_p extends its previous characteristic with the new fact.

$$Z_5 = \langle \{l_8, l_{13}\}, \{l_{13}, l_{14}\}, \begin{array}{c|cc} & l_{13} & l_{14} \\ \hline l_8 & true & false \\ l_{13} & true & W_{13,14} \end{array} \rangle,$$

where,

- $W_{13,14} =$ "the new fact is suitable for the current concepts for the educational content".

Token φ from place l_8 enters place l_{13} where it is merged with token δ_u and token δ_u extends its previous characteristic with the new fact.

When the predicate $W_{13,14}$ is valid, token δ_u splits to the same token δ_u which continues to stay in place l_{13} without a new characteristic and to one φ-token that enters place l_{14} without a new characteristic.

$$Z_6 = \langle \{l_9, l_{16}, l_{18}\}, \{l_{15}, l_{16}\}, \begin{array}{c|cc} & l_{15} & l_{16} \\ \hline l_9 & false & true \\ l_{16} & W_{16,15} & true \\ l_{18} & false & true \end{array} \rangle,$$

where,

- $W_{16,15} =$ "there is need for information about the current educational content".

Token φ from place l_9 enters place l_{16} where it is merged with token δ_l and token δ_l extends its previous characteristic with the new fact.

Token τ from place l_{18} enters place l_{16} where it is merged with token δ_l and token δ_l does not obtain a new characteristic, but on the next time-step, it is split into two tokens—the same token δ_l which continues to stay in place l_{16} without a new characteristic and to one τ-token which enters place l_{15} with the characteristic.

"the need from information about the current educational content".

The γ-token with criteria for evaluation the concepts enters the net via place l_{17}. This token has initial characteristic.

"Criterion or new criteria system".

$$Z_7 = \langle \{l_{14}, l_{15}, l_{17}, l_{19}\}, \{l_{18}, l_{19}\}, \begin{array}{c|cc} & l_{18} & l_{19} \\ \hline l_{14} & \mathit{false} & \mathit{true} \\ l_{15} & \mathit{false} & \mathit{true} \\ l_{17} & \mathit{false} & \mathit{true} \\ l_{19} & W_{19,18} & \mathit{true} \end{array} \rangle,$$

where,

- $W_{19,18}$ = "there is a new concept for the educational content".

The φ-tokens from place l_{14}, the τ-token from place l_{15}, and the γ-token from place l_{17} enter place l_{19} where they are merged with token δ_c and the token δ_c extends its previous characteristic with the current characteristics of these tokens.

The token δ_c is split into two tokens—the same token δ_c which continues to stay in place l_{19} without a new characteristic and to one τ-token which enters place l_{18} with the characteristic.

"a new concept for the educational content".

The presented GN model can be used for collecting information for education of medical students in diabetes mellitus and possibilities for venous thrombotic disease. In the educational sources, there is facts in a contradiction that diabetes is a separate risk factor for venous thrombotic disease [5, 8]. Diabetes mellitus is not an independent risk factor for venous thromboembolic disease. Patients with diabetes are often hospitalized for inpatient treatment or surgery, which is a risk factor for venous thromboembolic disease. If age, sex, age and complications of diabetes are taken into account, diabetes can be considered a risk factor for a venous thrombotic disease. If hospitalization for high-volume surgery, inpatient treatment or bed rest is also taken into account, diabetes mellitus and its complications are not separate risk factors for venous thromboembolic disease [7]. In patients with diabetes mellitus, the incidence of venous thrombotic events is relatively higher, and it is not clear whether this is due to diabetes mellitus and, consequently, elevated plasma glucose levels, or to risk factors for diabetes mellitus and its complications [11]. There are facts in contraversion on the correlation between diabetes mellitus and possibilities for venous thrombotic disease. In the university educational courses, the process needs to collect, update and confirm the information. For the collecting new facts for diabetes mellitus and venous thrombotic disease a closed-answer will be created and patients from University Hospital in Burgas will be asked to fill out a questionnaire.

3 Conclusions

In the paper a GN model that can be used in educational process for creating and renewal of educational content is proposed. The collecting the information about diabetes mellitus and venous thromboembolic disease parameters, as well as qualitative observations can be used for:

- creating the DBs (of sure/possible/useful facts, that can be used in the educational content/facts for current educational content/current concepts for the educational content);
- renewing the DBs;
- collecting medical information for scientific investigations;

for training of students.

Gender group data, for marital status data and by years with collected statistical data for registered patients with oncological diseases in Burgas for 2014–2018.

Acknowledgements This research was funded in part by the European Regional Development Fund through the Operational Program "Science and Education for Smart Growth" under contract UNITe No BG05M2OP001-1.001-0004 (2018-2023).

References

1. Atanassov, K.: Generalized Nets. World Scientific, Singapore (1991)
2. Atanassov, K.: On Generalized Nets Theory. Prof. M. Drinov Publishing House, Sofia, Bulgaria (2007)
3. Atanassov, K.: Intuitionistic Fuzzy Sets. Springer Physica-Verlag, Berlin (1999)
4. Atanassov, K.: Index Matrices: Towards an Augmented Matrix Calculus. Springer, Cham (2014)
5. Bell, E.J., Folsom, A.R., Lutsey, P.L., et al.: Diabetes mellitus and venous thromboembolism: A systematic review and meta-analysis. Diabetes Res. Clin. Pract. **111**, 10–18 (2016). https://doi.org/10.1016/j.diabres.2015.10.019
6. Bakardjieva, T., Gercheva, G.: Knowledge management and e-learning—An agent-based approach. World Acad. Sci. Eng. Technol. **76**, 663–666 (2011)
7. Heit, J.A., Leibson, C.L., Ashrani, A.A., Petterson, T.M., Bailey, K.R., Melton, L.J., III.: Is diabetes mellitus an independent risk factor for venous thromboembolism? A population-based case-control study. Arterioscler. Thromb. Vasc. Biol. **29**(9), 1399–1405 (2009)
8. Kensington-Miller, B., Novak, J., Evans, T.: Just do it: flipped lecture, determinants and debate. Int. J. Math. Educ. Sci. Technol. **47**(6), 853–862 (2016). Padberg Jr, F., Cerveira, J.J., Lal, B.K., Pappas, P.J., Varma, S., Hobson, R.W., II: Does severe venous insufficiency have a different etiology in the morbidly obese? Is it venous? J. Vasc. Surg. **37**(1), 79–85 (2003)
9. Melo-Pinto, P., Kim, T., Atanassov, K., Sotirova, E., Shannon, A., Krawczak, M. (2005). Generalized net model of e-learning evaluation with intuitionistic fuzzy estimations. In: Issues in the Representation and Processing of Uncertain and Imprecise Information, Warszawa, pp. 241–249 (2005)
10. Nedyalkova, A., Bakardjieva, T., Nedyalkov, Kr.: Social media tools on the eve of e-learning 3.0. Glob. J. Comput. Sci. Technol. **13**(6), 19–23 (2013). ISSN 0975-4172

11. Petrauskiene, V., Falk, M., Waernbaum, I., Norberg, M., Eriksson, J.W.: The risk of venous thromboembolism is markedly elevated in patients with diabetes. Diabetologia **48**(5), 1017–1021 (2005)
12. Shannon, A., Langova-Orozova, D., Sotirova, E., Petrounias, I., Atanassov, K., Krawczak, M., Melo-Pinto, P., Kim, T.: Generalized Net Modelling of University Processes. KvB Visual Concepts Pty Ltd, Monograph No. 7, Sydney (2005)
13. Shannon, A., Atanassov, K., Sotirova, E., Langova-Orozova, D., Krawczak, M., Me-lo-Pinto, P., Petrounias, I., Kim, T.: Generalized Nets and Information Flow Within a University, Warszawa (2007)
14. Shannon, A., Sotirova, E., Petrounias, I., Atanassov, K., Krawczak, M., Melo-Pinto, P., Kim, T.: Intuitionistic fuzzy estimations of lecturers' evaluation of student work. In: First International Workshop on Intuitionistic Fuzzy Sets, Generalized Nets & Knowledge Engineering, pp. 44–47. University of Westminster, London, 6–7 Sept 2006
15. Shannon, A., Sotirova, E., Petrounias, I., Atanassov, K., Krawczak, M., Melo-Pinto, P., Kim, T.: Generalized net model of lecturers' evaluation of student work with intuitionistic fuzzy estimations. In: Second International Workshop on Intuitionistic Fuzzy Sets, Banska Bystrica, Slovakia, 3 Dec 2006, Notes on IFS, vol. 12, no. 4, pp. 22–28 (2006)
16. Shannon, A., Sotirova, E., Atanassov, K., Krawczak, M., Melo-Pinto, P., Kim, T.: Generalized Net Model for the Reliability and Standardization of Assessments of Student Problem Solving with Intuitionistic Fuzzy Estimations, Developments in Fuzzy Sets, Generalized Nets and Related Topics. Applications, vol. 2, pp. 249–256. System Research Institute, Polish Academy of Science (2008)
17. Shannon, A., Dimitrakiev, D., Sotirova, E., Krawczak, M., Kim, T.: Towards a Model of the Digital University: Generalized Net Model of a Lecturer's Evaluation with Intuitionistic Fuzzy Estimations, Cybernetics and Information Technologies, vol. 9, no. 2, pp. 69–78. Bulgarian Academy of Sciences (2009)
18. Shannon, A., Sotirova, E., Hristova, M., Kim, T.: Generalized net model of a student's course evaluation with intuitionistic fuzzy estimations in a digital university. In: Proceedings of the Jangjeon Mathematical Society, vol. 13, no. 1, pp. 31–38 (2010)
19. Shannon, A., Sotirova, E., Atanassov, K., Vasilev, V.: Generalized net model for creating and evaluating of educational content. In: Proceedings of 2020 IEEE 10th International Conference on Intelligent Systems, pp. 517–520. Varna, 28–30 Aug 2020
20. Sotirova, E., Atanassov, K., Shannon, A., Krawczak, M., Melo-Pinto, P., Riečan, B., Kim, T.: Intuitionistic fuzzy evaluations for analysis of a student's knowledge of mathematics in the university e-learning courses. In: 2016 IEEE 8th International Conference on Intelligent Systems, pp. 535–537 (2016)

Virtualization of Things in a Smart Agriculture Space

Stanimir Stoyanov, Todorka Glushkova, Ivan Popchev, and Lyubka Doukovska

Abstract This article discusses an adaptation of ViPS (Virtual Physical Space) to build an intelligent infrastructure for smart agriculture. ViPS is a reference architecture aiming to support development of applications where AI enhanced with IoT (Internet of Things) and CPSS (Cyber-Physical-Social System) technologies are first class citizens. The main challenge is the virtualization of "things" from the physical world, as well as related events, time and spatial aspects. In this paper, two modeling approaches are presented that are used to virtualize the physical "things" we are interested in. To present the spatial aspects of things we use an Ambient-Oriented Modeling approach. By adapting of ambient-oriented interpreter, we created an animated simulator for modelling of some scenarios implementing basic services in the space. The second approach uses intelligent agents with an environment presented as artifacts. Both modeling approaches are demonstrated with a realistic example. Currently, a prototype implementation of the smart agriculture space is conducted in the region of Plovdiv.

Keywords Virtual physical space (ViPS) · Virtualization of things · Modelling · Ambient intelligence · Intelligent agents · Artefacts · Intelligent agriculture · Cyber-physical-social system · Internet of things (IoT)

S. Stoyanov · T. Glushkova (✉)
Faculty of Mathematics and Informatics, University of Plovdiv "Paisii Hilendarski", 236 Bulgaria Blvd., Plovdiv 4027, Bulgaria
e-mail: glushkova@uni-plovdiv.bg

S. Stoyanov
e-mail: stani@uni-plovdiv.bg

S. Stoyanov · I. Popchev · L. Doukovska
Intelligent Systems Department, Institute of Information and Communication Technologies, Bulgarian Academy of Sciences, Acad. Georgi Bonchev Str., Sofia 1113, Bulgaria
e-mail: ipopchev@iit.bas.bg

L. Doukovska
e-mail: lyubka.doukovska@iict.bas.bg

V. Sgurev et al. (eds.), *Advances in Intelligent Systems Research and Innovation*, Studies in Systems, Decision and Control 379,
https://doi.org/10.1007/978-3-030-78124-8_16

1 Introduction

The integration of the virtual and physical worlds opens up new interesting opportunities for developing intelligent applications in many application areas. This integration is a shared concept for modern technologies as IoT, Cyber-Physical Systems (CPS), and CPSS. A CPSS-like application, putting the user in the focus of events, stimulates the creation of distributed, autonomous, contex-aware, and intelligent software.

In this paper, we briefly describe a reference architecture known as a Virtual Physical Space to support the creation of CPSS-like applications. The main challenge is the virtualization of "things" from the physical world, as well as related events, time and spatial aspects. In the ViPS, the virtualization of physical objects ("things") is supported by combined modeling approaches. Two such approaches are presented in more detail in this article.

ViPS does not appear suddenly—it has a 15-year history. The first version known as Distributed eLearning Center (DeLC) has been used for years to support e-learning at the Faculty of Mathematics and Informatics at the University of Plovdiv has been used for years [1]. Although a successful project, the lack of integration between its virtual environment with the physical world caused problems, especially for disabled learners. For this reason, DeLC has been reengineered into a Virtual Education Space (VES) that operates as an IoT ecosystem [2]. Summing up the experience of constructing VES, we began developing the ViPS architecture [3] that could be adapted to various CPSS-like applications such as a smart seaside city [4] or a smart tourist guide [5]. Implementing a CPSS application, a major challenge is the virtualization of "things" from the physical world, which are of interest to us. Moreover, account has to be taken of related events, time and spatial aspects. To present the spatial aspects of "things", we have decided to use an ambient-oriented modeling approach [6].

Recently, we started an adaptation of ViPS to build an intelligent infrastructure for agriculture in the region of Plovdiv. In this paper, we intend to present briefly the ViPS architecture and to demonstrate two modelling approaches aiming to support the virtualization in a smart agriculture space. The rest of the paper is organized as follows: a short review of related works is presented in Sect. 2, which is followed by an over-all description of the ViPS architecture in Sect. 3. Section 4 presents the ViPS's adaptation for the agriculture domain known as Smart Agriculture Space. In Sect. 5, the virtualization in the Smart Agriculture Space by an ambient-oriented and an agent-artifact modeling is discussed. A demonstration scenario is shown in Sect. 6. Finally, Sect. 7 concludes the paper.

2 Related Works

Agriculture is becoming an increasingly relevant area for conducting research and practical developments in the field of CPSS, CPS and IoT. These three relatively new and interconnected technologies provide a variety of unthinkable opportunities to build infrastructures for intelligent and precise farming. As water supplies become scarce due to climate change, there is an urgent need to irrigate more efficiently in order to optimize water use. Effective pest control and, at the same time, minimizing the risk of environmental pollution by pesticides are the other major challenges of smart farming. In the same time, agriculture usually has a massive number of crops and machinery growing and operating in wide geographic areas in which they are located. They are subject to operational factors often beyond their control, such as the amount of sunlight or rainfall they receive, or the temperature fluctuations. In this context, agriculture is ripe for the adoption of new technologies to help monitor and manage assets on a granular level and the development of intelligent CPSS- and IoT-based agriculture infrastructures would be of great benefit [7]. In the specialized literature, a great number of examples of such solutions and systems can be found; in this section we cite briefly only some of them.

In [8], an agricultural cyber-physical-social system integrating three worlds that in turn are mapped to the cyber space and physical space is presented. The physical world consisting of crops, their environment and control devices, is mapped into cyberspace through various IoT techniques. The mental world includes a wide variety of data in the form of producers' experience and knowledge, customer preferences, timing, quantity and application of fertilizers. In the artificial world of the system, crop models are maintained and decisions are made that through the control devices affect the physical world. The functionality of the proposed architecture is demonstrated with a case study on the solar greenhouse. In [9] architecture for monitoring soil humidity and temperature in small farms divided in regions is described. Using wireless sensing networks and IoT methods the proposed system monitors soil moisture, humidity and temperature in the relevant areas and if needed it sends reports to the end users. The farmers can more effectively plan crop cultivation, harvesting, irrigation and fertilization by help of reports' information. In order to assists precision agriculture, an infrastructure consisting of five nodes with one or more sensors is proposed in [10]. The data acquisition from various sensors allows monitoring of climate elements such as soil moisture, air quality, temperature and humidity, precipitation intensity, precipitation level, wind speed and direction, light flux and atmospheric pressure. In addition, a web page has been developed where the user can view the climatic conditions of cultivation or afforestation. Another architecture and an application prototype for precision farming using a wireless sensor network with an IoT cloud is proposed in [11].

A useful review of current research and research projects in the field of big data to analyze and solve various significant issues in agriculture is presented in [12]. The review highlights the large capabilities of large data for intelligent and precision agriculture. Furthermore, it shows that the availability of hardware and software, techniques and methods for analyzing large data as well as the increasing openness of data sources will encourage further academic research, public sector initiatives and business initiatives in the agricultural sector.

3 Brief Overview of ViPS

With the creation of ViPS we aim for a genetic infrastructure that can be used to develop systems with CPSS-like behavior for different application areas. The main challenge to be taken into account is the integration of the virtual and physical worlds in both directions—from the physical world to the virtual and vice versa. The first bottom-up direction addresses the virtualization of physical objects ("things"). The top-down direction aims to "locate" the system intelligence as close as possible to the physical world. Special attention is paid to the support of various groups of users who will work with applications created with the help of adapted versions of the genetic infrastructure.

The IoT technology is built on the concept of "smart" objects that are able to autonomously and proactively collect, store and transmit data. According to this concept, ViPS can receive data from the physical world, as well as share knowledge, plan and make decisions in the virtual world.

An agent-oriented approach is applied to build the genetic infrastructure. The fundamental components of ViPS are implemented as rational practical reasoning agents. Personal assistants will assist users in working with domain-specific adaptations of ViPS.

Our idea is ViPS to operate as a space; i.e. the active components are autonomous and proactive and they interact through messages rather than through strictly defined interfaces. However, we can consider the individual components as located in three functional layers: ViPS User Interface, ViPS Middleware, and ViPS Physical World.

Usually, due to the inherent complexity it is hard for users to work in a heterogeneous space such as a ViPS-like application. For this reason, personal assistants operating on the behalf of the users are aware of their needs to select or prepare appropriate scenarios for the execution of users' requests. Furthermore, personal assistants can manage and control scenario's execution interacting with the ViPS middleware. Besides, the personal assistants exercise control of personal access in a ViPS-like space (Fig. 1).

The work of the personal assistants is closely related to the Event Engine. The personal assistant's lifecycle is supported by a user's personal timetable. The personal timetable presents the activities that the user has to perform in a certain period of time. The entries on the timetable are presented as domain-specific events correspond to a suitable activity. Furthermore, the events can be used for a specification of policies that in turn present space operative scenarios.

The personal assistants are implemented as BDI rational intelligent agents [13]. The BDI lifecycle incorporates two steps known as deliberation and means-ends reasoning. During the deliberation, a personal assistant generates its current goal while in the next step (means-ends reasoning) a plan for the goal's achievement is prepared. Usually, the plan is launched when a corresponding event occurs, i.e. by means of the Event Engine.

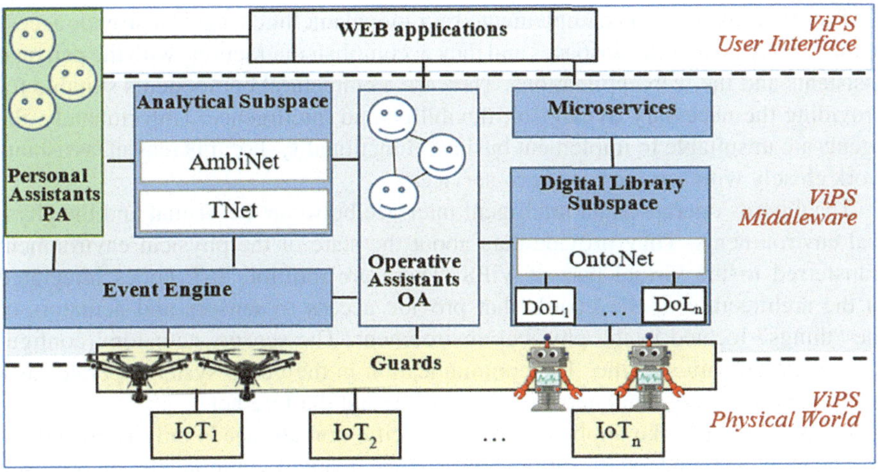

Fig. 1 Architecture of ViPS. © 2018 IEEE [3]

The personal assistants also have to be able to operate in an "overtaking action" mode, i.e. as an early warning system. This way, certain operations must be performed before the event occurs. Therefore, the personal assistants cannot rely on the Event Engine in the timeframe before the event occurs. Then, they are supported by the AmbiNet and TNet components. The interaction between the personal assistants and the Analytical Subspace is implemented by specialized interaction protocols.

Additionally, the user interface can be supplemented with a web application. The web application could offer public information resources of the space as well.

In the middle layer of the architecture are located several components extremely essential for the operation of ViPS. Practically, the virtualization of "things" is supported by this level. The middleware takes into account time and spatial aspects of the "things" to be virtualized as well the events binding "things" with user's activities. In this way, the Analytical Subspace provides means for the preparation of a domain-specific analysis based on the following formalisms and the supported tools:

- Interval Temporal Logics [14] and TNet (Temporal Net)—provide an opportunity to present and work with the temporal aspects of things, events and locations;
- Calculus of Context-aware Ambients [15] and AmbiNet—the spatial characteristics of the "things" and the events are modeled as ambients;
- Event Model [16] and Event Engine—model and interpret various types of events and their arguments such as identification, conditions for occurrence and completion.

The components of the Analytical Subspace are supported by the Digital Libraries Subspace operating as a central repository of ViPS. This subspace stores domain-specific background knowledge and documentation. The repository is implemented as a knowledge network including relational data bases and a hierarchy of ontologies known as OntoNet.

The operative assistants, implemented as rational intelligent agents, provide access to the resources of both subspaces and they accomplish interactions with the personal assistants and the web applications. They are architectural components suitable for providing the necessary dynamism, flexibility, and intelligence. Unfortunately, the agents are unsuitable to implement business functionality. For this reason, assistants work closely with services or micro-services.

The Guards operate as an intelligent interface between the virtual and the physical environments. They provide data about the state of the physical environment transferred to the virtual part of ViPS. There are multiple IoT nodes integrated in the architecture of the guards that provide access to sensors and actuators of the "things" located in the physical environment. The sensors-actuators' configurations are domain-specific. The communication in the guard system operates as a combination of a personal network (e.g. LoRa) and the Internet.

Usually, a ViPS-like application for a specific domain needs only some ViPS's components to be adapted. Subsequently, these adapted components are archived in the Domain Libraries. In this way, the genetic architecture can be continuously ended.

4 A Smart Agriculture Space

Currently, we are going to adapt the ViPS to build a smart agriculture infrastructure. The smart agriculture aims to help address three major challenges: an efficient use of water for irrigation, an effective pest control and, at the same time, minimizing the risk of environmental pollution by pesticides, and an early detection of plant diseases. The space (Fig. 2) consists of the following four components: Operative Center, DiLib "Smart Agriculture", Operative Assistants, and Guard Network.

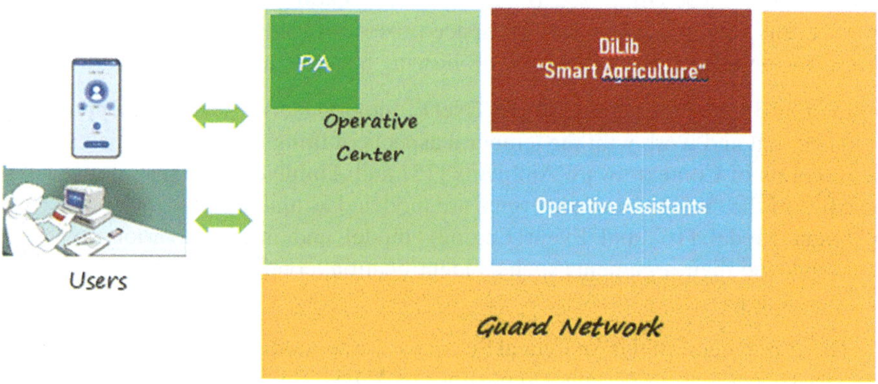

Fig. 2 Smart agriculture space

Operative Center. The Operative Center makes decisions to perform various types of operations, including planning and control. Furthermore, the operative center is responsible for the virtualization of physical "things". The core of the center are the agriculture operators' personal assistants. The personal assistants are the active and proactive components that initiate the overall functioning of the space. Usually, a Cyber-Physical-Social System, in our case a smart agriculture space, is a complex and multifaceted infrastructure. Operators of the smart agriculture space, who usually do not have specialized training, will encounter great difficulties in working with it. For this reason, the main task of the Operative Center is to support farmers easily and intuitively be able to operate in space and in this way to become fully integrated in the space.

A personal assistant (PA) is being developed as a multi-agent system. Significant problems to be solved are how the environment of the individual agents will be presented and how they will interact through it. Usually, the environment of a PA consists of active components (other assistants) and passive components. The interaction with the active ingredients is typically implemented by means of the ACL [17]. The passive components of the environment are a bigger challenge. Theoretically, the environment can be considered as part of the beliefs of agents. However, from a technological and program-technical point of view, this assumption can become a serious challenge when the environment is complex as in our case. It is desirable that agents' beliefs be implemented and managed as a not very complex system.

DiLib "Smart Agriculture". The DiLib "Smart Agriculture" functions as a centralized repository primarily for storing immutable (or infrequently modifiable) data characterizing physical things, which are interesting for virtualization of smart agriculture. Examples of such things are soils, whole blocks and individual plots, crops, water sources. The repository is implemented as a combination of shared relational databases and ontologies. The access to the information objects of the DiLib "Smart Agriculture" is carried out by the operative assistants.

Operative Assistants. The main responsibility of operational assistants is to support the PA providing it with the necessary data and services needed to make an informed decision. Specifically, for the smart agriculture space, the operative assistants extract data from the DiLib "Smart Agriculture" on request from the PA. Furthermore, the operative assistants are responsible for keeping the DiLib up to date. If necessary, they interact with the guards.

Guard Network. We started building a regional guard network for the region of Plovdiv. Currently, the guard network includes two sensor networks located on the territory of two agricultural institutes The sensor networks cover indoor and outdoor areas. The outdoor network is additionally extended by drones that operate as mobile guards. Drones are mainly used for visual observation. Specialized drones for spraying and pollination are under development. Communication within the Guards Network is primarily conducted in accordance with the LoRa specification. The main task of this component is to collect significant information from the physical space and to transmit it to higher levels of the system (fog and cloud levels). For the convenience of working with the above levels, we have a standardized interface for connection with about 200 types of sensors.

5 Virtualization of the Physical "Things"

As emphasized, supporting of physical things' virtualization is a significant responsibility of the Operative Center. The virtualization requires a complex approach including multifaceted modeling of the things interested in smart agriculture. Here we will present two modeling approaches that support virtualization and interaction in the smart agriculture space:

- Modelling of things as artifacts—artifacts represent the inherent properties (attributes) of things. They form the passive part of the PA's environment.
- Modelling of things as ambients—the spatial aspects of things are modeled. This modeling is also used to prepare plans that are stored in the Plan Library of the PA.

5.1 Modelling of Things as Ambients

AmbiNet is a subspace of ViPS whose task is to pre-model the scenarios from the respective domain. AmbiNet realizes its main functionalities on the basis of a hierarchical structure of "ambients". According to established definitions [18], the ambient is an identity with the following characteristics: restriction, inclusion and mobility. The first characteristic describes the limited space in which the calculations are performed; the second feature represents the possibility of one ambient to be included in another ambient; the third feature represents the ability of an ambient to move from one place to another along with its entire sub-ambient hierarchy. Thus, one ambient is presented through its following elements: identifier or name, corresponding computational processes and a number of sub-ambients that have identifiers, and its own sub-ambient structure.

Ambient Intelligence (AmI) provides the opportunity for the information that users need on a daily basis to become "more sensitive" to their needs, problems and expectations. This is realized by providing continuous context-aware sensory information from the surrounding world and the ability of certain triggers and actuators to respond and change this environment. All these devices are connected via a suitable communication network. The main characteristics of AmI technologies defined in [19] are: sensitivity, adaptability, transparency, versatility and intelligence.

The convergence of AmI and IoT [20] aims to build environments, in which intelligent Web-enabled objects cooperate by exchanging their own data and services, and thus they provide comprehensive services to the users. According to the AmI and IoT vision, the devices are widespread in the environment in the form of sensors, actuators, smart devices, active labeled objects or mobile robots. These devices are interconnected and can be accessed anytime anywhere through a variety of small portable mobile devices. Compared to classic web services, AmI and IoT services deliver not only computing and software resources, but also various physical devices and event control options [21]. Using IoT and AmI technologies, the number of user support services is steadily increasing, and these services are dynamic due to frequent changes occurring in the environment.

Ambients can model various physical or virtual objects in a unified way. The network of ambients can be parameterized and adjusted to the dynamically changing physical or virtual environment through the data that is constantly coming from IoT nodes. All this gives us a reason to use Ambient-Oriented Modeling (AOM) for scenarios and processes managed by the user's personal assistant (PA).

Ambients can be static or dynamic, depending on their location and their ability to change that location. Static ambiences have a permanent location in the physical world (ex. fields, Pump Station, etc.). Dynamic ambients have a changing location in the physical world (ex. smart tractors, drones, etc.).

Let A is the set of ambients. An ambient $a =$ < name, location, type, parent, $P(a)$, attr > $\in$ A where: ***name*** is the ambient identifier; ***location***—the location of the ambient in the physical world (for a virtual ambient: location = *null*); ***type*** is the type of the ambient (static, dynamic); ***parent*** is the parental ambient (if none exists, parent = *null*); ***P*(*a*)**—the plurality of the ambient processes; ***attr*** is a set of additional ambient attributes.

If an ambient ***b*** is a child of the ambient ***a***, we denote $a =$ parent (***b***). Let **a, b** $\in$ **A**, we define the following two operations:

- $\boldsymbol{b \downarrow a}$ (in)—ambient ***b*** enters into the boundaries of the ambient ***a*** and becomes its child. Then $a =$ parent(b).
- $\boldsymbol{b \uparrow a}$ (out)—ambient ***b*** goes out of the boundaries of ambient ***a*** and becomes its sibling. Then parent(a) = parent(b).

Let the ambients $a_1, a_2 \in$ A. They can communicate each other by sending and receiving messages as follows: $a_1 \xrightarrow{<message>} a_2$ as $a_2 =$ parent (a_1) or $a_1 =$ parent (a_2) or parent (a_1) = parent (a_2).

Virtual ambients don't have a physical location, so their type may be indefinite. They are mobile in the virtual space, but their physical location could be null. For example, $a_1, a_2, a_3, a_4 \in$ A:

- $a_1 =$ <Orchard, loc_a_1, static, null, $P(a_1)$, attr(a1)>;
- $a_2 =$ <Tractor, loc_a_2, dynamic, a_1, $P(a_2)$, attr(a2)>;
- $a_3 =$ <Drone, loc_a_3, dynamic, a_1, $P(a_3)$, attr(a3)>;
- $a_4 =$ <AmbiNet, null, virtual, null, $P(a_4)$, attr(a4)> .

In the example, the ambient a1 is static, a2 and a3 are dynamic; a1 has no parent in the current context; a2 and a3 are children of a1, i.e. a = parent (a2) = parent (a3).

All ambients can communicate with each other by sending and receiving messages. If ambient a2 wants to send a message to ambient a4, this cannot be done directly as a2 and a4 do not meet the above conditions. Sending and receiving messages can only be done by retransmission from parental ambients in the hierarchy. In this case:

$$a_2 \xrightarrow{<message>} a_1; a_1 \xrightarrow{<message>} a_4.$$

The formal semantics of the Calculus of Context-Aware Ambients (CCA) fully implements the π-calculus modelling [22] and provides better opportunities for context-aware modelling. The Ambients in CCA are defined as an identity that is used to describe an object or a component. Each CCA ambient has a name and a boundary; it may contain other ambients within itself, and it may be included in another one. Between two ambients there are three possible relationships—parent, child, and sibling. Each ambient can communicate with other ambients around it. The process of messaging is performed using the handshaking process. In the CCA notation, when two ambients exchange a message, we use the "::" symbol to describe an interaction between sibling ambients; "↑" and "↓" are symbols for interaction between parent and child; the symbols "< >" means sending, and "()"—receiving a message. The CCA ambient can be mobile. It has two ways to move from one location to another: inwards (in) and outwards (out). In CCA, four syntax categories can be distinguished: processes **P**, opportunities **M**, locations $\boldsymbol{\alpha}$ and context expressions **k**.

The personal assistant PA has a key role in the CPSS-like applications. The PAs interact mainly with the analytical sub-space to prepare a suitable plan. We distinguish two types of plans—a Preventive plan and a Plan for the event. Preparing the Preventative plan typically uses the interface to the physical world, and this determines the need for this plan to be generated using AmbiNet, working in modeling mode, i.e. assisted by AmbiNet planning. Generally, the plan is presented in the form of a sequence of actions that will lead to the goal being achieved in the period before the desired event is started. In many domains, we can look at this plan as a possible route that should ensure achieving the goal. This determines the need to create AmbiNet components to help modeling and to optimize the possible plans—the AmbiNet Route Generator and the AmbiNet Route Optimizer. In addition, already modeled plans can be stored in a specialized repository and used in similar situations from the AmbiNet Repository. In this sense, we can define two approaches to create the AmbiNet subspace: as a structure of individual components, the interactions between which are interpreted together or as a presentation of the individual components as virtual ambients, the interaction of which is interpreted by the existing interpreter.

We chose the second approach, and we decided to extend the existing ccaPL environment with an additional AmbiNet Editor to help simulate the implementation of different scenarios in a user-friendly manner and identify possible critical components or interactions. Synchronizing the work of individual AmbiNet components is assisted by operative assistants with specialized functions.

The main task of AmbiNet Editor is to enable visual modeling of context-relevant scenarios. It can be used to present the necessary ambients, to create the ambient hierarchical structure and to describe the ambient's location, type and attributes. In addition, AmbiNet Editor provides the ability to enter ambient processes, capabilities, and contextual expressions using parameters with dynamic input values from the IoT nodes of the surrounding world. The editor describes the process of sending and receiving messages between ambients according to the current scenario. The AmbiNet ccaPL Interpreter is an official cclPL modeling language interpreter based on CCA calculations. After describing the scenario with AmbiNet Editor, the Interpreter simulates the current scenario and visualizes it on a console (Console) or through a visual animator (Animator). The AmbiNet Route Generator is used to generate routes (plans) based on user-defined criteria. The route is a series of nodes that are modeled as ambients with their current location. Different search algorithms can be used. AmbiNet Route Optimizer aims to optimize the generated routes (plans) and the current state of the participating ambients. The AmbiNet Visualizer is used to present the scenario in the virtual reality. AmbiNet Repository is a library of already completed routes, plans and scenarios.

5.2 *Modelling of Things as Artifacts*

Why did we prefer modeling of physical things as artifacts? In ViPS, the focus is on the user, in this case the agricultural operator represented by the PA. For this reason, how the environment looks like in terms of PA is essential. Fundamentally, various technologies can be used to develop multi-agent systems. It is desirable that the technology chosen promotes the use of first-class abstractions that address three basic dimensions that characterize a multi-agent system, namely: the agents themselves, the agent environment, and the organization of agents as an integrated system. A key aspect is that all of these dimensions are maintained by the platform chosen. In this way, the implemented multi-agent system using such means can be dynamically changed and reorganized in different ways. Usually, for each specific application we consider some structure of the environment, which is highly dependent on the specifics of the problem to be solved. It would be good to be able to develop a genetic structure of the environment that can be parameterized or adapted to the desired domain.

According to our experience, we have chosen the JaCaMo framework [23] to develop the agents' environment as artifacts. JaCaMo integrates three environments:

- Jason [24]—a development environment for creating rational agents with BDI architecture.
- Cartago [25]—a platform that provides a general-purpose programming model for building shared working environments that agents can use to collaborate on a multi-agent system. Based on the A&A (agents and artifacts) conceptual model, CArtAgO work environments are modeled and implemented as a set of artifacts [26].

- Moise [27]—an organizational model for multi-agent systems based on concepts such as roles, groups and missions. Moise allows the organization of a multi-agent system to be specified explicitly. In this way, a systematized organization in a multi-agent system can be specified.

A Jason agent owns a library where pre-prepared plans are stored. Each plan is presented as a structure consisting of three components: triggering event, plan context, and plan body. In accordance with the life cycle of the Jason agent, the selection of a plan suitable for a particular situation is made in two steps:

- First, a subset of relevant plans is determined from the set of all plans—the choise of relevant plans is conducted by triggering events. To this end, the PA updates its event queue through up-to-date perceptions or through the processing of messages received from other agents in the shared environment. Only those plans whose triggering event unifies with the currently selected event are selected.
- Second, an applicable plan is selected from the set of relevant plans—the choice of applicable plans is managed by context of the plans. It is checked whether the context of a plan is a logical consequence of the actual assistant's beliefs base.

In JaCaMo, we are able to introduce an abstract layer, which makes it possible to encapsulate those services and resources that are not modeled as cognitive agents, i.e. are not active and autonomous components. From the agents' point of view, artifacts are identities representing resources and/or tools from a shared environment that they can dynamically create, share and use to support their individual and collective activities. If the agent abstraction supports an effective modeling of autonomous, proactive, target-driven computing objects, the artifact abstraction aims to an effectively modelling functionally oriented, non-autonomous, computational objects intended to be used (monitored or controlled) by agents to perform their work. The basic features of an artifact are observable properties, signals and feedbacks. When focusing an artifact, its observed properties are directly mapped to the beliefs base of the observer agent. Furthermore, each time a observed property of a focused artifact is updated, an event is generated that is automatically perceived by the agent. The signals are specific messages generated by the artifact that are processed asynchronously by the monitoring agent. The results of successfully completed artifact operations can be provided to the agent as feedbacks.

6 Demonstration Scenario

Let's take a look at the following scenario: a drone flies over an orchard or a vegetable garden, and by recognizing images and interacting with appropriate sensors it determines the need for watering in individual regions of it. In the field there is built an underground irrigation system with multiple aggregates, which allow the water jet to be redirected to different channels, according to the necessities. Static sensors are located in the field to monitor the soil moisture in the different soil layers.

6.1 Ambient-Oriented Modeling in AmbiNet

The information for this scenario is sent to AmbiNet for modeling. The result of AmbiNet's work can be a list of possible irrigation plans, and a proposal for an optimal plan or the information that because of inactivity of certain aggregates it is not possible to create an irrigation plan. A smart tractor with an irrigation system can also be used in modeling.

The ambients that will be involved in modeling the sample scenario are according to their type:

- Static Ambients:
  - Field = <Orchard, loc_Field, static, null, P(Field), attr_Field> ;
  - IrrU = <IrrU1_n, IrrU1_n, loc_IrrU1_n, static, Orchard, P(IrrU1_n), attr_IrrU> —irrigation units in the field.
- Dynamic Ambients:
  - ST = <Tractor, loc_ST, dynamic, Orchard, P(ST), attr_ST>
  - Drone = <Drone, loc_Drone, dynamic, null, P(Drone), attr_Drone> .
- Virtual Ambients:
  - ANet = <AmbiNet, null, virtual, AS, P(ANet), attr_ANet> for describing of AmbiNet sub-space;
  - RG = <RG, null, virtual, ANet, P(RG), attr_RG> for AmbiNet Route Generator;
  - RO = <RO, virtual, null, ANet, P(RO), attr_RO> for AmbiNet Route Optimizer;
  - OC = <OC, null, virtual, null, P(OC), attr_OC> for describing of Operative Center
  - PA = <PA, null, virtual, null, P(PA), attr_PA> for Personal Assistant in the Operative Center.

CCA modeling of the sample scenario considered and the main ambients are presented below. The personal assistant (PA) communicates with the Operative Center (OC) and it requests information on the need to include the irrigation system [see (1)].

$$P(PA) \triangleq (OC \uparrow < needPlan > .0 | OC \uparrow (plan).0) \quad (1)$$

The OC connects to the drone (Drone) that flies over the levels, and by a bidirectional communication with the IoT nodes and their sensing devices, it determines the need for watering each of the individual plots in it. In addition, the Drone collects information on the activity of the individual key aggregates in the irrigation system and sends it to the OC. Then OC activates a bidirectional communication with the AmbiNet environment with a request to generate an irrigation plan [see (2) and (3)].

$$P(OC) \triangleq \begin{pmatrix} !PA \downarrow (needPlan). \\ Dron ::< PA_i, getStatus\_IoT_i, getData\_IoT_i, getStatus\_Tractor > .0| \\ Dron :: (PA_i, Status\_IoT_i, Data\_IoT_i, Status\_Tractor). \\ ANet ::< PA_i, Status\_IoT_i, Data\_IoT_i, Status\_Tractor, loc\_ST > .0| \\ ANet :: (PAi, reply).(has(plan))?PA ::< plan > .0 \end{pmatrix} \tag{2}$$

$$P(Dron) \triangleq \begin{pmatrix} OC :: (PA_i, getStatus\_IoT_i, getData\_IoT_i, getStatus\_Tractor). \\ IrrU ::< needStatus\_IoT_i, needData\_IoT_i > .0| \\ ST ::< needStatus\_Tractor > .0| \\ IrrU :: (Status\_IoT_i, Data\_IoT_i).ST :: (loc\_ST, status\_Tractor). \\ OC ::< PA_i, Status\_IoT_i, Data\_IoT_i, loc\_ST, Status\_Tractor > .0 \end{pmatrix} \tag{3}$$

AmbiNet activates its Route Generation (RG) component in a modeling mode by providing information about the activity and the connectivity of the individual aggregates in the irrigation system. The RG, which has the structure of the static graph of the irrigation system, based on this information, creates a subgraph of the currently active nodes and, through a certain algorithm, it defines a list of separate irrigation plans. This list is transmitted to the Route Optimizer (RO) ambient to divide the optimal plan, which is transmitted back to the OC, with which the modeling stage is completed [see (4–6)].

$$P(ANet) \triangleq \begin{pmatrix} OC :: (PA_i, Status\_IoT_i, Data\_IoT_i, Status\_Tractor, loc\_ST). \\ RG \downarrow < PA_i, Status\_IoT_i, Data\_IoT_i, Status\_Tractor, loc\_ST > .0| \\ RG \downarrow (PA_i, reply).OC ::< PA_i, reply > .0 \end{pmatrix} \tag{4}$$

$$P(RG) \triangleq \begin{pmatrix} ANet \uparrow (PA_i, Status\_IoT_i, Data\_IoT_i, Status\_Tractor, loc\_ST). \\ has(ListPlans)?RO ::< PA_i, ListPlans > .0| \\ RO :: (PA_i, OptimalPlan).ANet \uparrow < PA_i, reply > .0 \end{pmatrix} \tag{5}$$

$$P(RO) \triangleq (RG :: (PA_i, ListPlans).RG ::< PA_i, OptimalPlan > .0) \tag{6}$$

In case of inactivity or lack of interconnection between individual irrigation units (IrrU), the plan includes the Smart Tractor (ST), which can replace the inactive aggregates and will be included in the irrigation plan [see (7) and (8)].

$$P(IrrU) \triangleq \begin{pmatrix} Dron :: (needStatus\_IoT_i, needData\_IoT_i). \\ Dron ::< StatusIoT_i, DataIoT_i > .0 \end{pmatrix} \tag{7}$$

$$P(ST) \triangleq \begin{pmatrix} Dron :: (needStatus\_Tractor). \\ Dron ::< loc\_ST, status\_Tractor > .0 \end{pmatrix} \tag{8}$$

The Sequence UML diagram for realization of Ambient-oriented modelling of the scenario is presented in Fig. 3.

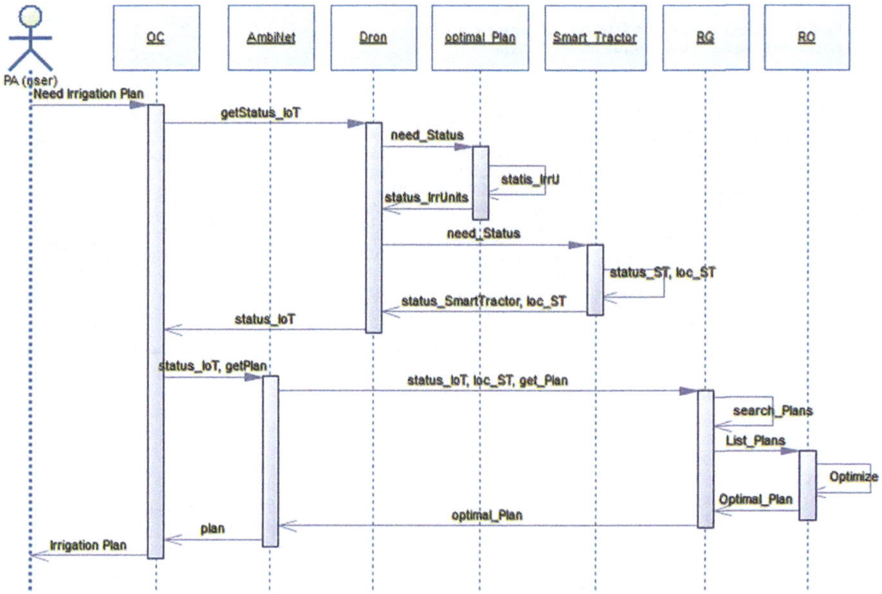

Fig. 3 UML sequence diagram of ambient-oriented modelling of the current scenario

Verifying the correctness of CCA modeling of certain scenarios requires the development of an interpreter that validates the processes of the ambients in the course of their interaction. Since the CCA notation contains symbols that cannot be directly interpreted, a special language ccaPL, which is a computer-readable version of the CCA syntax [28], has been developed. The ccaPL code of described scenario is presented here:

```
OC[PA
[OC@send(needPlan).0|
OC@recv(plan).0
]|
PA#recv(needPlan).
Dron::send(PAi,getStatus_IoTi,getData_IoTi, getStaus_ST).0|
Dron::recv(PAi,Status_IoTi,Data_IoTi,Status_ST).
ANet::send(PAi,Status_IoTi,Data_IoTi,Status_ST,
loc_ST).0|
ANet::recv(PAi,reply).PA#send(plan).0]|
Dron
[OC::recv(PAi,getStatus_IoTi,getData_IoTi,getStatus_ST).
IrrU::send(needStatus_IoTi,needData_IoTi).
ST::send(needStatus_ST).0|
IrrU::recv(Status_IoTi,Data_IoTi).
ST::recv(loc_ST,Status_ST).OC::send(PAi,Status_IoTi,
rrU::recv(S
IrrU[Dron::recv(needStatus_IoTi,needData_IoTi).Dron::
::send(Status_IoTi,Data_IoTi).0]|
ST[Dron::recv(needStatus_ST).Dron::send(loc_ST,Status_ST).0]|
```

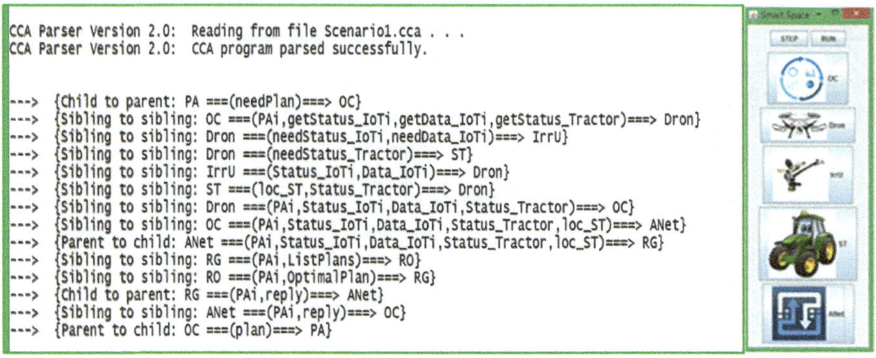

Fig. 4 CcaPL simulator in console and animator mode

```
ANet
[OC::recv(PAi,Status_IoTi,Data_IoTi,Status_ST,loc_ST).
RG#send(PAi,Status_IoTi,Data_IoTi,Status_ST,loc_ST).0|
RG#recv(PAi,reply).OC::send(PAi,reply).0|
RG[ANet@recv(PAi,Status_IoTi,Data_IoTi,Status_ST,loc_ST).
RO::send(PAi,ListPlans).0|
RO::recv(PAi,OptimalPlan).ANet@send(PAi,reply).0]|
RO[RG::recv(PAi,ListPlans).RG::send(PAi,OptimalPlan).0]]
```

Each row of outputs of the program execution is read as follows: Note "A === (X) ===> B" means Ambient "A" sends an "X" message to Ambient "B". "Child to parent", "Parent to child" and "Sibling to sibling" provide information about the relationship between the sender A and the recipient B in terms of the hierarchy of ambients. The Animator presents graphically the ambients, their location, the hierarchy relationships and the processes in the implementation of the modeled scenario (Fig. 4).

6.2 Agent- and Artifact-Oriented Modelling

In this scenario, a personal assistant supports the operation of the irrigation system. The responsibility of the personal assistant is to choose an applicable plan for irrigation of a parcel selected by the user. Exemplary tasks (presented as plans) that can be assisted by the personal assistant can be calculations of the required amount of water, of the slope of the terrain, of the evaporation level, of the maximum amount of the needed water, of the number of irrigation zones. In the ambient-oriented model, various specific details (e.g., inherent attributes of the physical "things" to be observed) were ignored. However, the details are essential to support decision-making process of the personal assistant. Instead, these kind of details can be considered in an artifact-oriented model.

In fact, the personal assistant will be implemented as a multi-agent system operating as a specific calculator called agroIrregator, which will select an applicable plan depending on the specific conditions in the physical world. The agroIrregator consists of the following components:

- Personal Assistant (PA)—manages and synchronizes the work of the agroIrregator. The PA saves in his own plan library pre-prepared plans that can be updated depending on the specific conditions.
- Operative Assistant (OA)—assists the PA; in particular, in having the PA update his beliefs base. For this purpose, it provides information from DiLib, necessary for the selection of an applicable plan.
- Guard—operates as an interface with the physical world delivering data from the sensors needed to select an applicable plan.
- Artifacts—constitute a shared environment of the assistants. Furthermore, the artifacts provide information (in form of triggering events) to generate a set of relevant plans.

We model the agents' environment as artifacts. In our case, essential artifacts are for example parcels, plants, irrigation system, sensor network. The observable properties of the artifacts in interest can be:

- For a parcel—identifier, size, location, soil type, soil moisture (at different depths), air humidity. Some of these data are mostly static and they are stored in the DiLib. Another part of the data are variable obtained from the sensor network (by the guards).
- Plant—name, identifier, species, root system.
- Irrigation system—identifier, type.

Let's look at a possible scenario for using agroIrregator. The PA has to select an applicable irrigation plan for a parcel equipped with a combined drip irrigation system. A drip irrigation system consists of pipes with drip valves mounted on them. Depending on various characteristics, such as diameter, flow, and pressure the drip valves are mounted on the pipes at different distances. Furthermore, depending on the root system of the plantation and the type of soil, it is advisable to provide for the possibility to switching to a suitable drip distribution configuration. The PA is also able to obtain information about the soil moisture (at different depths) by the guards. By the supporting OA, it receives information about the soil type of parcel to be irrigated and the type of root system respectively. The assistants and guard are implemented as rational BDI agents with built-in components event queue (EQ), beliefs base (BB), and plan library (PL).

An applicable plan is selected by implementing the following steps (Fig. 5):

- user chooses artefactParcel to be irrigated.
- artefactParcel generates a trigger event (pId) that is sent to Guard, Operative Assistant, and Personal Assistant.
- trigger event (pId) placed in the PA's event queue paEQ causes activation of the personal assistant to select relevant plans, i. e. to generate the set relPlans.

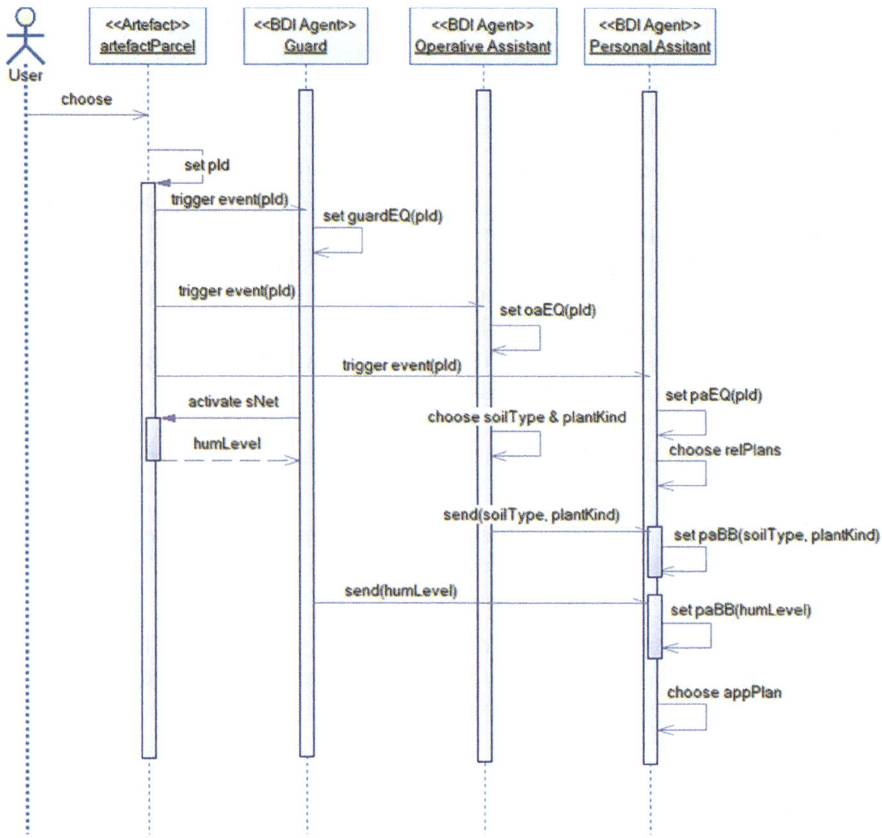

Fig. 5 UML Sequence diagram for realization of simple scenario

- trigger event (pId) placed in the guard's event queue guardEQ activates the guard to read data from the humidity sensor in the parcel pId. Then, the guard sends this data to the personal assistant to update its beliefs base paBB.
- trigger event (pId) placed in the OA's event queue oaEQ activates the operative assistant to extract information on the soilType and plantKind. Then, the OA sends this data to the personal assistant to update its beliefs base paBB.
- Now, the update of the PA's beliefs base paBB is completed and the PA is able to select an applicable plan appPlan from the set of relevant plans relPlans.

7 Conclusion

This article presents a reference architecture and its adaptation for an important domain, namely precise and intelligent agriculture. The virtualization of the physical objects of interest, which could be a crucial way for the integration of the virtual and physical worlds, is indicated as a serious challenge. In the Smart Agriculture Space, the virtualization is supported by combination of various modelling approaches; two of them are demonstrated in this paper.

Currently, a prototype regional version of the space is being built for the Plovdiv area. The guard system is planned to include three local sensor networks—two for outdoor and one for indoor (greenhouse) areas. Furthermore, a stable and efficient communication infrastructure at the edge, fog, and cloud levels is under construction. The DiLib "Smart Agriculture" is deployed on a private cloud built in the Faculty of Mathematics and Informatics of the University of Plovdiv, where the data collected by the sensor networks are also stored. The digital library repositories implemented mainly as relational databases and ontologies complement the modeling methods presented in this article.

Acknowledgements The research work reported in the paper is partly supported by the Bulgarian National Science Fund of the Ministry of Education under the project BG PLANTNET "Establishment of National Information Network GenBank—Plant genetic resources", Grant KP-06X36/2.

We acknowledge the provided access to the e-infrastructure of the Centre for Advanced Computing and Data Processing, with the financial support by the Grant No BG05M2OP001-1.001-0003, financed by the Science and Education for Smart Growth Operational Program (2014–2020) and co-financed by the European Union through the European structural and Investment funds.

References

1. Stoyanov, S., Popchev, I., Doychev, E., Mitev, D., Valkanov, V., Stoyanova-Doycheva, A., Valkanova, V., Minov, I.: DeLC educational portal. Cybern. Inf. Technol. Bulgarian Acad. Sci. **10**(3), 49–69 (2010). ISSN 1311-9702
2. Stoyanov, S., Orozova, D., Popchev, I.: Virtual education space—present and future. In: Proceedings of the Jubilee scientific conference with international participation, BFU, vol. II, pp. 410–418. ISBN 978-619-7126-28-0
3. Stoyanov, S., Stoyanova-Doycheva, A., Glushkova, T., Doychev, E.: Virtual physical space—an architecture supporting internet of things applications. In: Proceedings of the XXth International Symposium on Electrical Apparatus and Technologies, SIELA, Burgas, Bulgaria, pp. 251–252 (2018)
4. Stoyanov, S., Orozova, D., Popchev, I.: Internet of things water monitoring for a smart seaside city. In: Proceedings of the XXth International Symposium on Electrical Apparatus and Technologies, SIELA, Burgas, Bulgaria, pp. 253–254 (2018)
5. Glushkova, T., Miteva, M., Stoyanova-Doycheva, A., Ivanova, V., Stoyanov, S.: Implementation of a personal internet of thing tourist guide. Am. J. Comput. Commun. Control **5**(2), 39–51 (2018). ISSN 2375-3943
6. Glushkova, T., Stoyanov, S., Popchev, I., Cheresharov, S.: Ambient-oriented modeling in a virtual educational space. Comptes rendus de l'Academie bulgare des Sci. **71**(3), 398–406 (2018). ISSN 1310-1331
7. Kirkpatrick, K.: Technologizing agriculture. Commun. ACM **62**(2), 14–16 (2019)
8. Kang, M., Fan, X.-R., Hua, J., Wang, H., Wang, X., Wang, F.-Y.: Managing traditional solar greenhouse with CPSS: a just-for-fit philosophy. IEEE Trans. Cybern. **12**, 3371–3380 (2018)
9. Kiani, F., Seyyedabbasi, A.: Wireless sensor network and internet of things in precision agriculture. Int. J. Adv. Comput. Sci. Appl. **9**(6), 99–103 (2018)

10. Moreno-Moreno, C.D., Brox-Jiménez, M., Gersnoviez-Milla, A.A., Márquez-Moyano, M., Ortiz-López, M.A., Quiles-Latorre, F.J.: Wireless sensor network for sustainable agriculture. In: Proceedings of the Green Technology and Engineering International Conference EGTE IC'18, Caceres, Spain, Proceedings 2018, vol. 2, pp. 1302 (2018). https://doi.org/10.3390/proceedings2201302
11. Karim, F., Karim, F., Frihida, A.: Monitoring system using web of things in precision agriculture. Procedia Comput. Sci. **110**, 402–409 (2017)
12. Kamilaris, A., Kartakoullis, A., Prenafeta-Boldú, F.X.: A review on the practice of big data analysis in agriculture. Comput. Electron. Agric. **143**, 23–37 (2017)
13. Wooldridge, M.: An Introduction to Multi Agent Systems. Wiley (2009). ISBN 978-0-470-51946-2
14. Moszkowski, B.: Compositional reasoning using interval temporal logic and tempura. In: Lectures Notes in Computer Sciences, vol. 1536, pp. 439–464. Springer, Berlin
15. Siewe, F., Zedan, H., Cau, A.: The calculus of context-aware ambients. J. Comput. Syst. Sci. **77**(4), 597–620 (2010)
16. Guglev, Z., Stoyanov, S.: Hybrid approach for manipulation of events in the virtual referent space. In: Proceedings of the International Scientific Conference Blue Economy and Blue Development, pp. 197–203. Burgas Free University, Burgas (2018). ISBN 978-619-7126-57-0
17. FIPA Agent Communication Language. http://www.fipa.org/repository/aclspecs.html
18. Cardelli, L., Gordon, A.: Mobile ambients. Theoret. Comput. Sci. **240**, 177–213 (2000)
19. Cook, D.J., Augusto, J.C., Jakkula, V.R.: Ambient intelligence: technologies, applications, and opportunities. Pervasive Mob. Comput. **5**(4), 277–298 (2009)
20. Ashton, K.: That 'Internet of things' thing, in the real world things matter more than ideas. RFID J. **22**, 65–78 (2009)
21. Yachir, A., Amirat, Y., Chibani, A., Badache, N.: Event-aware framework for dynamic services discovery and selection in the context of ambient intelligence and internet of things. IEEE Trans. Autom. Sci. Eng. **13**(1), 85–102 (2016)
22. Milner, R., Parrow, J., Walker, D.: A calculus of mobile processes. Inf. Comput. Parts 1–2 **100**(1), 1–77 (1992)
23. Boissier, O., Bordini, R.H., Hübner, J.F., Ricci, A.: Multi-Agent Oriented Programming. Multi-Agent Systems Using JaCaMo. The MIT Press, Cambridge, Massachusetts, London, England (2020)
24. Bordini, R.H., Hübner, J.F., Wooldridge, M.: Programming Multi-Agent Systems in AgentSpeak using Jason. Wiley Ltd. (2007)
25. Cartago: http://jacamo.sourceforge.net/cartago/doc/
26. Ricci, A., Viroli, M., Omicini, A.: The A&A programming model and technology for developing agent environments in MAS. In: Dastani, M., et al. (eds.) ProMAS 2007, LNAI 4908, pp. 89–106. Springer, Berlin (2008)
27. Moise: http://moise.sourceforge.net/doc/.
28. Al-Sammarraie, M.H.: Policy-Based Approach For Context-aware Systems. Software Technology Research Laboratory. De Montfort University, Leicester, UK (2011)

Turbine Hill Chart Generation Using Artificial Neural Network

Arlindo Rodrigues Galvão Filho, Filipe de S. L. Ribeiro, Rafael Viana de Carvalho, and Clarimar José Coelho

Abstract The hill chart is an important tool for study of turbine performance, energy production as well as management and hydropower control. This paper propose a feed-forward Artificial Neural Network (ANN-FF) in two different configurations in order to respectively generate a hill chart and to estimate turbine efficiency. The dataset used for training ANN-FF model was retrieved from a small-scale test model of a hydroelectric turbine, provided by Jirau power plant situated on Madeira River in state of Rondônia, Brazil. Predicted values obtained by applying proposed ANN-FF model for each parameter is similar to values measured from small-scale test model of turbine. The training errors of proposed ANN-FF model have significant values from third decimal point. It is concluded that ANN-FF is a good strategy for generation of hill charts and estimate accurately hydroelectric turbine efficiency.

Keywords Energy · Turbine hill chart · Artificial Neural Network

1 Introduction

The hydraulic turbine characteristic curves (hill chart), formulated by model test, is widely used to describe energy conversion and other aspects of turbine performance, as well as properties at different operating conditions [1]. The hill chart is calculated

A. R. Galvão Filho (✉)
Institute of Informatics, Federal University of Goiás, Esperança Avenue, s/n, Chácaras de Recreio Samambaia, Goiânia, Goiás 74690-900, Brazil
e-mail: arlindogalvao@ufg.br

R. V. de Carvalho · C. J. Coelho
Scientific Computing Lab, Pontifical Catholic University of Goiás, First Avenue, Setor Leste Universitário, Goiânia, Goiás 74605-020, Brazil

F. de S. L. Ribeiro
Jirau Hidroeletric Power Plant, Energia Sustentável do Brasil, BR-364 Highway, KM 824, Jaci Paraná, Porto Velho 76840-000, Brazil
e-mail: filipe.ribeiro@esbr.com.br

V. Sgurev et al. (eds.), *Advances in Intelligent Systems Research and Innovation*, Studies in Systems, Decision and Control 379,
https://doi.org/10.1007/978-3-030-78124-8_17

as a function of water head, flow rate coordinate, runner blade angle, turbine power output and other characteristics of turbine. It is an important tool for study of turbine performance, energy production as well as management and hydropower control [2].

To generate hill chart, experimental measurements are performed on small-scale tested model of a turbine at test facilities specifically designed for this purpose [3]. The collected data still undergo normalization and interpolation to make it as representative as possible. Any changes in project of turbine, data from hill chart becomes erroneous, and constructing a new small-scale model or actual plant measurements is necessary but extremely costly. To that end, artificial neural networks (ANNs) can be used to predict turbine parameters and generate a hill chart based on measurement points of small-scale test models.

Several works propose use of ANNs for pattern recognition and prediction solutions applied to power generation problems. Shaw et al. [4] present an ANN-based method for determining optimal reservoir operations near Nashville, Tennessee, USA. Hammid et al. [5] propose a method based on ANN to simulate models of a small hydropower plant on Lake Himreen located in Diyala, Iraq. Božić and Jovanović [6] present an ANN-based method for predicting turbine characteristics considering small-scale turbine model dataset.

However, models presented so far in literature propose methods that do not group all parameters for construction of hill chart using ANN. Or when grouping, they use an ANNs for each predicted parameter, making solution more complex in terms of construction and parameterization. Moreover, a hill chart is used to analyze a turbine-generator set. The hill chart identifies a variation in yield values for each net height. Thus, it is possible to know which point minimizes losses of set operation. Outside this point, operation of generating unit causes a loss in production of electricity. The modeling process of production of electric energy, average yield is considered as a measure of performance of power generation. However, average yield is a simplification for problem of modeling of produced energy. Turbine efficiency is a function of drop height and instantaneous power (hill chart). The solution of hill chart in simulation models is a difficult task. The same yield value can be obtained from combination of several variables involved in model. This difficulty in obtaining prediction of values related to a some variables is typical of solutions using artificial neural networks.

In this context, this work proposes prediction of a hill chart using a feed-forward Artificial Neural Network (ANN-FF) grouping parameters and predicting solution for each parameter. Moreover, a different configuration is presented in order to estimate with more accuracy turbine efficiency. The proposed model was validated using data from two horizontal bulb Kaplan-type turbines from two different manufacturers. The results obtained were evaluated by predictive ability and visual similarity in relation to original hill chart. The data used were obtained from Jirau Hydroelectric Power Plant, installed on the Madeira River in the state of Rondônia, Brazil. The Jirau Power plant is managed by Sustainable Energy Consortium of Brazil S.A. (ESBR).

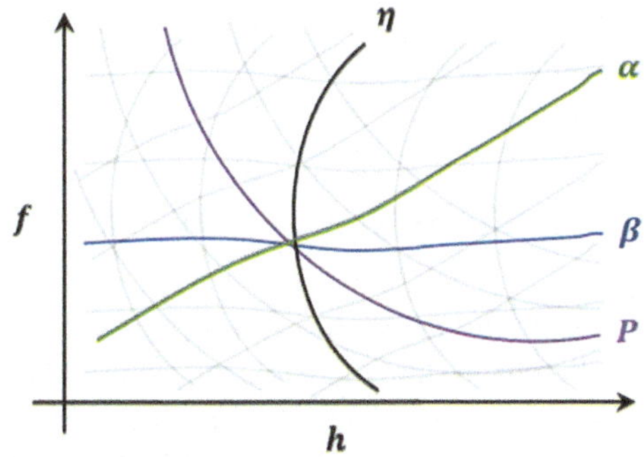

Fig. 1 Example of hill chart

2 Hydropower Turbine and Hill Chart

Turbine performance can be described as a complex dependence between multiple parameters [7]. Efficiency and turbine power output are information strongly related with gravity over dam, water density and water head, flow rate, wicked gate angle and runner blade angle, rotor speed, and other parameters of turbine [8]. This massive amount of geometric, energetic, cavitation parameters and their mutual dependencies makes it difficult to build a single and functional interface for turbine management. Modelling hill chart is a graphical alternative to control these parameters [9]. Such representation mathematically reduces dependencies by grouping them into a small number of parameters. A simplified model of turbine for hill chart can be defined as

$$\eta = f(h, f, p, \alpha, \beta), \tag{1}$$

where efficiency (η) is obtained by relation between water head (h) and flow rate (f), and turbine parameters of turbine power output (p), wicked gate angle (α) and runner blade angle (β). A hill chart example that represent Eq. (1) is depicted in Fig. 1.

The hill chart is generated through mathematical interpolation of data measures for each parameter. The values to generate hill chart are obtained experimentally by measurements in different scenarios on a small-scale test turbine model [10].

The most used approach is based on classical models of deterministic interpolation i.e. polynomial approximation [11], regression [12], splines [13] and statistics [14, 15]. The problem with classical models is approximation in which can result on a non representative model, where polynomial coefficients may not correspond with tested data.

An accurate prediction of hill charts are useful in an optimization procedure operation of turbine. The Hill Chart shows turbines characteristics and how it performs under different operating conditions. Accuracy performance measurements is estimated from small scale prototypes and simulation tests. However, efficiency measured may not be precise and an estimation of power production can be compromised. Therefore, a computational model using Artificial Neural Network can be used to generate a more accurate hill chart and improve estimation of turbine efficiency operation.

3 Artificial Neural Networks

ANN are human brain-inspired computational models that make it possible to build intelligent systems capable of generating predictive analytics [16]. ANNs can be defined as a set of processing units, characterized by artificial neurons, which are interconnected by a large number of connections. Such connections are associated by a certain weight, representing synapses generated by neurons. Learning process occurs by presenting a set of examples inserted in network input, and with information contained in these examples. Weight values adjustments are performed in order to approximate desired output of network. Therefore, process of adjusting synaptic weights aims to tune network for better approximation to expected result [17]. Therefore, an ANN can be considered an universal approximator between input and output variables [18].

The ANNs Feed-forward (ANN-FF) have an architecture consisting of one input layer, one or more hidden layers, and one output layer. Each layer contains one or more neurons, which are completely connected to sub-sequential layer and associated with an adaptive weight. Hidden layers may have different activation functions, which enable network to be a universal approximator.

The main advantage of ANN-FF is self-learning ability [19]. There are several algorithms for obtaining this learning ability [20]. In general they aim to minimize error between output and input, adjusting weights over epoch. The most used training algorithm applied for error minimization is called back-propagation [21]. Back-propagation uses descending gradient method to minimize cost function, which is usually defined by mean square error between expected output and network output. Then, training process occurs minimizing error by iterative update weights and bias, until there is convergence to an expected error or a maximum number of time is reached.

4 The Dataset

Jirau hydroelectric power plant, installed on Madeira River in state of Rondônia—Brazil, consists of fifty Kaplan type horizontal bulb generating units. Among these units, 28 are located on right margin and 22 on left margin. The right margin units were produced by a different manufacturer from left margin units. Model tests were performed in two small-scale test models for each manufacturer, at test facilities in Lausanne, Switzerland.

For right margin small-scale model, tests were performed with water head of 7.7 to 17.7 m ranging from 0.5 m. For each value of water head tested, a wicked gate angle variate from 36 to 106%. For each wicked gate angle, a turbine blade opening was tested so that assembly had as little cavitation as possible. For each set of measurements including water head, wicked gate angle and runner blade angle, turbine output, flow and efficiency were also applicable. For small-scale test model of left margin, similar test were performed considering water head from 8.1 to 19.5 m. Thus, there are two types of hill chart of turbines provided by their respective manufacturers, according to their small-scale test models.

Dataset used in this work were obtained from two small-scale test models, one of left margin and one of right margin. The overall set was divided into a set of training and tests to validate proposed neural network. For right margin, 1243 measurements were generated, divided into 994 for training and 249 for testing. For left margin, 342 measurements were generated, divided into 274 for training and 68 for testing. To ensure that no input dominates another due to numerical magnitude differences, data were normalized over a range of $[-1, 1]$ by a linear scaling function. Dataset were provided by Energia Sustentável do Brasil (ESBR)—JIRAU Hydroelectric Power Plant (Jirau HPP).

5 ANN-FF Proposed Model

The proposed ANN-FF is structured with three layers, consisting of an input layer, a hidden layer, and an output layer each. This structure was chosen because it is able to approximate any continuous function of interest with good precision [22]. For this model, two different sets of configuration was used: one to generate hill chart and another to estimate turbine efficiency. An schematic representation of two sets of proposed ANN-FF is depicted in Fig. 2. The first configuration model (Fig. 2a) is based only on river data as input to estimate all other parameters of turbine. The input layer contains two neurons where input data of water head (h) and flow rate (f) is inserted. The output layer has four neurons where output data for efficiency (η), turbine power output (p), wicked gate angle (α) and runner blade angle (β) are estimated and used to generate the hill chart. The second model configuration (Fig. 2b) uses river data and turbines parameters to estimate turbine efficiency. The input layer contains five neurons where input data of water head (h), flow rate (f), turbine power output (p), wicked gate angle (α) and runner blade angle (β) is inserted. As mentioned before, output layer has just one neurons, where output data for efficiency (η) are estimated with accuracy.

Since hydroelectric power plant has two different turbines, it is necessary to adjust quantity of neurons of hidden layer for each scenario. The quantity was empirically selected from 100 test for each configuration of 5–50 neurons. The best configuration found for both networks was 40 neurons for model that represents right margin and 10 neurons for the model that represents left margin. As dataset is small, training algorithm used was Levenberg-Marquart back-propagation. It was selected hyperbolic tangent function for differentiable activation function, required by algorithm.

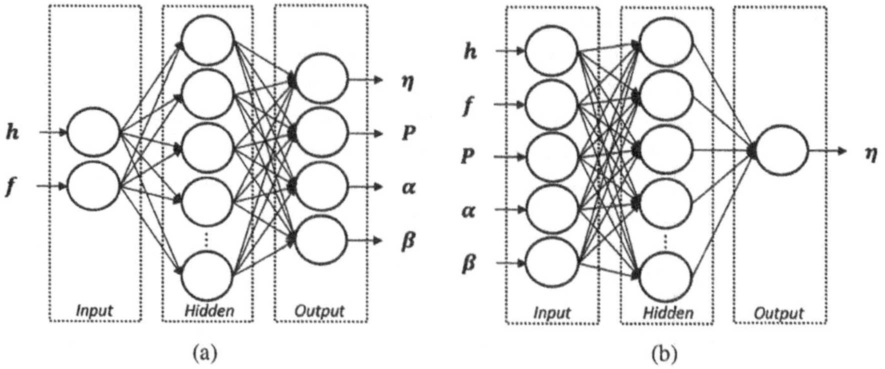

Fig. 2 Schematic representation of ANN-FF proposed for **a** hill chart generation and **b** efficiency estimation

5.1 Evaluation of ANN-FF Models

Predictive ability of proposed ANN-FF was evaluated by three different criteria. They were root mean square error ($RMSE$)

$$RMSE = \sqrt{\frac{1}{N}\sum_{i=1}^{N}(\hat{y}_i - y_i)^2}, \tag{2}$$

Mean Absolute Error (MAE)

$$MAE = \frac{1}{N}\sum_{i=1}^{N}|\hat{y}_i - y_i|^2, \tag{3}$$

and determination coefficient (R^2)

$$R^2 = 1 - \frac{\sum_{i=1}^{N}(y_i - \hat{y}_i)^2}{\sum_{i=1}^{N}(y_i - \bar{y}_i)^2}, \tag{4}$$

where N is amount of data, $\hat{y}_i$ is predicted value, y_i is measured value, $\bar{y}_i$ is average of measured values.

Moreover, a qualitative evaluation of results was performed, consisting of visually compare predicted hill chart with hill chart from small-scale tested model, provided by turbine manufactures.

6 Results and Discussion

This work proposes an ANNN-FF model that can be configured to estimate turbine parameters to generate hill chart based on water head and flow rate data of river. Moreover, same model in a different configuration, can estimate turbine efficiency based on data from water head, flow rate, power output, wicked gate angle, runner blade angle. This section presents the results for each configuration set of model.

6.1 ANN-FF for Hill Chart Generation

The predicted parameters of turbine output (p), wicked gate angle (α), efficiency (η) and runner blade angle (β) from first configuration set of ANN-FF model are compared with data from small-scale test models of right margin turbine. This results is depicted in Fig. 3.

One can notice that, curves from measured data of small-scale test models (blue) and predicted values (red) almost overlapping for all parameters compared. Similar result is found for left margin turbine of predicted data from ANN-FF model compared with data from small-scale test model. Figure 4 shows that, for all parameters, curves from measured data of small-scale test models (blue) and predicted values (red) also are overlapping for left margin. In both cases, test sample number (x axis of Figs. 3 and 4a–d) correspond to 20% of measured data from small-scale test model of turbines for each margin (which have different quantity).

The evaluation of first configuration of ANN-FF model for prediction errors of training and test dataset presents excellent results for all parameters ((η), (p), (α) and (β)) under study. Considering $RMSE$ and MAE criteria, errors are considerable small for both margins (left and right). Moreover, R^2 criteria presents results close to 1 which means that ANN-FF model has a good generalized fit measure. These results are presented in Tables 1 (for the right margin) and 2 (for left margin). Analyzing both tables, the $RMSE$ and MAE criteria from left margin are slightly bigger compared with predictive errors from right margin. The reason may be related with difference on quantity of measured data from small-scale models of turbines from left and right margins.

For qualitative evaluation, two hill charts from each margin of ANN-FF model was generated using complete dataset (training and testing). The hill charts are formed on a Cartesian plane with axes containing drop-in input data ($h \times f$), and curves containing efficiency values (η, black line), turbine output (p, magenta line), wicked gate angle (α, green line) and runner blade angle (β, blue line). Figure 5 shows original hill chart from small-scale test model of right margin turbine along with predicted curve of right margin. While Fig. 6 shows original hill chart from small-scale test model of left margin turbine along with predicted curve for left margin.

Both hill charts where evaluated by operating staff experts, from Jirau Hydroelectric Power Plant, which attested accuracy of result from proposed model and its

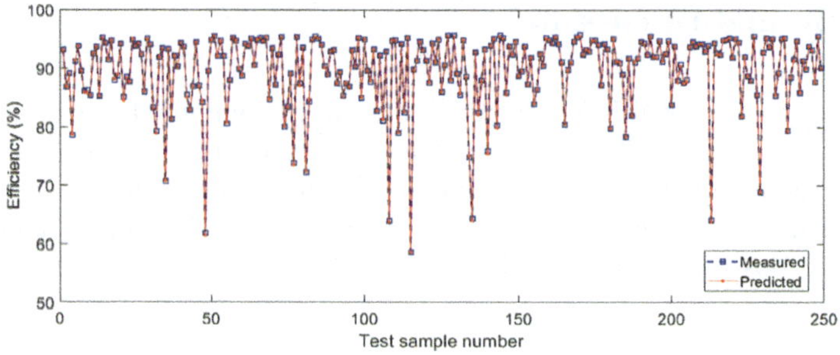

(a)

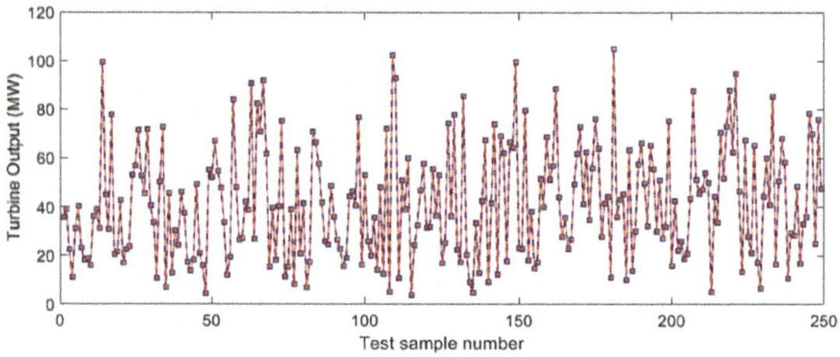

(b)

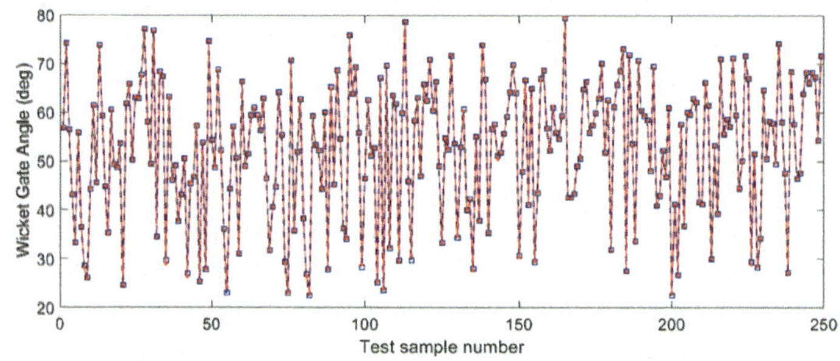

(c)

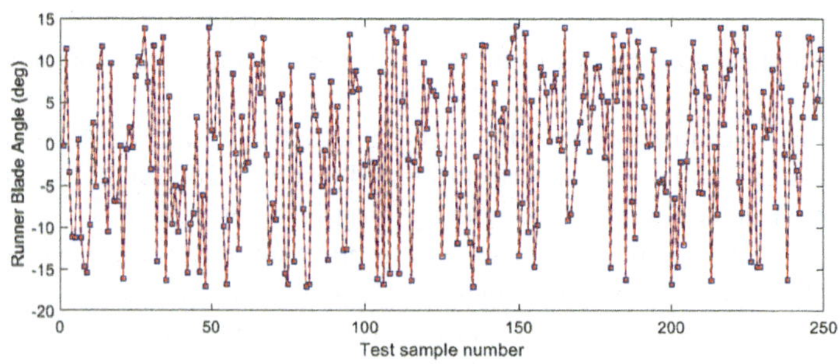

(d)

Fig. 3 Prediction results of **a** efficiency (η), **b** turbine output (p), **c** wicked gate angle (α) and **d** runner blade angle (β) using test data from right margin units

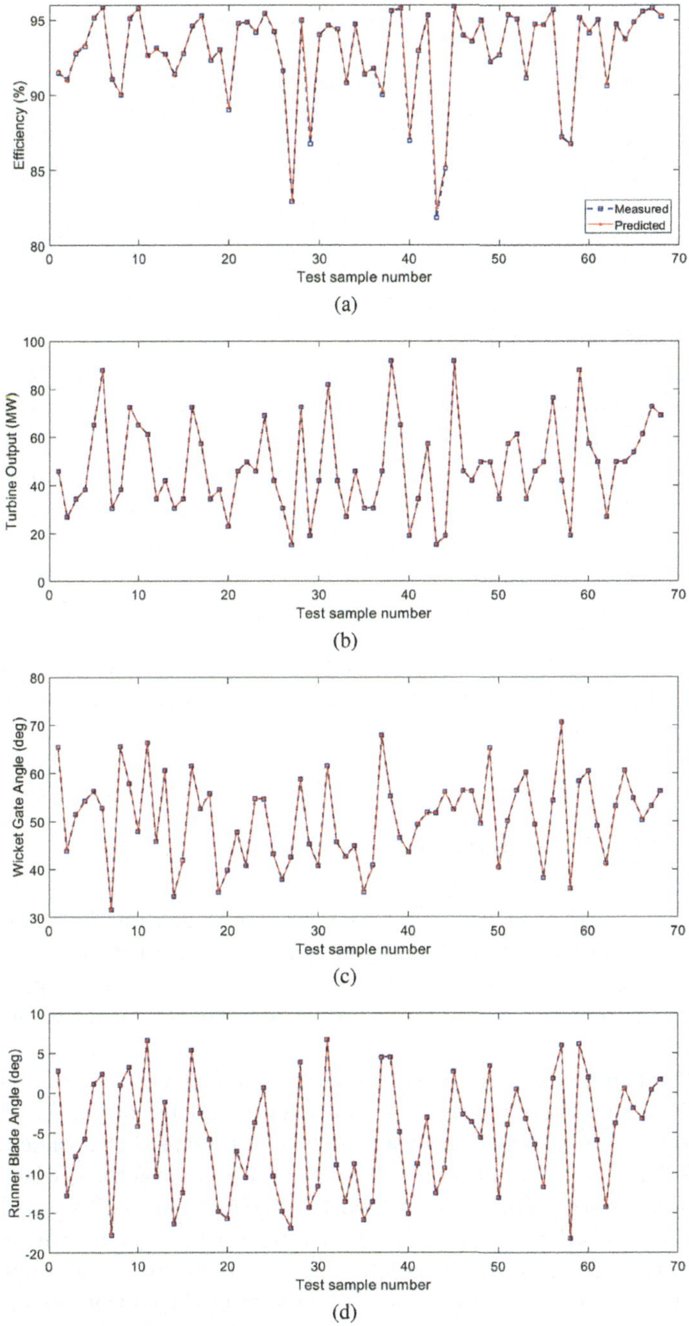

Fig. 4 Prediction results of **a** efficiency (η), **b** turbine output (p), **c** wicked gate angle (α) and **d** runner blade angle (β) using test data from left margin units

Table 1 Prediction errors for right margin

Dataset	Error	η	p	α	β
Train	RMSE	0.0403	0.0394	0.0649	0.0367
	MAE	0.0302	0.0289	0.0493	0.0296
	R^2	0.9999	0.9999	0.9999	0.9999
Test	RMSE	0.0494	0.0506	0.0744	0.0405
	MAE	0.0355	0.0338	0.0523	0.0329
	R^2	0.9999	0.9999	0.9999	0.9999

Table 2 Prediction errors for left margin

Dataset	Error	η	p	α	β
Train	RMSE	0.1168	0.0715	0.1556	0.0526
	MAE	0.0753	0.0556	0.1207	0.0422
	R^2	0.9989	0.9999	0.9997	0.9999
Test	RMSE	0.1302	0.0720	0.1564	0.0523
	MAE	0.0774	0.0570	0.1287	0.0407
	R^2	0.9982	0.9999	0.9997	0.9999

value for analysis of turbine performance, energy production as well as management and hydropower control.

6.2 ANN-FF for Efficiency Estimation

Despite proposed ANN-FF model set configuration to generate hill chart presents efficiency estimation (η), second configuration set was defined specifically to estimate (η) parameter with more accuracy. The objective is to provide a specific model that can use dataset from river and from turbine to predict efficiency without the hill chart.

To test this configuration set, model output is compared with data from small-scale test models of right margin turbine. This results is depicted in Fig. 7a. The curves from measured data of small-scale test models (blue) and predicted values (red) also almost overlap. Similar result is found for the left margin turbine when compare the efficiency estimation of ANN-FF model with data from small-scale test model. Figure 7b shows overlapping curves from measured data of small-scale test models (blue) and predicted values (red) for left margin. In both cases, test sample number (x axis of Fig. 7a and b) correspond to 20% of measured data from small-scale test model of turbines for each margin (which have different quantity).

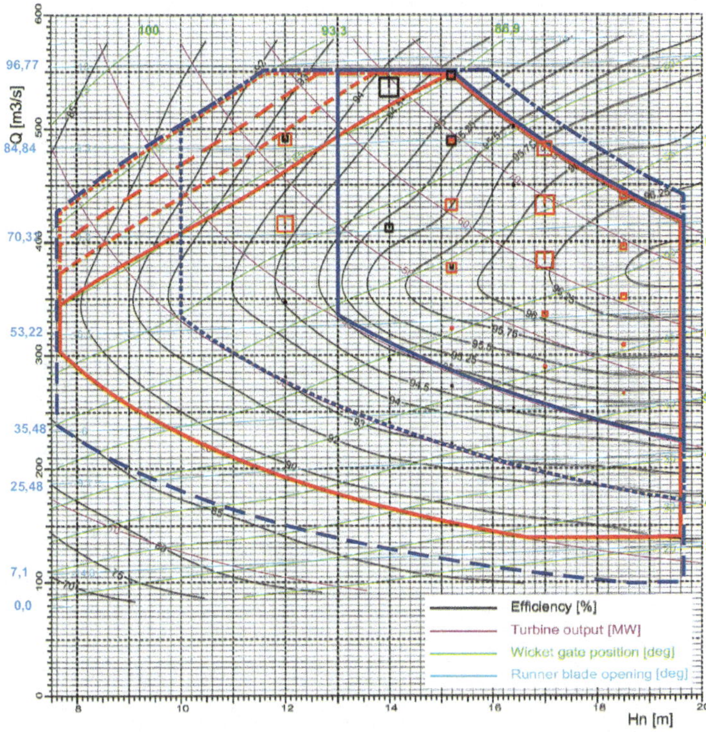

(a)

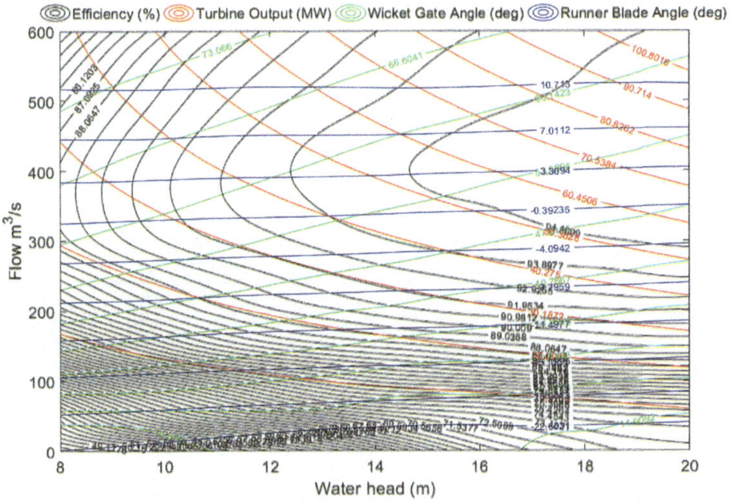

(b)

Fig. 5 The hill chart from small-scale test model of right margin turbine (**a**) and hill chart from respective predictive model using full dataset (**b**)

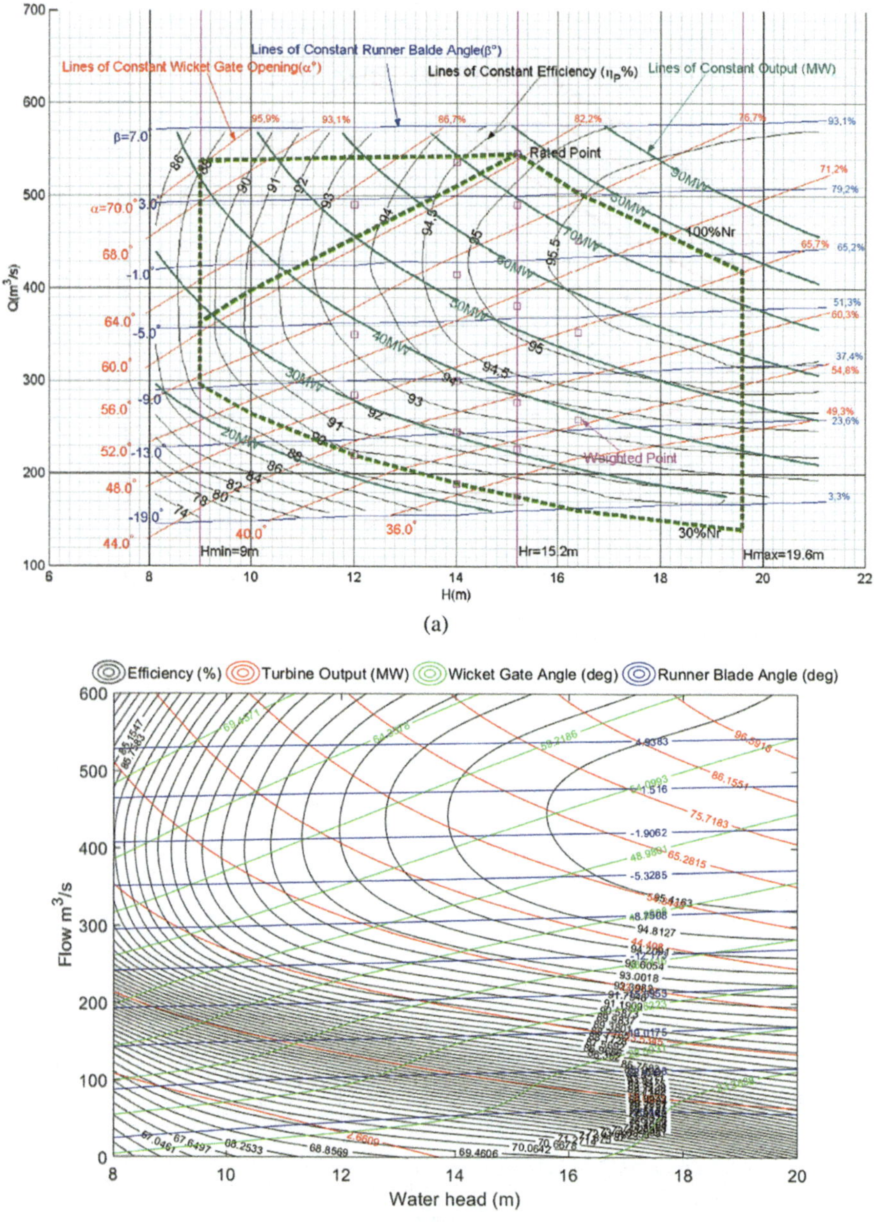

Fig. 6 The hill chart from small-scale test model of left margin turbine (**a**) and hill chart from respective predictive model using full dataset (**b**)

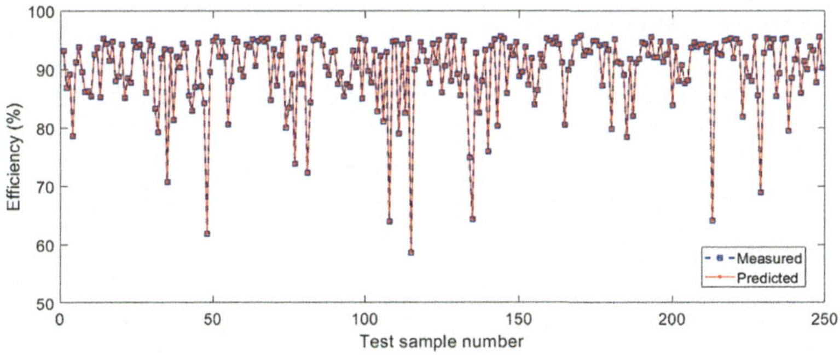

(a)

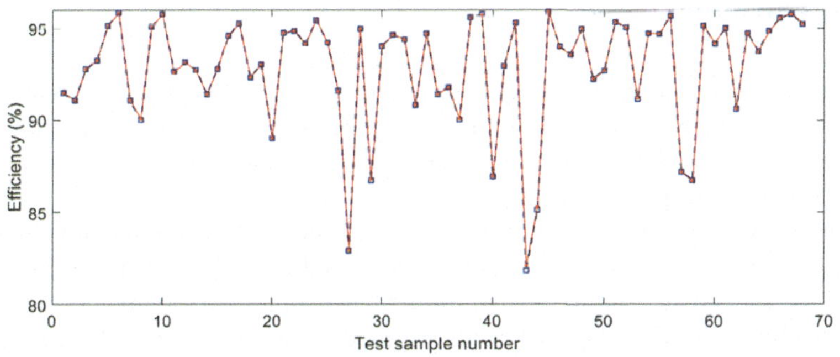

(b)

Fig. 7 Prediction results of efficiency (η) using test data from **a** right and **b** left margin units

Finally, the evaluation for prediction errors of ANN-FF model, considering training and test dataset presents better results for efficiency (η) estimation than the previous configuration for the generation of hill chart.

Considering $RMSE$ and MAE criteria errors are considerable small for both margins (left and right). Moreover, R^2 criteria presents results close to 1 which means that ANN-FF model has a good generalized fit measure. These results are presented in Table 3 (for both margins). Similar results are found for $RMSE$ and MAE criteria from left margin been slightly bigger compared with predictive errors from right margin. The reason again, may be related with difference on quantity of measured data from small-scale models of turbines from left and right margins. Comparing with Tables 1 and 2, the results are slightly better. This behavior can be explained by the numbers of input ANN-FF for this scenario, as shown in Fig. 2.

Table 3 Prediction errors for both margins

Dataset	Error	Right margin	Left margin
Train	$RMSE$	0.0079	0.0330
	MAE	0.0055	0.0201
	R^2	0.9999	0.9999
Test	$RMSE$	0.0125	0.0484
	MAE	0.0080	0.0275
	R^2	0.9999	0.9998

7 Conclusion

This paper presents an ANN model based on feed-forward to predict two hill charts of hydropower turbine. The comparison of predicted results with measured from small- scale test models of the two turbines are similar (overlapping) in terms of curves produced for predicted and measured values.

Root mean square error for proposed ANN-FF model has significant values from first decimal place for efficiency (η) parameter and third decimal place for turbine output (p), wicked gate angle (α) and runner blade angle (β) parameters. Moreover, mean absolute error for proposed ANN-FF model has significant values from third decimal place for all parameters of models from both margins of Madeira River. Whereas determination coefficient for proposed ANN-FF model are close to 1 for turbine models also for both margins of Madeira River.

The proposed ANN-FF model produced predicted values similar to values measured of small-scale test models, with minimal errors, for hydroelectric turbines installed on both margins of Madeira River. Therefore, it can be concluded that Artificial Neural Network based model is a promising approach for generation of hill chart for study of efficiency of hydroelectric turbines.

Acknowledgements Authors thank Energia Sustentável do Brasil for their support in conducting this study, contract Jirau 064/2018, ANEEL PD-06631-0007/018.

References

1. Krishna, H.C.R.: Hydraulic Design of Hydraulic Machinery. Gower Technical Press, Aldershot, Hants (1997)
2. Dixon, S.L., Eng, B.: Fluid Mechanics and Thermodynamics of Turbomachinery. Elsevier Butterworth-Heinemann, Boston (2005)
3. Kavurmaci, B., Celebioglu, K., Aradag, S., Tascioglu, Y.: Model testing of Francis-type hydraulic turbines. Meas. Control **50**(3), 70–73 (2017)
4. Shaw, A.R., Sawyer, H.S., LeBoeuf, E.J., Hadjerioua, B.: Hydropower optimization using artificial neural network surrogate models of a high-fidelity hydrodynamics and water quality model. Water Resour. Res. **53**(11), 9444–9461 (2017)

5. Hammid, A.T., Sulaiman M.H.B., Abdalla, A.N.: Prediction of small hydropower plant power production in Himreen Lake dam (HLD) using artificial neural network. Alexandria Eng. J. (2018)
6. Božić, I., Jovanović, R.: Prediction of double-regulated hydraulic turbine on-cam energy characteristics by artificial neural networks approach. FME Trans. 44, 125–132 (2016)
7. Brekke, H.: Performance and safety of hydraulic turbines. In: 25th IAHR Symposium on Hydraulic Machinery and Systems (2010)
8. Nechleba, M.: Hydraulic Turbines Their Design and Equipment. Artia, Prague (1957)
9. Subramanya, K.: Hydraulic Machines. Tata McGraw-Hill, New Delhi (2013)
10. Murgausová, M., Olusiak, R.: Design of Kaplan turbine parameters in relation to design point position in a turbine hill chart. AIP Conf. Proc. (2019)
11. Kulkarni, S.S.: Design study of a horizontal axis tidal turbine blade. Doctoral thesis, Faculty of Computing, Engineering, and the Built Environment. Birmingham City University (2016)
12. Stople, R. A.: Testing efficiency and characteristics of a Kaplan-type small turbine. Doctoral thesis, Department of Energy and Process Engineering. Norwegian University of Science and Technology (NTNU) (2011)
13. Piratelli-Filho, A., Araújo, J.A., Brasil Júnior, A.C.P.: Reverse engineering of hydraulic turbine runners using coordinate measuring arms and NURBS modeling. Rev. Tecnol. Fortaleza **30**(1), 114–122 (2009)
14. Konga, J., Skjelbreda, H.I., Fossob, O.B.: An overview on formulations and optimization methods for the unit-based short-term hydro scheduling problem. Electr. Power Syst. Res. **178** (2019)
15. Zhao, J., Wang, L., Liu, D., Wang, J., Zhao, Y., Liu, T., Wang, H.: Dynamic model of Kaplan turbine regulating system suitable for power system analysis. Math. Prob. Eng. **2015** (2015)
16. Hassoun, M.: Fundamentals of Artificial Neural Networks. A Bradford Book (2003)
17. Chapmann, J.: Neural Networks: Introduction to Artificial Neurons, Backpropagation Algorithms and Multilayer Feedforward Networks. Create Space Independent Publishing Platform (2017)
18. Hagan, M.T., Demuth, H.B., Beale, M.H., Jesús, O.D.: Neural Network Design (2014)
19. Du, K.L., Swamy, M.N.S.: Neural Networks and Statistical Learning. Springer, London (2019)
20. Binkhonain, M., Zhao, L.: A review of machine learning algorithms for identification and classification of non-functional requirements. Expert Syst. Appl. **1** (2019)
21. Rao, M.A., Srinivas, J.: Neural Networks: Algorithms and Applications. Alpha Science International, Oxford (2003)
22. Fine, T.L.: Feedforward Neural Network Methodology. Springer, Berlin (1999)

Stimuli-Based Control of Negative Emotions in a Digital Learning Environment

Rossitza Kaltenborn, Mincho Hadjiski, and Stefan Koynov

Abstract The proposed system for coping negative emotions arising during the learning process is considered as an embedded part of the complex intelligent learning system realized in a digital environment. By applying data-driven procedures on the current and retrospective data the main didactic-based stimuli provoking emotion generation are identified. They are examined as dominant negative emotions in the context of learning. Due to the presence of strong internal and output interconnections between teaching and emotional states, an intelligent decoupling multidimensional control scheme is accepted to overcome the lack of sample effective independent control actions that separately affect the states of learning and emotions. To avoid existing drawbacks in emotion-focused partial control, an approach with integrating emotions with a low-dimensional representation is accepted. Two-stage procedure is proposed to compensate the negative impact of emotions on the learning process.

Keywords Data · Emotion · Intelligent control · Learning · Stimulus

1 Introduction

The past half century is characterized by an extremely rapid development of three areas—computer science, computer technology and artificial intelligence. They have had a particularly strong influence on all aspects of public life in the present century. The information age marks a sharp increase in the generation of knowledge by mankind. The "big data" problem is no longer so much to find more, but mainly—the necessary information. In the transformation sequence "data-information-knowledge" not only the quantity but also the quality of the available data is crucial for the successful application. The information obtained as a result

R. Kaltenborn
Smart Infomanagement & Consulting, Frankfurt, Germany

M. Hadjiski (✉) · S. Koynov
Intelligent Systems Department, Institute of Information and Communication Technologies, Bulgarian Academy of Sciences, Acad. Georgi Bonchev Str, Sofia 1113, Bulgaria
e-mail: zdravkah@abv.bg

V. Sgurev et al. (eds.), *Advances in Intelligent Systems Research and Innovation*, Studies in Systems, Decision and Control 379,
https://doi.org/10.1007/978-3-030-78124-8_18

of modern data-driven procedures—collection, classification, clustering, multimodal analysis, is now being transformed into knowledge by combining human participation and technologies based on artificial intelligence (AI) [1–4]. The emergent technologies based on AI have made remarkable achievements in several major areas over the last 15 years—machine learning (ML), image processing and pattern recognition, natural language processing (NLP), problem solving (PS), decision making (DM), intelligent robotics (IR) [1, 2, 5, 6]. AI has acquired the status of an essential factor in public life through mobile communications, social networks, voice assistants, home and service robots [6–10]. These successes would be impossible without the anticipatory development of advanced computer technologies, methods and algorithms—supercomputers, PCs with graphical process units (GPUs), cloud technologies, Internet of Things (IoT), intelligent computer networks [11, 12]. Industry 4.0, are based on AI's significant innovation potential for the digitalization of industry and society. Estimates suggest that by 2030, 30% of existing routine jobs will be replaced by robots, and laid-off employees will need to be retrained [5, 6, 8].

The problem with training acquires new dimensions. In addition to solving the problem of unemployment and guaranteeing specialists for the traditional maintenance and improvement of existing systems, ways must be found for the training of specialists of a new type, prepared for jobs that do not currently exist and for activities that have not yet been discovered [13, 14]. These imperative requirements of business and society face a number of significant challenges [14, 15]:

- Finding a solution to the problem of the explosive increase of the amount of data, information and knowledge by transforming education from obtaining mainly fundamental knowledge, useful facts and necessary skills to the development of knowledge and skills for self-learning, innovation, critical thinking;
- Search for new approaches in education, aimed at opportunities for problem solving, active participation in the learning process, skills for team working;
- Student-oriented individualized training in order to ultimately develop the personal potential of each learner;
- Development of a synergistic approach in the educational system with assessment of all the main influencing factors by the teachers, the students, the learning technologies, the social environment;
- To make full use of the modern achievements of AI-based computer technologies to increase the intelligence of each element in the learning process—pedagogy, the curriculum synthesis, the assessment of academic achievements, the risk estimation, the g13
- blended teacher-robot education, the personal development;
- To study and avoid potential hazards and damages that emerging AI-based technologies could cause.

These challenges can be resolved only using AI-based learning-oriented technologies, developed and implemented with very promising results in the last two decades:

- Tools for data processing—learning data mining, learning analytics, context-oriented learning data bases [11, 13, 16];
- AI focused on learning:
  - Natural Language processing (NLP)—personal assistants, the transformations voice-paper and paper-voice, recommenders, question/answer systems;
  - Data-driven mathematical modeling of learning system elements—student, teacher, assessment, based on machine learning (supervised learning, unsupervised learning, reinforcement learning using deep learning and convolutional neural networks (CNN));
  - Image processing (face, scene, video) [17, 18];
  - Affective computing [19–21];
- Intelligent Tutoring Systems (ITS) [21–23] and Educational Maintenance Systems [24];
- New generation of learning resources—flipped classroom, Virtual Reality (VR), Extended Virtual Reality (EVR), video [25–27];
- Intelligent robots directed to service in learning systems (blended learning, autonomous learning assistants) [6, 10, 12, 25].

These systems, both functionally and algorithmically, are far from mature and much work remains until experimental and single examples become reliable elements of the learning process of the future.

Learning is a complex multidimensional dynamic process subject to continuous external and internal disturbances. Its effectiveness is widely determined by its robustness in terms of those disturbances that divert it from the optimal trajectory. Numerous theoretical and applied research in recent decades has shown that the emotional state of the learner has a significant impact on the course of the learning process—the outcomes and the academic achievements [17–20, 28, 29]. The works on lots of important aspects of the impact of emotions on the students' performance in learning, for instance the influence of emotions in different learning contexts in digital learning environments, the effect of multiple emotions simultaneously, the formation of individual emotional state in students' groups or especially formed clusters of students with different emotional profiles are still only a few [30–32]. Some of these investigations [33, 34] are focused only on the psychological aspects of the emotional behavior not related to the emotional state regulation in intelligent tutoring systems. Despite the fact that most studies on emotions in learning establish a general pattern that positive emotions have a positive impact on the learning process, and negative emotions—negative, there is a situation where there is still no common understanding of the impact of emotion valence and the specific influence of basic emotions in different combinations and learning contexts [35–38].

Over the last decade, a large number of methods and instrumental implementations for measuring and recognizing emotions based on the identification of facial expressions, assessment of the emotional content of human speech in monologue, dialogue or social environment, analysis and classification of gestures and movement have been discovered and required [39, 40]. Nevertheless, a number of problems remain in the field of emotion recognition: what is useful, meaningful and profitable to measure—with what models, methods and tools [37, 39].

Because in any learning system the process of knowledge generation in the learners is dominant in importance, the main indicator of its quality is the efficiency of the learning process, the outcomes and the academic achievements. Therefore, the emotional impact must be treated mainly in the context of the improvement of the main target indicators of the system as a disturbing factor. On the other hand, the emotional state of the students has the character of an element that determines the behavior of the system, i.e. it has both functional and structural and parametric impact. Thus, emotion (and especially negative emotion) cannot be treated simply as an external disturbing effect of the learning environment, but must be considered in the context of an unified learning emotion system with complex interrelationships, dependencies and limitations. Such a synergistic approach is only productive in the context of the main problem to be solved—innovative learning in a digital AI-based environment. Unfortunately, solving this problem encounters a number of difficulties and limitations—a very complex situation of multifunctionality, transformability from one state to another, the influence of pedagogical strategies, the significant presence of difficult to formalize factors, the significant internal evolution of the learning process, a great uncertainty—unfortunately not only parametric but also structural [35, 36, 41].

In the present research, an attempt has been made to generalize the influence of the emotional state of students on the effectiveness of the learning process. It is assumed that the main stimuli for emotional affects arise primarily from specific didactic actions, the learning environment, the learning interpersonal interactions, but they are inevitably accompanied by smaller in intensity and irnpact emotional effects on the living and social environment.

2 Characteristics of Emotions in the Context of a Learning Element

2.1 Basic Emotions

There is no consensus among psychologists on the classification of emotions. Various studies offer models of the emotional state, including from 2 generalized emotions to 32 specific emotions, which so far can be recognized with a probability of over 80% [37, 39].

2.2 Emotion Stimuli

Emotions are evoked by lots of stimuli. In a large part of the research, research situations are formed in which individual stimuli evoke certain emotions [28, 38, 51]. These results represent separate sections of the space of emotions, but the realistic staging presupposes a combined effect of many different stimuli. In the present study, the focus is on the emotions that arise in the learning process from different types of didactic and instrumental influences.

2.3 Emotion Recognition

Among the large number of methods for measuring and classifying emotions for the case of interest in learning systems, three methods are essential:

- Recognition of emotions based on analysis of the con-figuration of a system of facial features (facial expression recognition) via analysis of images;
- Assessment and classification of emotions based on analysis of speech signals, using advances in the study of natural language processes (NLP);
- Recognition of emotions through analysis of gestures and body language.

The reliable aggregation of this disparate information remains an unresolved issue.

This is primarily related to the detection of emotions with a level of confidence higher than a given threshold (e.g. 80%). The next problem consists in data fusion for the same type of emotion, detected from a different sensor system. The serious problem is to identify the dominant (or integral) emotion or dyads of two emotions with the highest level of confidence.

From a practical perspective the only these three computer-based emotion recognition systems are relevant because they are least intrusive during the learning process.

2.4 Mathematical Models of Emotions

A significant number of mathematical models of emotions have been proposed: Valence-Arowsal (ZFM), Affective Circumflex Model (ACM), Zone Flow Model (ZFM), Ortony, Clore, Clorince (OCC) Model and many others [28, 31, 43, 44]. They have a direct focus related to the purpose of the study—parallel learning processes, cognitive processes at different levels, decision making; emotion metrics, emotion generalization; emotion simulation, emotion regulation, emotion self-regulation, emotion personalization; emotion interfacing [28–31, 36, 40, 45].

2.5 Uncertainties and Limitations of Correct Estimation of the Stimuli—Emotional State Relations

The analysis of the bibliography allows outline a number of complexities and unclear relations between stimuli, emotional behavior and academic achievements in educational systems [31, 35, 36, 38, 39, 45, 48]. The main ones are:

- There are conflicting opinions about the optimal emotional state for the learning process, including the role of weak negative and positive emotions;
- The combined influence of emotion, mood and attitude in a dynamic perspective is not clear;
- The emotional reaction of the individual depends on a large number of dynamically changing factor—his/hers individual characteristics, his/hers assessment of the achievement of goals, the social impact of the investigated group;
- Significant complexity in the assessment of the emotional state in the presence of many stimuli—difficult to define boundaries between the attributes of individual emotions, the transition of one emotion to another, ambiguity between stimulus and emotional reaction;
- The emotional reaction strongly depends on the mental, cognitive and knowledge state, the degree of realization of the intermediate goals planned by the student and the personal efforts therefore;
- Predicted emotions via simulation modules can have an anticipatory influence on the emotional state;
- Self-regulation abilities and the current relation of the student strongly depend on the arising situation and it is temporarily evolving during the learning process.

Despite the significant achievements in affective computing, there is still a lack of consensus or even a common understanding of the fundamental influence of different types of emotions on the learning process, their relationships and interference, their joint impact on the overall emotional state of the student. Thus, the problem of simple generalized representation of emotion as a component of learning arises in theoretical and practical aspects, especially in the digital learning environment [28, 32–34, 36, 39, 41].

2.6 Emotion Integration

A lot of research has been devoted to the problem of emotion integration. Two different points of view are distinguished.

2.6.1 Formation of Integral Emotion in the Human Brain

Emotion Integration as a Conception

Investigations focus on the integration of emotions in the process of reasoning, decision-making, choice and generation of relevant action(s) and behaviors. There is a variety of hypotheses here:

- There is a key zone in the brain where emotion and cognition are integrated [50, 51];
- The effect of integration allows us to categorize our experience on the axis of "good-bad" [52];
- Deterministic rather than probabilistic judgment in making a decision based on integral emotion [53, 54];
- Combining integral and incidental emotion [40].

Gestalt-Based Approach

- Emotion is considered as a single state formed as a result of complexation of many variables—environment, cognitive and physiological variables;
- The whole is considered as an object that has different properties from the sum of its components;
- The emotional state is considered as a Parallel Constraint Satisfaction, which in [55] is modeled by a sequence of neural networks.

Dual Process

It is believed that the emergence of any new emotional state is the result of two processes:

- Emotional, which is intuitive, fast and performed without cognitive effort;
- Cognitive, which is a controlled process and is relatively slow, adaptive to the situation and rational.

In [56] neural networks are applied to integrate emotional and cognitive processes. The focus of this research is the search for a generalized model for the formation of the emotional state of man, but in the psychological context of consideration. For our task, dedicated to the management of negative emotions in the learning process, it is essential to form a generalized variable (or vector with a small dimension) based on objective sensory information (face, voice, gesture), which adequately describes the emotional state of the student in different learning situations.

2.6.2 Measured Emotion Integration.

Despite the growing intensity of research on emotion modeling in learning systems, this area is still far from mature [31, 35, 36, 45]. In the present research it is important for us not to regulate the emotional state of the student under the influence of one or more specific emotions, but to provide robust management of the learning process in the real conditions of current didactic influences, considered as dominant stimuli of the emotional state. Therefore, of considerable interest are the results of relevant prior research, which treats various aspects of the creation of integrated models of emotions:

- Affect Integrated Model of Decision Making (AIMD) [46]. There are two forms of influence on the learning process—from Integral Emotion and from Incidental Emotion in an attempt to search for a general model of emotion-based impact on learning systems;
- Two-dimensional "Circumplex Model of Affect" (CMA) [37], where the emotions are considered as combinations of valence and arousal;
- Dominant emotions [47] which define the main direction of emotion state dynamic changes;
- Integration of Incident and Integral Affect [40], where the mechanisms of their crossing effect is treated;
- A number of fuzzy logic-based models for emotion integration [18, 48, 49] are developed in order to take into account both the type of the emotion and their intensities as well;
- Acceptance of an integration approach as a framework for considering the learning process [43, 57] to aggregate the integral and incident emotion [40], to create an integrated psychometric model of *Em* [58];
- More direct methods for emotion information integration: averaging and summation [59], multimodal integration [60], integration via weighted integrated methods [61];
- Form a FL-based integration [18, 46, 49, 62–64];
- Integration in a system of EM registers [65, 66].

As can be seen, the problem of recognized learning-based emotions' reduction bas been studied from different aspects but without a common understanding. In this study, an attempt has been made to modify and develop further the existing results concerning integration of emotions provoked by the learning process in order to regulate them as a stage of partial copying of the whole emotion control system.

3 Computer-Based Learning Systems

3.1 Learning Process

Advanced education in AI-era is increasingly transformed to developments on its digital basis in the direction of using pedagogical strategies that realize synergy and synchronicity between its three main dimensions—(i) acquisition of knowledge, useful information and applied skills (K); (ii) development of cognitive abilities—understanding, innovation, decision making, generalization, critical thinking (C) and (iii) building the learner's personality with epistemological means (Ep), as shown in Fig. 1.

Assessments of academic achievement are key indicators of the quality of the learning process in different situations and disturbances from the external environment (d_L), internal changes and under the influence of management influences (U_L).

The learning process is considered as a multi-stage sequence of transition from one state (learning and emotional state) to another, in discrete moments of time n (Fig. 2). This discrete sequence of states allows information to be accumulated about the course of the unified learning/emotion process, which is used for data-driven procedures (tuning the model via ML, current decision making, control actions formation, adaptation). The sequence of states forms learning trajectories representing academic achievements and outcomes during the time for individuals or for clusters of students formed according to the accepted features of classification. An idealized learning trajectory is shown in Fig. 3.

Learning trajectories could have a variety of particular forms, because the forms depend on all factors influencing the learning process. Testing, assessment, estimation and evaluation give temporal transformations for different aspects of education (K, C, E_p) necessary for analysis, optimization, control and adaptation in the time.

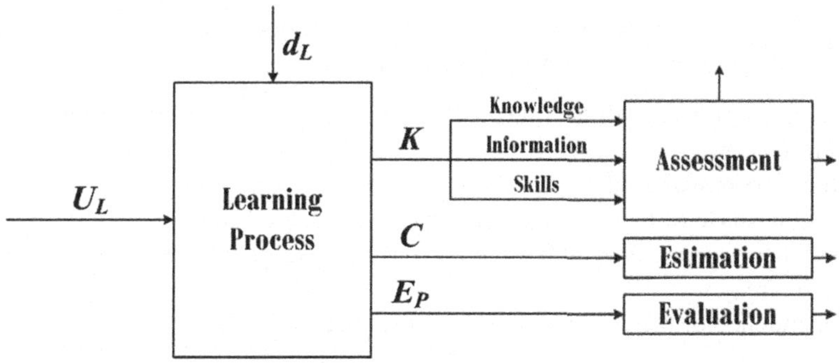

Fig. 1 Scheme of learning as a multilateral process

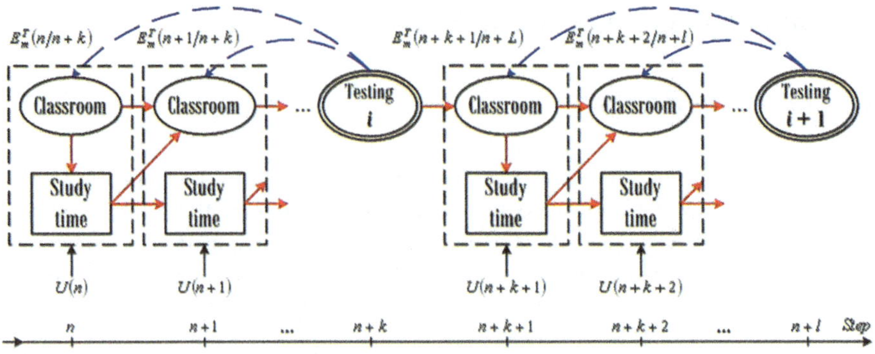

Fig. 2 Learning as a sequencing temporal process

Fig. 3 Learning trajectory

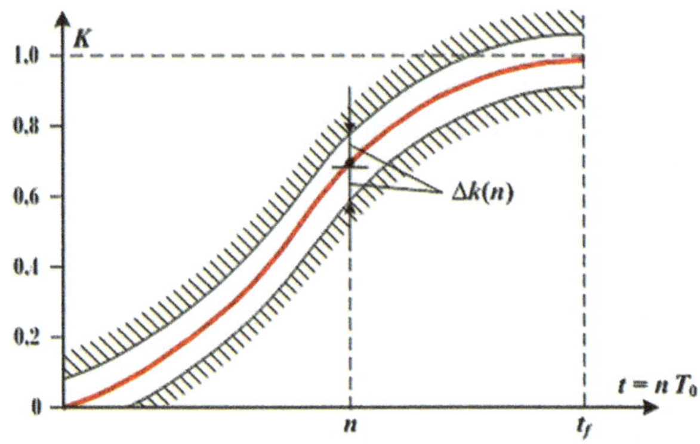

3.2 *Personalization of the Learning Process*

Traditional education resides on the application of a single learning trajectory for all students (Fig. 3). It is formed as an average and does not take into account the individual differences of students [30–32, 45]:

- Different cognitive abilities (perception, memorizing, creativity, critical thinking, generalization);
- Different personal goals;
- Individual style of learning (preference, abilities, attitudes);
- Broad dispersion of the individual current knowledge relevant to the considered domain area;
- Independence of the personal emotional profile.

The average learning trajectory reflects the routine experience not only of the particular lecturer, but also of the most often formed conservative tradition of the teaching staff for a given time interval.

Traditional average education is unsatisfactory for two categories of students:

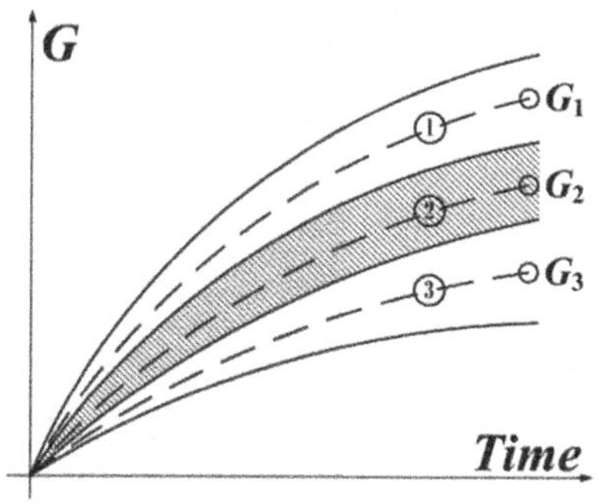

Fig. 4 Decomposition of an average learning trajectory

- The best ones because they are not motivated to develop their higher cognitive capabilities, goals, personal ambitions and motivation;
- The weaker ones, because they could not cover the imposed average requirements, exceeding their goals and capabilities, which is why they often find themselves in a state of lagging behind or dropping out.

The desire to create a learning system that maximally develops the personal potential of each student imposes their differentiation into at least three groups: stronger (1), medium (2) and weaker (3) (Fig. 4).

According to the average data, the number of students in these groups is 10%, 80%, 10%, respectively. As individual differences remain in the groups thus formed, the following successive degrees of individualization emerge:

A. Implementation of the traditional average learning trajectory (2) with an average achievable goal G_2;
B. Formation of 3 groups of students who study on learning trajectories 1, 2, 3 with goals G_1, G_2, G_3;
C. Creating clusters in each of the groups in order to use cluster-oriented learning trajectories;
D. Elaborating individual trajectories for each student.

Figure 5 shows the main characteristics in the implementation of the different degree of individualization (*DI*).

- *R*—efforts to build a computer system for personalized learning;
- *S*—the amount of the necessary supplementary learning material, which must be prepared for the respective basic trajectory (average, cluster, individual);
- *Em*—intensity of the negative emotions arising as a difference between the ones generated during the training on the average and the individualized optimal learning trajectory.

In the learning process, although for different reasons, the stronger and weaker students (groups 1 and 3) can form higher goals in their education by covering only the requirements set by the respective trajectories depending on the degree of personalization (A, B, C, D in Fig. 6).

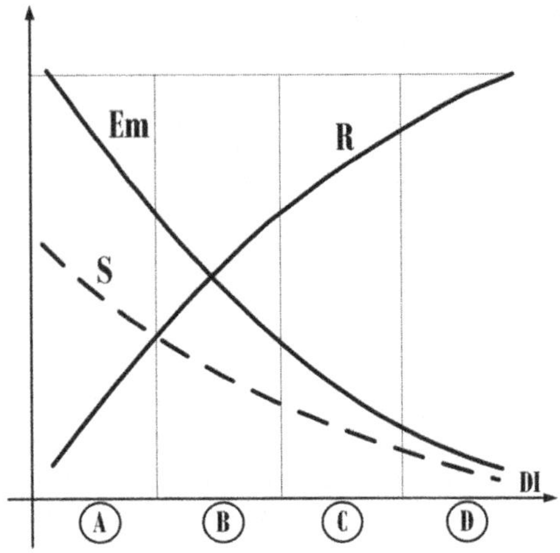

Fig. 5 Effects of different degrees of personalization

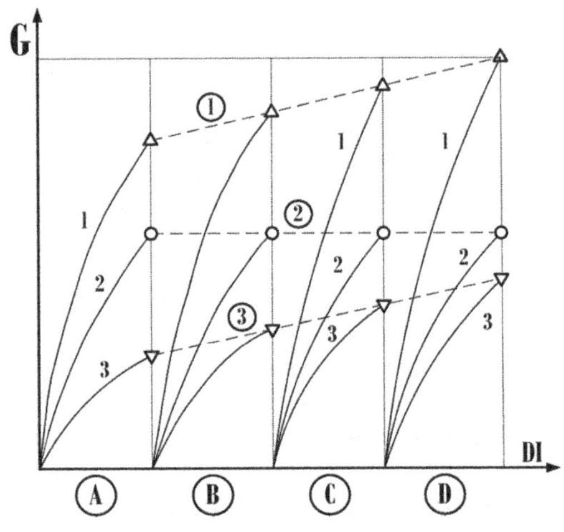

Fig. 6 Personal goal evolving in different degrees of personalization

With additional personal motivation and effort, students from all groups can significantly increase their goals and corresponding academic achievements (Fig. 7).

A variety of instruments can be used for personalization of the learning process:

1. Individualization of the learning content [14, 15, 67];
2. Taking into account the individual style of learning [45];
3. Using personalized pedagogic actions (change the rhythm and/or the sequence of the topics, the choice of supplementary learning material, individual advices) [69, 70];
4. Individualizing recommended learning resources (question/answer system, computer assistant, RV, ER) [71];
5. Individual training to receive practical skills [72];

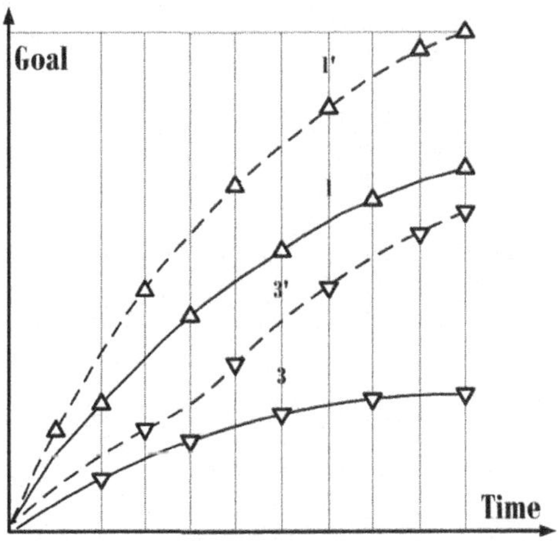

Fig. 7 Perspectives for further increase of the personal goal (1 → 1′), (3 → 3′)

6. Implementation of personalized learning assessment (form, frequency, complexity of tests, interactive evaluation) [73].

Figure 8 shows the scheme of a computer-based representation of the *j*-th topic of the *i*-th learning course with ranges for each feature of the content (volume, complexity and novelty), two generalized types of learners (experience or principle-oriented) and three different preferences in each type according to the individual learning style. As you can see, for each topic of the course, depending on the accepted degree of personalization, 3, 9, 18 or 54 variants must be developed. With an average of 10 topics per course, the volume increases significantly—30, 90 180 or 540 options. Assuming that preparation for each lesson requires 30 – 50 h (content, optimization, presentation, control questions), the task can be successfully solved at the level of modern domain knowledge only with the help of an AI-based intelligent tutoring system (ITS). The implementation of personalized training is associated with a number of other limitations and difficulties.

1. Some pedagogical contradictions need to be overcome:
   - More students have an individual learning style of the type: sensory, visual, inductive,
   - Almost all teachers belong to the opposite type: intuitive, verbal, deductive;
2. It is necessary to find a solution for optimization of some basic ratios in the learning process, which are decisive for the adoption of the learning strategy:
   - Degree of personalization (atomization or collaborative learning);
   - Fundamental/applied knowledge,
   - Transformation-oriented/traditional approach,
   - Optimal balance between classroom learning, self-study and project work;
3. Search for new student-oriented methods and tools for assessment of the academic performance, the achievements and the outcomes;

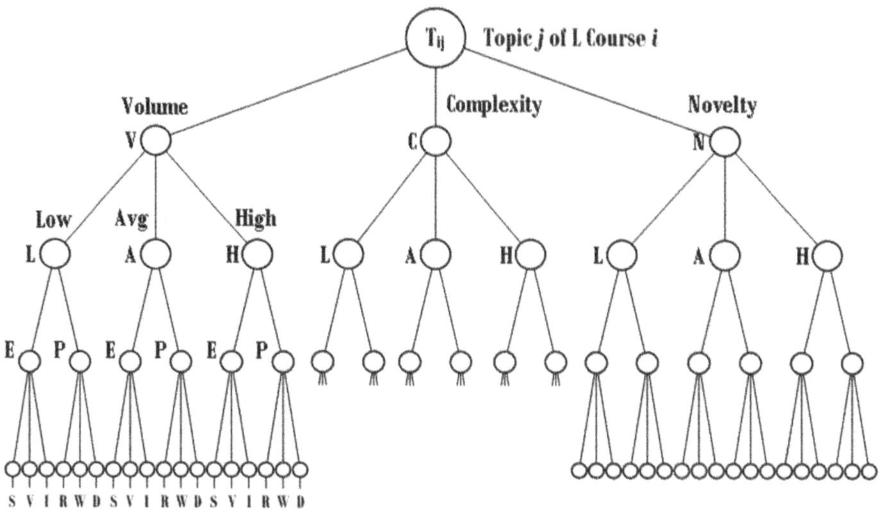

Style of Learning

Generalized	Experience-Oriented	E	Principle-Oriented	P
Detailized	Sensors	S	Reasoners	R
	Visual	V	Word	W
	Inductive	I	Deductive	D

Fig. 8 Tree-like structure of personalized curricula

4. Integration of the emotion regulation system with the learning system.

4 Learning Context-Based Framework of Emotional Regulation

In this research an attempt is made to regulate emotions arising in the learning process combining feedback with feedforward principles of control. Following the accepted approach the emotional state of each learner is treated as a combination of two main components:

Figure 9 presents schematically this interpretation of emotion formation, emotion detection and recognition via emerging AI-based technologies including emotion integration in a vector with a small dimension Z_i $(i \leq 3)$.

Unlike most of the publications devoted to the regulation of emotions in the learning systems based only as a rule on the measured variables of the emotional state in different approaches for the feedback formation of the control algorithm, in

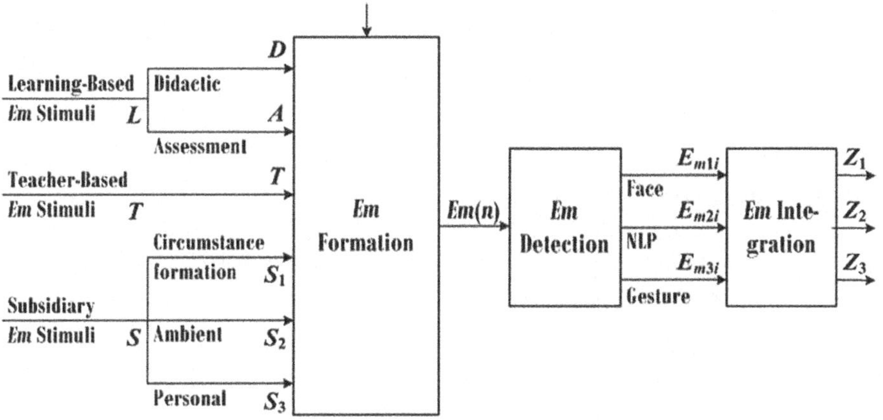

Fig. 9 Scheme of learning-based emotion formation

this research additionally all temporal variables connected with the learning process inputs are considered as feedforward factors contextually belonging to learning. They are considered as stimuli provoking corresponding emotions and it is assumed that they could be measured or inferred using secondary data or functional variables; they are available in a form of time series for each episode of the sequential learning process presented in Fig. 2.

The main input factor in Fig. 9 referred to the next functional groups:

1. Didactic—*D*:

$$D = (\ D_1,\ D_2,\ D_3,\ D_4,\ D_5)\ ,$$

where:
$D_1 = (\ d_{11}, d_{12}, d_{13}, d_{14}, d_{15})$ is the curriculum material for each stage of the learning process: Fig. 2 according Fig. 8;
d_{11}—content, d_{12}—complexity, d_{13}—novelty, d_{14}—volume, d_{15}—sequence of representation;
$D_2 = (\ d_{21}, d_{22}, d_{23})$ are pedagogical actions according to the accepted learning strategy and the current teacher's decision making;
d_{21}—relation on-site/on-line education, d_{22}—degree and scheduling collaborative work, d_{23}—temporal variables (rate of teaching on the *j*-th topic in the *i*-th course, Fig. 8, deadlines for homework);
$D_3 = (\ d_{31}, d_{32})$—all variants of learning styles (Fig. 8);
d_{31}—experiment-oriented (sensory, visual, inductive), d_{32}—principle-oriented (intuitive, verbal, deductive);
$D_4 = (\ d_{41}, d_{42})$ is the organization of practical training;
d_{41}—volume, dead time of the projects, d_{42}—scheduling practical exercises;
$D_5 = (\ d_{51}, d_{52})$—level and degree of learning resources using;
d_{51}—access to ITS, d_{52}—access to project materials;

2. Assessment (A):

$$A = (\ A_1,\ A_2,\ A_3,\ A_4)\ ,$$

where:

A_1 results of the assessment,
A_2 complexity of the test;
A_3 way of assessment (face-to-face, on-line, dialog-based);
A_4 frequency of assessment.

Teacher's emotions-based stimuli related to learning (1):

$$T = (\ T_1,\ T_2,\ T_{3,})\ ,$$

where:

T_1 recognized teacher's basic emotions,
T_2 integral teacher's emotion;
T_3 intensity of integral teacher's emotion;

Subsidiary emotions could be the result of a variety of reasons, unrelated to the learning process:

$$S = (\ S_1,\ S_2,\ S_{3,})\ ,$$

where emotions S_1 $(\ i = 1,\ 3\ )$ are caused accordingly by life circumstances, ambient environment and personal problems.

Subsidiary emotions quite often could be of a significant level as a component of the integral emotion and they can impact strongly on the learning process. Thus copying them in this case will be an unavoidable regulation task. This broad kind of emotions is studied in depth in psychology both as a phenomenon and as methods of regulation as well [28, 35, 44, 47, 50, 53], but they are out of the focus of our research. As it was mentioned above our assumption is that in the context of the learning process the negative emotion states are mainly the result of learning-originated stimuli.

The generalized functional teaching/learning scheme is presented in Fig. 10. The teacher's emotional reactions are included in the learning and in the emotional control realization. Detailed learning state formation as a dominant process but with interrelations with the emotional state formation is presented in Fig. 11. Learning state L in accordance with Fig. 1 comprises three main axes: knowledge acquisition K, cognition at a lower level C and also at a higher level (epistemological) *Ep*. The control vector

$$U_L = (\ U_{L1},\ U_{L2},\ U_{L3})$$

includes corresponding components which are complex and cross-correlated.

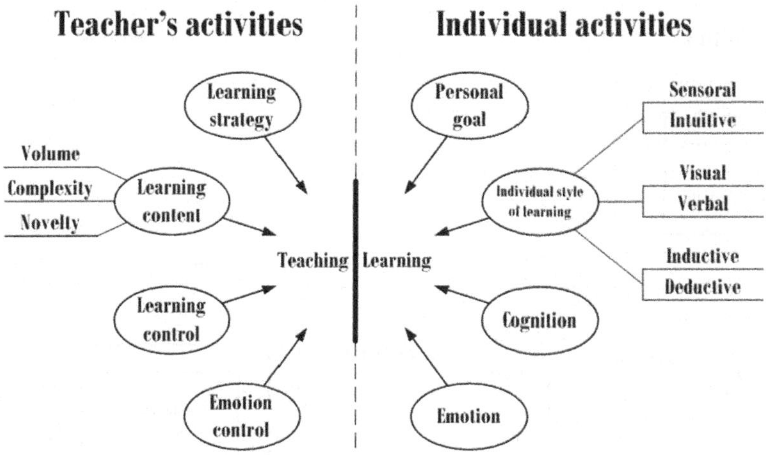

Fig. 10 Teaching/learning functional scheme

The main disturbances on the learning process represent the emotion state *Em*, the control actions for emotion regulation U_{Em} and also a variety of unpredictable and immeasurable disturbances d_L. In an AI-based computerized learning system (e.g. an Advanced Intelligent Tutoring System (ITS)) the implementation of an integrated mathematical model of knowledge acquisition is compulsory. The model outputs—the estimates of the learning state $\tilde{L}$ and the emotional state $\tilde{E}m$ are of key importance for the process of iterative improvement of the learning system performance [35, 39, 43, 47, 74–76].

The emotional state *Em* is a subject of a number of input actions: emotion-focused control action U_{Em}, the effect of control actions on the learning process U_L, the inference from the learning outcomes and the academic achievements *Ass*, the impact of emotional self-regulation *ESR*, the specific influence formed due to teacher's own emotional state T_{Em} and subsidiary emotions $D_L = S$ according to Fig. 9.

5 Control Problem Formulation

The task is to manage emotions in a learning system based on a computerized environment using AI emergent technologies. A control system is considered as complex, dynamic with multiple and multilayer interconnections. The system evolves due to the students' personal, cognitive, emotional and societal characteristic changes during the learning process. A number of different processes take part in the system—parallel learning processes, complex cognitive processes at different levels, multimodal learning (knowledge-oriented, project-oriented, skills-oriented).

A new strategy for emotional control in the learning process is presented with the following main peculiarities:

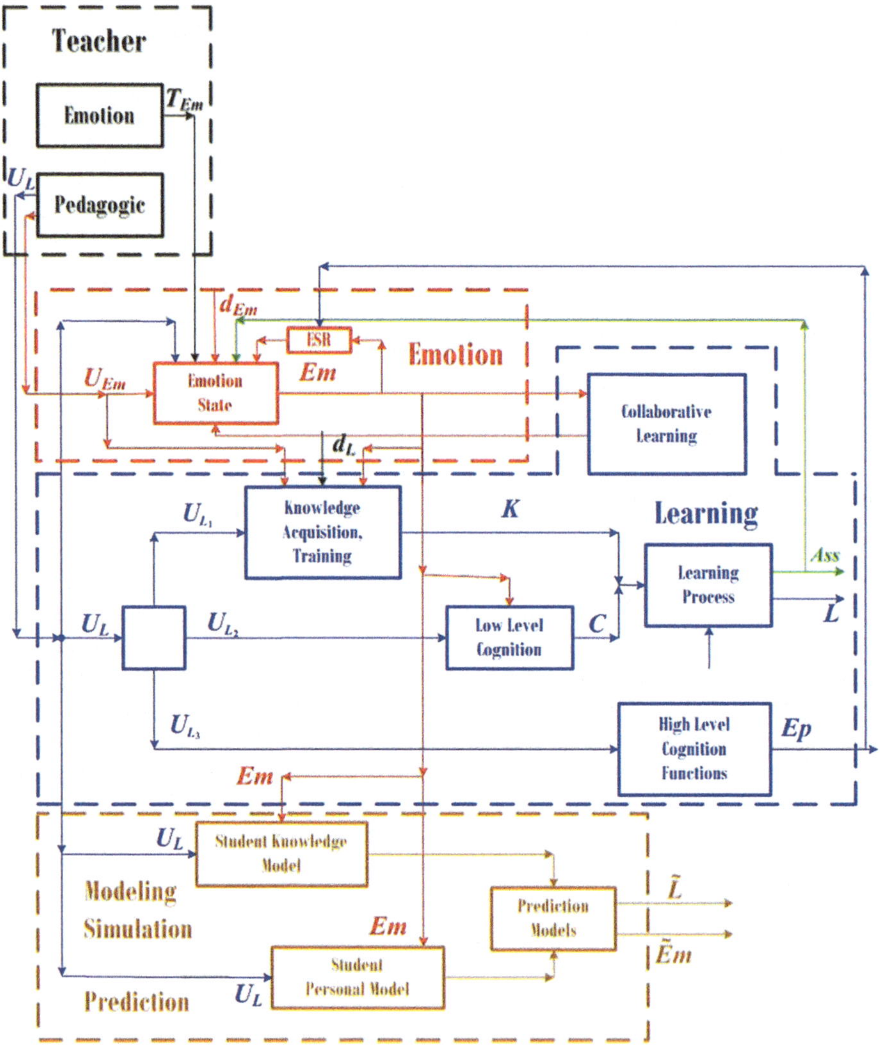

Fig. 11 Multilateral interconnected learning/emotion system

1. The learning process is dominant;
2. It is contextually focused on emotions originated from the learning process;
3. It is causality-based using antecedent-output relations;
4. It contains two functional stages to cope with the negative emotions: (i) first, regulating learning-based emotions, (ii) second, coping with all subsidiary emotions;
5. It proposes a new possibility to detect and interpret the part of emotions arising during the learning process.

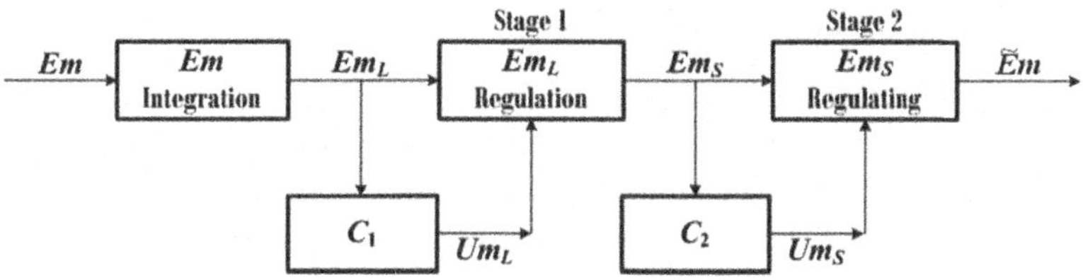

Fig. 12 Sequential regulation of negative emotions

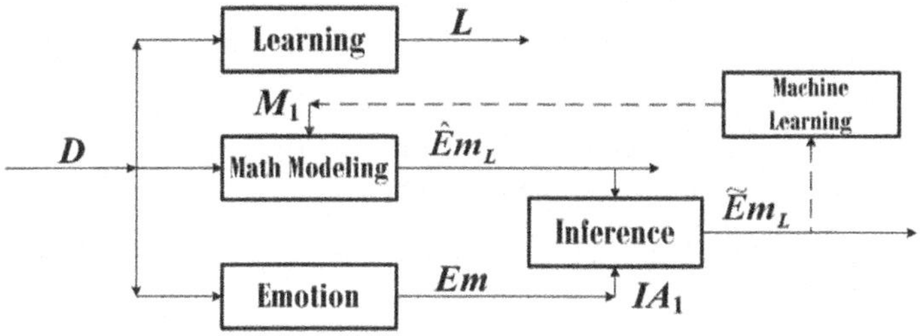

Fig. 13 Tuning the mathematical model for learning-caused emotion Em_L prediction

The two-stage produce of sequential emotion regulation is presented in Fig. 12.

The learning-based component of emotion Em_L is the input for the control algorithm C_1 which decreases iteratively Em_L up to a corresponding threshold $Em_L \leq Em^R$. The control actions Um_L are formed from the relevant didactic actions D. The rest of the negative emotion (the subcidiary Em) Em_S is directed to the control algorithm C_2 which realizes a well known action to cope with the negative in the common case [43, 47, 54] like reappraisal, suppression, amplification, substitution (Fig. 13).

In the research the emotion is considered as useful information. For the application of feedback-like control a prediction of $Em_L = \hat{Em}_L$ should be available. It is generated via the mathematical model M_1. In the intelligent agent IA_1 two information time series are processed: (i) the output of the mathematical model M_1 and an AI-based system for total emotion detection, recognition and integration. The resulting estimate $\hat{Em}_L$ is used in the machine learning procedure to improve the accuracy of the mathematical model M_1.

The proposed system has a number of advantages:

1. It is a realistic tool for applying the principles of personalization based on an advanced digital learning environment:
   - Overcoming the traditional principle "one-size-tits-are" of teaching;
   - Providing optimal learning trajectories for each student;
   - Making an available optimal pedagogical strategy and situation-provoked actions from the teacher;
   - In the different stages of learning personalization make it possible to optimize the learning trajectories for a number of clusters or the whole group;

- Reducing the failing students;

2. It can be a solid base for new methods of optimal curricula synthesis on different levels (for instance application of ontology engineering methods [77]);
3. The pedagogical model can be presented in the form of a set of rules via data mining and case-based reasoning techniques in the real classroom everyday practice;
4. It can guarantee the flexibility and a near-to-optimal behavior in changing circumstances—different learning contexts, changing personal characteristics of the students, development of teaching paradigms;
5. It will synchronize the current tasks and the evolution of long-term goals.

The proposed system is learning-oriented, not emotion-oriented. The emotion control is only a subsystem of the whole learning control system.

The system uses directly measurable variables:

- The parameters of objectively measured and consistent attributes of basic emotions,
- Current and retrospective data of all U_L and U_{Em} management impacts, measurable assessments of learning and the emotional state.

A combined approach has been adopted to compensate for the impact of emotional stimuli:

(a) Using feedback control,
(b) Using feedforward control after identifying most probable didactic disturbances on the learning process.

The present work is focused on computer-oriented Intelligent Tutoring Systems (ITS), which have been successfully developed and implemented for more than 20 years [22, 23, 37]. It assumes a certain level of development of the technological infrastructure of the learning process:

- Automatic measurement and recognition (face, speech, gestures) of emotions;
- Availability of mature enough ITS with the following possibilities:
  - To synthesize optimal learning trajectories for each individual learner,
  - An accessible set of alternative variants for each discrete step of the planned learning process,
  - A full set of didactic actions available in the ITS.
- Rich enough learning environment;
- Computer system for a dialog analysis "teacher-student";
- System for realization of a variety of assessment methods;
- Computer-based tools to regulate not learning process caused emotions e.g. [47].

6 Causality-Based Regulation of Emotions in an Advanced Learning System

The goal is to create a learning system that reduces the impact of negative emotions on the learning process and reduces the dropouts of students.

The training aims to follow for each student the optimal trajectory for the acquisition of knowledge, skills and cognitive skills within acceptable limits (Figs. 3 and 4).

The process of knowledge accumulation $K(n)$ can be summarized as:

$$K(n+1) = F_K(K(n), U_L(n), Em(n), A_L(n), P(n)), \tag{1}$$

where K is a vector of the elements of knowledge; U_L is a management influence in the learning process

$$U_L = (\ U_{L1},\ U_{L2})\ ; \tag{2}$$

U_{L1} are learning activities:

(a) Learning material (complexity, novelty, volume, content, arrangement);
(b) Pedagogical actions (individual learning style, time problems (pace, test and project deadlines);
(c) Use of learning resources (types,% on-line);

U_{L2} are evaluation results:

(a) Results of tests and inspections;
(b) Frequency of assessment;
(c) Evaluation method;

Em is the influence of the emotional state on the learning process:

$$Em = F_e(U_{L1}, U_{L2}, A_e(d_{Em})) = F_e(U_L, A_E); \tag{3}$$

d_{Em} is an external stimulus for generating emotions A,

$A_L,\ A_E$ are external disturbances on the learning process and the emotional state,

P are parametric short-term and long-term effects.

If a mixed model of emotions is accepted then a two-component model of emotions can be written in the form:

$$Em(n) = (Em_L(n), Em_A(n)). \tag{4}$$

Figure 14 shows the space of the accepted bifocal mixed emotional model in normalized form. The OAV triangle is of particular interest for this research because it is dominated by stimuli generated by different types of U_L learning management

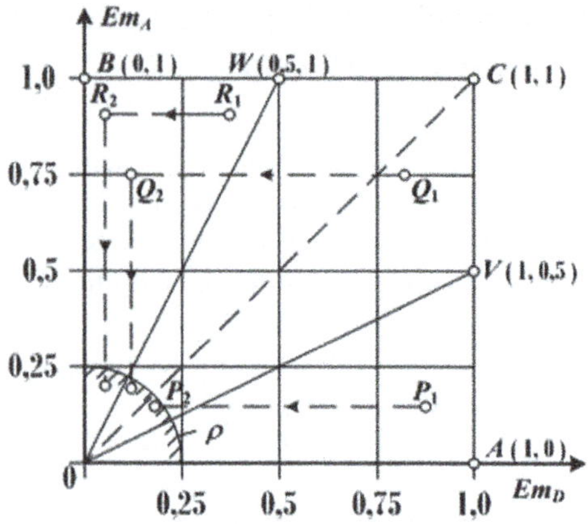

Fig. 14 Bidimensional representation of learning-based emotions

influences U_L (2). In zone 2, it is not the learning stimuli for emotion that predominate, but zone 3 is transitional with different dominance.

The main idea of the present work is shown in Fig. 15. Regardless of the starting situation P_1, Q_1 or R_1, an iterative process of reducing the negative emotion caused by the learning process (P_1–P_2, Q_1–Q_2, R_1–R_2) is initially realized. Then, if the residual emotion E_{mA} from extracurricular stimuli exceeds a set threshold value ρ, some of the established methods for coping the residual emotion developed in detail in psychology (e.g. [44, 47, 50]) are used. The transitions of the type $P_1 \rightarrow P_2$, $Q_1 \rightarrow Q_2$, $R_1 \rightarrow R_2$ are realized following the functional dependence (3). The following procedure is proposed for this purpose:

(1) An array of current and retrospective values of the emotion vectors *Em* didactic influences U_{L1} and assessments from texts and inspections U_{L2} is formed;
(2) The integral emotion $\tilde{E}m$ is determined;
(3) If $\tilde{E}m < \rho$, then the emotional state is normal and no control action is taken ($\Delta U_{L1} = 0$);

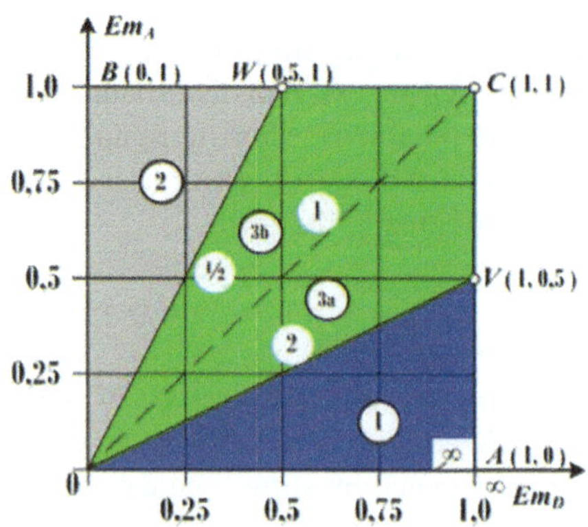

Fig. 15 Iterative regulation of learning-based emotions

(4) If $\tilde{E}m > \rho$, then a decision for relevant didactic actions has to be made, which is done in the following sequence:

(a) Differences are formed:

$$\Delta U_{L1i}(n/n-1) = U_{L1i}(n) - U_{L1}(n-1) \quad (i = \overline{1, k}), \tag{5}$$

$$\Delta U_{L2j}(n/n-1) = U_{L2j}(n) - U_{L2}(n-1) \quad (j = \overline{1, l}); \tag{6}$$

(b) Compare the differences with the corresponding thresholds:

- If all ΔU_{L1i}, ΔU_{L2j} are smaller than β_{ij}, the causes of the negative emotional state are not caused by the learning process, the situation is in zone 2 of Fig. 15 and the regulation of emotions must be done by one of the methods established in psychology (e.g. [47]);
- If some of the differences ΔU_{L2j} are significantly greater than ΔU_{L1i}, the probable cause of the negative emotions are the results of academic performance (test, incidental inspections) and the regulatory impact ΔU_L $(n+1)$ should be dominated by actions ΔU_{L2j} combined with actions ΔU_{L1i};
- If some of the differences $\Delta U_{L1m} (m = 1, 2)$ are significantly larger than the others $\Delta U_{L1i} (i = 3 - k)$ and ΔU_{L2j} $(j = 1, 2)$ the regulatory impact ΔU_L must compensate ΔU_{Lm} by appropriate single or combined didactic effects;

(c) The decision for management action $\Delta U_L(n+1)$ is taken on the basis of a system of rules R_r that take into account the attributes of the current situation;

(5) If after the undertaken regulatory influences of the type $\Delta U_{Ll}(n+1)$ and/or $\Delta U_{L1,2}(n+1)$ the emotional state is not brought to zone ρ (Fig. 15), the procedure is repeated iteratively;

(6) If the possibilities for reducing the negative emotional state with the means of di-dactic influences are exhausted$U_L = (U_{L1}, U_{L2})$ (zone 2 of Figs. 14 and 15), the general approaches developed in the theory of emotion management [22, 23, 39, 47] are used for regulation.

7 Integrated Learning/Emotion Control System

The structural scheme of the proposed Integrated Control System is shown in Fig. 16.

It contains five main functional blocks: Learning/Emotion Processes, Data Processing, Strategy, Control, Decoupling. P_L and P_E describe the learning and emotional states under the control inputs U_L and U_E and also the disturbances d_L and d_E. The operators P_{LE} and P_{EL} represent the input–output cross-connection

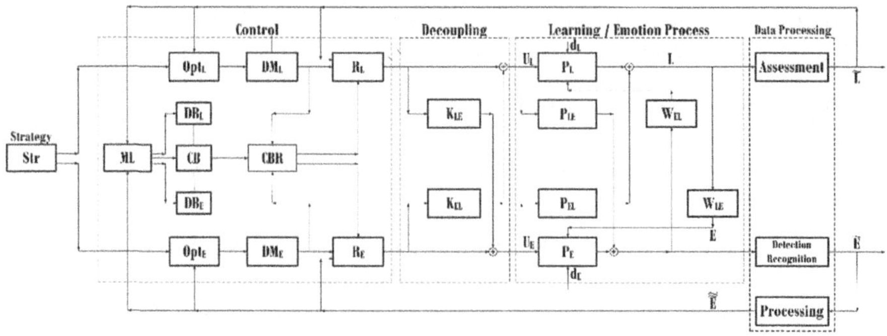

Fig. 16 Integrated learning/emotion AI-based control system

among learning and emotion processes; the operators W_{LE} and W_{EL}—the output-input inverse cross-connections. The mutual impact between learning and emotion processes is multidirectional and multimodal. The Data Processing block embraces all measurements and new data processing: the Assessment (tests, question/answer data acquisition and analysis, dialog-based estimation, results from basic emotions (face, voice, gesture); the processing involves integration and decomposing emotions. The Control block is AI-based and it includes a number of agents with different functionalities: DM_L and DM_E are modules for decision making which concerns learning and emotion control; R_L and R_E are executive elements which realize the accepted decisions as real actions. They are the main source for stimuli for both learning and emotion processes thus forming the content of the control variables U_L and U_E. the modules Opt_L and Opt_E are tools optimizing the candidate variants for decision making. The rest of the elements are: *ML*—module for machine learning, DB_L, DB_E—data bases with information about the states of learning and emotions, *CB* is a case-base as a part of the system of case-based reasoning (CBR).

Depending on the stage of education transformation, four different control principles can be accepted:

1. Control without using instrumental feedback,
2. Independent learning and emotion control with feedback,
3. Coupled control,
4. Decoupled control (Table 1).

Table 1 .

		R_L	R_E	K_{LE}	K_{EL}	DM_L	DM_S	ML	DB_L DB_E	CB CBR	Ass_L	Em R_{BC}
1	Open loop					✔					✔	
2	Not connected	✔				✔					✔	
3	Coupled	✔	✔			✔	✔		✔		✔	✔
4	Decoupled	✔	✔	✔	✔	✔	✔	✔	✔	✔	✔	✔

The proposed in this research system corresponds to the coupled variant and it can be performed in full volume only in an AI-based learning digital environment. The main problem arising in the design of this alternative consists of the strong cross-connections represented by the operators P_{LE}, P_{EL} and W_{EL}, W_{LE}. As a result, the effect of learning control actions U_L significantly influences the emotional state E and vice versa. To overcome this problem, data mining techniques are relevant to derive a set of rules for decoupling the control processes in both control loops. They are realized by operators K_{LE} and K_{EL}. As the control system will be virtual and personalized for each student the content of the decoupling operators must be adaptively changeable and available only applying data-driven procedures. In this regard, the realistic intelligent learning system with embedded emotion control is feasible only in the case if it realizes the full volume of the necessary measurements of all coordinates in the scheme presented in Fig. 16.

8 Decreasing Learning-Based Emotion Em_L in the Integrated Intelligent Control System

In the case when the integrated negative emotion Z (Fig. 9) is represented by two-component vector (Figs. 14 and 15) we can normalize Z_1 and Z_2—the coordinates of vector Z (Fig. 9), in the following way:

$$Z_1 = \frac{|Z_1|}{|Z_1| + |Z_2|}, \quad Z_2 = 1 - Z_1. \tag{7}$$

The emotional state can be presented as shown by point S in Fig. 17. In the marked triangle, zone 1 with coordinates (0.5, 1, 0.5), the negative emotion based on learning process Em_L is dominant. During the control performance that emotion will decrease following the hypotenuse AB from point A $(1, 0)$ to point B $(1, 0)$ as shown in Fig. 18.

In the case of presenting the integral emotion with three-component vector $Z = (Z_1, Z_2, Z_3)$ in Fig. 9, the normalization can be done in the next way:

$$Z_i = \frac{|Z_i|}{|Z_1| + |Z_2| + |Z_2|} \quad (i = 1, 2, 3), \sum_{i=1}^{3} Z_i = 1. \tag{8}$$

For the three-dimensional case we can use barycentric coordinates with a representative point S (Fig. 19).

The corresponding steps to decrease negative learning-based emotion Em_L are presented in Fig. 20.

Figure 21 shows several possible scenarios for dynamic behavior of the trajectory for the acquisition of knowledge and skills in the event of a negative emotional impact in point A.

- K_1 is a situation of returning to the basic trajectory K_0 with a given goal;

Fig. 17 Normalized integrated emotion representation in the two-dimensional case

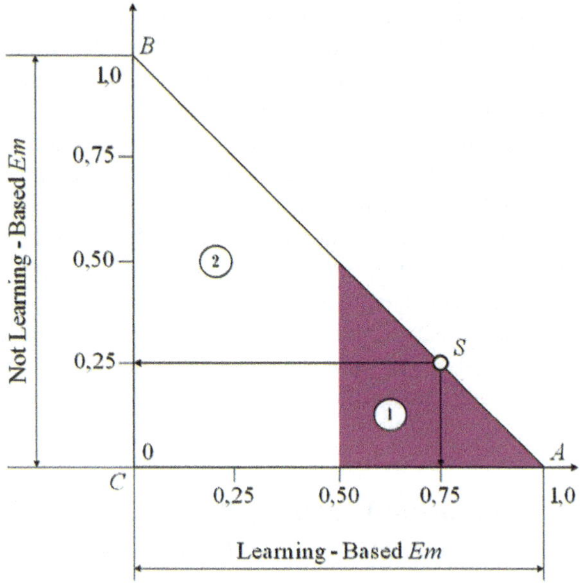

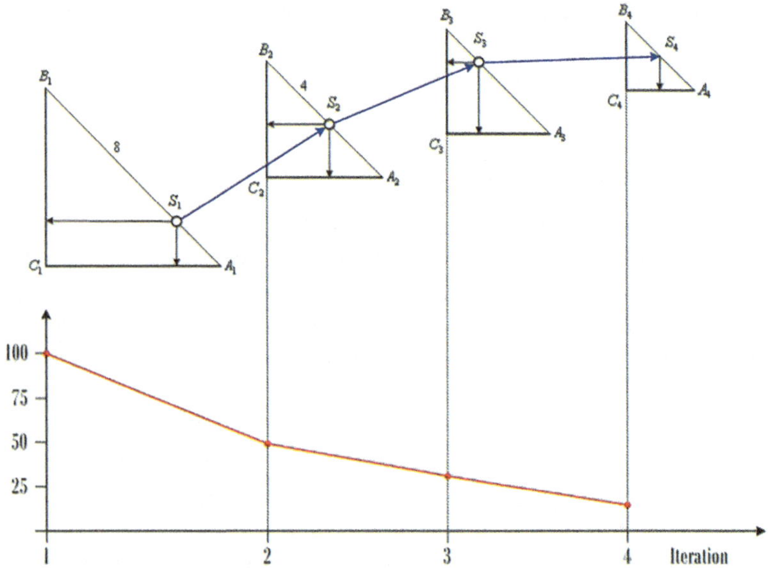

Fig. 18 Trajectory of decreasing negative emotion Em_L

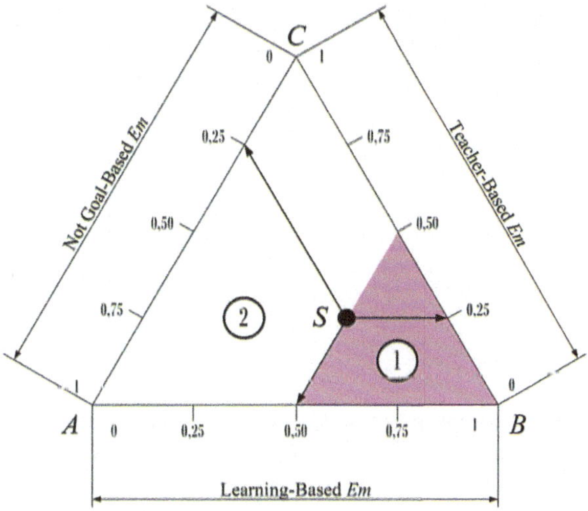

Fig. 19 Integral emotion I_{Em} represented in barycentric coordinates

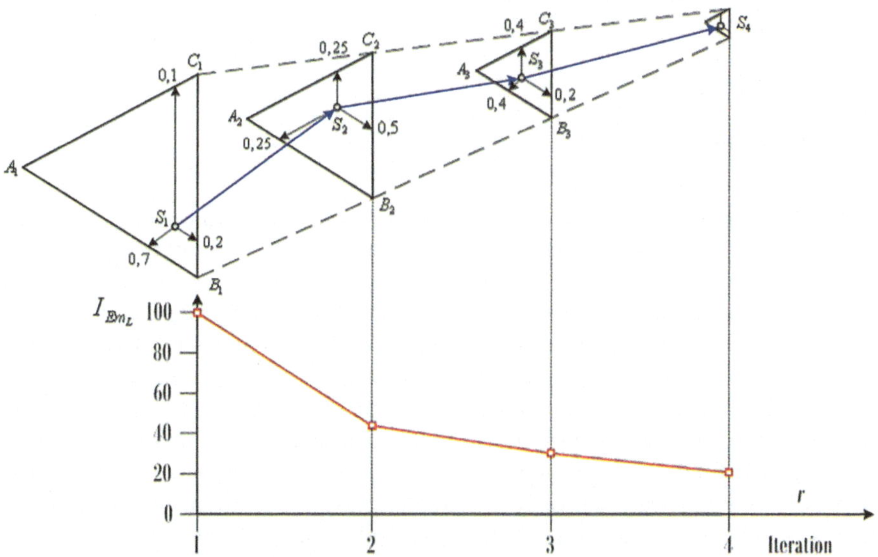

Fig. 20 Sequential decreasing learning-based emotion Em_L

- K_2 is mastering the emotion, but reducing the expected goal;
- K_3 is a case of dropping out of the student;
- K_4 is indicative of the possibility of a weak negative emotion (e.g. a negative test result) to move to a learning trajectory with a higher goal.

In this research the modules of decision making DM_L, DM_E, optimization Opt_L, Opt_E, all modules *ML*, *CBR*, *CB*, DB_L, DB_E in the AI-based intelligent part of the

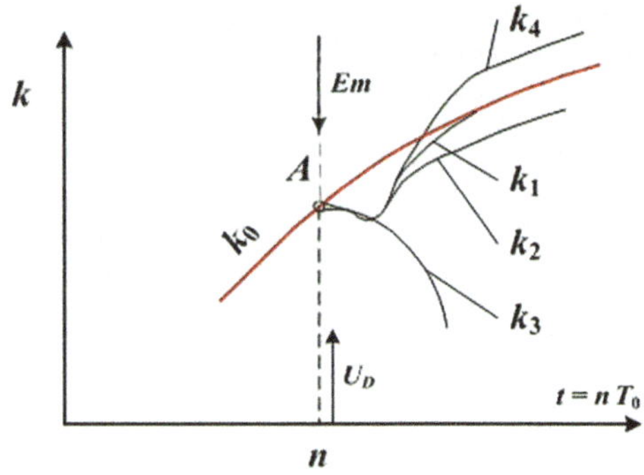

Fig. 21 Alternative learning trajectories after an Emotion disturbance

integrated learning control system as well as the problem of strategy formation and goal setting are not a subject of consideration.

As noted above the problem of regulating subsidiary integral emotion is realized as a stage of the whole control action in each emerging learning/emotional situation (Fig. 12) using traditional methods for coping with negative emotions in the common case [47, 57].

9 Conclusion

This study proposes a new combined feedback and feedforward control of negative emotions as an embedded part of complex systems aimed at significant innovation and transformation of learning processes using an emerging digital environment based on artificial intelligence. Current negative emotions are presented via integration as a combination of learning-based and subsidiary components. The algorithm is proposed to cope with them in the framework of a two-stage procedure where there are applied both knowledge of real life teaching practice and advanced achievements in cognitive psychology. A stimuli-based approach accepted in this work enables to aggregate data for the current emotion received from two different sources: one is the output of a contemporary sensor system for detection, recognition, classification and integration of emotions and the second is available as a result from data-driven ML-based mathematical model. Due to the strong interconnections between learning- and emotion- subsystems an intelligent decoupling control system is proposed to overcome the lack of effective independent control actions addressed to learning and/or emotion regulation. Because of the included advanced AI-based techniques the whole system meets the present-day requirements of scaling, flexibility and robustness in arbitrary learning and control situations.

References

1. Russell, S., Norvig, P.: Artificial Intelligence: A Modern Approach, 3rd ed. Pearson (2009)
2. Negnevitsky, M.: Artificial Intelligence: A Guide to Intelligent Systems, 3rd edn. Pearson Education, Canada (2014)
3. Baesens, B.: Analytics in a Big Data World. Wiley, New York (2014)
4. Berthold, M., Hand, D.: Intelligent Data Analysis, 2nd ed. Springer, Berlin (2007)
5. Kordon, A.K.: Applying Data Science: How to Create Value with Artificial Intelligence. Springer, Berlin (2020)
6. McKinsey: Global Institute, Artificial Intelligence: The Next Digital Frontier? (2017)
7. Pascual, D.: Artificial Intelligence Tools: Decision Support Systems in Conditions Monitoring and Diagnosis. CRC Press, Boca Raton (2015)
8. Chen, N., Christensen, L., Gallagher, K., Mate, R., Rafert, G.: Global Economic Impacts Associated with Artificial Intelligence. Study Analysis Group, Boston, MA, vol. 25 (2016)
9. Andreu-Perez, J., Poon, C.C., Merrifield, R.D., Wong, S.T., Yang, G.-Z.: Big Data for Health. IEEE J. Biomed. Health Inform. **19**, 1193–1208 (2015)
10. Kober, J., Bagnell, J.A., Peters, J.: Reinforcement learning in robotics: a survey. Int. J. Robot. Res. **32**, 1238–1274 (2013)
11. Sobh, T., Elleithy, K. (eds.): Emerging Trends in Computing, Informatics, Systems Sciences, and Engineering. Springer, Berlin (2013)
12. IEEE Computer Society's Top 12 Technology Trends for 2020, https://www.prnewswire.com/news-releases/300971863.html.
13. Crawley, E.F., Hosoi, A., Mitra, A.: Redesigning undergraduate engineering education at MIT—the new engineering education transformation (NEET) initiative. American Society for Engineering Education (2018)
14. Graham, R.: The Global State of the Art in Engineering Education. Survey Report commissioned by MIT, Phase I 2016, Phase II (2018)
15. Natarajan, R. (ed.): Proceedings of the International Conference on Transformations in Engineering Education (ICTIEE 2014). Springer, Berlin (2014)
16. Dahdouh, K., Dakkak, A., Oughdir, L., Ibriz, A.: Improving Online Education Using Big Data Technologies (2020). https://doi.org/10.5772/intechopen.88463
17. Vaessen, M., van der Heijden, K., de Gelder, B.: Decoding of emotion expression in the face, body and voice reveals sensory modality specific representations. bioRxiv preprint (2019)
18. Kleinjohann, B.: Fuzzy emotion recognition in natural speech. In: IEEE International Workshop on Robot and Human Interactive Communication (ROMAN 2005), pp. 317–322 (2005)
19. Yadegaridehkordi, E., Noor, N.F.B.M., Ayub, M.N.B., Affal, H.B., Hussin, N.B.: Affective computing in education: a systematic review and future research. Comput. Educ. **42**, (2019). https://doi.org/10.1016/j.compedu.2019.103649
20. Troussas, C., Virvou, M.: Affective computing and motivation in educational contexts: data pre-processing and ensemble learning. In: Advances in Social Networking-Based Learning, Intelligent Systems Reference Library, vol. 181. Springer, Berlin (2020)
21. Hernández, Y., Sucar, L.E., Arroyo-Figueroa, G.: Building an affective model for intelligent tutoring systems with base on teachers' expertise. In: Gelbukh, A., Morales, E.F. (eds.) MICAI 2008: Advances in Artificial Intelligence MICAI 2008, Lecture Notes in Computer Science, vol. 5317. Springer, Berlin (2008)
22. Han, J., Zhao, W., Jiang, Q., Oubibi, M., Hu, X.: Intelligent tutoring system trends 2006–2018: a literature review. In: 2019 Eighth International Conference on Educational Innovation through Technology (EITT), Biloxi, MS, USA, pp. 153–159 (2019)
23. Kulik, J.A., Fletcher, J.D.: Effectiveness of intelligent tutoring systems—a meta-analytic review. Rev. Educ. Res. (2015). https://doi.org/10.0132/0034654315581420
24. Pratap, M.: How Learning Management System Revolutionizing the Education Sector (2020). https://hackernoon.com/qor32bx
25. Popenici, S., Kerr, S.: Exploring the impact of artificial intelligence on teaching and learning in higher education. Res. Practice Technol. Enhanced Learn. **12**(1), 1–13 (2017)

26. Saltan, F.: The new generation of interactive whiteboards: how students perceive and conceptualize?. Participatory Educ. Res. **6**(2), 93–102 (2019)
27. van Alten, D.C.D., Phielix, C., Janssen, J., Kester, L.: Effects of flipping the classroom on learning outcomes and satisfaction: a meta-analysis. Educ. Res. Rev. **28**, 1–18 (2019)
28. Alexander, P.A., Pekrun, R., Linnenbrink-Garcia, L. (eds.) International Handbook of Emotions in Education. Routledge, London (2014)
29. Faria, R., Almeida, A., Martins, C., Gonçalves, R., Figueiredo, L.: Including emotion in learning process. In: Portuguese Conference on Artificial Intelligence, pp. 27–32. Springer, Berlin (2015)
30. Kuppens, P., Oravecz, Z., Tuerlinckx, F.: feelings change: accounting for individual differences in the temporal dynamics of affect. J. Personal. Social Psychol. **99**(6), 1042–1060 (2010)
31. Hamann, S., Canli, T.: Individual differences in emotion processing. Curr. Opin. Neurobiol. **14**, 233–238 (2004)
32. Vishkin, A., Hasson, Y., Millgram, Y., Tamir, M.: One size does not fit all: tailoring cognitive reappraisal to different emotions. In: Personality and Social Psychology Bulletin, Society for Personality and Social Psychology, pp. 1–16 (2019)
33. Lewis, M.D.: Bridging emotion theory and neurobiology through dynamic systems modeling. Behav. Brain Sci. **28**(2), 169–194 (2005)
34. D' Mello, S., Graesser, A.: The half-life of cognitive-affective states during complex learning. Cogn. Emotion (2011). https://doi.org/10.1080/02699931.2011.613668
35. Ekman, P.: What scientists who study emotion agree about. Persp. Psychol. Sci. **11**, 31–34 (2016)
36. Tiedens, L.Z., Linton, S.: Judgment under Emotional Certainty and Uncertainty: the effects of specific emotions on information processing. J. Pers. Soc. Psychol. **81**, 973–988 (2001)
37. Feidakis, M., Daradoumis, T., Caballe, S.: Emotion measurement in intelligent tutoring systems: what, when and how to measure? In: Third International Conference on Intelligent Networking and Collaborative Systems, pp. 807–813 (2011)
38. D'Mello, S., Lehman, B., Pekrun, R., Graesser, A.: Confusion can be beneficial for learning. Learn. Instr. **29**, 153–170 (2014)
39. Colombo, D., Fernandez-Alvarez, J., Palacios, A.G., Cipresso, P., Bottela, G., Riva, G.: New technologies for the understanding, assessment and intervention of emotion regulation. Front. Psychol. **10**(12) (2019)
40. Västfjäll, D., Slovic, P., Burns, W.J., Erlandsson, A., Koppel, L., Eand, A.: et al.: The arithmetic of emotion: integration of incidental and integral affect in judgments and decisions. Front. Psychol. **7**(325) (2016)
41. Feidakis, M.: A review of emotion-aware systems for e-learning in virtual environments in formative assessment. In: Learning Data Analytics and Gamification. Elsevier, Amsterdam (2016)
42. Cowena, A.S., Keltnera, D.: Self-report captures 27 distinct categories of emotion bridged by continuous gradients. PNAS, USA, Published online 5 Sept 2017
43. Harley, J.M., Pekrun, R., Taxer, J.L., Gross, J.J.: Emotion regulation in achievement situations: an integrated model. Educ. Psychol. **54**(2), 106–126 (2019)
44. Kort, B., Reilly, R.: Analytical models of emotions, learning and relationships: towards an affect-sensitive cognitive machine. In: Proceedings of the International Conference on Virtual Worlds and Simulation (VWSim), San Antonio, Texas (2002)
45. Ibrahimoglu, N., Unaldi, I., Samancioglu, M., Baglibel, M.: The relationship between personality traits and learning styles: a cluster analysis. Asian J. Manage. Sci. Educ. 93–108 (2013)
46. Lerner, J.S., Li, Y., Valdesolo, P., Kassam, K.: Emotion and decision making. Ann Rev. Psychol. (2014)
47. Gross, J.J.: Emotion regulation: current Status and future prospects. Psychol. Inq. **26**, 1–26 (2015)
48. Esau, N., Kleinjohann, L., Kleinjohann, B.: An adaptive fuzzy emotion model for emotion recognition. In: Proceedings of the 4-th European Society for Fuzzy Logic and Technology, Barcelona, pp. 73–78 (2005)

49. Zhou, R., Feng, J., Chang, H., Zhou, Y.: fuzzification of attribute information granules and its formal reasoning model. CAAI Trans. Intell. Technol. **2**(3), 116–125 (2017)
50. Damasio, A.R., Grabowski, T.J., Bechara, A., Damasio, H., Ponto, L.L., Parvizi, J., et al.: Subcortical and cortical brain activity during the feeling of self-generated emotions. Nat. Neurosci. **3**, 1049–1056 (2000)
51. Ozawa, S., Matsuda, G., Hiraki, K.: Negative emotion modulates prefrontal cortex activity during a working memory task: a NIRS study. Front. Hum. Neurosci. **8**(46) (2014)
52. Kahneman, D., Riis, J.: Living, and thinking about it: two perspectives on life. In: Kahneman, D., Riis, J.: The Science of Well-Being (2005). books.google.com
53. Shafir, R., Thiruchselvam, R., Suri, G., Gross, J.J., Sheppes, G.: Neural processing of emotional-intensity predicts emotion regulation choice. Social Cogn. Affect. Neurosci. 1863–1871 (2016)
54. Wortha, F., Azevedo, R., Taub, M., Narciss, S.: Multiple negative emotions during learning with digital learning environments—evidence on their detrimental effect on learning from two methodological approaches. Front. Psychol. **10** (2019). https://doi.org/10.3389/fpsyg.2019.02678
55. Thagard, P., Nerb, J.: Emotional gestalts: appraisal, change, and the dynamics of affect personality and social psychology review. **6**(4), 274–282 (2002)
56. Gregoire, M.: Is it a challenge or a threat? a dual-process model of teachers' cognition and appraisal processes during conceptual change. Educ. Psychol. Rev. **15**(2) (2003)
57. Pekrun, R., Erenzel, A., Goetz, T., Perry, R.: The control-value theory of achievement emotions: an integrative approach to emotions in education. In: Schutz, P., Pekrun, R. (eds.) Emotion in Education, pp. 13–36. Academic Press, Amsterdam (2007)
58. Lange, J., Dalege, J., Borsboom, D., van Kleef, G.A., Fischer, A.H.: Toward an integrative psychometric model of emotions. Perspect. Psychol. Sci. **15**(2), 444–468
59. Seta, J.J., Haire, A., Seta, C.E.: Averaging and summation: positivity and choice as a function of the number and affective intensity of life events. J. Exp. Soc. Psychol. **44**, 173–186 (2008)
60. Tsiourti, C., Weiss, A., Wac, K., et al.: Multimodal integration of emotional signals from voice, body, and context: effects of congruence on emotion recognition and attitudes towards robots. Int. J. Soc. Robot. **11**, 555–573 (2019)
61. Byun, S.-W., Lee, S.-P.: Human emotion recognition based on the weighted integration method using image sequences and acoustic features. Multimedia Tools Appl. **9** (2020)
62. Zhou, R., Feng, J., Chang, H., Zhou, Y.: Fuzzification of attribute information granules and its formal reasoning model. CAAI Trans. Intell. Technol. **2**(3), 116–125
63. Liliana, D.Y., Basaruddin, T., Widyanto, M.R., Oriza, I.I.D.: Fuzzy emotion: a natural approach to automatic facial expression recognition from psychological perspective using fuzzy system (2019)
64. Ray, T., Kordon, A., Wells, C.: Applied Data Mining in Forecasting. SAS Institute (2012)
65. Roth, G., Vansteenkiste, R., Ryan, M.: Integrative Emotion Regulation: Process and Development from a Self-determination Theory Perspective, pp. 1–12. Development and Psychopathology, Cambridge University Press (2019)
66. Roth, G., Shahar, B.H., Zohar-Shefer, Y., Benita, M., Moed, A., Bibi, U., Ryan, R.M.: Benefits of emotional integration and costs of emotional distancing. J. Pers. **86**, 919–934 (2018)
67. Melo, F.R., Flores, E.L., Carvalho, S.D.: Multilevel content's structure for personalization in conexionist intelligent tutor systems. In: 8th International Conference on Information Systems and Technology Mmanagement, 2011, Proceedings of 8th CONTECSI, TECSI, EAC FEA USP (2011)
68. Reigeluth, C.M., Aslan, S., Chen, Z., et al.: Personalized integrated educational system: technology functions for the learner-centered paradigm of education. J. Educ. Comput. Res. **53**(3), 459–496
69. Zhou, G., Azizsoltani, H., Ausin, M.S., Barnes, T., Chi, M.: Hierarchical reinforcement learning for pedagogical policy induction. In: Isotani, S., Millán, E., Ogan, A., Hastings, P., McLaren, B., Luckin, R.R. (eds.) Artificial Intelligence in Education AIED 2019, Lecture Notes in Computer Science, vol. 625. Springer, Berlin (2019)

70. Frijda, N.H.: Emotions, individual differences and time course: reflections. Cogn. Emotion **23**(7), 1444–1461 (2009)
71. Jevsikova, T., Berniukevicius, E., Kurilovas, E.: Application of resource description framework to personalise learning: systematical review and methodology. Inf. Educ. **16**(1), 61–82 (2017)
72. Connor, C.M.: Using technology and assessment to personalize instruction: preventing reading problems. Prev. Sci. **20**(1), 89–99 (2019)
73. Stewart, C.: Learning analytics: shifting from theory to practice. J. Empower Teach. Excel **1**(1) (2017)
74. Kaltenborn, R., Hadjiski, M., Koynov, S.: Intelligent control of negative emotions in a computer-based learning system. In: Proceedings of 2020 IEEE 10th International Conference on Intelligent Systems, Bulgaria, Varna, pp. 119–124, 28–30 Aug 2020
75. Kaltenborn, R.: Embedding the assessment of emotion in the learning process with ai-driven technologies. In: Petrov, V., Anderson, K. (eds.) Traditional Learning Theories, Process Philosophy and AI, pp. 145–166. Les Editions Chromatica (2019)
76. Hadjiski, M., Kaltenborn, R.: Learnability as an indicator for planning and control of learning systems. Inform. Technol. Control **4** (2020)
77. Chang, M., D'Anjello, G., Gaeta, M., Orciuoli, F., Sampson, D., Simonelli, C.: Building ontology-driven tutoring models for intelligent tutoring systems using data mining. IEEE Access **8**, 48151–48162 (2020)

Intelligent Network-Flow Solutions with Risks at Transportation of Products

Vassil Sgurev, Lyubka Doukovska, and Stanislav Drangajov

Abstract The paper proposes a method for making optimal decisions with risks in the very important task of transporting of commodities on a network of arbitrary configuration. One of the biggest flows of commodities in worldwide scale are the agricultural ones. This includes also the case, which happens often at harvest when the vehicles make circular course e.g. between the field and the granaries or grain silos. The method takes into account both the specific network properties and the probability of adverse events and the risks to the individual sections of the network in the implementation of these services. The proposed method minimizes not only the total cost of transporting commodities but also the insurance costs to cover the risks. A complete numerical modeling of decision-making and transportation of commodities on a network of arbitrary configuration is implemented, whereby both the cost of transport and the coverage of emerging risks are minimized by the proposed method. The possibilities for practical application of the proposed method for transportation of commodities as well as for solving other applied problems are pointed out.

Keywords Transportation of commodities · Risks · Minimization · Maximum flow · Minimum cut

1 Introduction

Decision-making models and methods emerge as a separate scientific area within the framework of artificial intelligence (AI) [1] in the middle of the previous century. At the same time, researchers in the field of Operations Research (OR) [2] and, later, intelligent systems [3] begin to address these issues. They are gradually clarifying that the use of specific methods and approaches from one scientific area—AI, leads to

V. Sgurev · L. Doukovska · S. Drangajov (✉)
Intelligent Systems Department, Institute of Information and Communication Technologies, Bulgarian Academy of Sciences, Acad. Georgi Bonchev Str., Sofia 1113, Bulgaria

L. Doukovska
e-mail: lyubka.doukovska@iict.bas.bg

V. Sgurev et al. (eds.), *Advances in Intelligent Systems Research and Innovation*, Studies in Systems, Decision and Control 379,
https://doi.org/10.1007/978-3-030-78124-8_19

effective results in the other direction—OR and vice versa. Moreover, the introduction of new real-world parameters into statistical models, based on statistical, stochastic and more recently, risk factors, can best be accounted for using mixed models from different scientific fields.

In recent years a new scientific direction emerges—Network Science [4] emerges, which has much to do both with AI and OR, and not only. In fact today's interlinked world is enveloped by networks—transport, communication, financial, information etc. and all these networks are crucial to the existence of the world.

A mixed network two-product model of resources and risks in conducting logistics operations is proposed in the present work.

The importance of transport and logistics for every social, military and economic activity is undeniable since the emergence of public relations to this day. Obviously, without transfer of loads, labor force, etc. the development of modern society is impossible. It is therefore important to evaluate the opportunities and risks involved in transferring a product unit, regardless of its type, on separate sections of a route.

One of the largest commodity flows in today's global society is that of agricultural production [5, 6]. Not in vain, the main problem of trade contradictions USA–China is the issue of import of agricultural products from the USA into China. The European Union also allocates more than €385 billion in agricultural subsidies for the next 2021–2027 planning period. Another important risk that arises in the transportation of agricultural products is the danger to human health and the transmission of contagion [7, 8], but this is not the subject of this work, since matter affects many social structures.

A method is proposed in the present work, for reducing the total transportation costs, risks and means of insurance for transportation of commodities, both operationally and in long-term prospective.

The type of commodities—agricultural e.g. plant-growing or livestock, perishable or durable, or industrial is not taken into account, neither is the mode of transport—land (road, rail), marine, air (in rare cases), etc. An attempt has been made a generalized scheme of risks and their optimal insurance security from the perspective of the producer and the recipient of production to be presented, on the basis of a network of arbitrary complexity.

Most often, large bulks of commodities are transported by land—rail and road on a continental scale, and large flows, especially from grain, are carried by waterways, sea and river, ultimately reaching the processing plants and finally to end users through combined logistics.

Another case, which is quite frequent, is when the vehicles travel to transport the commodities along a circular route from the production site to the processing plants or large storehouses and back again.

2 Problem Formalization and Denotations

Hereinafter, the plain graph with 7 vertices and 11 arcs, shown in Fig. 1 will be used

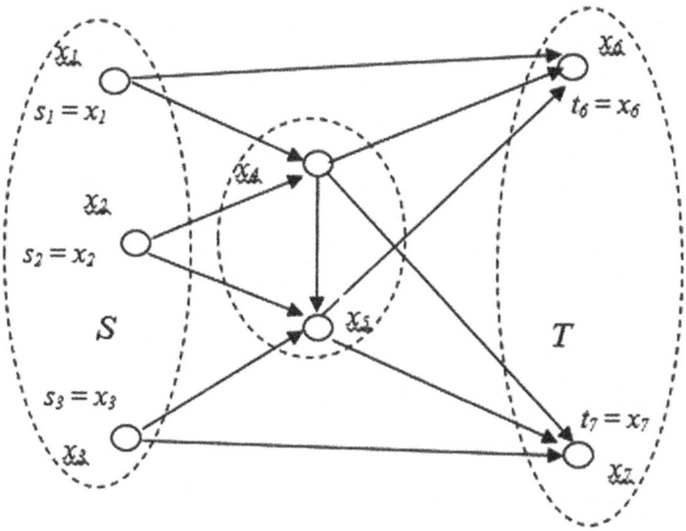

Fig. 1 Illustrative plain graph with 7 vertices and 11 arcs

as an illustration.

The following denotations will be introduced [9, 10]:

$G(X, U)$—directed graph with a set of vertices (nodes) $X = \{x_i/i \in I\}$ and a set of arcs (edges) $U = \{x_{ij}/(i, j) \in G\}$, where I is the set of indices of all vertices from X; $I = \{i/x_i \in X\}$, and G is the set of the pairs of indices of all arcs U of the graph—$G = \{(i, j)/x_{i,j} \in U\}$. For each arc $x_{i,j}$ vertex x_i is initial, and vertex x_j—end.

It is assumed that the entire volume of agricultural resource integrated as a flow v is generated in a set of source-vertices S and collected in a set of consumer-vertices T. The network flow conservation equations [11–13] are observed in all others vertices of the network $\{X\backslash(S\backslash\bigcup T)\}$, in which the amount of resource entering the vertex is equal to the amount of outgoing from it. Some arc flow (resource) passes through any arc and a price must be paid for both the passage of the resource and the corresponding risk, i.e. there are two arc flow functions on the network:

f_{ij}—function of the resource network flow on arc x_{ij};

r_{ij}—network flow function of risk on section x_{ij}.

As shown in [12, 14] a relation exists between f_{ij} and r_{ij}:

$$r_{ij} = p_{ij} f_{ij}, \tag{1}$$

$$0 \le p_{ij} \le 1; (i, j) \in G. \tag{2}$$

where p_{ij} is the probability an adverse event to occur at transportation of resource f_{ij} along arc x_{ij}. Risk r_{ij} in (1) is defined as a product of two measures—the amount of resource being transported along the arc and the probability for an adverse event during this operation.

The following denotations are accepted, concerning the directed graph $G(X, U)$:

$\Gamma_i^1 = \{j/(i, j) \in G\}$—direct mapping, consisting of the set of end nodes' indices of arcs outgoing from vertex x_i;

$\Gamma_i^{-1} = \{j/(j, i) \in G\}$—reverse mapping, representing the set of indices of all initial vertices of arcs whose end vertex is x_i; [10]

c'_{ij}—upper bound of capacity for resource, feasible along arc x_{ij};

c''_{ij}—upper bound of capacity for risk, feasible along arc x_{ij};

a'_{ij}—evaluation (cost) of passing a unit of resource along arc x_{ij};

a''_{ij}—evaluation (e.g. insurance) for a unit of risk along arc x_{ij};

v and $v_{\max}$—flow and maximal flow of resources;

v' and $v'_{\max}$—flow and maximal flow of risks.

The requirement of observing the constraints on the flows of risk $\left\{c''_{ij}\right\}$ and resource $\left\{c'_{ij}\right\}$ leads to the following generalized constraint for resource, coordinated with the constraint for risk:

$$c^s_{ij} = \min\left[c'_{ij}, \frac{c''_{ij}}{p_{ij}}\right]; \tag{3}$$

and keeping the constraints for resources $\left\{c'_{ij}\right\}$ in the constraints for risk $\left\{c''_{ij}\right\}$—to:

$$c^r_{ij} = \min\left[p_{ij}c'_{ij}, c''_{ij}\right]. \tag{4}$$

It follows from the two relations above, that if (3) and (4) are observed and their both sides are multiplied by p_{ij}, the following two equalities will be obtained:

$$c^r_{ij} = p_{ij}c^s_{ij}; \; c^s_{ij} = \frac{c^r_{ij}}{p_{ij}}. \tag{5}$$

The two coordinated values c^s_{ij} and c^r_{ij} will be considered as upper bounds of the arc capacities for resources and risks on the arc x_{ij}. Both network flows, resource—$\{f_{ij}\}$ and risk—$\{r_{ij}\}$, are upper bounded on the separate arcs in the following way:

$$f_{ij} \le C^S_{ij} \text{ for each } (i, j) \in G; \tag{6}$$

$$r_{ij} \le C^r_{ij} \text{ for each } (i, j) \in G. \tag{7}$$

Some clarifications should be made here regarding Fig. 1. Each of the source vertices of the set S can be considered as a set of sources of different agricultural products. E.g. x_3 can be a large holding of several units producing different types of product—grain, livestock, etc. to be transported to some of the distribution nodes, e.g. x_4 or x_5, or directly to the end consumer. The set of distribution vertices $\{x_4, x_5\}$ (surrounded by dotted line) also has a more complex topology. By rail or road, production can be transported to many end consumers T, also surrounded by dotted line. End users are e.g. processing plants, chain stores or ports. This implies, as the case may be, that the topology is precisely defined.

The optimal allocation of agricultural resources on the transport network, taking into account the associated risk, is not possible without the creation of network models that take into account the following two features:

(a) Network-flow interpretation of resource allocation;
(b) Effectively taking into account risk and its allocation on the network.

Each section of the network $x_{ij} \in U$ is of upper-bounded capacity c_{ij}, i.e.

$$f_{ij} \leq c_{ij};\ (i, j) \in G. \tag{8}$$

Physical constraints exist at the resource allocation

$$f_{ij} \leq c'_{ij};\ (i, j) \in G, \tag{9}$$

and the following is valid at the network risk allocation:

$$r_{ij} \leq c''_{ij};\ (i, j) \in G. \tag{10}$$

Besides, passing a unit of resource along arc $x_{ij} \in U$ costs a'_{ij} price per unit, and for a unit of risk r_{ij}—a''_{ij} units value, respectively.

For optimal allocation of resources and risks over the network, the following network-flow model is used, in which the objective function L and the corresponding constraints are defined as follows:

$$L = \sum_{(i,j)\in G} a_{ij} f_{ij} \to \max(\min), \tag{11}$$

subject to constraints: for each $i \in I$ and $(i, j) \in G$

$$\sum_{j\in\Gamma_i^1} f_{ij} - \sum_{j\in\Gamma_i^{-1}} f_{ji} = \begin{cases} v_i\ , \text{ if } x_i \in S; \\ 0, \text{ if } x_i \notin (S \cup T) \\ -v_i, \text{ if } x_i \in T \end{cases} \tag{12}$$

$$f_{ij} \leq c_{ij} \text{ for each } (i, j) \in G; \tag{13}$$

$$f_{ij} \geq 0 \text{ for each } (i, j) \in G; \tag{14}$$

$$\sum_{i\in I(S)} v_i = \sum_{j\in I(T)} v_j = v,$$

$$I(S) = \{i/x_i \in S\} \text{ and } I(T) = \{j/x_j \in T\}$$

v is the total sum flow from sources of S to consumers of T.

The correlation between the resource f_{ij} on the section $x_{ij} \in U$, the probability p_{ij}; $(i, j) \in G$ for an adverse event and risk r_{ij}; $(i, j) \in G$ is defined in formula (1) in the following way: for each $(i, j) \in G$

$$f_{ij} = \frac{r_{ij}}{p_{ij}}, \tag{15}$$

where

$$r_{ij} \geq 0; \; 1 \geq p_{ij} \geq 0. \tag{16}$$

It follows from (1) that the risk r_{ij} is expressed by the same measure as f_{ij}. Methods for finding optimal flows of resources and risks are proposed in [12], at which all requirements, concerning capacities $\{c'_{ij}\}$ and $\{c''_{ij}\}$ are observed. Further, the index s will denote the parameters relating to the resource flows, and the index r—the parameters of the risk flows. The two-product flow may be written down as:

$$f_{ij}a^s_{ij} = a'_{ij} f_{ij} + a''_{ij} r_{ij} = f_{ij}\left(a'_{ij} + p_{ij}a''_{ij}\right); \tag{17}$$

$$a^s_{ij} = a'_{ij} + p_{ij}a''_{ij}. \tag{18}$$

If in (5) and (7) is put: for each $(i, j) \in G$

$$c_{ij} = c^s_{ij} \text{ and } a_{ij} = a^s_{ij}, \tag{19}$$

then the optimal resource flow may be defined through relations (5)–(8) subject to requirements from (9) to (14).

In this case, the allocation of resources $\{f_{ij}\}$ to the network sections satisfies the requirements for network flow, while observing the flow conservation Eqs. (12). In the general case, this is not observed for $\{r_{ij}\}$ obtained from $\{f_{ij}\}$ by relation (15). This means that the model under consideration generally lacks "flowability" for the arc values $\{r_{ij}\}$. If all f_{ij}; $(i,j) \in G$ in (12) are replaced by $p_{ij}r_{ij}$; $(i,j) \in G$ from (15), then a generalized flow of risks with gains and losses will be received, for which the flow from sources in S is generally not equal to the similar total flow in consumers T. This leads to a variable distribution of the risk on the separate cuts of the network.

2.1 Numerical Example

The following three optimization problems may be formulated on the transportation network for agricultural production, shown in Fig. 1:

Problem 1 v is the total sum flow of a resource that is to be transported from the sources S to the consumers T. Such a realization of this flow $\{f_{ij}/(i,j) \in G\}$ is required, at which the requirements for feasible risk along the individual arcs of the network $\{x_{ij}\}$as well as the feasible physical capacity for resource are observed, while minimizing functional $\sum_{(i,j)\in G} a_{ij} f_{ij}$.

Problem 2 The total-sum flow $v = v_1 + v_2 + v_3$, as well as the arc flow functions are variable values and the maximum value v_{max} is sought, subject to the corresponding constraints, including those on capacities. Then the calculation of v_{max} will be reduced to the following problem of network flow programming:

$$L = \sum_{i\in I(S)} v_i \rightarrow \max;$$

subject to the requirements from (12) to (14), (3)–(5) and (17)–(19).

Problem 3 If the total-sum maximal flow v_{max} is determined by Problem 2. then, once $v_{max} = \text{const}$ is accepted, and v_1, v_2, and v_3 are given the values obtained for them after solving Problem 2, then Problem 1 is solved again. In such a way the maximum flow with minimum cost v'_{max} is determined. This follows from the fact that the minimum cuts can be more than one and the new realization searches for the maximum flow for different values of the functional $\sum_{(i,j)\in G} a_{ij} f_{ij}$.

Exemplary source data for solving the three optimization problems are shown in Table 1.

According to Fig. 1 and the data from Table 1 the relations from (9) to (14) can be written down in the following way:

(1) $f_{1,4} + f_{1,6} = v_1$;

Table 1 Exemplary source data for the numerical example

Param	Arcs										
	(1, 4)	(1, 6)	(2, 4)	(2, 5)	(3, 5)	(3, 7)	(4, 5)	(4, 6)	(4, 7)	(5, 6)	(5, 7)
a_{ij}	8	5	5	6	8	6	7	8	6	5	7
p_{ij}	0.2	0.1	0.15	0.1	0.2	0.25	0.1	0.2	0.2	0.3	0.15
c''_{ij}	8	9	6	4	3	5	3	4	5	3	4
c'_{ij}	25	15	25	10	12	23	35	25	20	25	15
c_{ij}	25	15	25	10	12	20	30	20	20	25	15

(2) $f_{2,4} + f_{2,5} = v_2$;
(3) $f_{3,5} + f_{3,7} = v_3$;
(4) $f_{4,5} + f_{4,6} + f_{4,7} - f_{1,4} - f_{2,4} = 0$;
(5) $f_{5,6} + f_{5,7} - f_{2,5} - f_{3,5} - f_{4,5} = 0$;
(6) $-f_{1,6} - f_{4,6} - f_{5,6} + v_6 = 0$;
(7) $-f_{3,7} - f_{4,7} - f_{5,7} + v_7 = 0$;
(8) $f_{1,4} \le 25$;
(9) $f_{1,6} \le 15$;
(10) $f_{2,4} \le 25$;
(11) $f_{2,5} \le 10$;
(12) $f_{3,5} \le 12$
(13) $f_{3,7} \le 20$;
(14) $f_{4,5} \le 30$;
(15) $f_{4,6} \le 20$
(16) $f_{4,7} \le 20$;
(17) $f_{5,6} \le 25$;
(18) $f_{5,7} \le 15$;
(19) $f_{i,j} \ge 0$ for each $(i,j) \in G$.

Based on the above equalities and inequalities, and the data in Table 1, Problem 1 can be solved with the following objective function with numerical values:

$$L_1 = 8\, f_{1,4} + 5\, f_{1,6} + 5\, f_{2,4} + 6\, f_{2,5} + 8\, f_{3,5} + 6\, f_{3,7} + 7\, f_{4,5} + 8\, f_{4,6} + 6\, f_{4,7} + 5\, f_{5,6} + 7\, f_{5,7} \to \min .$$

Let the values of agricultural production flows of different type from the three sources be 38, 32 and 30, respectively, i.e. $v_1 = 38$, $v_2 = 32$, and $v_3 = 30$, with the production being transported to distribution (reloading) points or directly to end users (processing plants or shipping vessels for export).

Determining the optimal values of $\{f_{ij}/(i,j) \in G\}$ is reduced to solving a linear programming problem [15] with the above constraints from (1) to (18) and with objective function pointed out above. For this purpose, the very user-friendly and freely distributed software package LPSolve IDE has been used.

The following optimal values of $\{f_{ij}\}$ have been received:

$$f_{1,4} = 23;\; f_{1,6} = 15;\; f_{2,4} = 22;\; f_{2,5} = 10;\; f_{3,5} = 10;$$
$$f_{3,7} = 20;\; f_{4,5} = 5;\; f_{4,6} = 20;\; f_{4,7} = 20;$$
$$f_{5,6} = 25;\; f_{5,7} = 0;\; v_6 = 60;\; v_7 = 40.$$

Based on these optimal values, through the respective relations from the previous section the values of the arc risks $\{r_{ij}/(i,j) \in G\}$ may be obtained, as well as the differences $\{c_{ij} - f_{ij}/(i,j) \in G$. These data are shown in Table 2.

Table 2 Differences between capacities and received flow functions

Arcs	(1, 4)	(1, 6)	(2, 4)	(2, 5)	(3, 5)	(3, 7)	(4, 5)	(4, 6)	(4, 7)	(5, 6)	(5, 7)
f_{ij}	23	15	22	10	10	20	5	20	20	25	0
r_{ij}	4.6	1.5	3.3	1	2	5	0.5	4	4	7.5	0
$c_{ij}-f_{ij}$	2	0	3	0	2	0	25	0	0	0	15

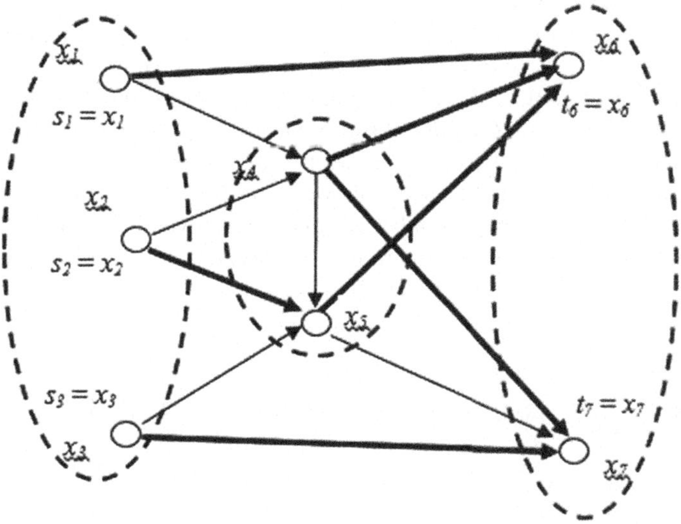

Fig. 2 Saturated arcs of the network in the numerical example

In the case being considered, the bulk of agricultural production being transported is 100 units and the total value of transport charges and insurance of transport risks is 1069 units. The risk assumed for these services is 37.4 units in total.

It follows from Table 2 and Fig. 2 that there are six saturated arcs for which $c_{ij} - f_{ij} = 0$, but there is no minimum cut between the sources S and the consumers T. Therefore, the transport network can absorb greater bulk of the agricultural production being transported. In Fig. 2 the saturated arcs for which no increase of production being transported is possible are marked with thick lines.

Problem 2 The maximum amount of agricultural production that can be transported between the sources S and consumers T is determined through it. In this case, the objective function L_2 is of the following kind:

$$L_2 = \sum_{i \in I(S)} v_i = v_1 + v_2 + v_3 \rightarrow \max.$$

At that in equations from (1) to (18), the parameters $\{v_i / i \in I(S) \cup I(T)\}$ are considered as variables. In the present Problem 2, the values of capacities and probabilities $\{p_{ij}\}$ from Table 1 are used.

Determining the maximum amount of L_2 of agricultural production that can be transported from S to T is reduced to a linear programming problem with constraints (1)–(18) and objective function L_2. For this purpose, the same software product has been used as in Problem 1. The optimal solutions obtained for $\{f_{ij}/(i,j) \in G\}$ are given in the second row of Table 3.

The maximum amount of agro-production being transported is equal to $L_2 = 107$ units and the total value of transportation amounts to 1200 units.

The third row of Table 3 shows the values of the risks assumed on the respective arcs. Their total-sum is 36.65 units. It follows from the same Table 3 that 9 arcs are saturated and only two, $x_{4,5}$ and $x_{5,7}$ are unsaturated. A transport network is shown in Fig. 3, in which saturated arcs are shown with thick lines. It follows from the saturation of the arcs that there is only one cut on the network, for which the equality of the minimum cut and the maximum flow is observed, i.e. the requirement of the mincut-maxflow theorem [16, 17]. That cut is

Table 3 Optimal solutions obtained for $\{f_{ij}/(i,j) \in G\}$

Arcs	(1, 4)	(1, 6)	(2, 4)	(2, 5)	(3, 5)	(3, 7)	(4, 5)	(4, 6)	(4, 7)	(5, 6)	(5, 7)
f_{ij}	25	15	25	10	12	20	10	20	20	25	7
r_{ij}	5	1.5	3.75	1	2.4	5	1	4	4	7.5	1.05
$c_{ij} - f_{ij}$	0	0	0	0		0	20	0	0	0	8

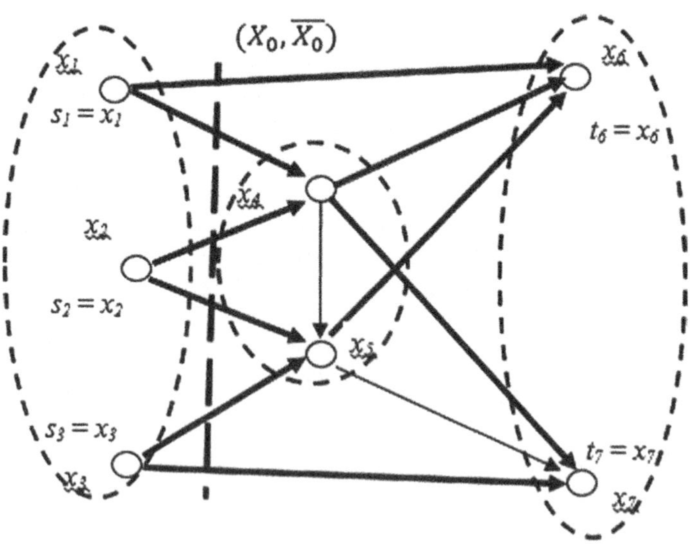

Fig. 3 Min cut

$$\left(X_0, \overline{X_0}\right) = \left\{x_{1,4}, x_{1,6}, x_{2,4}, x_{2,5}, x_{3,5}, x_{3,7}\right\};$$

at which $X_0 = \{x_1, x_2, x_3\}$; $\overline{X_0} = \{x_4, x_5, x_{6,} x_7\}$; $\left(X_0, \overline{X_0}\right) = \emptyset$, and $\emptyset$ is the empty set;

$$v\max = v_1 + v_2 + v_3 = C\left(X_0, \overline{X_0}\right) = f\left(X_0, \overline{X_0}\right); f\left(\overline{X_0}, X_0\right) = \emptyset.$$

The min cut is shown with a thick dotted line in Fig. 3.

Compared to Problem 1, the agricultural production being transported is by 7 units greater, the total risk assumed is one unit greater, and the value for transportation and insurance is 131 units greater. With the assigned capacity, it is not possible to increase the amount of the agricultural production being transported. This may be done by increasing the capacities mostly of the arcs of the min cut $\left(X_0, \overline{X_0}\right)$.

Problem 3 In practice, problems arise not only to find the maximum possible flow from the sources S to the consumers T, but also a problem for the maximum flow of a minimum cost (*mincost-maxflow* problem). This problem can also be solved by using the results of the previous Problem 2. For this purpose, it is necessary:

(a) In relations from (1) to (18), the values of $\{v_i / i \in I(S)\}$ obtained from Problem 2 should be assigned;

(b) The objective function L1 of Problem 1 should be used.

After solving the linear programming problem in compliance with the requirements (a) and (b) and using the specified software product, the required values for $\{f_{ij}/(i,j) \in G\}$, corresponding to the max flow of min cost are obtained. In this particular case at issue in Problem 3, they coincide completely with the results for the arc flow functions of Problem 2. This means that there is no other distribution of $\{f_{ij}/(i,j) \in G\}$ leading to lower total transportation costs of agro-products. Therefore, in Problem 3 the transportation costs and the minimum cut $\left(X_0, \overline{X_0}\right)$ are the same as in Problem 2 and Fig. 3.

The numerical examples described in Problems 1, 2 and 3 demostrate in a sufficiently convincing way the possibility of finding optimal solutions through the proposed method, taking into account the risks of transportation of agro-products on a complex network.

A possibility exists to use the proposed method and to find optimal intelligent solutions in other applied problems.

3 Circulation at Transportation

Very often vehicles travel on a circular route from the production sites to the processing plants or larger storehouses and back making a lot of courses, which is connected with serious expenses and risks.

Besides the denotations accepted in the previous section some more are needed for formalization of the transport circulation. They are as follows:

$I_s = \{i/x_i \in S\}$—indices of the set of loading points;

$$S = \bigcup_{i \in I_s} x_i; \tag{20}$$

d_i—capacity (performance) by resources of a loading point of index $i \in I_s$;

$I_d = \{i/x_i \in T\}$—indices in the set of loading points;

$$T = \bigcup_{i \in I_d} x_i \tag{21}$$

b_i—capacity (performance) by resources of a unloading point of index $i \in I_t$;

$I_p = \{i/x_i \in P\}$—indices of the set of loading points;

$$P = \bigcup_{i \in I_p} x_i; \tag{22}$$

The following requirement is to be observed for the product being transported [18]:

$$\sum_{i \in I_s} d_i = \sum_{j \in I_t} b_j \tag{23}$$

Let l_{ij} be the number of time units for traveling the distance from point x_i to point x_j, which includes the time for loading/unloading at the same point. Then k_{ij} it is the coefficient for transition from discrete to continuous flow, i.e.

$$k_{ij} = \frac{1}{l_{ij}}; \ (i, j) \in G. \tag{24}$$

It is expedient (15) to be observed, which means that

$$p_{ij} f_{ij} - r_{ij} \leq 0; \ (i, j) \in G. \tag{25}$$

The requirement above leads to the possibility to build a network risk flow, although with some increase in the risk on individual arcs. This allows the definition of a two-product network flow of resources and risks with a total minimum cost.

The total costs for resource and risks are equal to: $L_s = \sum_{(i,j) \in G)} a'_{ij} f_{ij}$.

and

$$L_r = \sum_{(i,j) \in G)} a''_{ij} r_{ij} \tag{26}$$

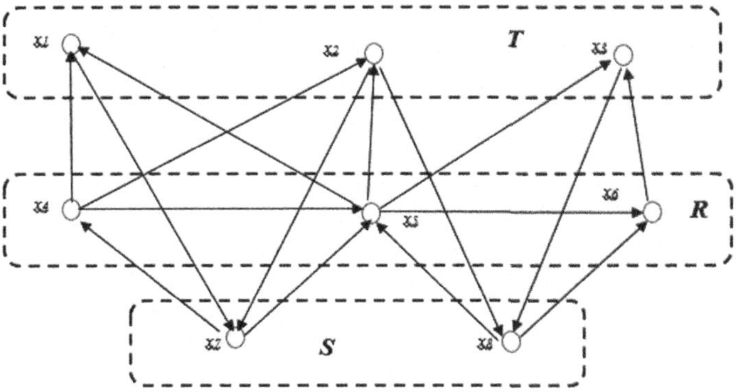

Fig. 4 Exemplary network for circulation

$$L = L_s + L_r = \sum_{(i,\,j)\in G} a'_{ij} f_{ij} + \sum_{(i,\,j)\in G} a''_{ij} r_{ij} \to \min\,|\,\max, \tag{27}$$

where L_s are expenses for transportation of resources, L_r—expenses for insurance of risks, L—total expenses $(L_s + L_r)$.

The value L can be considered as an objective function in the problems for optimization of the total costs of the two-product flow.

A transport network with two loading S, three unloading T and three intermediate R points is conditionally shown in Fig. 4. Circulation may be performed on the network. At that

$$X = S \cup P \cup T;\ S \cap P = \emptyset;\ S \cap T = \emptyset;\ T \cap P = \emptyset; \tag{28}$$

where Ø stands for the empty set.

If the transportation of the resource is carried out by discrete transport units and the travel times along the different arcs of the set U are different, then the transition to a continuous transport flow can be done using a generalized (with profits and losses) two-product network flow with capacities and with coefficients $\{k_{ij}/(i,\,j) \in U\}$. The flwoability requirements for $\{f_{ij}\}$ and $\{c_{ij}\}$ should be observed.

The denotations thus introduced allow for the formalization of the two-product network flow of resources and risks, including their circulation. Determining the optimal/minimum circulation of resources and risks in this two-product flow is reduced to the following linear programming problem:

$$L = \sum_{(i,\,j)\in G} a'_{ij} f_{ij} + \sum_{(i,\,j)\in G} a''_{ij} r_{ij} \to \min, \tag{29}$$

subject to constraints: for each $i \in I$ and $(i,\,j) \in G$

$$\sum_{j \in \Gamma_i^1} f_{ij} - \sum_{j \in \Gamma_i^{-1}} k_{ji} f_{ji} = 0,\ i \in I \tag{30}$$

$$\sum_{j \in \Gamma_i^{-1}} f_{ji} = d_i,\ i \in I_s \tag{31}$$

$$\sum_{j \in \Gamma_i^{-1}} f_{ji} = b_i,\ i \in I_t \tag{32}$$

$$p_{ij} f_{ij} - r_{ij} \le 0;\ (i, j) \in G \tag{33}$$

$$f_{ij} \le c_{ij}^s \tag{34}$$

$$f_{ij} \ge 0;\ (i, j) \in G \tag{35}$$

$$\sum\nolimits_{j \in \Gamma_i^1} r_{ij} - \sum\nolimits_{j \in \Gamma_i^{-1}} r_{ji} = 0,\ i \in I \tag{36}$$

$$r_{ij} \ge 0;\ (i, j) \in G \tag{37}$$

In order to have circulation, the requirement (23) should be observed.

After solving the linear programming problem (LP problem) above, an optimal distribution of resources $\left\{ f_{ij}^* / (i, j) \in G \right\}$ and arc risks $\left\{ r_{ij}^* / (i, j) \in G \right\}$ will be obtained along arcs $x_{ij} \in U$.

The value $\sum_{(i,j) \in G} f_{ij}^{*=} = f^*$ indicates the minimum number of transport units that can ensure the transportation of the produced resource $\sum_{i \in I_s} d_i$ from the loading points S to the unloading points T at a minimum total distance and with minimum transportation and risk insurance costs.

The optimal parameters obtained through the flow circulation f^* and $\left\{ f_{ij}^* / (i, j) \in G \right\}$ can and should be considered as an optimal shift plan—it can be used to compile an optimal schedule for the movement of each individual vehicle. However, since there is a real stochasticity in the movement of vehicles, the obtained optimal schedules must be performed through the dispatching computer system for operational control of transport units.

Problem 1 for circulation. Through equalities and inequalities (23) and from (29) to (36) a two-product flow of resources and risks is defined, which circulates on the network and leads to minimal value of the two-product flow at transportation of the resource

$$\sum_{i \in I_s} \sum_{j \in \Gamma_i^{-1}} f_{ji} = \sum_{i \in I_s} \sum_{j \in \Gamma_i^{-1}} f_{ji} = \sum_{i \in I_s} d_i \tag{38}$$

from loading points S through intermediate points R to unloading points T.

Problem 2 for circulation. The maximum possible value for the circulation of the generalized network flow of resources and risks can be determined by the following of linear programming problem:

$$L = v_{\max} = \sum_{i \in I_s} \sum_{j \in \Gamma_i^{-1}} f_{ji} \to \max; \tag{39}$$

subject to constraints (30), from (33) to (37), (40), and (41)

$$\sum_{i \in I_s} \sum_{j \in \Gamma_i^{-1}} f_{ji} = \sum_{i \in I_s} \sum_{j \in \Gamma_i^{-1}} f_{ji} = 0; \tag{40}$$

$$\sum_{i \in I_s} \sum_{j \in \Gamma_i^{-1}} r_{ji} = \sum_{i \in I_s} \sum_{j \in \Gamma_i^{-1}} r_{ji} = 0; \tag{41}$$

The obtained value for the maximum possible circulation can be used in a number of dependences of the classical network flow, including the Ford Fulkerson' theorem *mincut-maxflow* n [16], namely:

$$v_{\max} = f\left(\overline{X_0}, X_0\right) = c\left(\overline{X_0}, X_0\right) \tag{42}$$

$$f\left(\overline{X_0}, X_0\right) = 0; \tag{43}$$

$$(X_0, \overline{X_0}) = \left\{x_{ij} / x_i \in X_0;\ x_j \in \overline{X_0}\right\} \tag{44}$$

$$\left(\overline{X_0}, X_0\right) = \left\{x_{ij} / x_i \in \overline{X_0};\ x_j \in X_0\right\} \tag{45}$$

where $(X_0, \overline{X_0})$ is the minimal cut.

Problem 3 for circulation. If the maximum flow (39) for circulation is determined through Problem 2 for circulation, then the maximum flow with minimum value can be calculated by the following linear optimization problem: (29), (30), from (33) to (37) and the following two additional inequalities:

$$\sum_{i \in I_s} \sum_{j \in \Gamma_i^{-1}} f_{ji} = v_{\max};$$

$$\sum_{i \in I_s} \sum_{j \in \Gamma_i^{-1}} f_{ji} = v_{\max};$$

where $v_{\max}$ is the value received in (39) after solving Problem 2 for circulation.

After solving Problem 3, the obtained value of L in (39) is the minimum value for the maximum flow v_{max} in (39).

The model being developed makes it possible to minimize the number of circulating vehicles, which ensure the transportation of products with minimal total transportation costs and to cover the emerging risks.

The following example illustrates how the proposed two-product model with capacities and risks model can be used effectively.

3.1 Numerical Examples for Circulation

A graph $G(X, U)$ of five vertices $X = \{x_1, x_2, x_3, x_4, x_5\}$ and eight arcs $U = \{x_{15}, x_{25}, x_{31}, x_{32}, x_{41}, x_{42}, x_{53}, x_{54}\}$. There is one loading point $S = \{x_5\}$, two intermediate points $R = \{x_3, x_4\}$ and two unloading points $T = \{x_1, x_2\}$ that observe the requirement (28) is shown in Fig. 5.

Capacities of resource c_{ij}^s and in brackets—of the risk c_{ij}^r are shown in Fig. 1. Next in Table 4 the source data of the problem are given.

Problem 1 for circulation. The optimal two-product network flow with circulation and risks can be determined by the dependences from (29) to (37), which for the graph of Fig. 5 and the data in Table 4 are take the following form:

$$L = 16 f_{15} + 12\, f_{25} + 10\, f_{31} + 9\, f_{32} + 12\, f_{41} + 10\, f_{42} + 9\, f_{53} + 14\, f_{54}$$
$$+ 8\, r_{15} + 8 r_{25} + 12\, r_{31} + 8\, r_{32} + 8\, r_{41} + 12\, r_{42} + 8\, r_{53} + 6\, r_{54} \rightarrow \min;$$

subject to constraints and equalities:

1. $f_{15} - f_{31} - f_{41} = 0;$

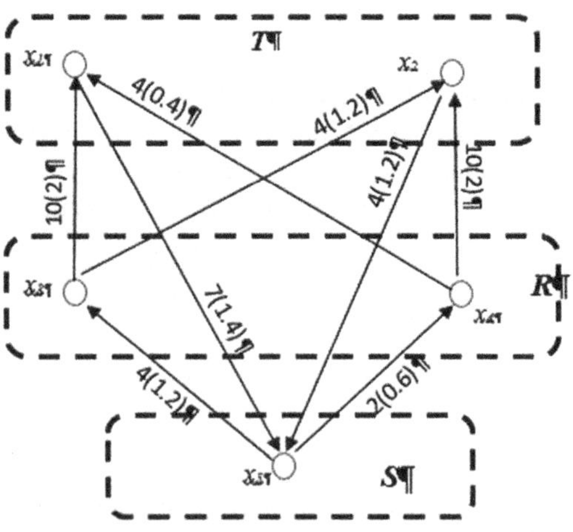

Fig. 5 Exemplary graph for circulation

Table 4 Source data for circulation

Parameters	Arcs—resource								Arcs—risks							
	(1, 5)	(2, 5)	(3, 1)	(3, 2)	(4, 1)	(4, 2)	(5, 3)	(5, 4)	(1, 5)	(2, 5)	(3, 1)	(3, 2)	(4, 1)	(4, 2)	(5, 3)	(5, 4)
c'_{ij}	7	4	10	8	4	10	8	4	–	–	–	–	–	–	–	–
c''_{ij}	–	–	–	–	–	–	–	–	1.8	1.2	2.2	1.2	1	2.2	1.2	0.6
c^s_{ij}	7	4	10	4	4	10	4	2	–	–	–	–	–	–	–	–
c^r_{ij}	–	–	–	–	–	–	–	–	1.4	1.2	2.0	1.2	0.4	2.0	1.2	0.6
a'_{ij}	16	12	10	9	12	10	9	14	–	–	–	–	–	–	–	–
a''_{ij}	–	–	–	–	–	–	–	–	8	8	12	8	8	12	8	6
p_{ij}	0.2	0.3	0.2	0.3	0.1	0.2	0.3	0.3	0.2	0.3	0.2	0.3	0.1	0.2	0.3	0.3
	Vertices—resource								*Vertices—risks*							
x_i	x_1	x_2	x_3	x_4	x_5	–	–	–	x_1	x_2	x_3	x_4	x_5	–	–	–
$d_i(b_i)$	0	0	0	0	5	–	–	–	0	0	0	0	1.7	–	–	–

2. $f_{25} - f_{32} - f_{42} = 0;$
3. $f_{31} + f_{32} - f_{53} = 0;$
4. $f_{41} + f_{42} - f_{54} = 0;$
5. $f_{53} + f_{54} - f_{15} - f_{25} = 0;$
6. $f_{15} + f_{25} = 5;$
7. $f_{31} + f_{41} + f_{32} + f_{42} = 5;$
8. $0.2f_{15} - r_{15} <= 0;$
9. $0.1f_{25} - r_{25} <= 0;$
10. $0.2f_{31} - r_{31} <= 0;$
11. $0.3f_{32} - r_{32} <= 0;$
12. $0.1f_{41} - r_{41} <= 0;$
13. $0.2f_{42} - r_{42} <= 0;$
14. $0.3f_{53} - r_{53} <= 0;$
15. $0.3f_{54} - r_{54} <= 0;$
16. $f_{15} <= 7;$
17. $f_{25} <= 4;$
18. $f_{31} <= 10;$
19. $f_{32} <= 4;$
20. $f_{41} <= 4;$
21. $f_{42} <= 10;$
22. $f_{53} <= 4;$
23. $f_{54} <= 2;$
24. $r_{15} - r_{31} - r_{41} = 0;$
25. $r_{25} - r_{32} - r_{42} = 0;$
26. $r_{31} + r_{32} - r_{53} = 0;$
27. $r_{41} + r_{42} - r_{54} = 0;$
28. $r_{53} + r_{54} - r_{15} - r_{25} = 0;$

$$f_{ij} \geq 0; \ (i, j) \in G, r_{ij} \geq 0; \ (i, j) \in G$$

It follows from the last two rows of Table 5 that at transportation of resources three sections—$\{x_{2,5}, x_{2,3}, x_{5,3}\}$ are saturated. The same sections are saturated also for risks (Table 5).

Figure 6 shows the graph from Fig. 5, along the arcs of which the obtained values of the resource flow are plotted, and in brackets—the flow of risks.

The saturated arcs of the graph of Fig. 6 are marked by bold lines. They form a cycle on the graph. A flow of a value of 4 circulates on the cycle. There is also a second cycle $\{x_{4,1}, x_{1,5}, x_{5,4}\}$, on which a resource flow of value of 1 and a risk flow of value of 0.3 circulate. This second cycle is with unsaturated arcs.

Problem 2 for circulation. To determine the maximum possible flow of resources and risks, almost the same relations will be used as in Problem 1, with minor modifications. In this case, the objective function is as follows:

$$L = v \rightarrow \max .$$

Table 5 Results for Problem 1 for circulation

Parameters	Arcs—resource								Arcs—risks							
	(1, 5)	(2, 5)	(3, 1)	(3, 2)	(4, 1)	(4, 2)	(5, 3)	(5, 4)	(1, 5)	(2, 5)	(3, 1)	(3, 2)	(4, 1)	(4, 2)	(5, 3)	(5, 4)
f_{ij}	1	4	0	4	1	0	4	1	–	–	–	–	–	–	–	–
r_{ij}	–	–	–	–	–	–	–	–	0.3	1.2	0	1.2	0.3	0	1.2	0.3
p_{ij}	0.2	0.3	0.2	0.3	0.1	0.2	0.3	0.3	0.2	0.3	0.2	0.3	0.1	0.2	0.3	0.3
$c_{ij}^{s} - f_{ij}$	6	0	10	0	3	10	0	1	–	–	–	–	–	–	–	–
$c_{ij}^{r} - r_{ij}$	–	–	–	–	–	–	–	–	0.9	0	2.2	0	0.7	2	0	0.3

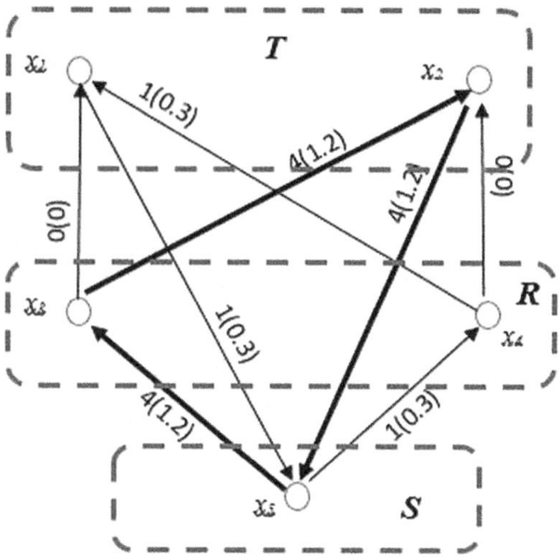

Fig. 6 Cycle with saturated arcs

Equations (6) and (7) in the constraints should be replaced by:

$$6'.\ f_{15} + f_{25} = 5;$$

$$7'.\ f_{31} + f_{41} + f_{32} + f_{42} = 5;$$

The results of solving Problem 2 thus formulated with the data from Table 4 are given in the following Table 6.

Figure 7 shows the graph from Fig. 5, along the arcs of which the obtained values of the resource flow are plotted, and in brackets—the flow of risks.

In Fig. 7 saturated arcs are bold, non-zero unsaturated arcs are dotted. The maximum flow obtained is equal to 6.

Problem 3 for circulation. The maximum value of the flow of resources obtained in the previous Problem 2 is used to obtain the maximum flow of a minimum cost. In this case, the objective function has the same coefficients and variables as in Problem 1, in the constraints the two Eqs. (6) and (7) have right-hand sides equal to the maximum flow, namely:

$$6 - a.\ f_{15} + f_{25} = 6;$$

$$7 - a.\ f_{31} + f_{41} + f_{32} + f_{42} = 6.$$

After solving the optimization problem, the maximum generalized two-project flow with circulation, risks with minimum price are determined.

In this case the value of the objective function is $L = 290.2$, i.e. the total amount to be paid for transportation and risk insurance for the maximum feasible flow.

Table. 6 Results for Problem 2 for circulation

Parameters	Arcs—resource								Arcs—risks							
	(1, 5)	(2, 5)	(3, 1)	(3, 2)	(4, 1)	(4, 2)	(5, 3)	(5, 4)	(1, 5)	(2, 5)	(3, 1)	(3, 2)	(4, 1)	(4, 2)	(5, 3)	(5, 4)
f_{ij}	6	0	4	0	2	0	4	2	–	–	–	–	–	–	–	–
r_{ij}	–	–	–	–	–	–	–	–	1.4	0.4	1.2	0	0.2	0.4	1.2	0.6
p_{ij}	0.2	0.3	0.2	0.3	0.1	0.2	0.3	0.3	0.2	0.3	0.2	0.3	0.1	0.2	0.3	0.3
$c^s_{ij} - f_{ij}$	1	4	6	4	2	10	0	0	–	−2	–	–	–	–	–	–
$c^r_{ij} - r_{ij}$	–	–	–	–	–	–	–	–	0	0.8	0.8	1.2	0.2	1.6	0	0

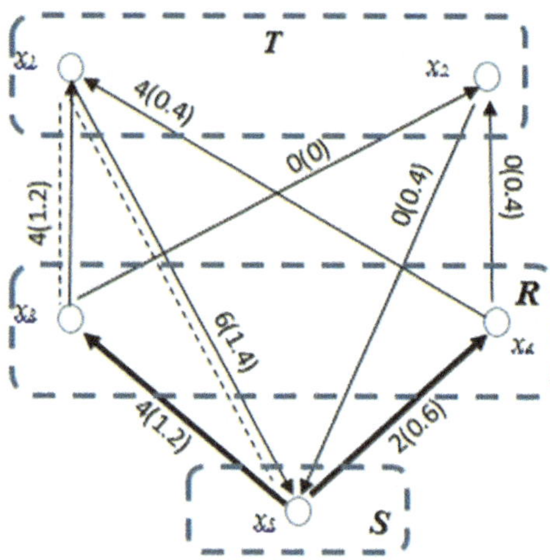

Fig. 7 Saturated and non-zero unsaturated arcs

4 Conclusion

1. The subject of this article is the problem of making optimal smart decisions, taking into account the risks at transporting commodities on transport networks of arbitrarily complex configuration.
2. A method is proposed for making optimal smart decisions, taking into account the network properties and transportation risks, and minimizing the total transportation costs and transport risk insurance along the separate sections of the network.
3. Numerical modeling of the possibility of using the proposed method has been carried out, and it has been shown sufficiently accurately that this method can provide optimal solutions in real conditions for transportation of commodities on a network of any configuration and various network risks.
4. The very frequent case when vehicles travel on circular route is considered, at which the expenses for transportation and risk insurance are minimized.
5. A method is proposed for determining the number of vehicles for transportation of the production on a circular route from the production sites to the processing plants or other consumers.

Acknowledgements The research work reported in the paper is supported by the Bulgarian National Science Fund under Grant № KP-06-X36/2, project BG PLANTNET "Establishment of National Information Network GenBank—Plant genetic resources".

References

1. https://builtin.com/artificial-intelligence
2. https://whatis.techtarget.com/definition/operations-research-OR
3. Gayle, S.: Intelligent Systems: The Big Picture The Future of Intelligent Systems is Exciting, But Making it Happen will Require Hard Work. https://doi.org/10.1080/08956308.2019.1613115
4. Albert-Lászlő, B.: Network Science, 456 pp. Cambridge University Press (2016). ISBN 1107076269, 9781107076266
5. Li, Q.: The study on the risk management of agricultural products green supply chain based on systematic analysis. In: 2012 Second International Conference on Business Computing and Global Informatization, 12–14 Oct. 2012, IEEE Xplore Digital Library. https://doi.org/10.1109/BCGIN.2012.71
6. Adamopoulos, T.: Transportation costs, agricultural productivity, and cross-country income differences. Int. Econ. Rev. **52**(2), 489–521 (2011). https://doi.org/10.1111/j.1468-2354.2011.00636.x
7. Li, P.-C., Shih, H.-C., Ma, H.-W.: Assessing the transfer of risk due to transportation of agricultural products. Chemosphere **120**, 706–713 (2015). https://doi.org/10.1016/j.chemosphere.2014.10.009
8. Wang, Y., Hao, H.X.: Research on the supply chain risk assessment of the fresh agricultural products based on the improved toptsis algorithm. Chem. Eng. Trans. **51**, 445–450 (2016). https://doi.org/10.3303/CET1651075
9. Christofides, N.: Graph theory: An Algorithmic Approach. Academic Press, London (1986)
10. Don Phillips, G.-D.: Fundamentals of Network Analysis, 474 pp. Prentice Hall, Englewood Cliffs, NJ (1981). https://doi.org/10.1002/net.3230120210
11. Jensen, P.A., Barnes, J.W.: Network Flow Programming. Wiley Inc., New York (1980)
12. Sgurev, V, Drangajov, S.: Intelligent control of flows with risks on a network. In: Proceedings of the 7th IEEE International Conference Intelligent Systems—IS' 14, September 24–26 2014, Warsaw, Poland, ISSN 2194-5357, ISSN 2194-5365 (electronic), ISBN 978-3-319-11309-8, ISBN 978-3-319-11310-4 (eBook). https://doi.org/10.1007/978-3-319-11310-4, Volume 2: Tools, Architectures, Systems, Applications, Springer International Publishing, Switzerland; Angelov, P., et al. (eds.), Advances in Intelligent Systems and Computing vol. 323, pp. 27–35 (2014)
13. Sgurev, V.: Network Flows with General Constraints. Publishing House of Bulg. Academy of Science, Sofia (1991).(in Bulgarian)
14. Sgurev, V., Drangajov, S.: Network Flows with Risks,. Publishing House of Bulg. Academy of Sci. "Prof. Marin Drinov. ISBN 978-954-322-990-1, Sofia (2019). (in Bulgarian)
15. https://www.amazon.com/Linear-Programming-Network-Mokhtar-Bazaraa/dp/0470462728/ref=pd_sbs_14_1/137-5260339-8864505?_encoding=UTF8&pd_rd_i=0470462728&pd_rd_r=ec3fe33b-fdd5-445d-b885-6dbdf1c5f435&pd_rd_w=RAQ8i&pd_rd_wg=1XoDV&pf_rd_p=ed1e2146-ecfe-435e-b3b5-d79fa072fd58&pf_rd_r=AJW1GP5D1WW8W448MNJA&psc=1&refRID=AJW1GP5D1WW8W448MNJA
16. Ford, L.R., Fulkerson, D.R.: Maximal flow through a network. Can. J. Math. **8**, 399–404 (1956)
17. http://pcwww.liv.ac.uk/~iken/teaching/optimisation/chapter6.pdf
18. https://courses.engr.illinois.edu/cs573/fa2013/lec/lec/14_flow_III.pdf

Embedded Intelligence in a System for Automatic Test Generation for Smoothly Digital Transformation in Higher Education

Pepa Petrova, Iva Kostadinova, and Majid H. Alsulami

Abstract Automated test generation is an important component in the digital transformation of education. Assessing knowledge using tests is an objective, fast and effective. Electronic tests are effective in verifying knowledge but they also can be used for self-learning. Generating tests by teachers is a laborious and time-consuming process. The development of an automatic test generation system is a key component for adapting education to the ever-changing digital environment. This paper presents a logical approach to assessing student success in the testing process. The approach satisfies the two goals of testing: knowledge testing and self-learning. A process for intelligent transition from one level of complexity of the testing system to another has been developed. This is a new and innovative approach to developing an intelligent automatic test generation system. None of our known training management systems has a self-test module with self-generated questions. The system we developed generates questions independently using templates. The system also navigates students through the different levels of complexity of the content, using the puzzle method and the curiosity function. Adding intelligence to an automated e-test generation system can be a milestone of digital transformation in education and can increase the efficiency of the teaching and learning process during a crisis. Today we are facing the crisis with COVID-19, which affects all areas of work, study and life. Our intelligent approach helps educational institutions to overcome atypical situations like the one we face now. It can also support the universities for a smooth digital transformation and better handling with future atypical situations.

P. Petrova (✉) · I. Kostadinova
Faculty of Information Science, University of Library Studies and Information Technologies (ULSIT), 1784 Sofia, Bulgaria
e-mail: p.petrova@unibit.bg

I. Kostadinova
e-mail: i.kostadinova@unibit.bg

M. H. Alsulami
Community College, Shaqra University, Shaqra, Saudi Arabia
e-mail: malsulami@su.edu.sa

V. Sgurev et al. (eds.), *Advances in Intelligent Systems Research and Innovation*,
Studies in Systems, Decision and Control 379,
https://doi.org/10.1007/978-3-030-78124-8_20

Keywords Automation · Curiosity · Intelligence · Logic · Puzzle method · System · Testing

1 Introduction

Higher education faces challenges in testing students to identify and develop their skills. All this comes with a digital transformation not only in higher education, but also in every other field of work and life. In the learning process, different skills are trained: reading, speaking, reproducing and applying. These skills are enshrined in a number of normative documents, including the European Reference Framework for Key Competences. Different methods and approaches are used for their build. One of them is the test of what has been learned. Entering the test as a form of knowledge testing proved to be the preferred way for both the teachers and the learners themselves. Global testing has been used successfully since the end of the last twentieth century. The introduction of the test as a form of testing in Bulgaria was imposed by the National Program for Development of School Education and Preschool Education and Training (2006–2015). By adopting this program, the test has become the basis of an "effective system for internal evaluation and national standardized external evaluation" in Bulgaria and is defined as "a leading form of ongoing control in the learning process and finishes each educational stage" [1].

Accepting challenges of the dynamically changing environment the good practices in training young people described in [2, 3] are followed by universities. The use of electronic tests not only as a means of verifying knowledge but also as a means of self-learning has led to the need to generate multiple training tests. Besides the well-known learning management systems (LMS), a lot of applications have been created in the space that offers the generation of training tests. Each one offers a set of algorithms to generate different types of test questions. In addition, each system offers a specific random selection mechanism for both the answers to a question and the order of the test questions in the final test. All the analysed LMS [4] offer test generation based on already existing in the system questions. This takes a significant amount of time for teachers to create their own questions. To reduce the time that teachers spend to make up test questions we have developed a unique module to automate the process. Based on templates, the module randomly changes the numerical attributes of the questions so as autonomously to generate unique question for each student.

Modern research and development for education apply intelligent approaches to adapt the learning process to the new technological reality. For example, AI is used to recognize emotions to eliminate negative emotions in a learning environment. In [5] a system for regulating the emotional state as a fragment of a modern Intelligent Learning System (ITS) has been proposed. The system uses direct sensory and retrospective information within the bifocal emotional model. The significant difficulties inherent in systems for automatic compensation of specific recognized emotions are avoided. Offering an approach for initial identification of possible stimuli for emotional state and subsequent compensation of their impact in the form of an iterative procedure. The proposed combined feedback and feedforward management is based on real measurements and reliable hypotheses. In a multi-connected system, control variables are defined to establish the connection "acquisition of knowledge and skills—emotional state". A generalization of the influence of the emotional state of students on the effectiveness of the learning process is made.

Modern research seeks different approaches to personalizing learning in order to increase competence and overcome differences between learners. In [6] dedicated devices for measuring learners' psychological and emotional characteristics are discussed. Learning logs, which are stored in the LMS, have become increasingly important in online learning for individual learning support. It is important that learners maintain personal motivation to learn in the online learning environment because, unlike classroom learning, there are no teachers to instruct or support them. Our system for e-test generation follow the same idea in its self-testing functionality.

This paper proposes an approach for embedding intelligence in an existing system for generating electronic tests. The aim is to incorporate logical data processing into the system and thus enable learners to self-test. With this work we present two algorithms for automatic transition from one level of complexity to another based on the learner's result.

The paper is organized in three main sections. Prerequisites for automatic test generation. The main problems and approaches in the field. Suitable techniques for embedding intelligence into systems for e-test generation. The second part explains the methodology. The system that we use as a fundament was developed by a team at the university. It is used by teachers to check student level of knowledge. To bring intelligence to the system and make it capable of self-testing, we apply a logical approach for data processing and analysis. There are some options for the trainees:

- Go through each category sequentially—statistical approach
- Demonstrate your knowledge—logical approach using levels of complexity.

Two algorithms are presented. One is focused on statistics because of the expected reliable result. The other is based on logical processing and dynamic assessment of the predicted success of the trainee. In the Results and discussion section an experiment conducted with 190 students is shown. Analysis of both algorithms is a basis for the discussion and conclusions.

2 Prerequisites for Authomatic Test Generation

The use of the test as a means of checking and controlling knowledge requires the preparation of multiple tests, and subsequently their verification and analysis of the results. This led to automation in the process of generating training tests and striving to prepare individual ones. Thus, the use of computer tests and automated knowledge testing systems greatly facilitate the learning process. Currently, a number of systems are available for generating training e-tests, and they can generally be grouped as e-test systems for training Learning Management Systems (LMS) and stand-alone e-test systems. The following systems for generating e-tests are considered: LMS ILIAS, LMS Moodle, etc. Stand-alone e-test systems are compared with existing LMSs in [4].

2.1 System for Automatic Generation of e-Tests

This research is based on an existing at the university stand-alone automated system for test generation and validation described in [7]. The automated system has a mechanism for generating e-tests on programming language, which is not found among the considered systems. The stand-alone system for generation of individual tests works using a template. The process of generating each e-test starts with placing a request and takes 1 s per question [8].

Some different approaches for automation are explored:

- Flexible approach for evaluation and ranking of students using multi-attribute decision making techniques [9].
- Approach for modeling of information systems. How data processing runs decision making for building of such systems [10, 11].

Part of the system that is used as a base for the proposed approach is shown on Fig. 1. As it was mentioned at the beginning the system is developed at the university and it is used for test generation by the teachers.

Figure 1 shows the part of the system when the selection of the template has happened. The system is automated in meaning that the teacher doesn't have to figure out every single question and its parameters. For instance, if a group of 20 students has to be tested whether they can calculate an equation the template approach will generate 20 equations with the same variables, but different numerical constants. That is how all students will calculate the same equation type but have to gain different results. This approach saves time to the teachers, but with limited application. To improve the applicability of the system intelligence is embedded. Thus, the system can be used for self-training. For the purpose of self-training, the system has to be able to navigate trainees between the different levels of complexity. To achieve the goal two algorithms for assessment and automatically switch form one level of complexity to another are presented. The test questions are designed as templates. In each template, there are separate zones for setting ranges of values. Based on these zones, a different test question is generated. Each template belongs to a group. Each group of templates covers a certain level of the study curriculum in a gradual way. For instance, the templates of the second level test the knowledge covering material of the second level, but also are based of the knowledge from the first level.

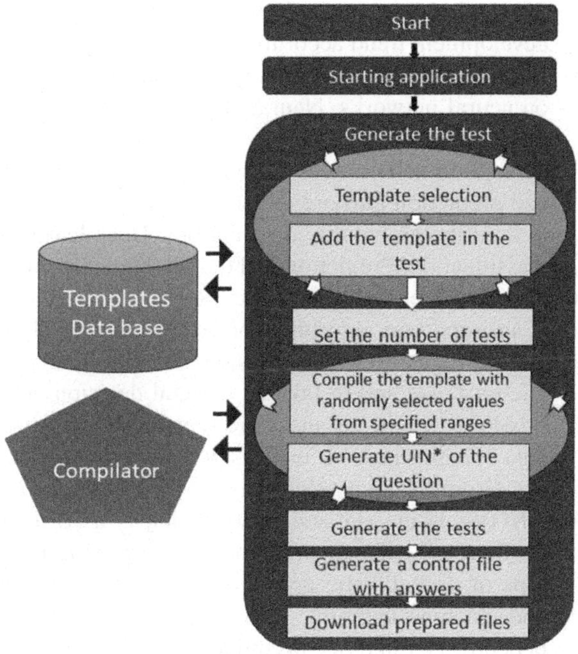

Fig. 1 Part of a system for e-test generation

2.2 Technologies that Brings Intelligence into Systems. Literature Review

As machines become increasingly capable, this replaces manual engineering and allows a machine to both learn the features and use them to perform a specific task. Algorithms for finding models, reasoning, machine learning, deep learning, knowledge representation are used to satisfy the autonomy of the machine. By adding reasoning to machine learning systems correlations and insights become much more useful. The system for automatic e-test generation can be improved and intelligence can be built-in. Approaches to be followed are:

- Data mining aims to discover the informative relationship, model, and trends of big data [12–14]. Data mining extracts implicit, unknown knowledge, and rules from a large number of noisy, fuzzy, or random data, which have potential value for decision-making, and can make result prediction for the non-occurrence behavior according to the existing information. Data mining technology is a new information processing technology in essence, which improves the application of data from low-level online query operation to higher-level application of decision support, analysis, and prediction. Datamining can deal with incomplete, redundant, and random data. Data mining technology has brought about the structural change of the traditional decision support system. The existing database system can achieve the effective input, modification, statistics and query functions, but it fails to find the relationships and rules in the data, and it cannot predict the future development trend according to the existing data [15, 16].

- Deep learning is part of a broader family of machine learning methods based on neural networks. Neural Networks are a set of algorithms that are designed to recognize patterns. They interpret sensory data through a kind of machine perception, labelling or clustering raw input. Can be used for both classification and prediction purposes as stated in [17, 18]. In [19] graphical models are used to describe probability distributions. Deep learning allowing future detection or classification. Prediction and qualification are on focus in this research.
- Decision Trees are predictive models as described in [20–22]. The decision tree is an important algorithm for predictive modelling and can be used to visually and explicitly represent decisions. It helps in making decisions under a particular circumstance and to take an optimal decision. The algorithm is well suited for problems where instances are represented by attribute value and when training data contains error. It is also applicable to the situation when the target function has discrete output value.
- ANOVA is the statistical model that is used to predict a continuous outcome on the basis of one or more categorical predictor variables [23]. They are a type of probabilistic graphical models lie at the intersection between statistics and machine learning. They have been shown to be powerful tools to encode dependence relationships among the variables of a domain under uncertainty.
- Unsupervised learning is a type of machine learning algorithm used to draw inferences from datasets consisting of input data without labelled responses. Cluster analysis as the most common algorithm of unsupervised learning is used for exploratory data analysis to find hidden patterns or grouping in data. Unsupervised learning through clustering is used for partitioning where the center of each cluster is displayed by the mean value of the objects [24].
- Puzzle methods are a logical approach applied usefully when we have weak and unreliable information [25, 26]. The effectiveness of the Puzzle methods in comparison with well-known statistical methods for prediction is presented in [27]. This research applies puzzle method and its constraints, as explained here [28].
- Collections of techniques for data fusion, data transformation, filtering, sampling are discussed in [29]. New Artificial Intelligence (AI) related data analysis architectures, methodologies, and techniques and their applications to various domains are presented.

The approaches are suitable for improving test generation systems by adding intelligence. For example, the classification capabilities of deep learning are combined with the binding constraints of the puzzle method to assess the level of knowledge of the test takers. The prediction modelling with decision trees in the presence of data with errors is supported by the curious function with which the statistical forecast becomes a logical causal result. The decision tree-based classification is very effective for small data-sets. To predict huge dataset decision trees are not efficient. The attributes of unidentified value can be easily estimated during the construction of the decision tree. It handles the data by evaluating the gain ratio for an attribute which is defined. Hence, the decision tree has an advantage of classifying the report with different probabilities [30]. Unsupervised learning and ANOVA approaches are appropriate when learner's statistical evaluation to be used as a basis for logical processing.

A base of the proposed approach in this research is the puzzle method. It allows deep modelling of different types of knowledge. Most often it serves as an upgrade to existing models. An upgrade that does not change other features, but only enhances existing ones [31, 32]. In this paper, it is used to upgrade the automation functionality of the test generation system to serve for self-testing purposes.

2.3 Digital Transformation in Higher Education

Traditional education is the most model used in educational institutions worldwide. This model is limited to fixed location, and students are required to attend to institutions. Each teacher is assigned to a group of students, and the number of students depends on the size of classroom. This in some cases make it so difficult to occupy all student who live close to the institution. The number of students is increasing, which require increasing the number of classrooms and teachers [33]. Therefore, digital transformation plays an important role in improving the education system.

Digital transformation can be defined as "the process of digitizing and making available what was previously analog/physical". it can use and depend on many emerging technologies such as internet of things, augmented and virtual reality, Artificial Intelligence (AI), cloud computing big data [34], social, mobile [35], and others. Digital transformation can change the way of teaching and improve the skills of teachers. Digital transformation enables students to access, learn, use, and attend classes at any time and from any place. Internet of Things (IoT) enables different technologies and devices to be connected form anywhere at any time. Students have the ability to use the tools to cooperate with other students, teachers, and utilize the digital materials provided for learning. Social allows students to work collaboratively. Mobile enables teachers to gather information about students, and be more interactive and dynamic [35]. Big data enables students find new techniques and methods that improve their skills of learning. AI enables teacher to follow up the progress of students and identify performance of students [36].

Digital transformation requires a change in operations, culture, technology, and other aspects. Institutions should reinvent themselves in order to benefit from digital transformation. They have to adopt new business model of learning and utilize the power of technologies to deliver products and service to their students and teachers [36].

3 Methodology

The system for automatic e-test generation based on templates follows some main steps: selection of the starting level; selection of a template from the level; generation of a question based on the template. In case of self-testing there are two options. First one is when a trainee takes a decision when to switch between levels, the second one is when the switch is done automatically. The focus of this research is the second case. To make the automatic switch the trainee has to be evaluated and a decision has to be made. To evaluate the result of the trainee in a current time a logical model for assessment and prediction is created. The model combines the puzzle method with its binding constraints and curiosity function which is a type of selection in the reprocessing of data or knowledge. We believe that these algorithms are suitable for training purposes and we based our hypothesis on this.

In our approach, we use three groups of questions to rule the logic. They are used to enhance the algorithms and to determine the complexity level. The groups of questions to determine the complexity:

- Multiple Choice Single Answers—closed test questions with a choice of one correct from several given answers. The complexity level of questions in the group is 1. In this type of complexity of the questions, there is no penalty scoring, i.e., no deduction of points for wrong or partially wrong answer.
- Multiple Choice Multiple Answers—closed type questions with more than one correct answer. Level of complexity of questions here is 2. This type of complexity does not have penalty scoring again. No deduction of points for wrong or partially wrong answer.
- Hot-Spot Questions—is a test question with an open short answer. These types of questions have a level of complexity of 3. They can apply penalty scoring.

Our two algorithms for logical evaluation of the success of the tested person are described as follows in the section. The first is with a fixed number of minimum questions to calculate the trainee's success and possibility to switch between levels. The second one determines dynamically the number of questions necessary to pass from one level of complexity to another.

Following characteristics are defined for the operation of the two algorithms:

Determining the speed of answering the questions:

- "quick answer"—answer in less than 30 s;
- "average answer"—answering between 31 and 1 min and 20 s;
- "slow response"—the response is given in more than 1 min and 20 s.

Calculate the average time to answer questions T_{mid}. The average time to answer is used as a binding constraint to assess the knowledge level of the trainee.

Determining the end of testing. The trainee can select an option "end of testing" or pass through all levels.

3.1 Algorithm with Fixed Number of Questions (Fixed Algorithm)

The minimum questions that the learner has to answer is set to a specific number. For instance, the learner has to answer at least 10 or 20 or other number of questions in a specific level to be assessed. Based on the answers, the knowledge level of the learner is assessed statistically. It is not possible to change the level earlier if the whole number of necessary questions is not answered. The questions are asked in portions of 10. Going to an upper level depends on the number of correct answers. If the trainee moves to the upper level, or stays at the same level, or goes to a lower level the next set of 10 questions should be answered. The statistics apply to each set of questions. It is needed as a base for statistical calculation. Here we have implemented two of the three group types of questions namely Multiple Choice Single Answer and Multiple Choice Multiple Answers. The steps of the algorithm are as follows:

- Step 1: Starting Level—there are three options. Start from first level. Start from last used level. Start from a level selected from the trainee.

  Select from.
  [case 1]: first level of complexity,
  [case 2]: last used level,
  [case 3]: trainee choice.
  Next step.

- Step 2: Selecting a template—templates are selected randomly from a class of templates according the level. There are not any limitation how many times a template will be used. This is a point for a future work. Labels on template can be used as additional information for the curiosity function, used in next algorithm.

  Random (template).
  Where template belongs to [case 1], [case 2] or [case 3].
  Next step.

- Step 3: Assessment of the answers—there are two possible ways. If the answer is correct and if the answer is wrong (1). There is no option for a partial answer.

Pseudocode:

Start

Calculate the trainee success for first 10 questions

assess the answer: f (x)=0 or 1.

$$f(x) = \begin{cases} 0, & x \ \textit{is incorrect answer} \\ 1, & x \ \textit{is correct answer} \end{cases} \tag{1}$$

$$calculate \quad S = \frac{\sum_{i=1}^{n} f(x)}{n}, \tag{2}$$

where S is the success of the trainee.

then apply puzzle binding constraints, to check if the questions were answered correctly by chance

add two new questions n=n+2,
if =1 for n+1 and n+2 and S ≥0.7 then go to next upper level,
if =1or 0 for n+1 and n+2 and S <0.7 then go to previous lower level, else ask next 10 questions at the same level and complexity,
Apply label "difficult" to all wrongly answered questions and use them again as difficult questions,
end.

Here mathematical model for selection of fewer questions with higher degree of difficulty is applied [37] supporting the decision whether to stay at this level.

3.2 Algorithm with Dynamic Determination of the Number of Questions (Dynamic Algorithm)

To accelerate the time of assessment and movement form one level to another we calculate dynamically the trainee success using puzzle method constraints [26] and curiosity function [27]. The steps are the same as in the "fixed algorithm", but the number of minimum questions that the trainee have to answer to move from one level to another is defined dynamically. Using the puzzle method constraints and the curiosity function we decrease the minimum number of questions to be answered. This allows the trainee to move faster across the levels. Steps 1 and 2 are the same as in A, but here we use all groups and can apply all types of question complexity. The e-testing system randomly selects the type of complexity of the questions.

- Step 1: Starting Level—as the Fixed Algorithm.

    Select from
    [case 1]: first level of complexity,
    [case 2]: last used level,
    [case 3]: trainee choice.
    [case 4]: randomly selected level.
    Next step.

- Step 2: Selecting a template.

    Random (template).

Where the template belongs to [case 1], [case 2], [case 3] or [case 4]. For each selected case, a question of the highest level of complexity: 3 is asked. Depending on the answer, the procedure below is performed.

If the answer is correct, the time to answer is considered:

- *If the answer time is less than 30 s, a question of complexity 3 from a higher category is asked.*
- *If the answer time is between 31 s to 1 min and 20 s, a question of complexity 3 from the same category is asked.*
- *If the answer time is more than 1 min and 20 s, a question of complexity 2 from a lower category is asked.*

If the answer is wrong—look at the time to answer:

- *If the time to give the wrong answer is less than 30 s, a question of complexity 3 from the same category is asked.*
- *If the time to give the wrong answer is between 31 s to 1 min and 20 s, a question of complexity 2 from a lower category is asked.*
- *If the time to give a wrong answer is more than 1 min and 20 s, a question of complexity 1 from a lower category is asked.*

Next step

- Step 3: Assessment of the answers. The success score is calculated when the level changes. The number of questions depends on both their complexity and the answer time of the learner. Level of success or failure is calculated based on Eq. (2). Moving to a higher level is possible with at least 70% correct answers. Moving to a lower level is necessary when 40% or less correct answers are given. The fact that the trainee answers complex questions quickly means that he has knowledge. It is a prerequisite for reducing the number of questions in the level and moving to another level of complexity.

Pseudocode:

Start

1. Calculate the trainee success (S)

$$assess\ the\ answer:\ f(x) = 0\ or\ 1,\ \ according \tag{3}$$

$$calculate\ \ \mathrm{S},\qquad according \tag{4}$$

2. Apply curiosity function – it is used to calculate the student's success in moving to a higher or lower level of complexity

$$ch_1^n \in \begin{cases} U_p - positive\ ch_i,\ where\ i = \{1, n\} \\ U_q - negative\ ch_j,\ where\ j = \{1, n\} \end{cases}, \tag{5}$$

where ch_1^n *is positive if the answer is correct and negative if the answer is incorrect. The goal is to identify whether a trainee is potentially likely to succeed or not. Curiosity* $Q(a_p)$*is calculated for each person as follows and links the trainee's score according the binding constrains with success or not:*

$$Q(a_p) = \frac{count\ (ch_i) - count\ (ch_j)}{count\ (ch_i) + count\ (ch_j)} \tag{6}$$

If $Q(a_p)$ >0.6 *"trainee will succeed". This is equivalent to 80% or more correct answers.*

3. Apply binding constraints
   if $Q(a_p)$*throws "trainee will succeed" in next three questions based on answers by now and the binding constraints options:*
   [option 1] for n_i, n_{i-1} *and* $n_{i-2}f(x) = 1$,
   [option 2] at least n_i *and* n_j*are labelled as "difficult questions" and* $f(x) = 1$.
   [option 3] the student chooses the option of the system "Show what you know", which includes test questions with the highest level of complexity 3—i.e. open questions and penalty scoring.
4. Change the level
   if [option p], $p = 1{:}2$ *and* $Q(a_p)$*throws "will succeed" then go to next upper level,*
   if $S \geq 0.7$ *and* $Q(a_p)$*labelled as "will not succeed" then stay on the same level,*
   if $S < 0.7$ *and* $Q(a_p)$*labelled as "will not succeed" then go to previous lower level,*
   repeat until the trainee exit or the highest level is finished.
   end.

To minimize the number of asked questions and to accelerate the movement from one level to another we calculate the probably success of the trainee dynamically using the curiosity function (3) with puzzle method [27]. To make the puzzle method constraints more effectiveness, we assume three more options, as follows, could be added to the dynamic algorithm:

[option 3] Greedy Algorithm [31]
[option 4] Loop Constraint Solver [38]

The options 3 and 4 have not been used in the experiment shown in this research, but the literature review shows it a potential and we assume it as future work.

3.3 Key Performance Indicators for e-Testing

Key performance indicators (KPIs) for e-testing can be easily shown as dimensions in 3D space of objects and their relationships. This helps to visualize space, movement, access, and so on. When a result of training is visualized in three-dimension it is not

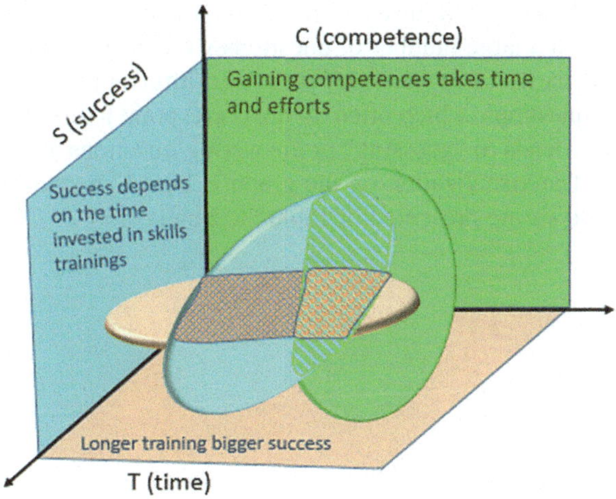

Fig. 2 Knowledge acquisition KPIs [40]

like to visualize a real object or surface [39]. We use the approach to visualize the success of the training and the intelligence of the system. Three main components to assess the success are shown on Fig. 2.

The three dimensions we define are the base of knowledge acquisition and building competences. Time (T) collect and represent how long the trainee was in the training process. Competence (C) is based on correlation between number of questions and the success in the following questions. Also, the value of time must be taken into account when assessing competence. It is explained in more detail in the next section. Success (S) is the score of the trainee.

4 Results and Discussion

The experiment was performed with the participation of 190 students in the field of informatics and computer science. The system for automated test generation has been improved with the proposed two algorithms: Algorithm with fixed number of questions (Fixed algorithm) and Algorithm with dynamic determination of the number of questions (Dynamic algorithm).

4.1 Results of the Experiment

The experiment has been performed with 190 participants—students in computer science field in their first year. We want to understand to what extent self-training affects the building of competence in learners. The time-success ratio can lead to an assessment of competence, but the correlation is not always correct.

As a result of the experiment with the fixed algorithm (Fig. 3) cases were found in which a large number of questions were answered in a very short time. This shows that the assessment of the Fixed Algorithm requires the answer to a very large number of questions, which often do not correspond to the level of knowledge of the learner. This leads to "guessing" or answering questions at random.

The visualization on the graph is based on students' data. Some results of data analysis are presented in Tables 1, 2 and 3.

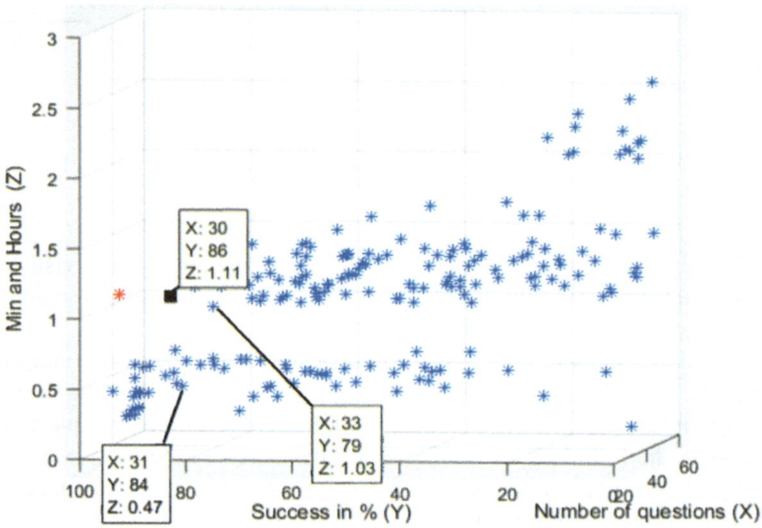

Fig. 3 Fixed algorithm—result of 190 students

Table 1 Distribution of learners according to the time of answering the questions—Fixed Algorithm

		Frequency	Percent	Valid percent
Valid	Lower than 40 min	17	8.9	8.9
	between 41 min and 1 h 20 min	101	53.2	53.2
	over 1 h 21 min	72	37.9	37.9
	Total	190	100.0	100.0

Table 2 Distribution of learners according to the number of correct answers to the questions—Fixed Algorithm

		Frequency	Percent	Valid percent
Valid	Lower than 50%	87	45.8	45.8
	between 51 and 70%	56	29.5	29.5
	over 70%	47	24.7	24.7
	Total	190	100.0	100.0

Table 3 Cross-analysis of the time for solving the test questions in relation to the success of the training—Fixed Algorithm

		Success			Total
		Under 50%	51–70%	Over 70%	
Time for question answers	Lower than 40 min	1	2	14	17
	between 41 min and 1 h 20 min	38	34	29	101
	over 1 h 21 min	48	20	4	72
Total		87	56	47	190

The data in Table 1 show the uneven distribution of the total time of answering the questions by the students. Relatively few learners answered their questions in less than 40 min—only 17 learners (8.9%). For the larger part of the trainees (101 learners, 52.3%) it took them up to 1 h and 20 min to complete the test. The remaining 72 students managed to answer the questions in more than 1 h and 20 min.

Table 2 shows data of the number of correctly answered questions given by the learners. Using the fixed algorithm, almost half of the sample (87 learners) gave up to 50% correct answers. This equates to a knowledge score of 3—low. Almost 30% (56 learners) gave between 51 and 70% correct answers, and only 24.7% have over 70% correct answers. This means that only 24.7% have acquired a very good or excellent knowledge of the material studied.

In Table 3, a cross-analysis was performed between the response time and the number of correct answers. It can be seen that only 14 learners showed very high success with an answer time fewer than 40 min. The learners who needed more time to respond respectively showed worse success. Only 4 of them had excellent or very good knowledge.

The result of the Dynamic Algorithm is presented on Fig. 4. It shows a balance between the number of questions and the time. This confirms that the number of questions needed to evaluate trainee when moving from one level of complexity to another is much more correctly determined. This also gives us reason to say that the quantitative indicators time, number of questions and number of correct answers have a direct correlation with competence. The result of the Dynamic Algorithm is shown in Fig. 4.

The analysis of the data collected by the Dynamic Algorithm is presented in Tables 4, 5 and 6.

The data in Table 4 show that when using the Dynamic Algorithm, the number of learners who responded in less than 40 min was higher than those with the fixed algorithm. 45 learners answered their questions correctly, compared to 17 learners with the fixed algorithm. The number of students who answered for an average time between 40 min and 1 h and 20 min is also higher—105 students using a dynamic algorithm, against 101 students using the fixed algorithm.

Table 5 shows that the use of Dynamic Algorithm leads to a higher number of learners who gave over 71% correct answers. While with the fixed algorithm they

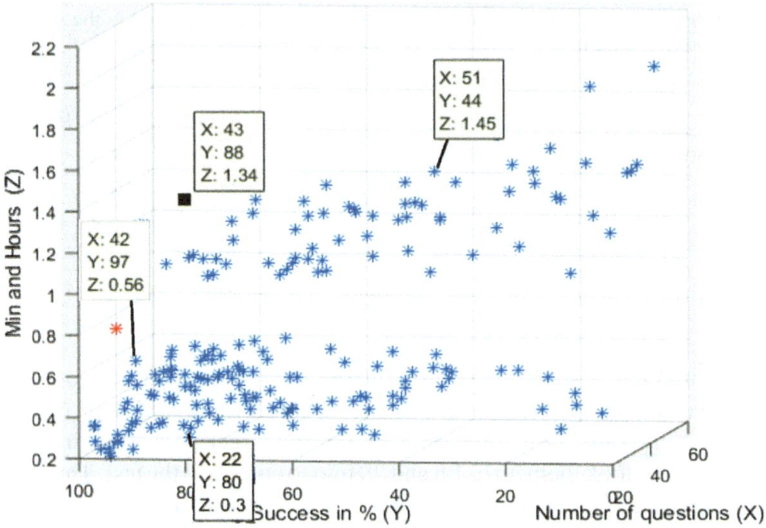

Fig. 4 Dynamic algorithm—result of 190 students

Table 4 Distribution of learners according to the time of answering the questions—Dynamic Algorithm

		Frequency	Percent	Valid percent
Valid	Under 40 min	45	23.7	23.7
	Between 41 min and 1 h 20 min	105	55.3	55.3
	Over 1 h 21 min	40	21.1	21.1
	Total	190	100.0	100.0

Table 5 Distribution of learners according to the number of correct answers to the questions—Dynamic Algorithm

		Frequency	Percent	Valid percent
Valid	Under 50%	56	29,5	29.5
	Between 51 and 70%	38	20.0	20.0
	Over 71%	96	50.5	50.5
	Total	190	100.0	100.0

are only 47 students, when using Dynamic Algorithm the number of trainees with more than 71% correct answers become 96 or 50.5%.

Cross-analysis of the data of Dynamic Algorithm shows an increase in learners with very good or excellent knowledge. In comparison 34 of the learners tested with Dynamic Algorithm gave more than 71% correct answers. Using Fixed Algorithm they are only 14 students.

Table 6 Cross-analysis of the time for solving the test questions in relation to the success of the training—Dynamic Algorithm

		Success			Total
		Under 50%	Between 51 and 70%	Over 71%	
Time for question answer	Under 40 min	4	7	34	45
	Between 41 min and 1 h 20 min	26	23	56	105
	Over 1 h 20 min	26	8	6	40
Total		56	38	96	190

The results show that the usage of Dynamic Algorithm is more effective than Fixed Algorithm. The results of the study clearly show that the intelligent approach meets the hypothesis. Applying the puzzle method constraints in combination with traditional intelligent methodologies improves the usage of the existing automated systems for an e-test generation. The second algorithm with a dynamically changing number of asked questions decreases the time for a pass from one level to another during the self-testing process. The dynamically changing number of questions, leads to a faster switch from one level of testing to another shown in Fig. 4. Binding constraints help the evaluation of the trainee not only based on correct and incorrect given answers. The additional information coming from curiosity function about whether the trainee answered difficult or easy questions, supports the puzzle method for decision making. The curiosity function makes clusters about questions complexity and supports the labelling process.

4.2 Discussion

The discussion in this paper is about whether to invest in the development and implementation of machine intelligence in automated test generation systems. Existing systems work under human control and do a very good job. Is it worth investing in algorithms for logical knowledge processing, complicating the system and making the system with higher consumption of computing power? Assessing the complexity of adding intelligence in the system, the time for data labelling and grouping, we created the two described algorithms. Logical data processing significantly reduces the time to find connections and the ability to group data. We know have to make more experiments with different machines and group of students to compare the efficiency of the algorithms. And it will be done as a future work.

What are the differences between the two proposed algorithms, why we have chosen such an approach for them and what opportunities and prospects we see? The choice for the first algorithm is dictated by the accuracy of the calculations based on statistical data. The guaranteed minimum number of questions to be answered before a knowledge assessment provides a basis for calculating a forecast. But skills assessment is a fuzzy approach. Using only Boolean values in the testing process is a weakness and leads to errors. Therefore, in order to improve and strengthen the evaluation, a clarification is added to the statistical calculation of the result of the fixed number of questions. The constraints of the puzzle check whether the trainee knows the terminology by answering special labeled questions correctly. Binding constraints bring accuracy to the statistical forecast and refine the learner's competence assessment. Whether success can be confirmed and maintained in the next assessment steps is a question to confirm the competence of the test takers. The second algorithm, which aims to speed up the movement through levels of complexity in the testing process. However, this requires the confidence that the trainee has the necessary knowledge. How to calculate this confidence? The Boolean nature of the test cannot give the required nuance. The statistics are also weak here, as there are external factors that it cannot take into account. The assumption of the existence of competence is achieved by applying the curiosity function. It gives additional information on the complexity of the questions asked and, through binding restrictions, gives the necessary confidence in the learner's competence.

The research provides a new direction in the automation of the assessment of knowledge and skills in students. The developed algorithms can be built into existing systems for automatic test generation to become self-testing systems, which will speed up the process of learning in students. The proposed approach is particularly suitable for application in areas of training in new technological areas, where the dynamics are high.

4.3 Limitations of the System with Embedded Intelligence

In our system with embedded intelligence we have observed some constraints:

- The more learners generate tests at the same time, the longer it takes for the system to generate a test. 1 s per question for each learner. To overcome this problem, it is necessary to install the system for generating e-tests on a PC with strong features—more RAM, more RAM.
- Only three types of test questions: Multiple Choice Single Answer, Multiple Choice Multiple Answer and Hot Spot questions. Larger e-testing systems offer a greater variety of types of test questions.
- The system does not take into account the student's previous experience. It makes a decision based on the knowledge shown in the process of solving the tests. We are thinking to develop and implement such functionality in our future work.

5 Conclusion

One of the ways to improve the process of knowledge acquisition is the frequent testing of students. Various training systems are used and improved to automate the test generation process. This paper discusses an approach to improve an existing automated system for e-tests generation to be suitable for self-testing. Automatic switch between levels of difficult is needed to navigate the trainee. To do so, an intelligence is embedded into the system. Two algorithms with fixed and dynamically changed number of questions have been proposed. The both of them use a logical approach that combines the puzzle method and the curiosity function with traditional intelligent methods to predict trainee's answers. The curiosity function groups the questions according to their complexity and in combination with the binding constraints of the puzzle method proposes the appropriate level of complexity to the learner. This approach is innovative and novel. There are some known approaches to predict learner success, but they are based on statistical calculations. When assessing a person's abilities and knowledge in a given field, the test is a very convenient tool, but it cannot assess hidden capability. The approach presented in this paper uses logical modelling to discover hidden capabilities if they exist in confidence. The opposite is also very useful. If the learner does not have enough knowledge and established competencies in the subject, then he is redirected to the study of more basic concepts. The results show that working towards the incorporation of intelligence into testing systems improves knowledge acquisition. This paper proposed a system for automatic generation of electronic tests which is also suitable for self-testing. No one of well-known learning management systems has the functionality to generate unique test for each student and to navigate the student through the different levels of complexity.

Acknowledgements The paper is realized within:

1. Project Bulgarian Supercomputer Centre: A High Performance Infrastructure for Computer Modelling, Simulation and Research with Application in Industry, Medicine, Pharmaceutics, Energetics, Transportation, Finances and Environment Protection, Contract № D01- 30/08.04.2011 with Ministry of Education, Youth and Science. 2011–2012.

2. National Science Program "Information and Communication Technologies for Unified Digital Market in Science, Education and Security".

References

1. National Program for Development of School Education and Preschool Education and Training. 2006–2015. Strategic documents. Council of Ministers
2. Petrova, P., Boyadzhiev, D.: Training young lecturers. In: XIV-th International Conference "Challenges in Higher Education and Research in the 21st Century", May 31–June 3, 2016, Sozopol, Bulgaria, Vol.14, 2016, p.23–26. Heron Press Ltd. ISBN: 978-954-580-356-9

3. Petrova, P., Kostadinova, I.: Higher education in the Era of digital transformation. In: 9th annual International Conference on Education and New Learning Technologies, 3–5 July 2017, Barcelona, Spain. IATED Academy, ISBN: 978-84-697-3777-4. ISSN: 2340-1117
4. Kostadinova, I., Rasheva-Yordanova, K., Garvanova, M.: Analysis of algorithms for generating test questions in e-testing systems. In: Proceedings of EDULEARN19 Conference 1–3 July 2019, Palma, Mallorca, Spain, pp. 1714–1719. ISBN: 978-84-09-12031-4
5. Kaltenborn, R., Hadjiski, M., Koynov, S.: Intelligent control of negative emotions in a computer-based learning system. In: Proceedings of 2020 IEEE 10th International Conference on Intelligent Systems, pp. 119–124. ISBN 978-1–7281-5456-5
6. Katsumata, M.: A multiple smart device-based personalized learning environment. In: Proceedings of 2020 IEEE 10th International Conference on Intelligent Systems, pp. 498–503. ISBN 978-1-7281-5456-5
7. Kostadinova, I., Rasheva-Yordanova, K., Ivanov, I., Petrova, P.: Automated system for generating and validation a learning tests. In: Proceedings of International Conference of Education, Research and Innovation (ICERI2017) Conference 14–16 November 2017, Seville, Spain, pp. 0414–0424, ISBN: 978-84-697-6957-7
8. Kostadinova, I, Getova, I., Dimitrov, G.P., Nikolova, B.: Model for building competencies through an electronic system for assessment of knowledge. In: Proceedings of International Conference of Education, Research and Innovation (ICERI2017) Conference 14–16 November 2017, Seville, Spain, pp. 0430–0436, ISBN: 978-84-697-6957-7
9. Borissova, D., Keremedchiev, D.: Group decision making in evaluation and ranking of students by extended simple multi-attribute rating technique. Cybern. Inf. Technol. **18**(3), 45–56 (2019)
10. Choi, J., Lee, Y., Lee, E.: Puzzle based algorithm learning for cultivating computational thinking. Wireless Pers. Commun. **93**, 131–145 (2017). https://doi.org/10.1007/s11277-016-3679-9
11. Panayotova, G., Dimitrov, G.P.: Modeling and data processing of information systems. In: International Conference on Artificial Intelligence and Pattern Recognition (AIPR), 19–21 Sept 2016. ISBN: 978-1-4673-9187-0
12. Cook, D.J., Holder, L.B.: Graph-based data mining. IEEE Intell. Syst. Appl. **15**(2), 32–41 (2000)
13. Keogh, E., Lonardi, S., Ratanamahatana, C.A.: Towards parameter-free data mining. In: Proceedings of the Tenth ACM SIGKDD International Conference on Knowledge Discovery and Data Mining, ACM, pp. 206–215 (2004)
14. Kumar, S., Mohbey, S.K.: A review on big data based parallel and distributed approaches of pattern mining. J. King Saud Univ.-Comput. Inf. Sci. (2019). ISSN 1319-1578
15. Yun, Y., Ma, D., Yang, M.: Human–computer interaction-based decision support system with applications in data mining. Future Gener. Comput. Syst. **114**, 285–289 (2021). ISSN 0167-739X
16. Mirzakhanov, V.E.: Value of fuzzy logic for data mining and machine learning: a case study. Expert Syst. Appl. **162**, 113781 (2020). ISSN 0957-4174
17. Geron, A.: Hands-on Machine Learning with Scikit-Learn & TensorFlow. O'Reilly Media (2017). ISBN: 9781491962299
18. Han, J., Kamber, M., Pei, J.: Data Mining: Concepts and Techniques, 3rd edn. Morgan Kaufmann (2011)
19. Bishop, C.M.: Machine Learning and Pattern Recognition. Springer, US (2007)
20. Tzirakis, P., Tjortjis, C.: T3C: improving a decision tree classification algorithm's interval splits on continuous attributes. Adv. Data Anal. Classif. **11**(2), 353–370 (2017)
21. Berthold, M.R.: Journeys to Data Mining. Springer, US (2012)
22. Foster, P., Fawcett, T.: Data Science for Business What You Need to Know About Data Mining and Data Analytic Thinking. O'Reilly (2013)
23. Sthle, L., Wold, S.: Analysis of variance (ANOVA). Chemometrics Intell. Lab. Syst. **6**(4), 259–272 (1989). ISSN 0169-7439
24. Kanellopoulos, Y., Antonellis, P., Tjortjis, C., Makris, C., Tsirakis, N.: Kattractors: a partitional clustering algorithm for numeric data analysis. Appl. Artif. Intell. **25**(2), 97–115 (2011)

25. Su, J., Kochan, O., Jotsov, V.: Methods of reducing the effect of the acquired thermoelectric inhomogeneity of thermocouples on temperature measurement error. Measur. Tech. **58**(3), 327–331 (2015)
26. Jotsov, V.: New proposals for knowledge and data driven applications in security systems. In: Innovative Issues in Intelligent Systems. Studies in Computational Intelligence, vol. 623, pp. 231–294. Springer, Berlin Heidelberg (2016). ISSN: 1860–949X
27. Jotsov, V., Iliev, E., Petrova, P.: Puzzle methods for data science applications. In: Intelligent Systems'2016: Proceedings of the 8th IEEE International Conference, September 04–06, 2016, Sofia, Bulgaria, vol. 1, pp. 322–332 (2016). ISBN 978-1–5090-1353-1. ISSN 15411672
28. Jotsov, V., Petrova, P., Iliev, E.: Learning through constraint applications. In: Learning Systems: From Theory to Practice. Studies in Computational Intelligence, vol. 756. Springer, Cham (2018)
29. Garcia, L.P.F., et al.: Boosting meta-learning with simulated data complexity measures. Intell. Data Anal. **24**(5), 1011–1028 (2020)
30. R. Santhosh, M. Mohanapriya, Generalized fuzzy logic based performance prediction in data mining, Materials Today: Proceedings, 2020, ISSN 2214–7853.
31. Pomeranz, D., Shemesh, M., Ben-Shahar, O.: A fully automated greedy square jigsaw puzzle solver. In: Proceedings of IEEE Conference on Computer Vision and Pattern Recognition (2011)
32. Cho, T.S., Avidan, S., Freeman, W.T.: A probabilistic image jigsaw puzzle solver. In: Proceedings of IEEE Conference on Computer Vision and Pattern Recognition (2010)
33. Balkin, J.M., Sonnevend, J.: The digital transformation of education. In: Education and Social Media: Toward a Digital Future. MIT Press (2016). ISBN-13: 9780262034470
34. Rodrigues, L.S.: Challenges of digital transformation in higher education institutions: a brief discussion. In: Proceedings of the 30th International Business Information Management Association Conference, IBIMA 2017—Vision 2020: Sustainable Economic Development, Innovation Management, and Global Growth, pp. 4490–4493 (2017)
35. Florez-Aristizabal, L., Cano, S., Collazos, C.A., Benavides, F., Moreira, F., Fardoun, H.M.: Digital transformation to support literacy teaching to deaf Children: From storytelling to digital interactive storytelling. In: Telematics and Informatics, vol. 38, pp. 87–99 (2019). ISSN 0736-5853
36. Abad-Segura, E., González-Zamar, M.-D., Infante-Moro, J.C., Ruipérez García, G.: Sustainable management of digital transformation in higher education: global research trends. Sustainability **12**, 2107 (2020)
37. Borissova, D., Keremedchiev, D.: Intelligent system for generation and evaluation of e-learning tests using integer programming. In: Modelling and Development of Intelligent Systems. MDIS 2019. CCIS, vol. 1126, pp. 97–110. Springer (2020)
38. Son, K., Hays, J., Cooper, D.B.: Solving square jigsaw puzzles with loop constraints. ECCV **6**(2014), 32–46 (2013)
39. Trenchev, I., Mavrevski, R., Traykov, M., Stoykov, D.: Mathematical approaches for creation of complex 3D models and their application in education. In: ICERI2019 Proceedings, 2019, pp. 4908–4914, ISBN: 978-84-09-14755-7. ISSN: 2340-1095
40. Petrova, P., Kostadinova, I.: An approach for embedding intelligence in a system for automatic test generation and a 3D result model. In: Proceedings of 2020 IEEE 10th International Conference on Intelligent Systems, pp. 358–363. ISBN 978-1-7281-5456-5

Simulation Algorithms to Assess the Impact of Aging on the Reliability of Standby Systems with Switching Failures

Kiril Tenekedjiev, Natalia Nikolova, Guixin Fan, Mark Symes, and Oanh Nguyen

Abstract We analyze the influence of aging on the reliability of a two-component standby system with switching failures (2SBSF). When the backup component is aging in standby as if it is working, we have a 2SBSF with full aging. Typically, such a system is modelled with a rate diagram (where the failure rates are time-dependent) which has an adjacent system of 3 ordinary differential equations (ODEs). The reliability characteristics of the 2SBSF with full aging can be obtained from the numerical solution of the formulated Cauchy problem. Alternatively, simulation modelling can produce the same results, but without using the rate diagram. In that way we can adapt the simulation algorithms to treat the cases when either the backup component does not age in standby (2SBSF with no aging), or when the backup component ages in standby slower than in operation (2SBSF with partial aging). In the last two cases there is neither rate diagram, nor a system of ODEs since the 2SBSF is not state dependent when the full aging assumption does not hold. We develop ten algorithms to facilitate the simulation solutions of the 2SBSF with no aging and with partial aging. We emphasize on the ways to deal with the numerical problems created by working with truncated distributions and with random sampling

---

K. Tenekedjiev · N. Nikolova (✉) · G. Fan · M. Symes · O. Nguyen
Australian Maritime College, University of Tasmania, 1 Maritime Way, Launceston, TAS 7250, Australia
e-mail: Natalia.Nikolova@utas.edu.au

K. Tenekedjiev
e-mail: Kiril.Tenekedjiev@utas.edu.au

G. Fan
e-mail: gfan@utas.edu.au

M. Symes
e-mail: mark.symes@utas.edu.au

O. Nguyen
e-mail: o.nguyen@utas.edu.au

K. Tenekedjiev · N. Nikolova
Nikola Vaptsarov Naval Academy, 73 V. Drumev St, 9026 Varna, Bulgaria

V. Sgurev et al. (eds.), *Advances in Intelligent Systems Research and Innovation*, Studies in Systems, Decision and Control 379,
https://doi.org/10.1007/978-3-030-78124-8_21

from conditional CDFs. Extensive numerical examples in two groups demonstrate the ideas of our procedures.

Keywords Reliability · Rate diagram · Truncated distributions · State probability functions · Aging in standby

1 Introduction

When a system or a component fails excessively, it is often necessary to improve the overall system reliability so that repair costs can be reduced, and scheduled maintenances can be arranged. A widespread way to ease the problem is to create redundancy by duplicating components. The primary and the additional components can be arranged in three different schemes. The first is parallel, where all duplicates virtually operate at the same time [10], the second is "*k* out of *n*", where not all duplicated components operate [1], and the third is standby, where only one component operates at a time while the others are on standby [4]. However, the duplicated components may also fail during standby, and there could also be failures during the switch from the failed working components to the standing by backup components. From this point onwards, we will be focusing on the standby systems.

Take the example below of a two-component standby system with switching failures (2SBSF). In the 2SBSF, the primary component operates till failure and then it is substituted with the backup component through a switch. The backup component and the switch could also fail. In each moment of time, the system falls under one of the five states given in Fig. 1.

In State 1 (Fig. 1a), the primary component is under normal operation, and the backup component is ready in standby, so the system functions normally. In State 2 (Fig. 1b), the primary component has failed, and the backup component is successfully switched on, so that the system would still operate in backup mode. In State

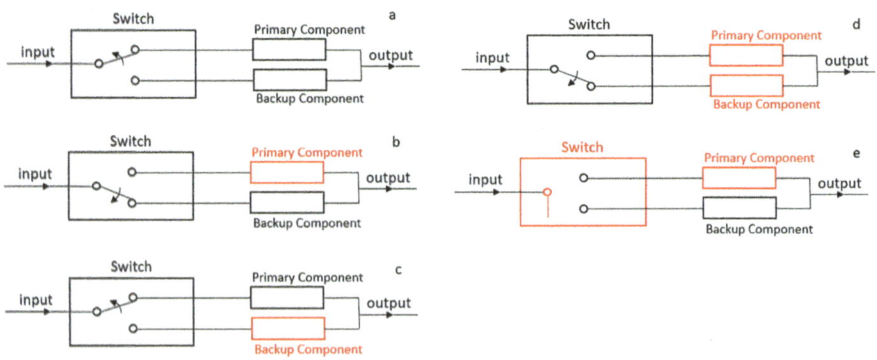

Fig. 1 Block diagrams of 2SBSF: **a** State 1: normal operation; **b** State 2: backup operation; **c** State 3: failure on standby; **d** State 4a: component system failure; **e** State 4b: switching system failure

3 (Fig. 1c), the primary component functions normally, but the backup has failed while in standby. Although the system still functions, it has lost its backup. To make matters worse, it would be unknown to operators that the backup component has failed. In State 4, the system will not operate, which can be caused by two scenarios: in scenario one (Fig. 1d), the primary component has failed, the switch has switched to backup mode successively, but the backup component also fails; in scenario two (Fig. 1e), the primary component has failed, but the switch fails to switch on the backup component. In either scenario, the system fails.

We can conclude that the overall reliability of any standby system depends on how the backup components will age during standby. According to [6], the duplicates in standby systems can be categorized into three types: hot reserve, cold reserve, and warm reserve. Hot reserve assumes that a backup component experiences the same failure rate as the active primary component, which means even when the backup component is inactive, it still decays and fails just as quickly as the working primary component. Cold reserve means the backup component has 0 failure rate and remains in pristine condition while in standby, it simply does not fail when it is not switched on. Warm reserve is an in-between state, in which the backup component still decays during standby, but not as quickly as the active primary component.

For this chapter, we would abandon the categories of hot, cold, and warm reserves and use the more precise terms of full aging, no aging, and partial aging. There are three reasons for such a substitute. First, those terms can cause confusion because the parallel systems are a.k.a. systems in hot reserve, and the standby systems are a.k.a. systems with cold reserve [15]. Second, the system may not experience a standby failure, but the backup component could still age during standby. Third, although rarely, a component could have decreasing failure rate, and in that case, the greater the aging the better the reliability.

To calculate the reliability characteristics of the system shown in Fig. 1, it is necessary to know:

(a) the reliability function of the primary component, $R_1(t)$,
(b) the reliability function of the backup component in operation, $R_2(t)$,
(c) the reliability function of the backup component in standby, $R_3(t)$,
(d) the switching failure probability, p.

Here, $R_i(t)$ (for $i = 1, 2, 3$) is the probability that the specific component under known operational conditions will not fail up to time $t \geq 0$. The distribution of each of the three continuous random variables "time to failure of the same specific component under known operational conditions" can be described with three more functions:

(a) the Cumulative Distribution Function (CDF), $F_i(t) = 1 - R_i(t)$ (for $t \geq 0$ and $i = 1, 2, 3$);
(b) the Probability Density Function (PDF), $f_i(t) = \mathrm{d}F_i(t)/\mathrm{d}t$ (for $t \geq 0$ and $i = 1, 2, 3$);
(c) the failure rate, $\lambda_i(t) = f_i(t)/R_i(t) = f_i(t)/[1 - F_i(t)]$ (for $t \geq 0$ and $i = 1, 2, 3$).

The α-quantile of any distribution, t_α, is defined as the time when the CDF is equal to α, i.e. $F(t_\alpha) = \alpha$. The R-design life of any component, $t_{des,\alpha}$, is defined as the time when the reliability function of the component is equal to R, i.e. $R(t_{des,\alpha}) = R$. In reliability engineering $t_{des,0.9}$ is a.k.a. B10 life, and $t_{des,0.99}$ is a.k.a. B1 life [27].

Let us define the probability function $P_r(t)$ (for $t \geq 0$ and $r = 1, 2, 3, 4$) to be the probability of the system being in state i at time t. Since at any instant of time, the system will always be in exactly one of the four states, the sum of the probability functions for any moment of time will be 1 [12, p. 200]. The system fails only in state four, which means that the reliability function of the system, $R_{sys}(t)$ (for $t \geq 0$), is the complement of the $P_4(t)$ function. Solving 2SBSF means to identify the probability functions and the reliability of the system.

The backup component's aging in standby by an equivalent time T means that the failure of the backup component in operation would follow not the given reliability function $R_2(t)$, but the conditional reliability $R_2(t|T)$, which assumes that the standby component has survived to time T without failure. Normally, the increased "wear-out" of the backup component will increase its failure rate with aging. However, the aging may also decrease the component's failure rate by avoiding the "infant mortality" section on the bathtub curve [14]. The "full aging assumption" means that the backup component will age in standby with the same intensity as when it is operating. Such scenario is, however, the most unrealistic and seldom occurs. A more realistic assumption would be the "no aging assumption", where the backup component does not age at all while in standby. This assumption is on the other extreme of the full aging assumption, and also does not occur frequently. The most realistic assumption is the "partial aging assumption", where the backup component ages in standby, but not as quickly as in operation.

The intention of this chapter is to look beyond the assumption of "full aging" and calculate the reliability of a two-component backup system with switching failure with the assumptions of "partial aging" and "no aging". Since one cannot establish ODEs with the above-mentioned two assumptions, it is only feasible to establish a simulation model.

A similar problem was formulated and solved in [22], where the influence of three different aging assumptions was investigated over the performance of the two-component standby system using simulation. This chapter is an extended version of [22], dealing with switching failures and with some additional numerical problems affecting the simulation solution.

2 Analytical Solutions Without Aging Assumptions

First, we will concentrate on the easiest case when in the 2SBSF given in Fig. 1, each of the three distributions has constant failure rate (CFR). In that case we will denote $\lambda_i(t) = \lambda_i$ (for $t \geq 0$ and $i = 1, 2, 3$). In other words, the failure distribution of the primary component during operation is exponential with parameter λ_1. Similarly, the failure distribution of the backup component during operation is exponential with

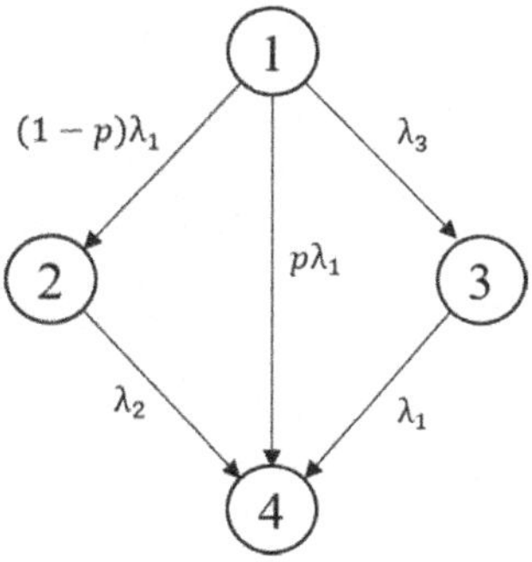

Fig. 2 Rate diagram of the constant failure rate 2SBSF

parameter λ_2 and the failure distribution of the backup component during standby is exponential with parameter λ_3.

In each moment of time, the system will be in either one of the four states, and the probability to transition between the states only depends on the current state. The transition between the states is instantaneous with time-independent transition probabilities. Therefore, it is possible to perform Markov analysis on this specific problem using a rate diagram [24]. The rate diagram shows the possible transitions that could happen between states with the corresponding transition rates. Thus, the rate diagram for the 2SBSF is shown on Fig. 2.

The resulting system of ODEs is:

$$\begin{cases} \dfrac{dP_1(t)}{dt} = -[\lambda_1 + \lambda_3]P_1(t) \\ \dfrac{dP_2(t)}{dt} = (1-p)\lambda_1 P_1(t) - \lambda_2 P_2(t) \\ \dfrac{dP_3(t)}{dt} = \lambda_3 P_1(t) - \lambda_1 P_3(t) \end{cases} \tag{1}$$

with initial conditions:

$$P_1(0) = 1,\ \ P_2(0) = 0,\ \ P_3(0) = 0 \tag{2}$$

The $P_4(t)$ can be written as:

$$P_4(t) = 1 - P_1(t) - P_2(t) - P_3(t) \tag{3}$$

Therefore, the reliability of the system is:

$$R_{sys}(t) = 1 - P_4(t) = P_1(t) + P_2(t) + P_3(t) \tag{4}$$

We can trivially solve the formulated Cauchi problem using Laplace transformation [28] and get the following analytical solutions:

$$P_1(t) = e^{-(\lambda_1+\lambda_3)t}, t \geq 0$$
$$P_2(t) = \frac{(1-p)\lambda_1}{\lambda_1+\lambda_3-\lambda_2}\left[e^{-\lambda_2 t} - e^{-(\lambda_1+\lambda_3)t}\right], t \geq 0$$
$$P_3(t) = e^{-\lambda_1 t} - e^{-(\lambda_1+\lambda_3)t}, t \geq 0$$
$$P_4(t) = 1 - e^{-\lambda_1 t} - \frac{(1-p)\lambda_1}{\lambda_1+\lambda_3-\lambda_2}\left[e^{-\lambda_2 t} - e^{-(\lambda_1+\lambda_3)t}\right], t \geq 0$$
$$R_{sys}(t) = e^{-\lambda_1 t} + \frac{(1-p)\lambda_1}{\lambda_1+\lambda_3-\lambda_2}\left[e^{-\lambda_2 t} - e^{-(\lambda_1+\lambda_3)t}\right], t \geq 0 \quad (5)$$

We derived the solution (5) of 2SBSF under the CFR assumption. However, the exponential distribution is memoryless in a sense that $R_2(t|T) = R_2(t) = \exp\{-\lambda_2 t\}$ for $t \geq 0$. It follows that any type of aging will never impact the solution. That is why we can think of this special case as a solution without aging assumption.

3 Numerical Solution with a Full Aging Assumption

The CFR model is rarely met in practice. It is more realistic to consider cases with time-dependent failure rates. If we relax the assumption of CFR, then the three failure rates in the rate diagram (Fig. 2) can be arbitrary non-negative functions $\lambda_i(t) \geq 0$ of the time (for $t \geq 0$ and $i = 1, 2, 3$). In other words, the failure distribution of the primary component during operation has failure rate $\lambda_1(t)$. Similarly, the failure distribution of the backup component during operation has failure rate $\lambda_2(t)$, and the failure distribution of the backup component during standby has failure rate $\lambda_3(t)$. The rate diagram in this case is shown on Fig. 3 and the corresponding system of three ODEs is given in (6).

$$\begin{cases} \frac{dP_1(t)}{dt} = -[\lambda_1(t) + \lambda_3(t)]P_1(t) \\ \frac{dP_2(t)}{dt} = (1-p)\lambda_1(t)P_1(t) - \lambda_2(t)P_2(t) \\ \frac{dP_3(t)}{dt} = \lambda_3(t)P_1(t) - \lambda_1(t)P_3(t) \end{cases} \quad (6)$$

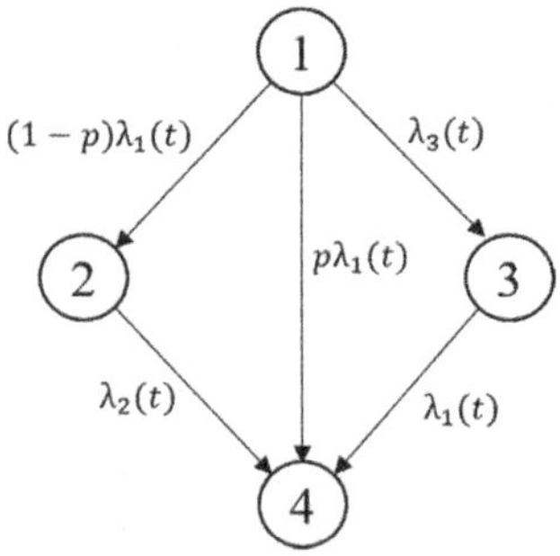

Fig. 3 Rate diagram of the time-dependent failure rate 2SBSF

The initial conditions of (6) are the same as in the CFR case and are given in (2). The so formulated Cauchi problem cannot be solved analytically, but a numerical solution is relatively easy to find. The numerical solution can be facilitated by the analytical Jacobian matrix of the RHS of (6) as a function of time:

$$J(t) = \begin{bmatrix} -\lambda_1(t) - \lambda_3(t) & 0 & 0 \\ (1-p)\lambda_1(t) & -\lambda_2(t) & 0 \\ \lambda_3(t) & 0 & -\lambda_1(t) \end{bmatrix} \tag{7}$$

The Cauchi problem in (6) and (2) together with the Jacobian (7) has to be solved numerically within the domain $t \in [0,\ t_{end}]$. The t_{end} is to be selected as low as possible, but so that the reliability at t_{end} is almost 0, i.e. $R_{sys}(t_{end}) \approx 0$.

However, the formulated solution in this section is making an implicit assumption that there is full aging in the system, i.e. the backup component ages while in standby, under the same rate as if it were operating. As one can see in (6), $\lambda_2(t)$ will depend only on the actual time t, but not from the moment when the system moves into State 2. That is true only when the system works under full aging. In fact, relaxing the restrictive CFR assumption using the rate diagram and the numerical solution of a Cauchi problem with 3 ODEs created another problem. The numerical solution in this section is true for any time dependent failure rates, but only for the case of full aging. That is why, the stated numerical solution is valid only if we assume that the backup component ages in standby as if it is operating.

4 Simulation Solution with a Full Aging Assumption

Alternatively, the solution from Sect. 3 for $t \in [0,\ t_{end}]$ can be obtained using simulation modeling [30]. The main part of the simulation is the generation of N pseudo-realities, where N is a large predetermined natural number (typically 1000, 10000, or 100000). In the jth pseudo-reality, the event chain EC_j (describing all the events in the pseudo-reality) is to be generated and remembered. The k-th event in the j-th pseudo-reality happens at time $time_j(k)$ and is the transition into the state $state_j(k)$. Therefore,

$$EC_j = \{[time_j(k), state_j(k)] | k = 1, 2, \ldots, q_j\} \tag{8}$$

The time values in (8) are in ascending order. Theoretically EC_j can be generated with algorithm Alg A given below.

Alg A: Theoretical generation of the event chain in the *j*th pseudo-reality for 2SBSF with full aging

1. Start the simulation from State 1 by setting $state_j(1) = 1$ and $time_j(1) = 0$
2. Sample a random variate failure time of the primary component, t_1, from $R_1(t)$

3. Sample a random variate failure time of the backup component in standby, t_3, from $R_3(t)$
4. If $t_3 = t_1$ and $t_1 \geq t_{end}$, then set $q_j = 1$ and stop the simulation (no system failure)
5. If $t_3 = t_1$ and $t_1 < t_{end}$, then set $state_j(2) = 4$, $time_j(2) = t_1$, $q_j = 2$ and stop the simulation (system failure)
6. If $t_3 < t_1$ and $t_3 \geq t_{end}$, then set $q_j = 1$ and stop the simulation (no system failure)
7. If $t_3 < t_1$ and $t_3 < t_{end}$, then:

   7a. Set $state_j(2) = 3$, $time_j(2) = t_3$ (move to State 3)
   7b. If $t_1 \geq t_{end}$, then set $q_j = 2$ and stop the simulation (no system failure)
   7c. If $t_1 < t_{end}$, then set $state_j(3) = 4$, $time_j(3) = t_1$, $q_j = 3$ and stop the simulation (system failure)

8. If $t_3 > t_1$ and $t_1 \geq t_{end}$, then set $q_j = 1$ and stop the simulation (no system failure)
9. If $t_3 > t_1$ and $t_1 < t_{end}$, then:

   9a. Generate a random number *RN* from the standard uniform distribution in the component interval (RN will determine the future state)
   9b. If $RN < p$, then set $state_j(2) = 4$, $time_j(2) = t_1$ and stop the simulation (switching failure)
   9c. If $RN \geq p$, then set $state_j(2) = 2$; $time_j(2) = t_1$ (move to State 2)

10. Sample a random variate failure time period of the backup component in operation, τ_2, from $R_2(t|t_1) = R_2(t + t_1)/R_2(t_1)$
11. If $(t_1 + \tau_2) \geq t_{end}$, then set $q_j = 2$ and stop the simulation (no system failure)
12. If $(t_1 + \tau_2) < t_{end}$, then set $state_j(3) = 4$, $time_j(3) = t_1 + \tau_2$, $q_j = 3$ and stop the simulation (system failure).

Using the event chain EC_j, it is then possible to build a state function, $St_j(t)$, which points the state of the 2SBSF for any time t in the domain $[0, t_{end}]$:

$$St_j = \begin{cases} state_j(k) \text{ if } \begin{cases} time_j(k) \leq t < time_j(k+1) \\ \text{for } k = 1, 2, \ldots, q_j - 1 \end{cases} \\ state_j(q_j) \text{ if } time_j(q_j) \leq t \leq t_{end} \end{cases} \tag{9}$$

Select n equally distributed times in the interval $[0, t_{end}]$:

$$T_i = \frac{i-1}{n-1} t_{end}, \text{ for } i = 1,2,\ldots,n \tag{10}$$

Knowing the N state functions (9), it is trivial to estimate the probability of the 2SBSF being in the State r at time T_i:

$$\hat{P}_r(T_i) = \frac{1}{N} \sum_{\substack{j=1 \\ St_j(T_i)=r}}^{N} 1 \text{ , for } r = 1, 2, 3, 4 \text{ and } i = 1, 2, \ldots, n \tag{11}$$

Formula (11) uses the frequentist definition of probability [29]. The continuous probability function for the r-th state of the 2SBSF, $\hat{P}_r(t)$ (for r = 1, 2, 3, 4), can be estimated as linear interpolation over the set of the n nodes estimated in (10) and (11):

$$\left\{\left(T_i - \hat{P}_r(T_i)\right) | i = 1, 2, \ldots, n\right\} \tag{12}$$

The system failure is observed only when the 2SBSF is in state 4. So, the reliability function of the system at time T_i can be approximated as:

$$R_{sys}(T_i) = \hat{P}_1(T_i) + \hat{P}_2(T_i) + \hat{P}_3(T_i), \text{ for } i = 1, 2, \ldots, n \tag{13}$$

The continuous reliability function of the system, $R_{sys}(t)$, can be estimated as linear interpolation over the set of the n nodes estimated in (10) and (13):

$$\left\{(T_i - R_{sys}(T_i)) | i = 1, 2, \ldots, n \right\} \tag{14}$$

Using the reliability function (14), it is possible to estimate all the important numerical characteristics of the failure process. The mean time to failure of a system ($MTTF_{sys}$), being the most widely used reliability characteristic, traditionally is calculated as the improper integral of the reliability function with bounds zero and plus infinity [3]. Its numerical approximation can be found by applying the trapezoidal rule [2]:

$$MTTF_{sys} = T_1 \left(\frac{R_{sys}(T_1) + R_{sys}(T_n)}{2} + \sum_{i=1}^{n-1} R_{sys}(T_i) \right) \tag{15}$$

To calculate the R-design life of the system, $t_{des,R}$, for any $R \in \left(R_{sys}(T_n), 1\right)$ the inverse of the reliability function is needed. However, $R_{sys}(t)$ is not easy to be inverted. The reliability function estimate $R_{sys}(t)$, being linearly interpolated on the nodes (14), is a continuous and decreasing function, but does not always have an inverse one because more often than not, some of the nodes in (14) are with equal reliability. In fact, for the nodes in (14), it is known that: $T_1 < T_2 < \cdots < T_n$ and $1 = R_{sys}(T_1) \geq R_{sys}(T_2) \geq \cdots \geq R_{sys}(T_n) \geq 0$. The last relations come from the nature of simulation, used for this system. One way of going around the non-existence of an inverse function is to form a new set of nodes by either keeping, or purging, or modifying some of the nodes in (14) to get:

$$\left\{\left(T_k^{pu} - R_k^{pu}\right) |k = 1, 2, \ldots, n^{pu} \right\} \tag{16}$$

In fact, T_1^{pu} is the greatest T_i from (14) that has $R_{sys}(T_i) = 1$, and $T_{n^{pu}}^{pu}$ is the least T_i from (14) that has $R_{sys}(T_i) = R_{sys}(T_n)$. Whenever several internal nodes from (14) form a horizontal platform, then all of them are removed and the center of the platform is added as a new node.

The nodes in (16) satisfy $T_1^{pu} < T_2^{pu} < \ldots < T_{n^{pu}}^{pu}$ and $1 = R_1^{pu} > R_2^{pu} > \cdots > R_{n^{pu}}^{pu} = R_{sys}(T_n) \geq 0$. Now we can define a new purged estimate of the reliability function $R^{pu}(t)$ to be linearly interpolated over the nodes in (16). The new function $R^{pu}(t)$ is continuous, strictly decreasing, and therefore invertible in its domain $\left[T_1^{pu}, T_{n^{pu}}^{pu}\right]$.

The procedure to transform the nodes (14) into nodes (16) can be described as "start with the end of the first horizontal segment, end with the beginning of the last horizontal segment, and substitute any other horizontal segments with their centroid". The procedure is embodied in algorithm Alg B.

Alg B: Purging the reliability function into an invertible reliability function

1. Set $i = 1, j = 1, k = 1$
2. If $R(T_i) > R(T_j)$, then go to step 5
3. Set $j = j + 1$
4. Go to step 2
5. If $k = 1$, then set $T_k^{pu} = T_j$, $R_k^{pu} = R_{sys}(T_j)$, $i = j + 1, j = j + 1$
6. If $k > 1$ and $R(T_j) > R(T_n)$, then set $T_k^{pu} = (T_i + T_j)/2$, $R_k^{pu} = R_{sys}(T_i)$
7. If $k > 1$ and $R(T_j) = R(T_n)$, then set $T_k^{pu} = T_i$, $R_k^{pu} = R_{sys}(T_i)$, $n^{pu} = k$ and stop the algorithm
8. Set $k = k + 1$ and go to step 2.

Figure 4 shows in black a reliability function $R_{sys}(t)$ linearly approximated on $n = 21$ nodes (14). The $R_{sys}(t)$ was transformed into the invertible reliability function $R^{pu}(t)$ using Alg B. Figure 4 shows in red an invertible reliability function $R^{pu}(t)$ linearly approximated on $n^{pu} = 8$ nodes (16).

The R-design life of the system can be assessed using the inverse of $R^{pu}(t)$, which here is readily available:

$$t_{des,R} = T_k^{pu} + \left(R_k^{pu} - R\right)\frac{T_{k+1}^{pu} - T_k^{pu}}{R_k^{pu} - R_{k+1}^{pu}}, \text{ for } R_k^{pu} \geq R > R_{k+1}^{pu} \tag{17}$$

The median, the B1 life, the 0.95 design life, the B10 life, and the interquartile range of the 2SBSF system are important reliability characteristics [31]. All of them depend on $t_{des,R}$:

$$\begin{cases} Median_{sys} = t_{des,0.5} \\ B1\_life = t_{des,0.99} \\ 95\%\_design\_life = t_{des,0.95} \\ B10\_life = t_{des,0.9} \\ IQR = t_{des,0.25} - t_{des,0.75} \end{cases} \tag{18}$$

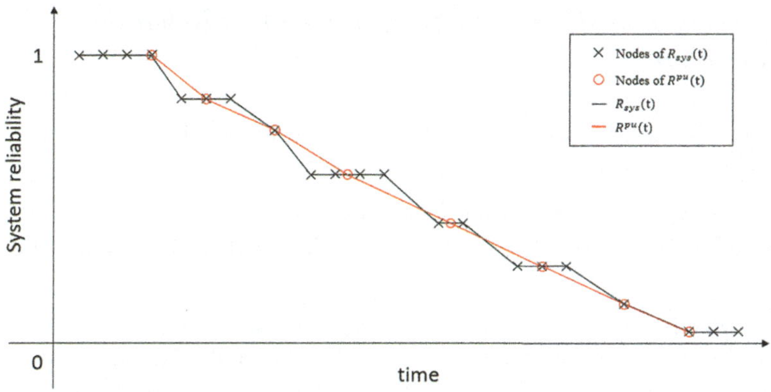

Fig. 4 Illustration of transforming a reliability function into invertible reliability function using Alg B

If we compare the numerical solution in Sect. 3 with the simulation solution in Sect. 4, we can conclude the following: (a) both solve a task under full aging hence results should be the same; (b) the simulation solution does not use the rate diagram in Fig. 3, but the more general block diagrams in Fig. 1. That allows the simulation solution to be modified to adapt to other aging assumptions; (c) the simulation results in each pseudo reality allow to calculate the characteristics of the system that would not be accessible under the numerical solution. Such characteristics are, for example, the probability for the backup component to work, which in fact is the probability to pass state 2. Another example is the probability for the backup component to fail in standby, which in fact is the probability to pass state 3.

5 Conditional and Unconditional Sampling

In each pseudo-reality, Alg A needs to sample one failure period variate from a conditional reliability (in step 10) and two failure times from two unconditional reliability functions (in steps 2 and 3). The sampling can easily be done with the inverse transformation method [9]. However, technical and numerical difficulties arise in its realization especially when the distribution is conditional and/or is truncated. This section develops practical procedures for random failure time sampling from distributions with arbitrary time dependent failure rates.

5.1 Inverse Transformation Sampling from a Reliability Function

For unconditional sampling from an arbitrary invertible reliability function $R(t)$, we can apply the inverse method:

(1) we generate a uniform pseudo-random variate r' in the unit interval (0, 1);
(2) transform r' into a random variate as the r'-design life of the reliability function: $t' = t_{des,r'} = R^{-1}(r')$.

Although other sampling methods can be faster for some specific distribution (e.g. the ziggurat method for normal variates sampling [19]), the inverse transformation sampling is the nowadays method of choice for sampling from generic invertible reliability functions describing continuous random variables.

Assume that there is no failure up to time T. The time interval τ between the T and the failure is a continuous random variable complying with the conditional reliability function $R_{cond}(\tau|T)$ which according to [25] is:

$$R_{cond}(\tau|T) = R(\tau + T)/R(T), \text{ for } \tau \geq 0 \tag{19}$$

For conditional sampling we can use the inverse method in two modifications—direct and indirect. The direct sampling method of a random interval variate from a conditional reliability function $R_{cond}(\tau|T)$ is straightforward:

(1) approximate $R_{cond}(\tau|T)$ with (19) using the unconditional reliability function $R(t)$;
(2) sample a uniformly distributed variate r' in the unit interval (0, 1);
(3) sample the time period variate τ' as the r'-design life from the reliability function $R_{cond}(\tau|T)$.

The direct sampling method from a conditional reliability function is illustrated on Fig. 5. The unconditional reliability function $R(t)$ is given on the upper panel together with the time T. The lower panel shows the conditional reliability function $R_{cond}(\tau|T)$ as well as the direct sampling of τ'.

It is much faster and convenient to apply the inverse transformation method over the unconditional reliability function, which is the essence of the indirect modification. The mentioned uniform random variate r' is equated to the RHS of (19). Then:

$$\begin{cases} R(\tau' + T) = r'R(T) \\ r'_{cor} = r'R(T) \\ \tau' + T = R^{-1}(r'_{cor}) \\ \tau' = R^{-1}(r'_{cor}) - T \\ \tau' = t_{des,r'_{cor}} - T \end{cases} \tag{20}$$

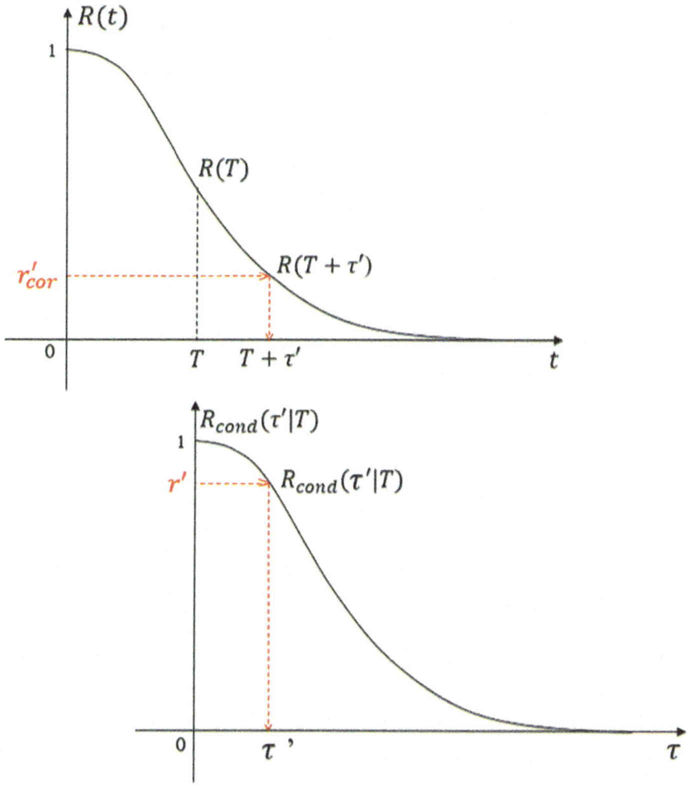

Fig. 5 Sampling variates from a conditional reliability function. On the upper panel the indirect sampling procedure utilises only the unconditional reliability function, $R(t)$. On the lower panel the direct sampling procedure uses the conditional reliability function, $R_{cond}(\tau\ |T)$, calculated from the unconditonal $R(t)$

The indirect sampling method of a random time interval variate from a conditional reliability function, $R_{cond}(\tau|T)$, is given below:

1. Estimate $R(T)$ as the unconditional reliability at time T
2. Sample a uniformly distributed variate, r', in the unit interval
3. Estimate $r_{cor}^{'} = r'R(T)$
4. Calculate $t_{des,r_{cor}^{'}}$ as the $r_{cor}^{'}$-design life from the unconditional reliability function $R(t)$
5. Estimate the time period variate as $\tau' = t_{des,r_{cor}^{'}} - T$.

Since any reliability value is not negative or does not exceed one, it follows that $0 \leq R(T) \leq 1$. On the other hand, r' is also a number between 0 and 1, i.e. $0 \leq r' \leq 1$. Using (20), it follows that

$$r_{cor}^{'} \in [1,\ R(T)] \tag{21}$$

Because of (21), the generated $t_{des,r'_{cor}}$ in step 4 is not less than T and therefore the time period variate is nonnegative ($\tau' \geq 0$). The procedure of the indirect sampling method is illustrated on the upper panel of Fig. 5.

5.2 *Inverse Transformation Sampling from a CDF*

In most of software implementations of the sampling algorithms in the previous subsection the reliability functions are not derived directly. In practice, for each component we know the CDF, $F(t)$ which is simply $[1 - R(t)]$. The reliability function a.k.a. survival function in the general statistics [7] has an auxiliary role, because of two main reasons: the lack of non-parametric distributions and the available software.

The primary source of information for deriving a failure distribution is an experimental data sample consisting of completely observed failure times and of right censored failure times [16]. However, the completely observed data points, as a rule, are concentrated in the early failure domain and rarely exceed a point where $F(t) = 0.8$. It is next to impossible to use the non-parametric distributions because the empirical CDF is approximated in a domain where $F(t)$ rarely reaches values more than 0.7–0.8 [23]. As a result, almost always $F(t)$ takes the form of some specific theoretical distribution (e.g. Weibull, Gamma, LogNormal, etc.) [8, p. 399].

The location, the scale, and the shape parameters of the particular theoretical distribution are estimated to fit best to the failure times in the experimental data sample (in "Least-Square" sense or in "Maximum Likelihood" sense). In almost all programing languages, the specific theoretical distributions are associated with library of functions performing all basic operations such as parameter identification, CDF calculation, sampling, inverse CDF identification, PDF calculation, numerical characteristics estimation, etc. On top, the MATLAB Statistical Toolbox [17] has a library to perform those basic operations using generic probability distribution objects which are identified only once in the software implementation (so from this time on, the simulation code does not depend on the particular distribution and its characteristics). As a result, the simulation software product can be quickly and reliably modified to work with other theoretical distributions. The library functions as a rule are used as black boxes, hence the necessity for substituting the reliability functions with CDFs.

The conditional reliability function $R_{cond}(\tau|T)$ given in (19) can be modified to depend on the unconditional CDF, $F(t)$:

$$R_{cond}(\tau|T) = [1 - F(\tau + T)]/[1 - F(T)], \quad \text{for} \tau \geq 0 \tag{22}$$

The conditional CDF, $F_{cond}(\tau|T)$ follows from (22):

$$F_{cond}(\tau|T) = 1 - [1 - F(\tau + T)]/[1 - F(T)], \quad \text{for} \tau \geq 0 \tag{23}$$

Since there are no advantages of using (23) instead of (22), the direct sampling method of a random time variate from a conditional CDF, $F_{cond}(\tau|T)$ is simple:

(1) approximate the function $R_{cond}(\tau|T)$ with (22) using the unconditional $F(t)$
(2) sample a uniformly distributed variate r'' in the unit interval (0, 1)
(3) sample the time period variate τ' as the r'-design life from the reliability function $R_{cond}(\tau|T)$

The direct sampling method with conditional CDF is illustrated on Fig. 6. On the upper panel the unconditional CDF, $F(t)$, is shown together with the time T. On the lower panel the conditional reliability function $R_{cond}(\tau|T)$ is constructed and the direct sampling of τ' is shown.

As with the conditional reliability, it is much faster and convenient to apply the inverse transformation method over the unconditional CDF. The mentioned uniform random variate r' is equated to the RHS of (22). Then:

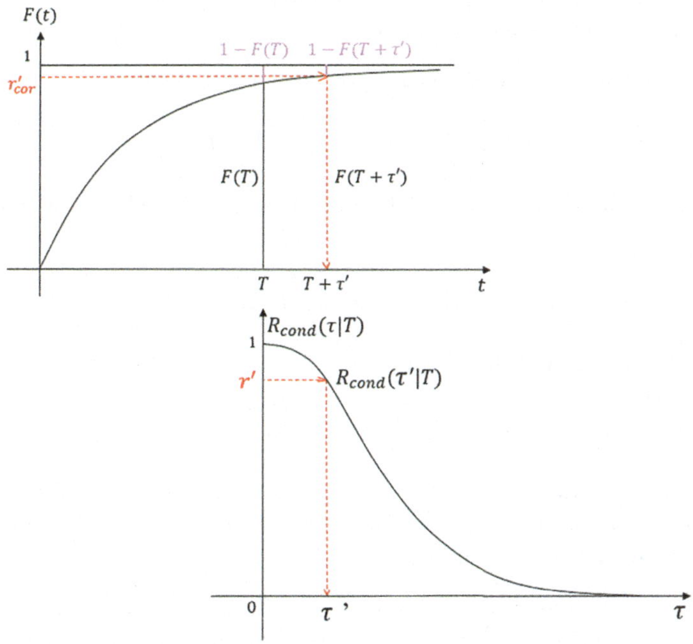

Fig. 6 Sampling variates from a conditional CDF. On the upper panel the indirect sampling procedure utilises only the unconditional CDF, $F(t)$. On the lower panel the direct sampling procedure uses the conditional reliability function, $R_{cond}(\tau\ |T)$, to calculate from the unconditonal $F(t)$

$$\begin{cases} 1 - F(\tau' + T) = r'[1 - F(T)] \\ F(\tau' + T) = 1 - r'[1 - F(T)] \\ r'_{cor} = 1 - r'[1 - F(T)] \\ \tau' + T = F^{-1}(r'_{cor}) \\ \tau' = F^{-1}(r'_{cor}) - T \\ \tau' = t_{r'_{cor}} - T \end{cases} \tag{24}$$

The indirect sampling method of a random time variate from a conditional CDF, $F_{cond}(\tau|T)$, is given below as algorithm Alg C:

Alg C: Theoretical indirect sampling method from a conditional CDF

1. Estimate $F(T)$ as the unconditional probability of the event $(t \le T)$
2. Sample a uniformly distributed variate, r', in the unit interval (0,1)
3. Estimate $r'_{cor} = 1 - r'[1 - F(T)]$
4. Calculate $t_{r'_{cor}}$ as the r'_{cor}-quantile of the unconditional CDF, $F(t)$
5. Estimate the time period variate as $\tau' = t_{r'_{cor}} - T$

Since r' is a number between 0 and 1, it can be proven that:

$$r'_{cor} = 1 - r'[1 - F(T)] \ge 1 - 1 \cdot [1 - F(T)] = F(T) \tag{25}$$

On the other hand, since any CDF value does not exceed one, it follows that $F(T) \le 1$, So,

$$r'_{cor} = 1 - r'[1 - F(T)] = 1 - r' + r'F(T) \le 1 - r' + r' \cdot 1 = 1 \tag{26}$$

From (25) and (26) it follows that:

$$r'_{cor} \in [F(T), 1] \tag{27}$$

Because of (27), the sampled $t_{r'_{cor}}$ in step 4 is not less than T and therefore the sampled time period variate is nonnegative ($\tau' \ge 0$). The procedure in Alg C is illustrated on the upper panel of Fig. 6.

5.3 Numerical Problems in the Implementation of Alg C

Alg C should be always in position to sample a random time period from a conditional CDF. However, in some pseudo-realities, Alg C will have a severe numerical problem when T is comparatively big, hence for some distributions the corresponding reliability $R(T)$ can be positive, but several orders less than the machine epsilon ε. The latter is defined as the smallest positive quantity for which the computer floating point arithmetic holds $1 + \varepsilon > 1$. For example, in MATLAB the reliability can be as low as (2.2×10^{-308}) which is the smallest positive floating point number but the machine epsilon ε is (2.2×10^{-16}) [26]. When the reliability is in the interval $(2.2 \times 10^{-308}, 2.2 \times 10^{-16})$ which is positive but smaller than ε, the corresponding CDF will be precisely 1 due to the machine epsilon definition:

$$\text{if } 0 < R(T) < \varepsilon, \text{ then } F(T) = 1 - R(T) = 1 \tag{28}$$

Then from (24):

$$r'_{cor} = 1 - r'[1 - F(T)] = 1 - r'[1 - 1] = 1 \tag{29}$$

Because of (29), the generated $t_{r'_{cor}}$ in step 4 of Alg C will be

$$t_{r'_{cor}} = t_1 = F^{-1}(1) = +\infty \tag{30}$$

Finally, the sample time period variate in step 4 of Alg C will be infinite:

$$\tau' = t_{r'_{cor}} - T = t_1 - T = \infty - T = \infty \tag{31}$$

It is obvious that the simulation in such a pseudo-reality will be unsuccessful.

On the other hand, certain distributions, like the Normal and the Generalized Extreme Value, have their time domains defined on $(-\infty, +\infty)$ instead of $(0, +\infty)$ [5, 12, p. 251]. It is very rare to create a model where the distributions are fixed. Rather than that, the software codes use generalized objects representing probability distributions that can be modified by the user (e.g. the MATLAB function *makedist* [18]). So, we can safely assume that any given distribution could have negative support in some stage of the software system life. To ensure that the sampled times or the sampled time periods are positive, it is necessary to work with truncated distributions. The CDF and the PDF of the truncated distribution can be expressed as functions of the initial CDF and PDF of the distribution as follows:

$$F_{tr}(t) = 1 - R_{tr}(t) = \begin{cases} 0 & \text{if } t < 0 \\ 1 - \frac{R(t)}{R(0)} = 1 - \frac{1-F(t)}{1-F(0)} = \frac{F(t)-F(0)}{1-F(0)} & \text{if } t \geq 0 \end{cases} \tag{32}$$

$$f_{tr}(t) = \frac{\mathrm{d}F_{tr}(t)}{\mathrm{d}t} = \begin{cases} 0 & \text{if } t < 0 \\ \frac{f(t)}{1-F(0)} & \text{if } t \geq 0 \end{cases} \tag{33}$$

The initial and truncated PDFs and CDFs are shown in Fig. 7.

The truncated function support is from to $(0, \infty)$, but it is difficult directly to apply the formulae (31) and (32) because the basic programming languages do not possess available libraries for most of the truncated distributions. That is a problem for both conditional and unconditional sampling.

The procedure for conditional sampling in Alg C can be modified to overcome the two problems discussed above. Let us define for a specific CDF, $F(t)$, the domain end

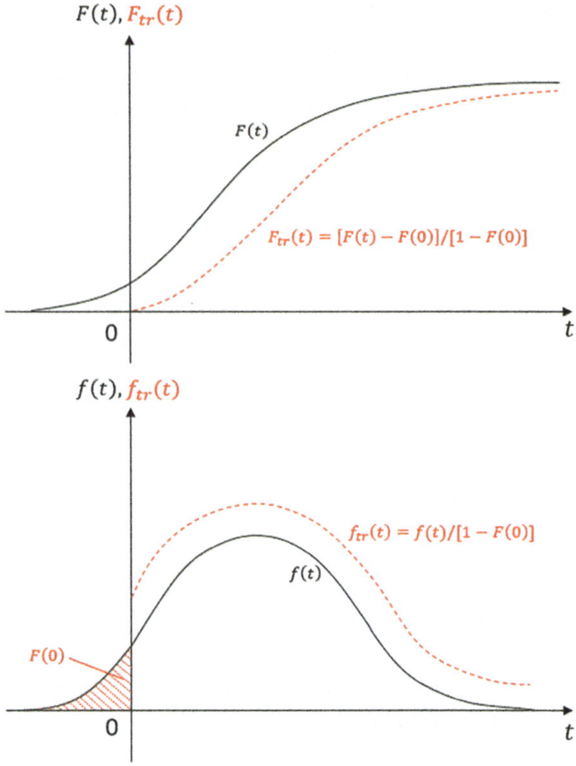

Fig. 7 Truncated distribution ensuring the non-negativity of the random time

time, T_{end}, as the greatest positive quantity for which the computer floating point arithmetic holds $F(T_{end}) < 1$. We do not need the precise T_{end}, but a reasonable estimate of some lower bound. It is convenient to identify T_{end} as a value for which (34) holds:

$$F(T_{end}) \in [1 - 200\varepsilon, 1 - 100\varepsilon] \tag{34}$$

It is obvious that for any $t \leq T_{end}$ the estimated CDF will be strictly smaller than 1. The introduction of T_{end} restricts the time domain of the CDF in order to mitigate the possibility that for some sampled simulation time, t' will have $F(t')$ precisely 1.

If it is known that no failure has happened up to time T, we can substitute T in Alg C with the cut-off time T_{cut} where:

$$T_{cut} = \begin{cases} T & \text{if } T < T_{end} \\ T_{end} & \text{if } T \geq T_{end} \end{cases} \tag{35}$$

All above mentioned modifications have been implemented in Alg D.

Alg D: Practical indirect sampling method from a conditional CDF, $F(t|T)$

1. Estimate T_{end} such that (34) is satisfied
2. Estimate the cut-off time T_{cut} using (35)
3. Estimate $F(T_{cut})$ as surrogate for the probability of the event ($t \leq T$)
4. Sample a uniformly distributed variate, r', in the unit interval (0,1)
5. Estimate $r'_{cor} = 1 - r'[1 - F(T_{cut})]$
6. Calculate $t_{r'_{cor}}$ as the r'_{cor}-quantile of the unconditional CDF, $F(t)$
7. Estimate the time period variate as $\tau' = t_{r'_{cor}} - T_{cut}$.

Alg D will work perfectly well with any truncated distribution for any $T \geq 0$, because the conditional truncated CDF will coincide with the conditional non-truncated CDF according to (19) and (32):

$$F_{tr}(t|T) = 1 - R_{tr}(t|T) = 1 - \frac{R_{tr}(t+T)}{R_{tr}(T)} = 1 - \frac{R(t+T)/R(0)}{R(T)/R(0)}$$

$$= 1 - \frac{R(t+T)}{R(T)} = 1 - R(t|T) = F(t|T) \quad (36)$$

By comparing (19) and (32) we can see that the truncated CDF, $F_{tr}(t)$, is in fact the conditional non-truncated CDF, $F(t|0)$:

$$F_{tr}(t) = 1 - R_{tr}(t) = 1 - \frac{R(t+0)}{R(0)} = 1 - R(t|0) = F(t|0) \quad (37)$$

Formula (37) is trivially true for non-truncated distributions, because then $R(0) = 1$. It means that we can use Alg D for generation from unconditional CDF by setting $T = 0$. As a result, Alg A could be improved so that the sampling should be done with Alg D. The resulting procedure is embodied in algorithm Alg E.

Alg E: Practical generation of the event chain in the *j*th pseudo-reality for 2SBSF with full aging

The steps of Alg E are the same as the steps of Alg A, except for the following three modifications:

***step2:**

2. Sample a random variate failure time of the primary component, t_1, from $F_1(t|0)$ using the sampling method in Alg D

***step3:**

3. Sample a random variate failure time of the backup component in standby, t_3, from $F_3(t|0)$ using the sampling method in Alg D

***step10:**

10. Sample a random variate failure time period of the backup component in operation, τ_2, from $R_2(t|t_1)$ using the sampling method in Alg D.

6 Aggregated Algorithm For Simulation Solution of a Full Aging System

With the event chain generation specified in Alg E, we construct the aggregated simulation procedure for the 2SBSF with full aging. The procedure is embodied in algorithm Alg F.

Alg F: Aggregated simulation solution of 2SBSF with full aging

1. Select the count N of pseudo-realities to be created
2. Generate the j-th event chain EC_j using Alg E, for $j = 1, 2, \ldots, N$
3. Select n equally distributed times T_i in the interval $[0, t_{end}]$ using (9)
4. Estimate the probability of the 2SBSF being in the state r at time T_i for $r = 1, 2, 3, 4$ and $i = 1, 2, \ldots, n$. Use formula (11) with the state functions $St_j(t)$ defined in (8), for $j = 1, 2, \ldots, N$
5. Estimate the nodes (14) of the reliability function $R_{sys}(t)$, using (10) and (13)
6. Estimate the nodes (16) of the invertible reliability function $R^{pu}(t)$ by Alg B
7. Estimate some system numerical characteristics of the 2SBSF reliability:

   7a. Find the mean time to failure, $MTTF_{sys}$, using (15)
   7b. Find the median time, $Median_{sys}$ using (17) and (18a)
   7c. Find the B1 life, *B1_life*, using (17) and (18b)
   7d. Find the 0.95-design life, *95%_design_life*, using (17) and (18c)
   7e. Find the B10 life, *B10_life*, using (17) and (18d)
   7f. Find the interquartile range, *IQR*, using twice (17) and (18e).

7 Simulation Solution with Partial Aging and No Aging

The ODE system (6) with initial conditions in (2) only describes a 2SBSF with full aging. No similar system of ODEs capturing the "no aging assumption" has been created to the best of our knowledge. The same applies for 2SBSF with partial aging. So, there is no numerical solution for either 2SBSF with no aging or for 2SBSF with partial aging. The simulation solutions for 2SBSF with no aging and for 2SBSF with partial aging are the only available options at this stage. Alg F for the 2SBSF with full aging can serve as a basis for modification to produce simulation solutions for both 2SBSF system with no aging and for 2SBSF system with partial aging.

7.1 2SBSF with No Aging

With no aging, the back-up component does not age while in standby. So, the conditional reliability of $R_2(t|t_1)$ can be substituted with the unconditional reliability $R_2(t)$, since component 2 would start operating like a brand-new component. We need to modify Alg E for the case of no aging to reflect the fact that after the failure of the primary component, the failure time period of the backup component should be determined from the failure distribution $F_2(t)$ starting from zero, but not from t_1. This means that Alg E for the event chain generation of 2SBSF with full aging should be modified only in step 10. That modification is embodied in Alg G below.

Alg G: Practical generation of the event chain in the *j*th pseudo-reality for 2SBSF with no aging
The steps of Alg G are the same as the steps of Alg E, except for the following modification:

***step*10:**

10. Sample a random variate failure time period of the backup component in operation, τ_2, from $R_2(t|0)$ using the sampling method in Alg D.

It is only natural that Alg *F* must be modified accordingly to Alg H.

Alg H: Simulation solution of 2SBSF system with no aging
The steps of Alg H are the same as the steps of Alg F, except for the following modification:

***step*2:**

2. Generate the j-th event chain EC_j using Alg G, for $j = 1, 2, \ldots, N$

Although trivial, the no aging simulation model (Alg H), together with the full aging simulation model (Alg F), will provide the two margins for the system reliability function in case of any model for 2SBSF with partial aging. This means that any position characteristics (such as MTTF, median, B1 life, B10 life) for the case of partial aging should lie between the respective characteristics of no aging and of full aging.

7.2 2SBSF with Partial Aging

In 2SBSF, aging has to be taken into account only when the backup component starts operation at time t_1 in case of primary component failure combined with no switching failure. To model 2SBSF with partial aging we can consider the time T in Alg D not as the actual time of the primary component failure, but as the equivalent aging time of the backup component. In the case of full aging, T is t_1 (Alg E). In the case of no aging, T is 0 (Alg G). So, when modeling 2SBSF with partial aging the equivalent aging time T should be somewhere between 0 and t_1. Similar idea for load sharing systems can be found in [8, pp. 167–169]. Let us denote with t^* the equivalent aging time for the backup component in the moment t_1 when the primary component fails. The novel idea we propose is to assume that the wearing of the backup component in standby in the time interval from 0 to t_1 is the same as the wearing of the standby component at the beginning of operation t_1. The assumption is only natural because

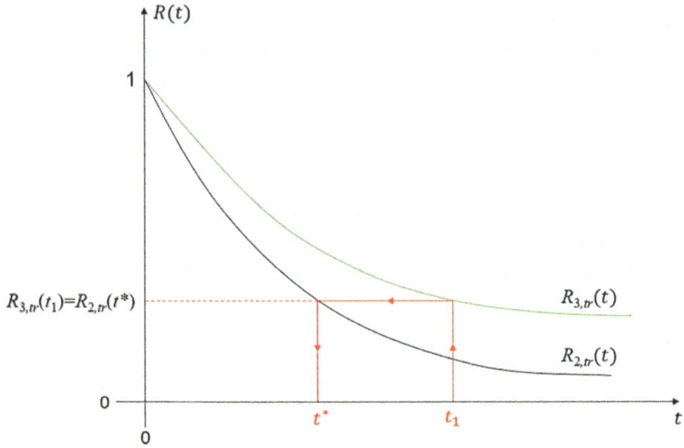

Fig. 8 Determining the equivalent aging time t^* for 2SBSF with partial aging

both types of wearing refer to the backup component in one and the same moment t_1. Since the reliability of a component at time t can serve as a proxy variable for the wearing at time t, we can determine t^* from the equation:

$$R_{3,tr}(t_1) = R_{2,tr}(t*) \tag{38}$$

This process of t^* identification is illustrated on Fig. 8.

In fact, (38) can be rewritten to use the available non-truncated distributions:

$$\begin{aligned}
&R_3(t_1)/R_3(0) = R_2(t*)\big/R_2(0) \Rightarrow \\
&R_2(t*) = [1 - F_3(t_1)][1 - F_2(0)]\big/[1 - F_3(0)] \Rightarrow \\
&F_2(t*) = 1 - [1 - F_3(t_1)][1 - F_2(0)]\big/[1 - F_3(0)] \Rightarrow \\
&t* = F_2^{-1}\big\{1 - [1 - F_3(t_1)][1 - F_2(0)]\big/[1 - F_3(0)]\big\}
\end{aligned} \tag{39}$$

This means that Alg E for the event chain generation of 2SBSF with full aging should be modified only in step 10. That modification is embodied in Alg I.

Alg I: Practical generation of the event chain in the *j*th pseudo-reality for 2SBSF with partial aging

The steps of Alg I are the same as the steps of Alg E, except for the following modification:

***step* 10:**

10. Identify the Failure Time Period of the Backup Component in operation, τ_2:

    10a. Estimate the reliability R^* of the backup component at the end of the standby period t_1: $R* = R_{3,tr}(t_1) = [1 - F_3(t_1)]\big/[1 - F_3(0)]$.

10b. Estimate the equivalent aging quantile number α^* according to (39): $\alpha* = 1 - R * [1 - F_2(t)]$.

10c. Find the equivalent aging time t^* as α^*-quantile, t_{α^*}, of the non-truncated $F_2(t)$: $t^* = t_{\alpha^*}$

10d. Sample a random variate failure time period of the backup component in operation, τ_2, from $R_2(t|t^*)$ using the sampling method in Alg D.

Again, it is only natural that Alg *F* has to be modified accordingly to Alg J.

Alg J: Simulation solution of a 2SBSF system with partial aging
The steps of Alg J are the same as the steps of Alg H, except for the following modification:

***step2:**

2. Generate the *j*-th event chain EC_j using Alg I, for $j = 1, 2, \ldots, N$

8 Numerical Examples

We shall present two groups of examples, each with three parts.

8.1 Numerical Example Group 1

In this group, we shall discuss the behavior of a 2SBSF system from 0 h to $t_{end} =$ 8000 h. The probability for switching failure is $p = 0.12$. The failure time of the primary component has Rayleigh distribution [20] with scale parameter 1600 h. The distribution has a PDF

$$f_1(t) = t/1600^2 \exp[-t^2/(2 \times 1600^2)], \text{ for } t \geq 0 \quad (40)$$

The failure time of the backup component in operation has truncated normal distribution [11] with location parameter 1000 h and scale parameter 900 h. The non-truncated distribution has a PDF

$$f_2(t) = \frac{1}{900\sqrt{2\pi}} \exp\left[-0.5\left(\frac{t - 1000}{900}\right)^2\right], \text{ for } t \in (-\infty; +\infty) \quad (41)$$

The failure time of the backup component in standby has Weibull distribution [13] with characteristic life 4500 h and shape parameter 2.2. The distribution has a PDF

$$f_3(t) = \frac{2.2}{4500}\left(\frac{t}{4500}\right)^{1.2} \exp\left[-\left(\frac{t}{4500}\right)^2\right], \text{ for } t \geq 0 \quad (42)$$

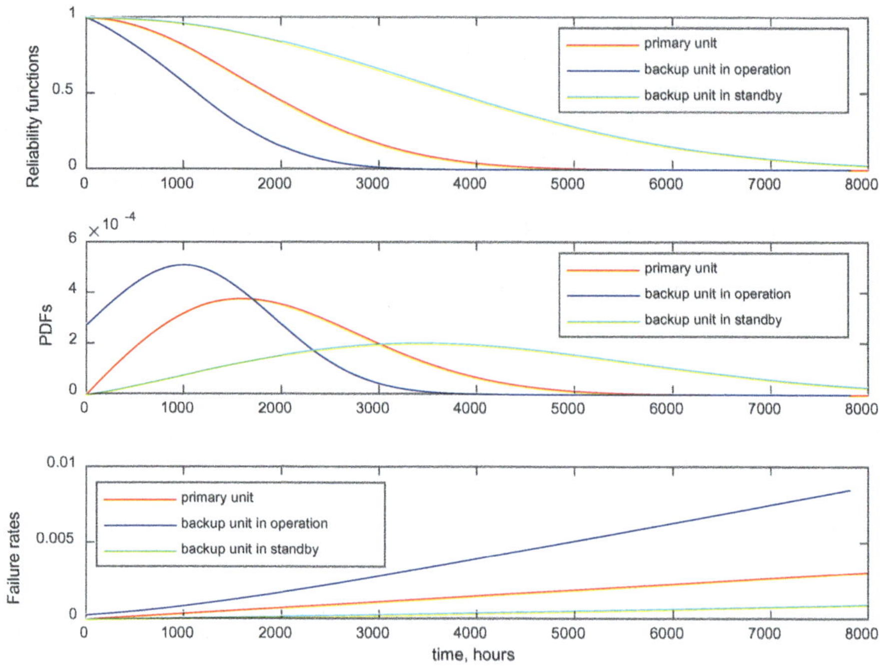

Fig. 9 Reliability functions, PDFs, and failure rates in example group 1

The reliability functions, the PDFs, and the failure rates of the three truncated distributions resulting from (40)–(42) are shown on Fig. 9.

The domain end times, $T_{1,end}$ and $T_{3,end}$ are bigger than $t_{end} = 8000$ h. The domain end time $T_{2,end} = 7808$ h. In the numerical solutions, for any $t > T_{2,end}$ the failure rate $\lambda_2(t)$ is approximated with $\lambda_2(7808)$. Similarly, in the simulation solution, for any $T > T_{2,end}$ the conditional distribution $F_2(t|T)$ is approximated with $F_2(t|7808)$.

The four state probability functions and the system reliability function for the 2SBSF with full aging are given on Figs. 10 and 11. On those figures, both the numerical and the simulation solution are presented. The numerical solution is performed with the MATLAB m-function *ode15s* using control for the non-negativity of the three components of the solution $P_1(t)$, $P_2(t)$, and $P_3(t)$ (MATLAB 2019b). The analytical Jacobian function (7) is provided, and the precision of the solution is the default one. The simulation solution is obtained using Alg E with $N = 10{,}000$ pseudo realities. The numerical characteristics of the system reliability are provided in the third column of Table 1.

From the graphical results in Figs. 10 and 11, it is evident that the numerical and the simulation solutions are practically identical to the extent that the numerical curves cannot be seen as they are covered by the simulation ones. This expected result is an additional empirical proof of the claim in the last paragraph of Sect. 3 that the ODE system (6) describes only 2SBSF with full aging. However, the simulation

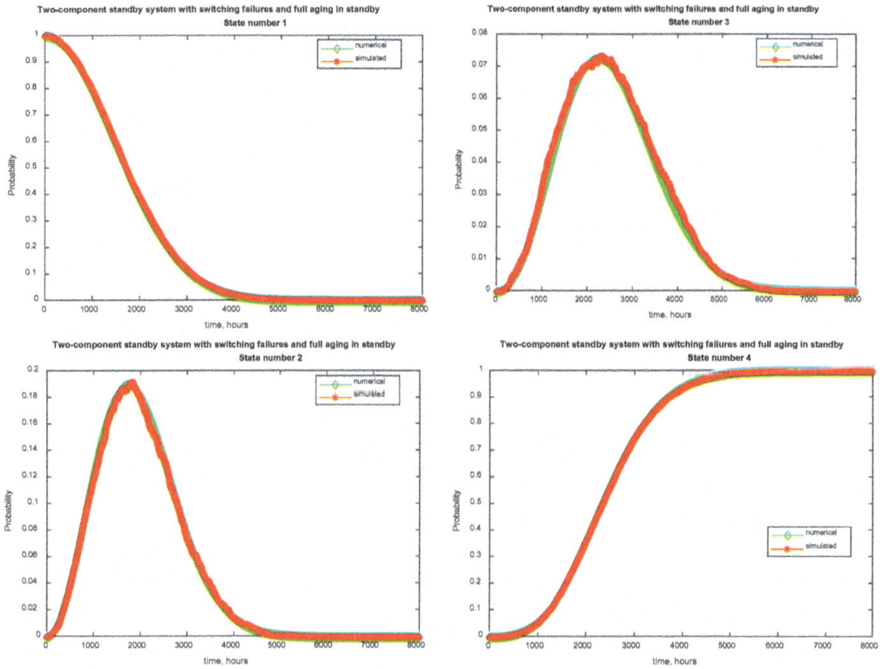

Fig. 10 State probabilities for 2SBSF with full aging from example group 1

Fig. 11 Reliability of 2SBSF with full aging from example group 1

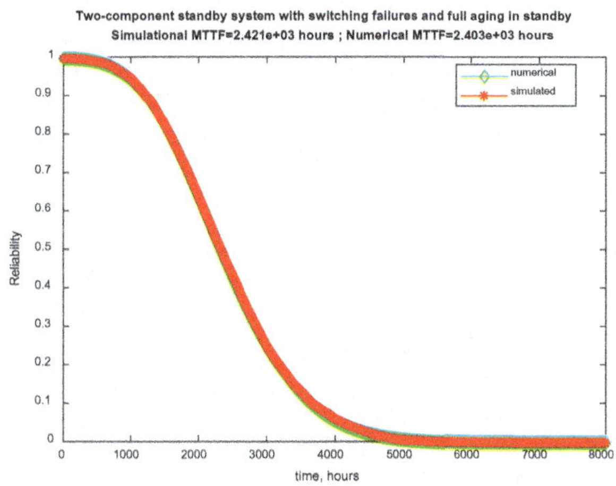

solution estimated that there is approximately 72% chance for the system to ever be in state 2, and 19% chance for the system to ever be in state 3. Such information is impossible to obtain from the numerical solution.

Table 1 Numerical characteristics for the system reliability of 2SBSF, estimated with $N = 10{,}000$ pseudo realities for example groups 1 and 2

Properties	Example group 1			Example group 2		
	No aging	Partial aging	Full aging	No aging	Partial aging	Full aging
$MTTF_{sys}$ (h)	2891	2723	2421	2867	2899	3464
$Median_{sys}$ (h)	2823	2679	2343	2533	2506	2645
B10_life (h)	1385	1374	1233	1157	1112	1076
95%_design_life (h)	1069	1041	988.9	881.0	843.0	779.3
B1_life (h)	584.3	556.3	544.3	477.6	443.0	389.9
IQR (h)	1623	1452	1285	1765	1817	2393

Using the same distributions, the 2SBSF was solved under the no aging assumption utilizing Alg H. The different type of aging affects only states 2 and 4, that is why the state probabilities 1 and 3 are practically the same as on Fig. 10.

The state 2 and 4 probability functions and the system reliability function for the 2SBSF with no aging are given on Figs. 12 and 13. On those figures, the numerical solution of 2SBSF with full aging is provided for reference. The numerical characteristics of the system reliability are provided in the first column of Table 1. There is approximately 72% chance for the system to ever be in state 2, and 18% chance for the system to ever be in state 3.

From the graphical results in Figs. 12 and 13 and from the analytical results in Table 1 it is evident that the reliability with no aging is better than the reliability with full aging. The later result is expected because the failure time of the backup unit in operation has an increasing failure rate (IFR) (see Fig. 9).

Using the same distributions, the 2SBSF was solved under the partial aging assumption utilizing Alg J. As in the no aging case, the state probabilities 1 and 3 are practically the same as on Fig. 10.

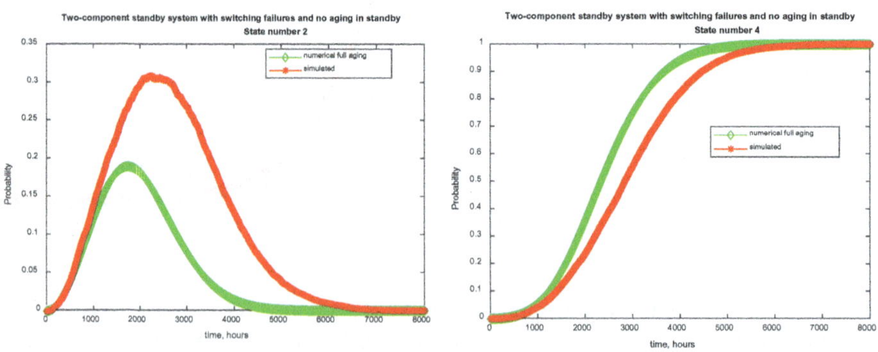

Fig. 12 State probabilities for 2SBSF with no aging from example group 1

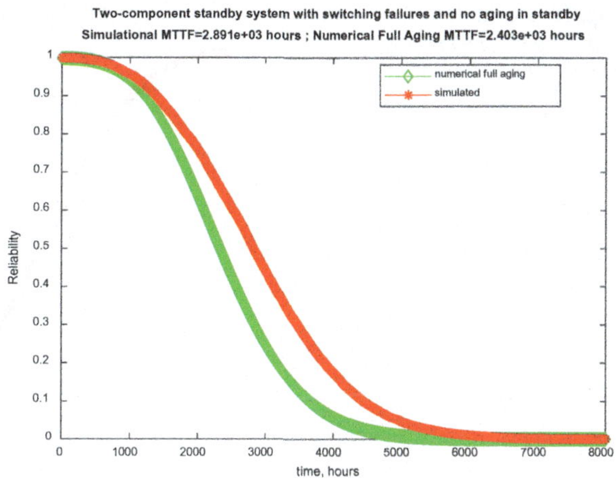

Fig. 13 Reliability of 2SBSF with no aging from example group 1

The state 2 and 4 probability functions and the system reliability function for the 2SBSF with partial aging are given on Figs. 14 and 15. On those figures, the numerical solution of 2SBSF with full aging is again provided for reference. The numerical characteristics of the system reliability are provided in the second column of Table 1. There is approximately 72% chance for the system to ever be in state 2, and 18% chance for the system to ever be in state 3.

From the graphical results in Figs. 14 and 15 it is evident that the reliability with partial aging is better than the reliability with full aging and worse than that with no aging. Table 1 shows that the positional numerical characteristics of the partial aging model fall between those of the no-aging and full-aging models. The reported results are expected because of the IFR of the backup unit in operation failures (see Fig. 9). Such phenomenon reflects the fact that the "no aging" and "full aging" models are the two margins of the "partial aging" model. The IFR of $F_{2,tr}$ (t) shows that the component 2 is deteriorating in a sense that the sampled failure time periods, τ_2, tend to decrease with the increasing of the equivalent aging time, t^*. As expected,

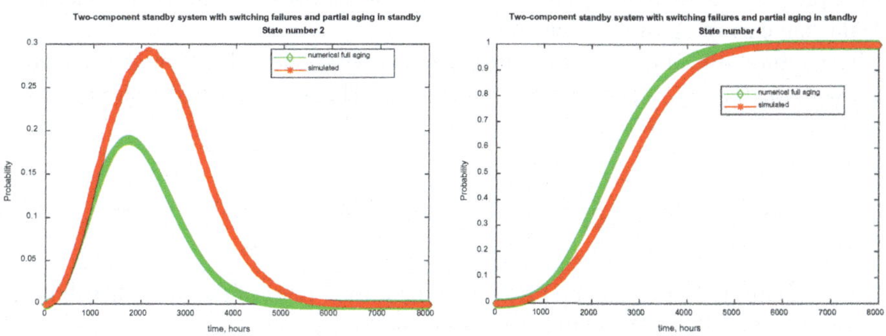

Fig. 14 State probabilities for 2SBSF with partial aging from example group 1

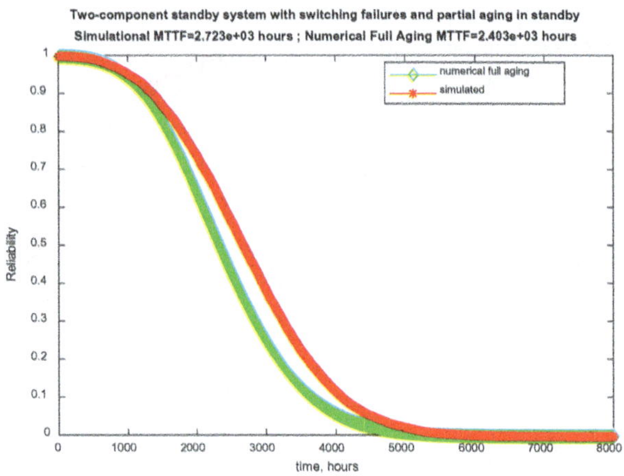

Fig. 15 Reliability of 2SBSF with partial aging from example group 1

the modeled situation is "older the backup component, bigger the chance for failure in operation". In the model with IFR, the lower bound of the positional numerical characteristics in the partial aging case will be given by the full aging model (the worst-case scenario). On the contrary, the upper bound of the positional numerical characteristics in the partial aging case will be given by the no aging model (the best-case scenario).

We will conclude this example group 1 with the remark that the variance of B1 life and the 95% design life are very high because the precision of the simulation in the beginning of the reliability curve is limited.

8.2 Numerical Example Group 2

In this group, we shall discuss the behavior of a 2SBSF system from 0 h to $t_{end} =$ 14,000 h. The probability for switching failure is again $p = 0.12$. The failure time of the primary component and the failure time of the backup component in standby are the same as in example group 1.

The failure time of the backup component in operation has lognormal distribution [21] with median 537 h and shape parameter 1.3. The non-truncated distribution has a PDF

$$f_2(t) = \frac{1}{2.2\sqrt{2\pi}t}\exp\left[-\frac{1}{2\times(2.2)^2}\left(\ln\frac{t}{537}\right)^2\right], \text{ for } t \geq 0 \tag{43}$$

The reliability functions, the PDFs, and the failure rates of the three distributions resulting from (40), (43) and (42) are shown on Fig. 16.

The domain end times, $T_{2,end}$ and $T_{3,end}$ are bigger than $t_{end} = 14{,}000$ h. The domain end time $T_{1,end} = 12{,}683$ h. In the numerical solutions, for any $t > T_{1,end}$

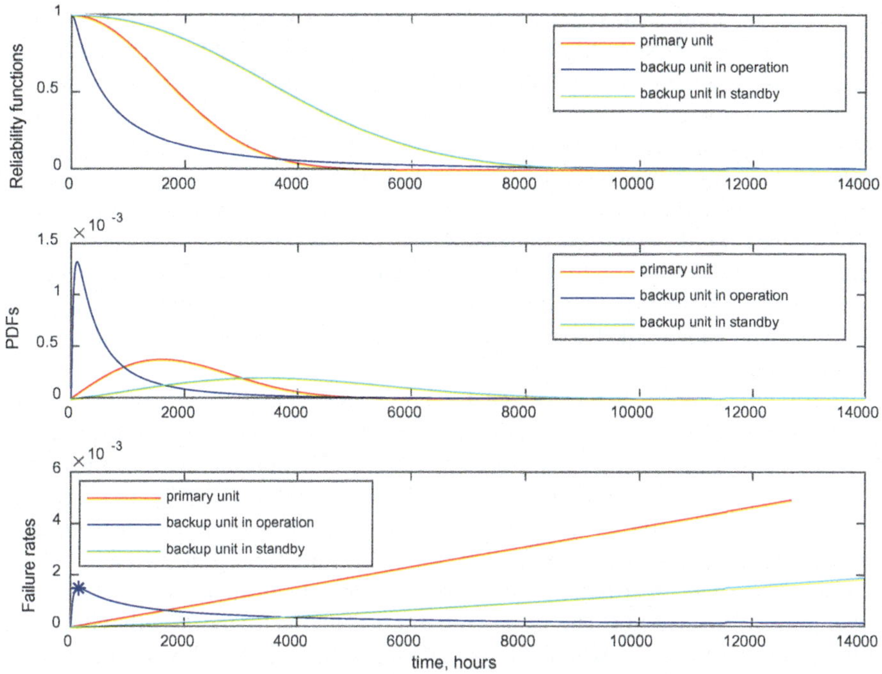

Fig. 16 Reliability functions, PDFs, and failure rates in example group 2

the failure rate $\lambda_1(t)$ is approximated with $\lambda_1(12{,}683)$. Similarly, in the simulation solution, for any $T > T_{1,end}$ the conditional distribution $F_1(t|T)$ is approximated with $F_1(t|12{,}683)$.

The system reliability function for the 2SBSF with full aging is given on Fig. 17,

Fig. 17 Reliability of 2SBSF with full aging from example group 2

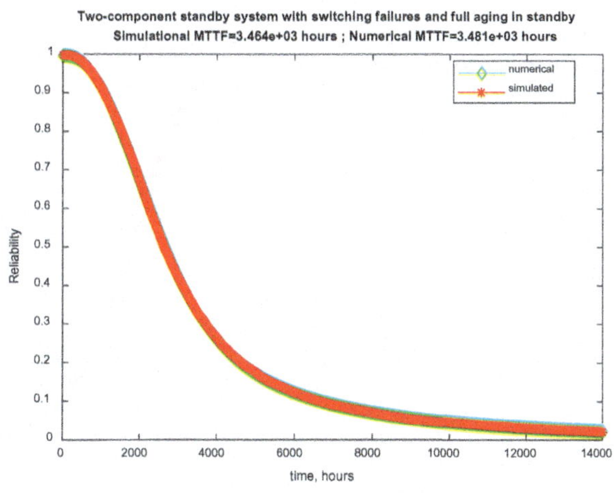

where both the numerical and the simulation solution are presented. The simulation solution is obtained using Alg E with $N = 10{,}000$ pseudo realities. The numerical characteristics of the system reliability are provided in the sixth column of Table 1.

From the graphical results in Fig. 17, we again see that the numerical and the simulation solutions are practically identical to the extent that the numerical reliability cannot be seen as it is covered by the simulation one, which is a second empirical proof of the claim that the ODE system (6) describes only 2SBSF with full aging.

Using the same distributions, the 2SBSF was solved under the no aging assumption utilizing Alg H. The system reliability function for the 2SBSF with no aging is given on Fig. 18, where the numerical solution of 2SBSF with full aging is provided for reference. The numerical characteristics of the system reliability are provided in the fourth column of Table 1.

The graphical results in Fig. 18 and the analytical results in Table 1 show a strange behavior. The reliability with no aging is worse than the reliability with full aging. The later result can be explained because the failure time of the backup unit in operation has a decreasing failure rate (DFR) for $t > 140$ h (see Fig. 16). Only B1 life and 95% design life are better for the no aging case compared to the full aging one. This can be explained with the IFR of $F_2(t)$ for the time up to 140 h.

Using the same distributions, the 2SBSF was solved under the partial aging assumption utilizing Alg J. The system reliability function for the 2SBSF with partial aging is given on Fig. 19, where the numerical solution of 2SBSF with full aging is again provided for reference. The numerical characteristics of the system reliability are provided in the fifth column of Table 1.

From the graphical results in Fig. 19 it is evident that the reliability with partial aging is worse than the reliability with full aging and better than that with no aging. Table 1 shows that the positional numerical characteristics of the partial aging model fall between those of the no-aging and full-aging models, with the exception of the median, which is practically the same as in the no aging case. Such phenomenon reflects the fact that the "no aging" and "full aging" models are the two margins of the

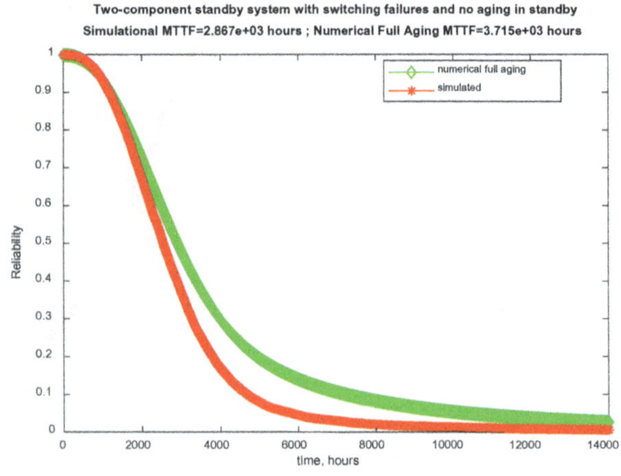

Fig. 18 Reliability of 2SBSF with no aging from example group 2

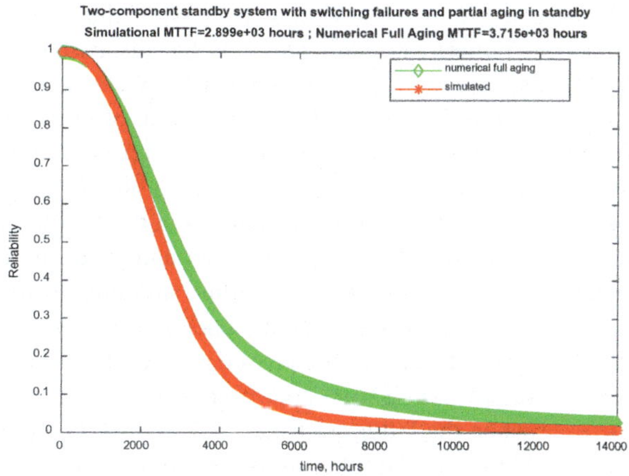

Fig. 19 Reliability of 2SBSF with partial aging from example group 2

"partial aging" model. The DFR of $F_2(t)$ shows that the component 2 is improving in a sense that the sampled failure time periods, τ_2, tend to increase with the increasing of the equivalent aging time, t^*. As expected, the modeled situation is "older the backup component, less the chance for failure in operation". In the model with DFR, the upper bound of the positional numerical characteristics in the partial aging case will be given by the full aging model (the best-case scenario). On the contrary, the lower bound of the positional numerical characteristics in the partial aging case will be given by the no aging model (the worst-case scenario).

We will conclude this example group 2 with the remark that the DFR of $F_2(t)$ is a rare case. On the other hand, the model with partial aging based on assumption (38) gives results very close to no aging for most of the distributions.

9 Conclusions

In this chapter, we discussed the impact of aging on the reliability of a 2SBSF under three assumptions for the standby component behavior—full aging, no aging and partial aging. In Sect. 1, we formalized the states of the 2SBSF with five block diagrams and setup the problem to calculate the reliability of the system under the three aging conditions of the standby component. In Sect. 2 we discussed the analytical solutions for the 2SBSF under CFR, which implies the lack of aging assumptions. We defined a system of 3 ODEs and formulated the reliability of the system as a Cauchi problem that has an analytical solution. Section 3 proposed the numerical solution of our task under time dependent failure rates. We showed that the formulated system of ODEs can only have a numerical solution. Yet, we discussed in detail that the solution had an implicit limitation as it only applied to 2SBSF with full aging. In Sect. 4, we discussed a theoretical simulation solution under the full aging assumption, formulated as Alg A. We showed how we obtain the reliability function of the system without the use of the rate diagram and how to calculate some key numerical characteristics. We discussed in detail the difficulties around finding an inverse for R_{sys}, which is usually constructed as a linear interpolation of a series of nodes. To solve this, we needed to offer Alg B for purging of the reliability function into an invertible one. We also offered a comparison between the numerical and simulation procedures from Sects. 3 and 4. Section 5 was entirely devoted to the technical and numerical difficulties in the sampling procedures from conditional and/or truncated distributions that surrounded the simulation solutions. We proposed practical procedures for random failure sampling from conditional and unconditional distributions with arbitrary time dependent failure rates, embodied into algorithms Alg C, Alg D, and Alg E. We proved that Alg E is a suitable universal tool to generate random samples both from conditional and from unconditional distributions even in the case when the latter are truncated. In Sect. 6, we proposed the aggregated algorithm—Alg F—for simulation solution of the 2SBSF under full aging in seven steps. Those approaches were further expanded on the assumptions of no aging and partial aging in Sect. 7. We proved that the ODE system is non-existent for any aging condition but for full aging, hence the simulation is the only option for a solution. We modified Alg F to derive the simulation solution for the 2SBSF under the no aging (Alg G, Alg H) and the partial aging (Alg I and Alg J) assumptions. In Sect. 8, we presented two groups of three numerical examples each, to demonstrate the ideas of our procedures. In group 1, we analyzed 2SBSF with IFR of the backup component in operation and showed that the partial aging characteristic is bounded between the best-case scenario of no aging conditions and the worst-case scenario of full aging conditions. In group 2, we concentrated on 2SBSF with generally DFR of the backup component in operation. We observed some strange aging behavior, where the reliability with no aging was worse than that under full aging, explained by the positive effect of increasing the equivalent aging time of the backup component. In the two groups of numerical experiments we identified two additional empirical proofs for the claim that the rate diagram and the adjacent ODE system implicitly assume full aging conditions. We also demonstrated that the results from partial aging will be very close to those for no aging.

We discussed that while the "full aging assumption" is rare in practice, it is popular in engineering practice due to the existing models and numerical solutions. Our discussions here concentrated to the extreme of full aging—the no aging assumption, as well as on the most realistic engineering case of partial aging. For those conditions, we offered simulation solutions realized in MATLAB (for ease of calculation). While there may be other forms of partial aging, the no aging and the full-aging systems should be utilized as limits for the reliability in any specific case. As a direction for future development, our algorithms will seek implementation is real world engineering reliability tasks.

References

1. Aghaei, M., Zeinal Hamadani, A., Abouei Ardakan, M.: Redundancy allocation problem for k-out-of-n systems with a choice of redundancy strategies. J. Ind. Eng. Int. **13**, 81–92 (2017). https://doi.org/10.1007/s40092-016-0169-3
2. Aigo, M.: On the numerical approximation of volterra integral equations of second kind using quadrature rules. Int. J. Adv. Sci. Tech. Res. **1**(3), 558–564 (2013)
3. Alavi, O., Viki, A.H.: Influential factors on switch power losses of a buck converter: a reliability approach. J. Telecommun. Electron. Comput. Eng. **8**(5), 1–6 (2016)
4. Amari, S.V., Dill, G.: A new method for reliability analysis of standby systems. In: 2009 Annual Reliability and Maintainability Symposium, Fort Worth, TX, pp. 417–422 (2009)
5. Bücher, A., Segers, J.: On the maximum likelihood estimator for the generalized extreme-value distribution. Extremes **20**, 839–872 (2017)
6. Cha, J., Mi, J., Yun, Y.: Modelling a general standby system and evaluation of its performance. Appl. Stoch. Model. Bus. Ind. **24**, 159–169 (2008)
7. Chakraborti, S., Jardim, F., Epprecht, E.: Higher-order moments using the survival function: the alternative expectation formula. Am. Stat. **73**(2), 191–194 (2019)
8. Ebeling, C.E.: An Introduction to Reliability and Maintainability Engineering, 3rd edn, pp. 167–169 & 399. Waveland Press Inc., Long Grove, IL (2019)
9. Grzelak, L.A., Witteveen, J.A.S., Suarez-Taboada, M., Osteerlee, C.W.: The stochastic collocation Monte Carlo sampler: highly efficient sampling from 'expensive' distributions. Quantitative Finance **19**(2), 339–356 (2019)
10. Hausken, K.: Strategic defense and attack for series and parallel reliability systems. Eur. J. Oper. Res. **186**(2), 856–881 (2008)
11. Horrace, W.C.: Moments of the truncated normal distribution. J. Prod. Anal. **43**, 133–138 (2015)
12. Keller, G.: Statistics for Management and Economics, 11 edn, pp. 200, 251. Cengage (2008)
13. Kizilersu, A., Kreer, M., Thomas, A.W.: The Weibull distribution. Significance **15**(2), 10–11 (2018)
14. Klutke, G.A., Kiessler, P.C., Wortman, M.A.: A critical look at the bathtub curve. IEEE Trans. Reliab. **52**(1), 125–129 (2003)
15. Kwiatuszewska-Sarnecka, B.: Reliability improvement of large multi-state series-parallel systems. Int. J. Autom. Comput. **2**, 157–164 (2006)
16. Li, Z.: Kong, Z.: A generalized procedure for monitoring right-censored failure time data. Qual. Reliab. Eng. Int. **31**, 695–705 (2015)
17. MATLAB R2019a and Statistics and Machine Learning Toolbox 11.5. The MathWorks, Inc., Natick, Massachusetts, United States (2019a)
18. MATLAB R2019b. The MathWorks, Natick (2019b)
19. McFarland, C.D.: A modified Ziggurat algorithm for generating exponentially and normally distributed pseudorandom numbers. J. Stat. Comput. Simul. **86**(7), 1281–1294 (2016)
20. Merovci, F., Elbatal, I.: Weibull Rayleigh distribution: theory and applications. Appl. Math. Inf. Sci. **9**(5), 1–11 (2015)
21. Mouri, H.: Log-normal distribution from a process that is not multiplicative but is additive. Phys. Rev. E **88**(4), 042124 (2013)
22. Nikolova, N., Fan, G., Symes, M., Tenekedjiev, K.: Simulating state-dependent systems with partial aging in standby. In: Proceedings of the IEEE 10th International Conference on Intelligent Systems (IS), 28–30 August, Varna, Bulgaria, pp. 51–60 (2020a)
23. Nikolova, N., Toneva, D., Tsonev, Y., Burgess, B., Tenekedjiev, K.: Novel methods to construct empirical cdf for continuous random variables using censor data. In: Proceedings of the IEEE 10th International Conference on Intelligent Systems (IS), 28–30 August, Varna, Bulgaria, pp. 61–68 (2020b)
24. Norris, J.: Differential equation approximations for Markov chains. In: Darling, R. (ed.) Probability Surveys, p. 71. Cambridge: s.n. (2008)

25. O'Connor, A.: Probability Distributions Used in Reliability Engineering, p. 13. University of Maryland, USA (2011)
26. Rajaraman, V.: IEEE standard for floating point numbers. Resonance **21**, 11–30 (2016). https://doi.org/10.1007/s12045-016-0292-x
27. Ryu, D., Chang, S.: Novel concepts for reliability technology. Microelectron. Reliab. **45**(3–4), 611–622 (2005)
28. Schwartz, L.: Mathematics for the physical sciences, Chapter VI: The Laplace transform, pp. 215–241. Dover books on Mathematics, Dover Publications, NY (2008)
29. Spanos, A.: A frequentist interpretation of probability for model-based inductive inference. Synthese **190**, 1555–1585 (2013)
30. Wang, W., Loman, J. M., Arno, R. G., Vassiliou, P., Furlong, E. R., Ogden, D.: Reliability block diagram simulation techniques applied to the IEEE Std. 493 standard network. IEEE Trans. Ind. Appl. **40**(3), 887–895 (2004)
31. Zhou, D., Wang, H., Blaabjerg, F.: Mission profile based system-level reliability analysis of DC/DC converters for a backup power application. IEEE Trans. Power Electron. **33**(9), 8030–8039 (2018)

The manufacturer's authorised representative in the EU is Springer Nature Customer Service Centre GmbH, Europaplatz 3, 69115 Heidelberg, Germany. If you have any concerns regarding our products, please contact ProductSafety@springernature.com

Printed and bound by CPI Group (UK) Ltd, Croydon, CR0 4YY
30/09/2026
02225287-0017